THE

ELEMENTS

OF

INTELLECTUAL SCIENCE.

A MANUAL FOR SCHOOLS AND COLLEGES.

ABRIDGED FROM "THE HUMAN INTELLECT."

By NOAH PORTER, D.D., LL.D.

PRESIDENT OF YALE COLLEGE

NEW YORK:
CHARLES SCRIBNER & COMPANY.
1871.

JAS. B. RODGERS CO.,
ELECTROTYPERS,
PHILADELPHIA.

JOSEPH J. LITTLE,
PRINTER,
108 to 114 Wooster St., N. Y.

PREFACE.

IN accordance with the wishes of many instructors and friends of education, the author has prepared an abridged edition of his work entitled, *The Human Intellect*, which was first published in 1868. In doing this, he has retained all the leading positions of the original work, with many of the illustrations, occasionally condensing the language, and not infrequently changing the order and method of the argument. Many important topics, less adapted to an elementary work, have been omitted altogether. The controversial and critical observations, have to a large extent been dropped, or greatly abridged. The historical matter has been in part retained, so far as seemed appropriate to a strictly elementary manual. In order, however, to meet the wants of schools, as well as of colleges, some of the matter which is less adapted to beginners, has been printed in smaller type. This may be reserved for a review, or omitted altogether. The author did not feel at liberty, however, to forego for the sake of beginners, a thorough discussion of the important speculative questions which occupy the concluding part of the treatise. For the convenience of those teachers and pupils who may wish to consult the larger work the leading divisions and titles in both volumes are the same. With many thanks for the favor with which the previous treatise has been received, this manual is now offered to the public, and especially to teachers and pupils in schools and colleges.

N. P.

YALE COLLEGE, July, 1871.

TABLE OF CONTENTS.

INTRODUCTION.

PSYCHOLOGY AND THE SOUL.

PAGE

I.—PSYCHOLOGY DEFINED AND VINDICATED. 1

§ 1. Psychology and kindred terms. § 2. Psychology is a science. § 3. Its relations to physiology and anthropology. § 4. Its phenomena known by consciousness. § 5. Its phenomena impel to scientific study. § 6. Value of Psychology. It promotes self-knowledge and moral culture—Disciplines to moral reflection. § 7. Trains to the knowledge of human nature. § 8. Is indispensable to educators. § 9. Disciplines for the study of literature. § 10. Psychology the mother of the sciences which relate to man. § 11. Its special relation to logic and metaphysics. § 12. Is a discipline to method.

II.—THE RELATIONS OF THE SOUL TO MATTER. . . . 11

§ 13. Psychology is a branch of physics. § 14. Reasons why its facts are at first distrusted by the student. § 15. Material phenomena are the earliest known. § 16. Materialistic misgivings and impressions. § 17. These should be set aside. In what way. § 18. The arguments of the materialist. (1). The soul is connected with a body—2. The soul is developed with the body—3. Is dependent on the body for its knowledge and enjoyment—4. Also for its energy and activity—5. It terminates a series of material existences—The conclusion of the materialist. § 19. Counter arguments. (1). Its phenomena are unlike material phenomena—2. The soul distinguishes itself from matter—3. The soul is self-active—4. Is not dependent on matter in its highest activities—5. Gradation of existence does not prove the soul to be material. § 20. The phenomena of the soul real. § 21. Phenomena of one sort cannot be judged by those of another. § 22. The phenomena, and language in which they are described. § 23. Misleading influence of language.

III.—THE FACULTIES OF THE SOUL. 24

§ 24. Question concerning the faculties. § 25. Faculties not parts or organs—Each faculty does not act at a separate time. § 26. States of the soul are like and unlike one another—Their elements are like and unlike in quality—They are dependent on one another—One element is preponderant in each state. § 27. Faculty defined. General authority—Special authority. § 28. These faculties common to all men. § 29. The faculties not independent of one another. § 30. The unity of the soul—Different kinds of unity—Psychical unity is the highest of all. § 31. Unity does not exclude complexness. § 32. Powers of the soul, threefold. § 33. Faculty, power, capacity. § 34. Function, state, phenomenon.

IV.—Is Psychology a Science, and what are its Principles and Methods? 34

§ 35. Materials of psychology; It is an inductive science, and the science of induction. § 36. Some hold psychology too vague to be a science. § 37. The materialistic view of psychology. § 38. The cerebralist theory. § 39. The phrenological theory. § 40. The Associationalist theory—Usually materialistic. § 41. Metaphysical or *a priori* Psychology.

THE HUMAN INTELLECT:

ITS FUNCTION, DEVELOPMENT, AND FACULTIES.

A Preliminary Chapter 42

§ 42. Knowledge defined. What is it to know? § 43. The process which prepares objects of knowledge. § 44. To know, implies the certainty of being. § 45. Also the reality of relations. § 46. When is the process of knowledge complete? § 47. The act of knowing is diverse in its energy. Attention. § 48. The psychological and logical relation of processes and products—The critical, or speculative, application of knowledge. § 49. Order of intellectual development, growth and studies. § 50. Principles of classifying the powers of the intellect. § 51. The presentative faculty. § 52. The representative faculty. § 53. Thought, or intelligence—Two aspects or forms of thought.

THE HUMAN INTELLECT.

PART FIRST.

PRESENTATION AND PRESENTATIVE KNOWLEDGE.

PAGE

I.—CONSCIOUSNESS—NATURAL CONSCIOUSNESS. . . . 61

§ 54. Consciousness defined. Variously applied. § 55. Two forms of consciousness. § 56. Natural consciousness *as an act*. § 57. Consciousness the object. § 58. Relation of consciousness to each of the elements of a psychical state. § 59. The activity may be chiefly noticed. § 60. Consciousness of the *ego*. § 61. The relation of consciousness to the objects of psychical activity. § 62. The object of consciousness is a state of being—Special sense of *cogito, ergo sum*. § 63. The validity of relations is also established. § 64. The development and growth of consciousness. § 65. Latent modifications of consciousness.

II.—THE REFLECTIVE, OR PHILOSOPHICAL CONSCIOUSNESS. . 78

§ 66. The reflective consciousness defined—The abnormal consciousness in children and adults—The ethical consciousness. § 67. The scientific reflective consciousness. Characterized by persistent attention. § 68. It attends to all the psychical phenomena. § 69. Compares and classifies them. § 70. Interprets and explains them by powers and laws. § 71. Relations of the philosophical to the natural consciousness. § 72. Office of language in respect to each—The language of common life sometimes the most trustworthy. § 73. The actions of men also an important test of truth. § 74. Conditions of reaching the decisions of consciousness. § 75. Uncertainty and slow progress of psychology explained. § 76. Peculiar difficulties in the study of the soul.

III.—SENSE-PERCEPTION: THE CONDITIONS AND THE PROCESS. 93

§ 77. Sense-perception defined and distinguished—Is developed earliest of all the powers—Seems to be the most familiar. Is not the most easily understood—Distinguished from other mental acts—Knowledge of matter not gained by sense-perception—What are acts of sense-perception?—Knowledge that is gained by sense-perception—Results of analysis. Eight topics proposed. § 78. The conditions enumerated. The first condition—The nervous system. The sensorium—The reflex action of the nerves. § 78 *a*. The second condition is an object or excitant. § 79. The third condition. Its action on the sensorium. § 80. The process of sense-perception in the simplest form; what is it?—It is psychical, not physiological—It is complex of two elements—The elements unequal in energy; in the same, and the different senses. § 81. Sensation proper pertains to the soul. § 82. Yet experienced by the soul connected with an organism. § 83. The sensations localized. § 84. Differ from one another in quality and definiteness. § 85.

Perception proper, an act of pure knowledge. § 86. Its object a non-*ego*. What kind of a non-*ego*. § 87. An extended non-*ego*. § 88. Perception attends all the sensations—The extension and externality of all objects not given with equal clearness. § 89. The varying relations of sensation and perception proper—In different sensations of the same sense—In the different senses.

IV.—CLASSES OF SENSE-PERCEPTIONS. 109

§ 90. Three classes of sense-perceptions. The muscular. § 91. The organic. § 92. The special sense-perceptions. Smell; its organ, conditions, and objects. § 93. Taste: organs and objects. Variety of the sensations. § 94. Hearing; its organ and objects—The sensations various. In what respects distinguishable—Sounds in succession and combination. Melody and harmony—The condition of oral language. Expressive of feeling. § 95. The sense of touch. Its organ—Essential condition of touch. § 96. Variety of sensations involved in touch. Sensations of gentle touch—Sensations involving violence or injury—Sensations of temperature—Sensations of pressure and weight—The muscular sensations—Sensations localized. § 97. Perceptions proper of touch—Extension perceived by touch. Not explicable by extension in the organism. § 98. The perception of externality by touch—Two meanings of externality—Externality in the first signification—Objection answered. § 99. Sense of touch the leading sense—Furnishes intellectual terms. § 100. Sight; its organ and the conditions of vision—Function of the image on the retina—Sensations proper of vision. § 101. Perception proper in vision. The object of vision—Is always extended—Visible extension superficial only—A single object seen with two eyes—Original place of the visible percept. § 102. Dignity of the eye.

V.—THE ACQUIRED SENSE-PERCEPTIONS. 132

§ 103. The sense-perceptions, as original and acquired—Importance and time of gaining the acquired perceptions. § 104. The acquired perceptions of smell and hearing. § 105. Acquired perceptions of sight. Distance judged by size—Judgments of magnitudes by distance—Judgments of distance by color, outline, clearness, etc.—Judgments of size by other equidistant objects—Influence of Intermediate objects. § 106. Judgments of form, etc., by sight. § 107. Acquired sense-perceptions of place, motion and expression within the body—The provisions of nature for these ends—The control by the intellect of these arrangements—How we learn to talk and to walk—Feats of dexterity. Expressional effects. § 108. The errors of the senses explained. § 109. The acquired perceptions as forms of knowledge—They involve induction—Objections from the cases of animals—Reasons why the perceptions of animals and of man should differ.

VI.—DEVELOPMENT AND GROWTH OF SENSE-PERCEPTION. . 150

§ 110. Nature, interest, and difficulty of the problem. § 111. The intellect, condition of the, before sense-perception begins—The beginnings and development of attention. § 112. The order in which the perceptions are developed. § 113. The development of touch—Hamilton's theory of the perception of the extra-organic. § 114. Development of vision. § 115. Combination of touch and vision—Observations upon infants. § 116. The blind from birth, upon the recovery of sight.

PAGE

VII.—The Products of Sense-Perception; or, The Perception of Material Things. 165

§ 117. Material things and sense-percepts. § 117 *a*. The first stage of perception; limited to coincidence in space and time. § 118. The second stage: The relation of substance and attribute—This relation supposes reflex and indirect knowledge. § 119. The conditions of complete perception. § 120. Can we attend to more than one thing at a time?

VIII.—Activity of the Soul in Sense-Perception . . 180

§ 121. Sense-perception held by many to be passive only—Grounds on which the theory rests. § 122. Evidence that the soul is active. § 123. Different modes of this activity—Is elementary, and easily exercised.

Sense-Perception: Summary and Review. . . . 187

IX.—Theories of Sense-Perception. 189

§ 125. Interest of the theories and their history. § 126. The early Greek philosophers—Aristotle. § 127. The schoolmen. Their doctrine of species. § 128. Descartes, Malebranche, and Arnauld. § 129. John Locke. § 130. Bishop George Berkeley, David Hume. § 131. T. Reid, Dugald Stewart, Dr. T. Brown. § 132. Sir William Hamilton. § 133. Immanuel Kant, and the German school.

PART SECOND.

REPRESENTATION AND REPRESENTATIVE KNOWLEDGE.

I.—The Representative Power Defined and Explained. 206

§ 134. Representation defined and illustrated. § 135. Appellations for the power. § 136. Objects of the representative power. § 137. These objects involve relations. § 138. Conditions and laws of representation considered. § 139. Representation divided into several varieties. § 140. Interest and importance of the representative power.

II.—The Representative Object—Its Nature and Importance. 215

§ 141. Why the *object* of representation needs special discussion. § 142. It is a pyschical object. § 143. It is a transient and short-lived object. § 144. It is an intellectual object. § 145. The relation can be compared to no other. § 146. Representative ideas of objects of consciousness and sense-perception do not resemble them. Memory. § 147. Positive characteristics of mental pictures. § 148. In thought, we prefer ideas to realities. § 149. Ideas especially useful in comparison and generalization. § 150. Images prepare for and aid to action.

PAGE

III.—THE CONDITIONS AND LAWS OF REPRESENTATION—THE ASSOCIATION OF IDEAS. 225

§ 151. Association of ideas. Importance and interest of the subject. § 152. Laws of association. § 153. Association not explained by bodily organization. § 154. The laws of association cannot be referred to any attractive power in ideas as such. § 155. Nor into the force of relations as such—Are not other relations supposable? § 156. The law of redintegration. § 157. The real explanation. How enounced—Associations with home. § 158. The secondary laws defined—How far reducible to the same principle with the primary. § 159. Apparent exceptions to the law of association. § 160. Representation unceasingly active. How it can be interrupted. § 161. Law of association and law of habit. § 162. Higher and lower laws of association.

IV.—REPRESENTATION.—(1.) THE MEMORY, OR RECOGNIZING FACULTY. 254

§ 163. The elements essential to an act of memory. § 164. Memory technically defined. Relation of memory to representation. § 165. The spontaneous memory. § 165 *a*. The intentional memory. § 166. Memory as the power to retain, and to lose. § 167. Dependence on the body. § 168. Varieties of memory; how explained—Development of memory. Its characteristics in the several periods of life. § 169. The education of the memory. § 170. The cultivation of the memory; mnemonics.

V.—REPRESENTATION.—(2.) THE PHANTASY, OR IMAGING POWER. 278

§ 171. Phantasy defined and illustrated. § 172. The interest of its problems—The power of association is operative in them all—Deviations accounted for. (1.) By changes in the relative proportion of the powers. § 173. Sleep physiologically considered. § 174. Sleep considered psychologically. § 175. Somnambulism, or abnormal sleep. § 176. Hallucinations, apparitions, etc. § 177. Insanity.

VI.—REPRESENTATION.—(3.) THE IMAGINATION OR CREATIVE POWER. 295

§ 178. Conditions and materials common to the imagination. § 179. The power of the imagination to create new products. § 180. The combining and arranging office of the imagination. § 181. The idealization of the relations of space and time in art, and mathematical science. § 182. 3. The formation of an ideal standard for psychical acts and states. § 183. The imagination is capable of growth and culture. § 184. Is developed from the earliest till the latest periods in life. § 185. Special applications of the imagination. The poetic imagination. § 186. Its medium is language. § 187. The philosophic imagination. § 188. The practical and ethical imagination. § 189. Relation of the imagination to religious faith.

PART THIRD.

THOUGHT AND THOUGHT-KNOWLEDGE.

PAGE

I.—THOUGHT-KNOWLEDGE DEFINED AND EXPLAINED. . . 319

§ 190. Thinking and the thought power defined. § 191. Appellations for the power of thinking, etc. § 192. Relation of thought to the lower powers. § 193. Concrete and abstract thinking. § 194. Relations of thought to language.

II.—THE FORMATION OF THE CONCEPT OR NOTION. . . 327

§ 195. The processes involved in forming the concept. § 196. The product, its nature and appellation. § 197. Concepts as concrete and abstract, as simple and complex; their content and extent. § 198. Classification, its origin and different species. § 199. How much do we gain by knowing by concepts. § 200. Relation of knowledge by concepts and by intuitions.

III.—THE NATURE OF THE CONCEPT.—SKETCH OF THEORIES. 339

§ 201. The doctrines of Socrates and Plato, and Aristotle. § 202. Porphyry's questions and the scholastics. § 203. Modern Philosophers—G. W. Leibnitz. § 204. G. W. F. Hegel.

IV.—THE NATURE OF THE CONCEPT.—GENERAL NAMES.—LANGUAGE. 45

§ 205. Essential characteristics of the concept. § 206. How far the conceptualist and nominalist are both right. § 207. The imaging of concepts—Different images illustrate the same concept. § 208. The truth represented by realism—The classifications of Botany. § 209. Value of naming and of language. § 210. The relation of symbolic to intuitive knowledge.

V.—JUDGMENT, AND THE PROPOSITION. 358

§ 211. Judgment implied in the formation and use of the concept. § 212. Judgments are psychological and logical—How the subject of a judgment is expressed in language. § 213. The Signification of the copula. § 214. Classes of judgments. Judgments of content. § 215. Judgments of extent. § 216. Scientific and common knowledge.

VI.—REASONING.—DEDUCTION OF MEDIATE JUDGMENT. . 366

§ 217. Nature and importance of reasoning. § 218. Reasoning, inductive and deductive. § 219. The forms of deduction. § 220. The syllogism not *a* but *the* form of deduction. § 221. The dicta or formula of the syllogism. § 222. Deduction rests on the relation of reason to consequent. § 223. The relation of logical reasons to cause and laws. § 224. Geometrical reasons.

VII.—REASONING.—VARIETIES OF DEDUCTION. . . 378

§ 225. The varieties are three; these subdivided. § 226. Probable reasoning. § 227. Mathematical reasoning, materials of. § 228. Definitions and axioms. § 229. The construction of geometrical figures. Auxiliary lines, etc. § 230.

Geometrical reasoning explained by an example. § 231. Immediate syllogisms. § 232. Two elements in most acts of deduction. § 233. Deduction adds to our knowledge. In what sense?

VIII.—INDUCTIVE REASONING OR INDUCTION. . . . 391

§ 234. Inductions properly and improperly so-called. § 235. Inductions of common life and inductions of science. § 236. Why are the indications of science more difficult? § 237. The a priori relations assumed in induction. § 238. The three rules of induction. § 239. The conditions of a successful hypothesis and discovery. § 240. The choice between hypotheses. § 241. The place of experiment.

IX.—SCIENTIFIC ARRANGEMENT.—THE SYSTEM. . . 416

§ 242. Scientific arrangement. System in its lower import. § 243. System in its higher significance.

PART FOURTH.

INTUITION.—THE CATEGORIES.—FIRST PRINCIPLES.

I.—THE INTUITIONS DEFINED AND ENUMERATED. . . 419

§ 244. The critical and speculative stage of our studies—They have been referred to a separate faculty—The appellations by which they are known. § 245. Not *first* in the order of time, but in logical importance—They are, in fact attained last in the order of time. § 246. Various significations of a principle. § 247. The relation of intuition to experience. § 248. The Three Criteria of First Truths. § 249. They are independent of one another—Hegel's development of the categories. § 250. Divided into three classes.

II.—THEORIES OF INTUITIVE KNOWLEDGE. . . . 433

§ 251. The theory of a direct mental vision of first truths. § 252. The theory that they are discerned by the light of nature. § 253. That they are innate or connate. § 254. The views of Locke and his school. § 255. Dr. Reid and the Scottish School. § 256. Kant and his School. § 257. Hamilton's Positive and Negative Necessity. § 258. The theory of Faith as contrasted with knowledge. § 259. J. G. Fichte. § 260. Schelling's view of the categories. § 261. Hegel's theory of pure thought. § 262. Herbart's theory.

III.—FORMAL RELATIONS OR CATEGORIES. . . . 446

§ 263. The category of being.—In what sense fundamental. § 264. The most abstract of all the categories. § 265. Is indefinable and indeterminate. § 266. Relationship. Diversity and similarity.—Relative notions. Negative notions. § 267. Substance and attribute formally conceived. § 268. The logical axioms of identity, etc.

PAGE

IV.—MATHEMATICAL RELATIONS: TIME AND SPACE. 454

§ 269. Development of the several relations of extension. § 270. Duration, how related to the acts of the soul. § 271. The mind discerns extended and enduring objects together. § 272. Limitations of sense-perception. § 273. Beyond these we use the imagination. § 274. Measures of time-objects are imaginary.—Different capacities in different men.—Differences in the estimates of time.—Whence standards for both space and time are derived. § 275. How the relations of space and time objects are generalized. § 276. Two classes of mathematical concepts. The geometrical.—Postulates of geometrical quantity. § 277. The concepts of number. § 278. The application of number to magnitude. § 279. Why, and how mathematical concepts are applicable to material objects. § 280. Time and space relations can be still further generalized. § 281. Extended and enduring objects are limited. § 282. Extension and duration distinguished from, but related to space and time. § 283. They limit objects and events. § 284. In what sense space and time are unlimited. § 285. Space and Time cannot be generalized under higher concepts. § 286. They are known as the conditions of their limited correlates. § 287. What are space and time? Conclusion.

V.—CAUSATION AND THE RELATION OF CAUSALITY. . 480

§ 289. Causation as a law, and as a principle. § 290. Event defined. § 291. Cause distinguished from conditions. § 292. The relation cannot be resolved into a time-relation. § 293. The principle of causality intuitively evident. § 294. Counter theories. The belief not acquired by induction or association. § 295. Not resolvable into outward or inner experience, or both. Locke and De Biran. § 296. The theory which resolves causality into a relation of concepts. § 297. Hamilton's theory of causation.—Conclusion. Our position re-affirmed.

VI.—DESIGN OF FINAL CAUSE. 498

§ 298. Terms explained. Formal, material, efficient, and final causes. § 299. Design and adaptation, how related. § 300. The relation assumed as necessary and *a priori*. § 301. Reasons. The mind impelled to connect objects by this relation. § 302. The relation is higher than that of efficient causation. § 303. The principle has been of essential service in scientific discovery. § 304. The foundation of the inductive philosophy. § 305. Required to explain the phenomena of organic existences. § 306. Relation of final to efficient causes in the higher orders of being. § 307. Objections: (1.) Men mistake in their judgments about final causes. § 308. (2.) Our interpretations can neither be tested nor confirmed. § 309. (3.) This relation derived from conscious experience.—The relation unquestioned in some applications. § 310. (4.) Two principles introduced into philosophy which may possibly conflict. § 311. (5.) The search after final causes has hindered discovery. § 312. (6.) The adaptations of nature are only the conditions of existence. § 313. (7.) Adaptation is limited to organic existence. § 314. (8.) We are not warranted in affirming it of all kinds of existence. § 315. (9.) Adaptation cannot be affirmed of an unlimited Being. § 316. The principle is illustrated and confirmed by its application to metaphysics. § 317. Applied in geometrical construction and deduction. § 318. Applied in geology, etc. § 319. Applied in geography and history; § 320. Also in comparative an tomy and physiology. § 321. Applied in anthropology;—In the provisions for and the capacities of language. § 322. Application to psychology. § 323. Applied and assuumed in ethics. § 324. Application to theology.—The common argument for the Divine existence.

VII.—SUBSTANCE AND ATTRIBUTE: MIND AND MATTER. . 524

§ 325. Uses and etymology of the terms. § 326. Substance and attribute in the abstract. § 327. Spiritual or mental substance. § 328. Material substance defined. § 329. Space occupation and identity of matter. § 330. The production of new substances.—The real Essence or *Thing in itself*. § 331. A material substance not necessarily independent.—Dogmas that seem to deny permanence. § 332. The reciprocal relations of material and spiritual substance.—Mind and matter directly and indirectly known. § 333. The qualities of matter as primary and secondary. § 335. Real and phenomenal or relative knowledge.

VIII.—THE FINITE AND CONDITIONED.—THE INFINITE AND ABSOLUTE. 541

§ 336. To know, a limiting process. § 337. The finite universe; how conceived. § 338. The import of the terms infinite and absolute. § 339. The unconditioned is the non-conditioned.—Applied tó quality and quantity. § 340, The absolute, several causes of. The Hegelian sense. § 341. What is not true of the absolute, etc. § 342. The absolute, etc., are knowable.—Views of Kant, Hamilton, and Mansel.—Herbert Spencer dissents in part. § 343. The absolute apprehended by the intellect. § 344. Not know exhaustively or adequately.—The finite universe infinite to our knowledge.—The absolute a thinking agent. § 345. Must be assumed to explain thought and science.

INTRODUCTION.

PSYCHOLOGY AND THE SOUL.

I.

PSYCHOLOGY DEFINED AND VINDICATED.

Psychology, and kindred terms.

§ 1. PSYCHOLOGY is *the science of the human soul.* The appellation is of comparatively recent use by English writers, but is now generally accepted as the most appropriate term to denote the scientific knowledge of the whole soul, as distinguished from a single class of its endowments or functions. The terms in frequent use—*mental philosophy, the philosophy of the mind, intellectual philosophy,* etc.—should be strictly limited to a single power of the soul, *i. e., its power to know,* and should never be extended to its capacity to feel and to will, or to all its endowments collectively. The terms *metaphysics* and *philosophy,* when used without an adjunct, cannot designate any special science, but only one which is general and fundamental to all the sciences, both material and psychical.

Psychology is a science.

§ 2. Psychology is a *science.* It professes to exhibit what is actually known or may be learned concerning the soul, in the forms of science—*i. e.,* in the forms of exact observation, precise definition, fixed terminology, classified arrangement, and rational explanation.

It is the science of *the soul; i. e.,* the science which has the soul for its subject-matter. *Soul* differs from spirit as the species from the genus; soul being limited to a spirit that either is or has been connected with a body or material organization; while *spirit* may also be applied to a being that has at present no such connection, or is believed never to have had any.

The term *soul* originally signified the principle of life or motion in a material organism. It was especially appropriated to the vital principle which was supposed to animate the body, whether in man or the lower animals. This signification is apparent in the threefold division of man into body, soul, and spirit, in which the soul occupies the place between the corporeal element, and the spiritual. This intermediate part was sometimes called the animal soul, and was believed to perish with the body. Hence, the term *spirit* was applied to a nature that had never been fixed in a body, or soiled and degraded by connection with it. But in the New Testament, *ψυχικὸς*—*psychical*—is often applied to the body in the sense of animal, to distinguish it from the spiritual body. We recognize somewhat of the earlier and lower meaning in the phrases, "*The soul* of the universe," "The soul of a plant," "The soul of an enterprise or interest;" *i. e. the animating principle* of the universe, etc., etc.

Its relations to physiology and anthropology.

§ 3. Psychology is distinguished from *physiology* and *anthropology*. Both these sciences have man as their subject. *Physiology* studies man as a material organism; distinguishing the several organs of which it is composed, the special functions of each, and the combined activity of all in a living being. It is true the structure and arrangement of some of these organs cannot be explained without a distinct recognition of their relations to a spiritual agent. But while physiology must recognize the functions of the soul, it need only consider those phenomena which are familiarly known. For all its purposes, the knowledge and the terminology of common life are entirely sufficient; as when physiology explains the structure of the eye, the ear, and the hand, by their relations to human vision and hearing, to tactual or mechanical skill. Its principal and almost exclusive sphere is the bodily structure and functions, as phenomena that can be explained with reference to the animal economy, and the conditions of bodily development and life.

Anthropology, as the term imports, treats of the whole man, as body and soul. It differs from psychology in treating of these factors when combined so as to form one product in many varieties. Of this product it gives the natural history. It investigates man as this complex whole, as varied in temperament, race, sex, and age; and as affected by climate, employment, or a more or less perfect civilization. It inquires how man is formed and changed in body and soul by inherited peculiarities and accidental circumstances. It discusses the influence of the soul upon the body and the influence of the body on the soul, whether in the normal or the abnormal states and functions of each. But it notices and records these phenomena, only so far as they are open to general observation and require no scientific analysis or explanation. To psychology it leaves the special and profound study of the soul; to physiology, the more thorough examination of the functions of the body.

§ 4. Psychology is distinguished still further from physiology in that the phenomena with which it has to do are apprehended by *consciousness;* while the phenomena of physiology are discerned by *the senses.* Psychology proceeds on the assumption that certain facts or phenomena may be known by the soul concerning itself. The power of the soul to know itself and its own states is termed *consciousness.* How the soul gains this knowledge, and what are the nature, the varieties, and the aids of consciousness, will be considered in the proper place.

Its phenomena known by consciousness.

That the soul does know itself, and confides in the knowledge thus attained, will be acknowledged by every one. The facts differ greatly from those which we observe by hearing, seeing, and touching. They are very numerous and various in their quality, differing from each other in important features, and yet having this feature in common, that they are known by the soul to which they pertain, and known to belong to itself.

§ 5. These phenomena, so numerous and peculiar, excite the desire and effort to reduce them to the exactness and symmetry of scientific knowledge. That they actually occur, cannot be questioned. No one doubts, or cares to deny, that he thinks and remembers, that he hopes and fears. They are the most interesting of all events to the individual who experiences them. The knowledge and the imaginings, the hopes and fears, the joys and sorrows of each person, make up the most important part of his being. They also go very far in deciding our success or failure in life. What we accomplish in our acts and achievements, depends most of all on what we are in our thoughts and aspirations, in our plans and

Its phenomena impel to scientific study.

energy. The mind, which we know so well, is ever at our hand as the instrument with which we execute our purposes and direct our acts. The soul within us is the well-spring ever open at our door and springing up at our feet, from which we draw our most satisfying joys and our bitterest sorrows. Phenomena like these are the legitimate objects of those scientific inquiries to which we are so powerfully impelled. The phenomena which are so near us at all times, which intrude themselves upon our attention even when we desire to exclude them,—which constitute the world within, to which the man himself alone has access, but which is yet, to him, more important than all the world without—deserve to be studied, and, if possible, to be scientifically classified and accounted for.

Value of Psychology. It promotes self-knowledge and moral culture.

§ 6. It may seem needless to dwell upon the *value* of psychological studies. They are peculiar in this, that, to whatever power of the soul they are directed, they require and strengthen the habit of self-knowledge. No real knowledge of the soul can be gained except by turning the gaze inward. Each student must do this himself, for no one can do it for another. Books and instructors, essays, poetry and the drama, cannot describe or teach that which is not confirmed by the researches of the learner within his own spirit. For the man who is disposed to reflect, they can do much, by instructing him where and how to look; but to him who will not converse with himself, they can impart no instruction; they must speak in an unknown tongue. They cannot create conceptions in the mind that will not verify them in its own experience.

This discipline to reflection, with the habits which it forms, is the condition of self-control. He that studies his own powers, may learn how to direct and use them. It also lays the foundation for moral self-improvement. He that would improve his character, must first know what his character is. He must discover what are his better and what his worse impulses; what are the points at which he is most easily assailed, and by what sensibilities or emotions he can most readily rally his forces and overcome their assailants. With self-improvement, self-government is intimately associated. He that would make himself better, must learn to set himself over against himself as his own master, repressing the evil, and educing and encouraging the good. But he that would rule himself, must first know himself. "Know thyself," was written over the portal at Delphi. It was inculcated by Socrates, that preëminent teacher of practical ethics, who, measuring every species of knowledge by its tendency to make man better, regarded this maxim as the summary of wisdom.

Disciplines to moral reflection.

We ought not to omit the peculiar grace and charm which is imparted to the character by that moral reflection which is the natural result of self-acquaintance. To learn to put ourselves in the condition of others, by imagining what would be our expectations and what our feelings were we in their place, not only disciplines and guides to that common justice which the laws enjoin, and to that unselfish morality which the Golden Rule prescribes, but it is the secret of that considerate sympathy and refined courtesy which invest with a peculiar attractiveness a few superior natures. It is by this process that we learn to clothe the severe form of allegiance to duty with the graceful robe of unselfish, sympathetic, and divine charity.

Dr. Thomas Arnold was accustomed to make much of what he called "moral thoughtfulness," as the trait of character which he desired most of all to perfect in his pupils, and which he defined as "the inquiring love of truth going along with the divine love of goodness." This "moral thoughtfulness" is fostered by self-acquaintance, when prosecuted with the honest purpose of self-improvement. It leads to a wider sympathy with man than is bounded by the circle of acquaintances, of countrymen, or even of those now living. It conducts the thoughts backward along the history of the past, and forward among the problems of the future. From this enlarged sympathy arise more hopeful and tolerant views of present evils, a firmer faith in the purposes of Providence and the prospects and progress of man, and a more cautious and candid estimate of the excitements and prejudices which attend the partisan conflicts of the passing hour. Superior natures, in all situations in life, have ever been reflective natures. When the opportunity has been furnished, they have been attracted by psychological studies and fascinated by the mysteries which these attempt to unveil and resolve.

Trains to the knowledge of human nature.

§ 7. The self-knowledge which psychology fosters, and to which it insensibly trains, is the one instrumentality by which we learn to understand our fellow-men. The sharp and searching look by which one man sees through another, and reads the secret which he is unwilling to confess, is attained only by the fine and subtle analysis of one's self. What is perceived, is only external signs; as words, looks, or gestures. To the thought, the feeling, the purpose which they suggest, there is no direct access. The only thoughts and feelings which the interpreter can know directly, are his own; and it is by a close and habitual study of these that he is able to connect them with the signs through which the thoughts and feelings of other men are revealed.

Is indispensable to educators.

§ 8. If, also, we would know our fellow-men to do them good, we must first know ourselves. This suggests the important service which psychology may render to teachers of every class. It is the office of the teacher to communicate knowledge. But to communicate is to *impart, i. e.*, to awaken in the mind of another—the thoughts which exist in the mind of the teacher. Hence, skill in the method or art of teaching, as distinguished from the possession of knowledge, depends almost entirely upon the power of a man to measure and judge of the effect of his instructions. The clear, methodical, and satisfactory communication of knowledge follows from often asking, What truths are most easily and naturally received at first, or as the foundations for others? What illustrations and examples are most pertinent and satisfactory? What degree of repetition and inculcation is required in order to cause the instruction to remain? How can individual peculiarities of intellect be successfully addressed, and, if need be, corrected? Such questions can only find answers through the habits and knowledge which come from intelligent self-study.

Education is even more than the communication of knowledge. It includes the training of the sensibilities, which are the springs of action, and the forming and fixing of the character. To this the knowledge of the feelings is as requisite as the knowledge of the intellect, and it is attained by a similar method.

Disciplines for the study of literature.

§ 9. We name another advantage from psychological study—the training which it ensures for the appreciation and enjoyment of literature, and the increased facility it imparts in writing that which may be worthy to be read. The great masters in literature, especially in poetry, fiction, and the drama, have sounded the depths of the human soul. They have studied man in the several phases which his being assumes, and as moved by the many varieties of human feeling and passion. They may not have learned the technical names which are given to his capacities, or the theories which have been formed of the essence and powers of the soul; but they have studied its thoughts and feelings to the most effectual purpose, and have exhibited the results of their studies in characters of surpassing interest, and by words of wondrous power. From their works the student of psychology may find most valuable aid, and, to enjoy and appreciate them, there is no study which is so useful as the systematic study of the human soul, with the habits and tastes which this study engenders. No fact is better attested by the history of literature, than that those trained by such studies enjoy with especial zest the best literary productions, and appreciate them more keenly than any other class of men. Other things being equal, they are better qualified to criticise them fairly and intelligently.

Psychology the mother of the sciences which relate to man.

§ 10. Psychology either furnishes or makes known *the first principles* for all those sciences which either directly or remotely relate to man—which concern his being, his aspirations and wants, the products of his genius, his institutions, his studies, or his destiny. It is from psychology that all these sciences derive their definitions, and it is in psychology that they find the evidence for their truth. They all begin with certain propositions, which they assume to be true. If their truth is questioned, the final appeal is made to the science of the human soul, as the highest court, beyond which there can be no resort.

Thus *ethics*, or the science of human duty, sets off with certain positions in respect to the nature of man, which assert that he is fitted for moral action, and that to right or virtuous activity he is impelled by the most sacred obligations. It defines conscience and duty, and the several relations of man, and from its definitions derives, by logical analysis and inference, the rules and maxims of practical ethics. But is man a moral being? What is it to be capable of moral activity and obligation? Is he endowed with conscience? What is conscience? These questions are all questions of fact, and can be answered only by the psychological study of man.

Political and social science also assumes that man is a social being, and that he is formed for and must exist in organized society. It defines the rights and obligations which grow out of this constitution. But is man thus endowed? and what is he as a social and political being? Psychology alone can answer.

Law, or the science of justice, lays down as its axioms certain assumptions in respect to the authority and limits of government, for the truth of which it must appeal to the consciousness of every one who consults his own inner life. This science is therefore carried back step by step, till its last footstep is firmly fixed in psychology.

Æsthetics, or the science of criticism, assumes that man is pleased with the beautiful and elevated by the sublime; and that he can form distinct conceptions of what is fitted to attract him in both. From these conceptions he can derive rules by which to try and measure whatever interests him in literature, nature, or art. The canons of taste are in the last analysis resolved by facts of psychology.

Theology is the science of God, of man's relations to God, and of the will of God as made known to man. But this science, whatever else is true of it, must assume that man is, in his nature, capable of religious emotion; as also that he believes in God, and can in some way understand His character and His will. What man believes, and how he comes to believe it, are in great part to be explained by psychology. Theology must go to psychology to vindicate its primary conceptions and justify its elementary principles. The science of religious faith and feeling must, so far as it is a science, rest on psychology.

By these considerations, psychology is shown to be the common parent of many of the sciences. To every one of these sciences the study of psychology furnishes the necessary groundwork, and is itself the necessary and appropriate introduction for the thorough understanding and orderly development of their teachings.

§ 11. To *logic and metaphysics*, psychology stands in a peculiar and most intimate relation, to understand which special consideration is required. Psychology, in one aspect, is like all the sciences of nature, a science of observation; and is subject to those rules of investigation and evidence which *logic* prescribes as common to them all. We study the soul aright when we collect and resolve its phenomena according to the inductive method; when we reason from premises to conclusions; when we infer, by analogy with similar phenomena; and when we arrange our products in the order and beauty of a complete and consistent system. Hence it follows that psychology though necessarily, as we have seen, the parent and director of many sciences, is itself in a most important sense subjected to logic as its guide and lawgiver.

Its special relation to logic and metaphysics.

But logic is itself subject to another science, viz., *metaphysics*, or speculative philosophy, inasmuch as this is the science of those necessary conceptions and fundamental relations on which the rules and the processes of logic are founded. Such are the conceptions of substance and attribute, of cause and effect, of means and ends, and the relations of inherence, causation, and design. Unless these are assumed, the concept, the judgment, the syllogism, the inductive process and the system, can have no meaning and no application. Psychology is therefore subject to logic as its lawgiver, and logic to metaphysics as its voucher.

But though, in the order of thought and methodical construction, psychology is subject to these sciences, yet, in the order of time and of acquisition, psychology is before both of them, though they are fundamental to itself and to all the other sciences. We must, in a certain sense, go through psychology in order to reach the logic by which we study psychology. Logic teaches the laws of right thinking. But what is it to think? What are the processes which it involves? We must ask these questions, in order to discover and prescribe the rules of thinking. We can answer them only by resorting to the facts which psychology discloses. Metaphysics involves the original conceptions which appear in all science, and the ultimate relations which are assumed in the language and inquiries of all the special philosophies. But what are these original conceptions, these prime relations, these categories, of which every particular assertion and every actual belief is only a special exemplification? Psychology only can answer, as, by her analysis, she shows that in all the processes which man performs, he necessarily originates and applies these conceptions and relations. By studying the mind, we discover the laws by which both mind and matter can be studied aright. By studying the mind, we unveil and evolve the necessary conceptions and primary beliefs, by which the mind itself interprets, or under which it views the universe of matter and spirit. It is, then, through psychology that we reach the very sciences to which psychology itself is subject and amenable. Psychology is the starting-point from which we proceed. Psychology is also the goal to which we must return, if we retrace the path along which science has led us. In synthesis we begin, in analysis we end, with this mother of all the sciences.

This special relation of psychology to these fundamental sciences explains why psychology is itself so often called philosophy and metaphysics, while it is neither, but simply a science of observation and of fact. It does, however, lead to philosophy and to metaphysics, as we have seen, by the discoveries which it evolves and the habits to which it trains. It is the natural introduction to metaphysical or philosophical studies, for its own investigations will conduct the mind step by step to those inquiries which will bring into view all the conceptions and relations, concerning the authority of which speculative intellects have disputed in all the schools. These conceptions and relations are employed in all the special sciences of nature, or, in the language of the ancients, in all *physics*, whether the τὰ φυσικά are material or spiritual. Hence it may be that all inquiries concerning them were called metaphysical, as beyond, or preliminary to, the physical, and the science was called *metaphysics*. Hence psychology itself was called philosophy, as it conducted to philosophy *par éminence*, the *prima philosophia*, which is fundamental to all the special and applied sciences.

Is a discipline to method.

§ 12. It is obvious that, if psychology holds these relations to so many special sciences, the study of it must of itself be a most efficient discipline to *method* and logical power.

"What is that," says Coleridge, (*The Friend*, Sec. II., Ess. 4,) "which first strikes us, and strikes us at once in a man of education? And which, among educated men, so instantly distinguishes the man of superior mind, that (as was observed, with eminent propriety, of the late Edmund Burke) we cannot stand under the same arch-way during a shower of rain, without finding him out? Not the weight or novelty of his remarks; not any unusual interest of facts communicated by him, * * * * * It is the unpremeditated and evidently habitual arrangement of his words, grounded on the habit of foreseeing, in each integral part, or (more plainly) in every sentence, the whole that he intends to communicate. However irregular and desultory his talk, there is method in the fragments."

It is impossible for a person to be accustomed to reflect upon his own psychical states, to analyze them into their elements, to trace his practical maxims and his scientific axioms to their fundamental principles, or to evolve them from their psychological beginnings; it is impossible that a man should be thus disciplined without acquiring the power of thinking clearly, rationally, and by orderly processes, and without also gaining the power to express his thoughts in a lucid and convincing manner. To whatever subject of investigation or business in life such a student may apply the discipline thus acquired, he will bring to it a mind capable of mastering the subject with satisfaction to himself and to others, and of gaining that supremacy which the man who thinks with order will always secure over those who think superficially, or who think with lack of method.

II.

THE RELATIONS OF THE SOUL TO MATTER.

Psychology is a branch of physics.

§ 13. Psychology is properly a branch of physics, in the enlarged signification of the term; *i. e.*, the science of the soul is one of the many sciences of nature. Whatever may be thought of the substance of the soul, its phenomena are unquestioned facts. They are facts which are as real and as potent as the phenomena of gravitation or electricity. As such, they assert their place in that vast system of beings which we call Nature, or the Universe, and they claim to be considered by the methods of inquiry which are appropriate to scientific investigation.

Reasons why its facts are at first distrusted by the student.

§ 14. The true philosopher will admit the justice of this claim, and will proceed to consider these phenomena in the light of scientific methods. But when he begins seriously to study them, he finds, perhaps to his surprise, that they are very unlike the phenomena to which he has been accustomed. He discovers that the subject-matter of investigation, in its manifestations, forces and laws, is strikingly and strangely peculiar. The inquirer is surprised, disturbed, and perhaps offended. His first impulse is, to question the reality and trustworthiness of the facts themselves; the next, to doubt whether they can be successfully analyzed and accurately defined. If it be conceded that they are actual, and worthy to be investigated, it is at once presumed that they may be attributed to some material substance or agent, or explained by material laws, or at least can be illustrated by material analogies. This tendency to resolve the soul into matter, or to judge the soul by matter, is very strong; at times it is almost irresistible, and has in all ages exerted over the most candid and truth-loving minds a powerful and unconscious influence. It has become, therefore, almost a necessity, in an Introduction to the study of this science, to consider this influence distinctly, so as to account for its existence and to guard against its effects. For the same reason also it is desirable, by a preliminary discussion, to determine what are the relations of the soul and its phenomena to the essence, powers, and laws of matter.

Material phenomena are the earliest known.

§ 15. We would first account for the existence of this tendency. By the natural course of development and training, we are for a long period exclusively occupied with material phenomena and material laws. What the man sees and hears and smells and tastes, first attracts and absorbs the attention. When he begins to reflect, the objects which he compares and distinguishes, which he classifies and arranges, are almost exclusively sensible objects. When he rises to scientific knowledge, it is to the science of material things. The laws of mechanics, of fluids, of light, of chemical union, of vegetable and animal life, are the laws which he first studies, masters, and learns to apply and to trust. It is in the order of nature, therefore, that the sciences of matter should be studied before the sciences of the soul. It follows, by a natural and almost necessary consequence, that the conceptions and methods of investigation which are appropriate to material objects, should so control the mind's habits and associations, as to take almost exclusive possession of them.

§ 16. When we pass over from the study of matter to the study of spirit, the prepossessions which we have thus derived remain with us. We ask, Are the phenomena real? Can they be actual and substantial when so unlike those phenomena which we see and hear, which we handle and taste? But allowing that they are actual, can they be definitely known? Can we compare and class them? When we ask, To what substance do they pertain? the readiest answer is, To some material substance; and the soul is readily resolved into some form of attenuated matter. Its functions, also, are explained by the action of the animal spirits, or by chemical or electrical changes in the nervous substance. Perception is resolved into impressions on the eye and the ear, which impressions are referred to motions in a vibrating fluid without, which in turn are responded to by motions aroused in a vibrating agent within. Memory and association are explained by the mutual attractions or repulsions of ideas similar to those to which the particles of matter are subjected by cohesion or electricity. Generalization and judgment, induction and reasoning, are resolved into the frequent and often-repeated deposits of impressions that have some mechanical affinity for one another.

Materialistic misgivings and impressions.

The mind that is trained by the most liberal culture, or that is schooled to the most complete self-control, cannot easily divest itself of the prejudices and prepossessions which have been contracted by previous studies. Indeed, the man devoted to a single class of studies or department of science is liable to stronger and more inveterate prejudices than he whose views have not been strengthened by reflection, tested by experiment, and enforced by authority. The man confirmed in his associations by means of a familiar mastery over some physical science, is the man of all others to whom the phenomena of the soul seem most novel and the conceptions most unfamiliar.

§ 17. But it is not enough to be forewarned of these influences, we need also to be forearmed against them. In order to this, it is wise to take a general and preliminary view of the relations of the soul to matter. We propose to present—first, those considerations which may fairly be urged by and conceded to the materialist, or the materialistic psychologist; and second, those which indicate and prove that the

These should be set aside. In what way.

soul has an activity that is independent of material agents, and follows laws that are peculiar to itself.

The arguments of the materialist. (1). The soul is connected with a body.

§ 18. The materialist urges, 1. That we know the soul only as connected with a material organization; that the agent called the soul exerts all its activities and manifests all its phenomena by means of the human body. Of a soul which acts or manifests its acts apart from the body, we have no experience, either by personal observation or through credible testimony.

2. The soul is developed with the body.

2. The powers of the soul are developed along with the powers and capacities of this organized structure. As these powers and capacities are severally called into action and reach their full perfection, the powers of the soul appear, one after another, and attain the full measure of the energy which nature has assigned them. The lower organs of the body act first in order, and these are developed and matured at the earliest period. Afterwards the higher organs are gradually matured and brought into action. After the body is completely developed for all its functions, it passes through certain stages of growth, increasing in size and strength. During these periods of development and growth the soul is also unfolded and matured. One power after another is made ready to act, and the capacity for the action of each is enlarged and strengthened. Because the soul is unfolded as the body is developed, and the soul grows with the growth of the body, it is urged that what we call the soul is but a name for the capacities to perform certain higher functions which belong to a finely organized and fully developed material organism.

3. Is dependent on the body for its knowledge and enjoyment.

3. The soul is dependent on the body for much of its knowledge and for many of its enjoyments. It is through the eye only that it perceives and enjoys color, and through the ear only that it apprehends and is delighted with sound. It is only as a material organ is affected by a material object, that the mind makes a single new acquisition concerning matter. Should these organs cease to exist, or cease to be acted on, all new acquisitions and new enjoyments would cease to be possible. Even the so-called higher kinds of knowledge and feeling have a nearer or remoter reference to the objects of sense with which we are brought in contact through the bodily organs.

Moreover, so far as we know, the soul begins to act and to enjoy, only when these organs are aroused by their appropriate material excitants or *stimuli;* and it would never act or enjoy at all, either in higher or lower forms, if these organs were not first called into action.

4. The soul is dependent on the body, and on matter, for its energy and activity. It sympathizes most intimately with every change in the body. The capacity to fix the attention so as to perceive clearly, to remember accurately, and to comprehend fully, varies with the condition of the stomach and the action of the heart. A slight indisposition is incompatible with the performance of the simplest functions of the intellect, and with the exercise of those emotions to which the soul is most wonted. An active disease disorders the imagination, filling it with offensive and incongruous phantasies, which the will can neither exclude nor regulate. The suffusion of the brain with blood or water, disqualifies the soul for action of any kind, or stupefies it into entire unconsciousness. A change in the structure or in the functions of the brain, or some lesion of the nervous system, induces that suspension of the higher and regulating functions which we call insanity. This state is permanent when its cause is permanent; and the soul may even relapse from this into the condition of idiocy, from which it is never known to emerge. That state of the body which we call fainting takes away all conscious perception and enjoyment, and causes the soul to sink into blank inaction. Another state of the body in sleep induces another kind of activity, in which the usual laws of perception, judgment, and memory, as well as the usual conditions of hope and fear, seem to be deranged or reversed. When the organization of the body is destroyed, the soul ceases to act, and, for aught we can observe, it ceases to exist.

4. Also for its energy and activity.

5. The soul is the termination of a series of material existences, which rise above each other in orderly gradation, each preparing the way for the other; and all are represented in that form of organized matter which manifests and sustains the highest of all, *i. e.*, the so-called phenomena of the soul. The lowest form of matter obeys mechanical laws. The form next higher is seen in bodies endowed with chemical properties and capable of chemi-

5. It terminates a series of material existences.

cal combinations. Here masses and molecules unlike each other unite in such a way as to form a third unlike either. In the form next higher, matter disposes its particles in crystalline arrangement, according to the law of which the elements arrange themselves in forms more or less symmetrical, after the rules of a natural geometry. Next we find the lowest types of organized existence, of which the crystal is the mute prophecy. In these, from the highest to the lowest, there are separate organs, each of which performs a special function, necessary to the existence and functional activity of every other organ and to the whole structure, which is made up of all the organs together. The plant, when the requisite conditions are present of nourishment, moisture, and light, expands into a developed organism, thrusts out the bud and leaf, opens the flower by which its beauty is perfected, and its seed and fruit are formed and matured. The animal requires material conditions of food and air and light. It comes into being by peculiar processes, it grows into a complicated structure of bone, muscle, viscera, nerves, and brain, each separate organ fulfilling its special duty, and all acting together so as to form a completed whole. As the animal structure becomes more perfectly and delicately organized, the phenomena of the soul begin to appear, requiring as their condition all the lower forms of nature, with the presence and action of mechanical, chemical, and organic powers and laws. So far also as we observe the various grades of animal life, as is the perfection of the material structure so is the perfection of the soul. The more simple the organization, the fewer are the instincts and the more limited is the intelligence. The more complex and delicate the structure, the wider is the range and the richer the capacities for knowledge, enjoyment, and skill. The human being also so far as its development can be traced, seems to pass in succession through the lower up to the higher grades of organic life. It is first, as it were, a plant, having only vegetative existence, in the capacity for nourishment and growth; then it becomes an animal, passing through the lowest to the highest forms of animal existence; last of all, it emerges into that which is still higher, viz., the special forms of activity, which are intelligent, sensitive, self-conscious, and rational. It would seem, it is argued, that the soul and the body are one organic growth. The one is perfected with the other, the one depends on the other, the one results from the other. To this is added the consideration already noticed, that organic or nervous force, and psychical or mental force, go hand in hand in energy.

The conclusion of the materialist.

From these analogies it is concluded that the soul is only a convenient term for the higher forms of activity which matter exerts in its more highly organized forms of existence. Or, in other words, the soul, in its essence and its acts, is dependent on organization; and when the organism is disintegrated, the activity of the soul must terminate. Its existence separately from organized matter, or as transferred to another and a new organism, involves an absurd and impossible conception.

Counter arguments (1). Its phenomena are unlike material phenomena.

§ 19. The considerations which may be urged in proof that the substance of the soul is not material, are the following: 1. The phenomena of the soul are in kind unlike the phenomena which pertain to matter. All material phenomena have one common characteristic—that they are discerned by the senses. They can be seen, felt, touched, tasted, and can also be weighed and measured. But the phenomena of the soul, at least, are known by consciousness, and, as thus known, are directly discerned to be totally unlike all those events and occurrences which the senses apprehend. The phenomena discerned by the senses are also known to have some relation to space. Motion, color, taste, sound, combustion, breathing, circulation, secretion, galvanic agency, chemical combination, growth, decomposition—every kind and form of material activity—require extension in the substance on which they operate, or in the effect or activity itself. But feeling, will, thought, memory, joy, sorrow, purpose, resolve, admit of no such relation to space. Even those agents in nature which are most imponderable and impalpable, as the electric force or fluid and the vital or organic force in the animal or plant, both require a certain portion of matter as the active or potent substance which exhibits electrical or vital activity. On the other hand, the phenomena of the soul are by consciousness not only not necessarily referred to any such portion of matter, but they are referred to another agent, the acting or suffering *ego*, which is not known by consciousness to have any sensible or material attributes. These peculiarities clearly and sharply distinguish the two classes or species of phenomena.

2. The soul distinguishes itself from matter.

2. The acting *ego* is not only not known to be in any way material, but it distinguishes its own actings, states and products, and even itself, from the material substance with which it is most intimately connected, from the very organized body on whose organization all its functions, and the very function of knowing or distinguishing, are said to depend. First, it distinguishes from this body every other material thing and object, asserting that the one is not the other. Second, it just as clearly, though not in the same way or on the same grounds, distinguishes itself and its states from the material objects which it discerns. It knows that the agent which sees and hears is not the matter which is seen and heard. Third, the soul also distinguishes itself and its inner states from the organized matter—*i. e.*, its own bodily organs—by means of which it perceives and is affected by other matter. Fourth, it resists the force and actings of its own body, and, in so doing, most emphatically distinguishes itself as an agent from that which it resists. By its own activity it struggles against and opposes the coming on of sleep, of faintness, and of death. Even in those conscious acts in which it feels itself most at the disposal and control of the body, it recognizes its separate existence and independent energy.

3. The soul is self-active.

3. The soul is self-active. Matter of itself is inert. The soul is impelled to action from within by its own energy. Matter only takes a new position, or passes into a new state, as it is acted upon by a force from without. True, the soul must begin its activities with the awakening of the senses; but when it is once awakened, it never sleeps, so far as we can observe or infer. If the senses should furnish it no new objects, it might go on without intermitting its action, busying itself with the materials already furnished under laws of its own. We grant also that what it perceives and desires and does, is determined, to a very great extent, by the objects which present themselves from without; but these direct the course of its action by furnishing it objects; they do not cause it to act. We concede even that its energy in action is dependent on material conditions, as the tension and healthful harmony of the nervous sys-

tem. When the nerves are relaxed or disturbed, as in fainting or disease, the force of the soul is greatly weakened or frightfully disordered; but there is no proof that any bodily conditions can arrest psychical activity after it has once been aroused. In this respect the contrast is striking between matter and spirit.

4. Is not dependent on matter in its highest activities.

4. To very many of the states of the soul no changes or affections of the organism can be observed or traced, as their conditions or prerequisites. It is argued that the soul and body are one material organism, because we know that in many instances some affection of the one is necessary as the condition of a correspondent affection of the other; *e. g.*, the soul cannot see unless the retina is painted by the light, nor can it hear unless the ear vibrates through sound. It ought greatly to weaken the force of this argument, to observe that the change in the soul is in its nature wholly unlike the conditions which go before it. The impression on the eye or the ear has no affinity with or likeness to the perception which follows. Moreover, the condition in the organism is often a condition only so far as to furnish an object which the soul apprehends, *i. e.*, the eye sees rather than hears, and sees this object rather than another, because the excited organism furnishes the object matter or occasion. The conclusiveness of the argument is entirely broken, when we reflect that no changes whatever in the organism are known to precede or to condition the most numerous and the most important psychical states and affections. We grant that the landscape which we see must first be pictured on the retina. But what change or affection of the material organism occurs, when the soul, at the sight of this landscape, images another like it, calls up by memory a similar scene, which had been seen years before a thousand miles distant, or, by creative acts of its own, constructs picture after picture that are more beautiful and varied than the one it is beholding? Or what bodily changes precede desire and disgust, hope and fear, at these memories and creations? No such changes have ever been discerned. No ground is furnished for surmising that they ever occur. They must occur in every instance, to justify *the theory* of the materialist. That they do occur is simply assumed. They have never been observed.

5. Gradation of existence does not prove the soul to be material.

5. The regular gradation in the arrangement of the several kinds of material existences, and the progressive development from the lower to the higher forms of organized matter, do not of themselves prove that the soul is matter in a more highly organized form. Nor does the fact that the transition from the highest forms of organized matter to the lowest types of psychical activity cannot be readily discriminated; nor that the body, which is organized for the uses of the soul, seems in its development to assume in successive order all the lower types of organization, force us to believe that a common substance, obeying material laws, is capable of rising into that refinement of organization which can perform the functions of knowledge and affection.

These facts can only be regarded as proof by the man who assumes that the existence of immaterial or spiritual being is impossible, and the belief of it is unphilosophical. This assumption involves the inference that there is no spiritual Creator. If there be a creating Spirit, who originated and controls matter, then it is not unphilosophical to believe that there may be a created spirit, which is intimately connected with and affected by a material organism, or which, perhaps, is itself the organizing agent.

To those who assume that there can be no creating Spirit, it is useless to attempt to prove that there may be spirit that is created. To those who admit that there is or may be a creating Spirit, or even to those who believe that design has a place in the universe, the regularity of development and progressive transition from one being to another simply indicate a unity of plan in the creation more clearly and more satisfactorily rather than prove a unity of substance in the agent. It may be impossible for us to draw the line where material organization ends and spiritual agency begins, where unconscious reaction ceases and conscious activity emerges. But we know enough to affirm that if spiritual existence is possible, and if it be necessary from its constitution or important to its destiny that it be developed with or organize matter, then all those phenomena by which it seems to rise by a natural evolution from the higher forms of matter, and to crown the series which it terminates as "the bright consummate flower," are fully explained by the unity, the beauty, and the harmony of the Creator's plan, and do not require a unity of substance.

This is all that needs to be determined at the present stage of our inquiries. What the substance of the soul is, and what its destiny, can be fully defined and vindicated by the philosophy and theology to which psychology is the appropriate introduction.

The phenomena of the soul real.

§ 20. It is important to remember, however, that whatever views we accept of the nature of the soul, its phenomena are as real as any other, and their peculiarities are entitled to a distinct recognition by the true philosopher. Whatever psychical properties or laws can be established on appropriate evidence, they all deserve to be accepted as among the real agencies and laws of the actual universe. Perception, memory, and reasoning are processes which are as real as gravitation and electrical action. In one aspect their reality is more worthy of confidence and respect, as it is by means of perception and reasoning that we know gravitation and electricity.

Phenomena of one sort cannot be judged by those of another.

§ 21. The analogy of the physical sciences establishes the principle, that facts of one sort are not to be distrusted because they differ in kind or quality from those of another class. Phenomena of one description are not to be resolved by laws that hold good of those of another. Chemical facts and laws are not disputed because they cannot be explained by mechanical properties and powers. The functions by which the plant is nourished and grows are not to be doubted because they cannot be explained by the laws which regulate the rise of water in a pump, or those which unite an acid or an oil with an alkali into a salt or a soap. The circulation of the blood, and the digestion of the food, are not to be questioned, or violently explained by laws which do not solve them, because they exhibit special and novel activities, and must be interpreted by peculiar methods. We are indeed prompted to reduce all our knowledge to unity, and we therefore seek to explain two events and two classes of phenomena, if it is possible, by a single agency and after a single law. We must prefer the well-known and the familiar to the unknown and the untried; but if we do not succeed, we may not for this reason doubt the facts or misconstrue their laws. If any of the phenomena concerning man which are discerned by consciousness alone happen to be newly observed,—if their truth is established through consciousness—then they are to be received as real, whether they are or are not like the phenomena of matter, or whether they can or cannot be explained by the laws or analogies which material phenomena illustrate and exemplify. To deny them, is unphilosophical. To attempt to explain them by any resort to physical analogies which fail to solve them, or which destroy their integrity or essentially alter their character, is to be more unphilosophical still. If either class of phenomena should take precedence of and give law to the other, the spiritual should stand before the material, for the reasons which have been already given.

The phenomena, and language in which they are described.

§ 22. We ought also to distinguish between the powers and laws which consciousness discovers, and the medium by which these discoveries are recorded and made known. This medium is language, in the large acceptation of the term—the language of signs, of looks, and of words. The most superficial inspection of the words which describe the thoughts and feelings, reveals the fact conclusively that they were all originally appropriated to material objects and to physical phenomena, The words *perceive, understand, imagine, disgust, disturb, adhere,* and a multitude besides, were all originally applied to some material act or event. It is only by a secondary or transferred signification that they stand for the states or acts of the soul.

Locke well observes on this point, (*Essay*, B. iii., c. 1, § 5):—"It may lead us a little toward the original of all our notions and knowledge, if we remark how great a dependence our words have on common, sensible ideas; and how those which are made use of to stand for actions and notions quite removed from sense, have their rise from thence, and from obvious sensible ideas are transferred to more abstruse significations, and made to stand for ideas that come not under the cognizance of our senses; *e. g.*, to *imagine, apprehend, comprehend, adhere, conceive, instil, disgust, disturbance, tranquillity,* etc., are all words taken from the operations of sensible things, and applied to certain modes of thinking. *Spirit,* in its primary signification, is breath; *angel,* a messenger; and I doubt not but if we could trace them to their sources, we should find in all languages the names which stand for things that fall not under our senses to have had their first rise from sensible ideas."

Misleading influence of language.

§ 23. The physical analogon which led to the selection of the word often lurks behind its psychical import, and is ready suddenly to spring out before the eyes, and not unfrequently to suggest erroneous and mischievous conclusions. Let the word *impression* be used, as it frequently is, for some affection of the intellect or the emotions, and it is conceived and reasoned of as involving some pressure or impulse. A mental *image* is taken to be a literal drawing or picture that is painted on the 'presence-chamber' of the soul, or can be restored or re-illuminated by the memory. The objects of the external world are said to be *out* of the mind, while the image or remembrance is said to be *in* it; as though the soul filled a portion of space, and disposed its thoughts within its walls or limits. The memory is conceived as a storehouse of facts, dates, or principles, all ready to be taken down or drawn out when required. Consciousness is thought and reasoned of as though it were an inner light, which illumines by its radiance the dark and winding recesses of the world within. Conscience is the voice of God, speaking with the distinctness and authority of audible speech.

When we reflect on the import of such terms in their application to the soul, we readily assent to the proposition that they are metaphors, either fresh or faded. But we do not always observe, nor do we always guard against the insidious influence of the image from which the metaphor was taken. When we are occupied with the thought, and not with the word—when we are reasoning earnestly, or seeking a solution which evades us, the material image may supply a suggestion which is more plausible than valid, and this will lead to a conclusion which is misleading. In such cases we reason and infer, not from what we think or know, but from what we *say;* and the very language which we use to define and steady our thinking, confuses and distracts it. Inasmuch as all the language which we use is materialistic in its origin and structure, it will incidentally favor those views of the soul which are materialistic, either as professed theories or insensible associations. The history of psychology is a perpetual testimony to the truth, that materialistic conceptions and theories find their readiest justification in the terms which the most thorough spiritualist is forced to employ, and that a quasi materialism seems to spring out of the very language by which it is confuted. Hence it becomes so important that the conceptions which we form should be sharply distinguished from the language in which they are uttered; and that the student of psychology should place himself ever on his guard against the influence of the images and associations which are continually put into his mouth by the language which the necessities of his being force him to use; which language, however high it may soar into the spiritual, can never free itself from the matter in which all its terms have their origin.

III.

THE FACULTIES OF THE SOUL.

Question concerning the faculties.

§ 24. We assume, as has been already stated, that the soul is endowed with the capacity to know its own phenomena. Reserving for future consideration the nature, the development, and the authority of this power, we proceed to apply it in inquiring what consciousness finds to be true of the soul, as to its phenomena, their conditions and laws.

The inquiry which comes first in order is the following: Do we find by consciousness that the soul is endowed with separate faculties or powers? This question is preliminary to all others, for it must be answered, to direct our classification, and fix our terminology.

Faculties not parts or organs.

§ 25. We answer, first, negatively. The soul is not divided into separate parts or organs, of which one may be active while the others are at rest. The plant and the animal have distinct and separate organs, of which each performs its appropriate and peculiar function, which none of the others can fulfil. The root, the bark, the leaf, the flower, in the one, and the stomach, the heart, the skin, and the eye, in the other, each performs an office which is peculiar to itself. While one of these organs is active, the others may be as yet undeveloped or in a state of comparative repose. There is no evidence of the division of the soul into any such organs. The whole soul, so far as we are conscious of its operations, acts in each of its functions. The identical and undivided *ego* is present, and wholly present, in every one of its conscious acts and states.

This peculiarity of the soul has not always been noticed as it should be; certainly it has not always been kept in mind. The so-called faculties have often been conceived and described as separable organs or parts of the soul's substance, any one of which might act of itself—nay, one or another of which might be conceived as added to or superinduced upon another, giving so much enlarged and diverse capacity. Sometimes the faculties have been represented as acting not only apart from one another, but apart from the conscious soul itself; the soul being conceived now as an arena or show-place within which the faculties prosecute their work or play, the soul being impassive and incognizant; or now as a spectator of their doings, more or less indifferent or interested. These representations are all derived from the analogies furnished by matter and its actings; they find no warrant in our conscious experience.

Each faculty does not act at a separate time.

Again, we do not find it true that the soul can only act with one of its so called faculties at the same instant of time. Some suppose,—perhaps inferring from a misconstruction of the doctrine of the faculties,—that when we know, feel and decide, or when we perceive, remember and judge, we must perform each of these separate acts in a definite and distinctly separable instant of time. Consciousness does not allot to each distinguishable kind of activity a separate interval or moment of duration, but before its eye many such are united in one undivided act.

States of the soul are like and unlike one another.

§ 26. We ask next, what facts authorize the conception and the use of the term faculty? We assume that the identical *ego*, or *I*, is not only distinguishable from its own states, but that each of these states is separated or individualized from every other, by occupying a separate portion of time. Each of these states is known by the soul's consciousness to be individually different from every other. But though they are thus divided, they are united by other relations, as follows—

Their elements are like and unlike in quality.

First, their prominent elements are known to be alike or unlike in the immediate experience of the soul. The person who is the subject of each, knows that what he calls his acts of knowledge are alike, and also that they differ from his states of feeling and of will, as readily and as clearly as he distinguishes blue from red, or green from violet, or hard from soft, or bitter from sweet.

They are dependent on one another.

Second, the elements which are the grounds of the classification of the several states are not only recognized as like or unlike, but each has a relation of dependence with respect to the others. Not only is one state different from another, as a so-called state of knowledge, feeling, or will, but the element of knowledge is known to be the necessary condition of the element of feeling, and the element of feeling the condition of that of will. A man does not feel, except he knows or apprehends some object which excites feeling. He always feels about or with respect to something cognized.

When he would increase or intensify an emotion, he applies the intellect to the appropriate object with greater energy and a more exclusive concentration. When he would excite the feeling anew, he brings the object before the attentive intellect a second time. When he would rid himself of an emotion, or prevent its return, he occupies the attention with some other object, so as to excite an emotion that shall exclude or displace the first. There is a similar dependence in the acts or states of the will. To choose, we must not only know, but we must also feel. If an object could be simply known, and excite no feeling, it could neither be chosen nor rejected.

One element is preponderant in each state.

Third, each act or state of the soul is characterized and distinguished by the presence and predominance of some one of the single elements which we have named. That is, each state of the soul is more conspicuously and eminently a state of knowledge, feeling, or will; some one of these elements being prevailing and predominant. It is natural and normal for the soul to blend all in one, and by the laws of its self-active nature, to spring at once into all these forms of its appropriate energy. At every instant of its being it should leap as by a single bound, along the completed curve of its several capacities. Sometimes its course seems to be arrested; often it seems to be detained in a single element; most usually, we may almost say invariably, one only is prominent to the eye of consciousness, the other elements being scarcely noticed as present at all. We distinguish, remember, and name such a state by the predominating feature or element. We think of and call it a state of knowledge, feeling, or will. We observe, too, the appropriate characteristics of the function which prevails, because a single element is conspicuous in each particular state.

Faculty defined. General authority.

§ 27. These considerations prove that the several states of the soul are strikingly distinguished as like or unlike. The capacity of the soul for any one of these distinguishable kinds of activity we call a faculty. We do this for the same reason that we ascribe or refer any material effect or phenomenon to a special power as its source or cause. One ore of iron exhibits magnetic agency, and produces magnetic effects. To another these are wholly wanting. To the one we ascribe, to the other we deny the magnetic power. On the same ground, if there were no other, we might interpret psychical effects by referring each to a special psychical power, which we call a faculty.

Special authority.

But we have higher authority for recognizing special faculties in the sphere of spirit, than for admitting determinate powers in the world of matter. Of material agencies we perceive nothing but the effects. Of the states and effects of the soul, we are conscious that we are the producers. In the one case, we stand before the curtain and see the result, which we ascribe to an agency whose arrangement and working we cannot directly inspect. In the other case, we are ourselves behind the scenes, and observe the working, if, indeed, we do not ourselves work the machinery. To certain of these actions, issuing in certain results, we are prompted by no effort at all. We cannot by any effort prevent ourselves from performing them, and we ascribe them to the nature or constitution of the soul. Hence, with eminent propriety, we connect with such acts the term faculty, from *facilitas*, as explained by Cicero: "*Facultates sunt, aut quibus facilius fit, aut sine quibus aliquid confici non potest.*"—*Cic. Inv.*, 1, 27, 41.

These faculties common to all men.

§ 28. We call the faculties thus ascertained, the *human faculties.* We do so, because certain states of the soul, and certain elements of these states, are believed to be alike in all human beings. No soul is truly human in which they are not present. The exercise and experience of them is necessary to every perfectly constituted and fully developed human being. They may not all be active in an infant of a few days old, but they are sure to become so, if the infant lives and nothing interferes with its normal development. But when we say that the soul must possess these powers in order to be human, we do not assert that any two human beings possess them in the same proportion, or exercise them with the same energy. All men perceive, remember, and reason; but all men do not perceive with the same quickness and accuracy, nor do all men remember with the same readiness and reach, nor do they reason with equal certainty and discrimination. The sensibilities of some men are obtuse, and of others are acute. The choices and practical impulses of men differ most of all. In these, each man is preëminently himself, sharing in no sense his individuality with any other human being.

The faculties not independent of one another.

§ 29. In these natural and original differences, the faculties are not altogether independent one of another. A powerful intellect, to be developed into its normal attainment, needs to be stimulated by strong feelings and to be held and directed by a determined will. Nature usually provides for the possibility of such a development, by proportioning the several endowments of the soul to one another. Hence, a man superior in intellect is usually superior in the capacity for energetic feeling and effective decision. If there be a marked disproportion between any one and the others, we observe it as irregular and unnatural.

This truth needs to be observed in the development of the soul, by special methods of discipline or plans of education. The whole soul must be educated in the harmony of its powers, or it cannot be successfully educated in any single one. The intellect cannot be trained to superior activity or successful achievement except as the feelings are stimulated to a strong interest for the objects to which the intellect is applied, or the ends for which it acts. The will must be taught to concentrate and fix the energies, and to direct them to harmonious and successful activity. We cannot, if we would, train a single power alone. When we seem to bestow all our power upon one only—as the intellect—in the education of ourselves or of others, we are always, in fact, acting upon the whole soul, in exciting new habits or kindling new aspirations.

§ 30. These truths also strikingly illustrate the organic unity and the eminent individuality of the soul. We need ever to be mindful of this. Science seeks after resemblances, and thus is continually impelled to overlook differences. Or, if science notices differences, it is the differences by which species are distinguished, not those by which individuals are separated. With those individual peculiarities which refuse to be classed with any other under some common conception, science disdains to concern itself. All objects in Nature have in some sense an individual unity, which science cannot wholly master and resolve; but the soul is more intensely and eminently one and individual than any other.

The unity of the soul.

We say a piece of iron, or any mere aggregate or mass, is one, when its constituent particles or atoms are permanently held together by adhesive attraction. The law of chemical affinity makes two unlike substances into a third unlike either, which is eminently *one* by the completeness of the interpenetration and combination. A plant is one, so long as its several organs act together, and the functions of each conspire with the functions of every other to the common existence and the developed growth of the whole. The unity of the plant arises from the action of each of these organs with and upon every other, and the united action of the whole through the integrity of an undivided structure. The same is true, only more strikingly and eminently, of the living animal. The animal ceases to be one when its structure is divided, because the reciprocal action of its several organs is thereby forever rendered impossible.

Different kinds of unity.

But the soul is one in a higher sense than the plant or the animal can be. It has, indeed, no material structure, the visible and tangible bond of its material organs. Its faculties are dependent on one another by a union so intimate, that the soul cannot act with one except as it

Psychical unity is the highest of all.

also acts with the others. It cannot grow in the capacity or energy of one except as it grows in the energy of the others. Above all, the soul, in all its conscious activity, refers these various forms of action, thus interdependent on each other, to a single central agent. It knows its unity, in a large portion of its direct experience. It is not more certain that it acts in various ways, each intimately related to one another, than it is that one person, the undivided and self-conscious *ego*, acts in all these ways. First, this ego *knows*, in all its varieties of cognition, and all the variety of objects which it can apprehend. It also *feels*, as variously in the quality and intensity of this kind of subjective experience as its subjective and objective conditions allow. But it is by its actings in *choice*, or as the will, that its individuality is preëminently known *to itself and by itself* to be *one*, as it makes itself to be what it is by its individual acts.

It is true that each soul is like every other soul in those powers by which it is human. It is unlike every other, not only in the proportion of the faculties and attainments which are comparable to those minuter shadings of form and properties in the individual plant or animal which are beyond the reach of the classifying power, but also in the conscious and necessary reference of every action to the individual *ego*. It is preëminently one, as by its own self-activity it gives to each act of its voluntary and rational life a direction and energy which it shares with no other being and no other act of its own being.

Unity does not exclude complexness.

§ 31. But though the soul in these respects is preeminently one, it is not thereby single in the sense of excluding a complex organization. Rather do its unity and individuality depend upon and require a complex organism of faculties and powers. We observe that, in all organisms, the more complicated the structure is, the more conspicuous is the individuality. Just in proportion as the structure is complex in its organs and in the variety of its possible functions, just in that proportion is there the possibility of an unshared individuality, by means of the greater number of particulars in which no other single being can be like this one. But the more largely complex the soul is in the wealth of its known and its yet unrevealed endowments, the more strikingly is its unity illustrated in the working of these endowments with one another to the progressive development and increasing power of a single living being. But its unity is conspicuous in the circumstance, that the being refers this increase of knowledge, skill, and moral capacity to itself, through its conscious knowing, feeling, and choosing. The dignity of the soul is shown by its varied adaptations to the universe of matter, life, and spirit, and by its capacity to respond to and interpret this complex universe by its answering powers, but most of all, in that it can distinguish itself, as the one agent and patient, from everything which it observes or cares for.

Powers of the soul, threefold.

§ 32. The powers of this complex yet individual soul with which our science is concerned, are those only which are manifested through *its conscious acts or states*. All the other powers are left unconsidered, except so far as they incidentally relate to these conscious exercises or experiences. These conscious acts or states are separated into three broad and general divisions of *states of knowledge, states of feeling,* and *states of will. To know, to feel,* and *to choose,* are the most obviously distinguishable states of the soul. These are referred to three faculties, which are designated as *The Intellect, The Sensibility,* and *The Will.*

This threefold division of the powers of the conscious *ego* is now universally adopted by those who accept any division or doctrine of faculties. It has taken the place of the twofold division which formerly prevailed, into *the understanding* and *the will;* according to which the sensibility, or the soul's capacity for emotion, was included under the will, and the affections, as they were usually called, were regarded as phenomena of the will.

Aristotle divided the powers of the soul into the vegetative, the perceptive (including the phantasy), the locomotive, the impulsive or orectic (including the affectional and emotional), and the noetic. All these, except the noetic, were shared by the brutes. The Νοῦς was divine, perhaps preëxistent and imperishable. Cf. *De Gen., et Cor.* ii. 3; *De An.* iii. 5. The distinction of body, soul, and spirit, as we have already noticed, was nearly coincident with this, though more general, and recognized under the Πνεῦμα special relations to the Divine Spirit. The schoolmen retained this division, and distinguished three classes of souls, as follows: the vegetative, of plants, the vegetative and perceptive, of animals, the vegetative, perceptive and rational, of man. Each of the last two have the impulsive and locomotive.

The moderns, throwing out of their classification the powers not apprehended in consciousness, reduced the remainder to two: the intellectual and impulsive, or the powers of the understanding and the powers of the will. This classification was a long time current.

Aristotle had recognized under the orectic, or impulsive powers—the powers of the will, which we have noticed—a threefold subdivision: ἐπιθυμία, θυμός, βούλησις. Theologians had for a long period distinguished the affections and the will and zealously discussed the relations of the one to the other. Locke carefully and earnestly distinguished will from desire, without, however, proposing a threefold division of the powers. (*Essay* B. II. c. 21, §§ 6, 30, 31.) Reid does substantially the same inasmuch as he retains the received division in its accepted import in his *Intellectual Powers, Essay* I., c. 7; but in his *Active Powers, Essay* II., c's 1 and 2, he limits the will to the capacity to determine or choose, excluding from it the capacity for both emotion and desire. Dugald Stewart (*Active and Moral Powers*), following Reid, adopted a threefold classfiication without its formal nomenclature. But Dr. Thomas Brown goes backward from all, distinctly asserting that the will is a modification of desire, and a volition is only the strongest or prevailing desire. *Inquiry*, etc., p. 1, § 3. *Kant* subdivided the impulsive and orectic into two, viz., *feeling* and *desire*. *Kritik d. Urtheils-Kraft*, *Einleitung* and *Anthropologie*. Prof. T. C. Upham distinguishes the power of the soul formally, as *intellect, sensibility*, and *will*.

Hamilton divided the powers of the soul into the faculties of knowledge, capacities of feeling, and powers of conation—*i. e.*, of desire and will. Desire and will he distinguished respectively as a blind or fatal, and a free or deliberate tendency to act. (*Met. Lect.* XI.)

Modern opponents of faculties.

Among modern writers, Herbart and his school have made themselves conspicuous by rejecting the doctrine of faculties of the soul in general, and of the intellect in particular, as inconsistent with the essential unity of the soul, and as self-contradictory in both conception and statement. But Herbart insists most earnestly that the soul possesses a capacity for self-assertion, and that these self-assertions vary both in kind and degree with the conditions which call them forth. His doctrine is not unlike that of Leibnitz respecting monads of all classes, and preëminently of the conscious monads, that they represent or reflect all other objects, and that in this individual capacity lies their individual being. But diverse capacities for these varying self-assertions, or, in modern terminology, for 'reactions,' involves all that is essential, and we may add, all that is objected to in the doctrine of 'faculties;' the one being no more incompatible with the soul's unity than is the other.

Herbart, moreover, affirms of the ideas—'*Vorstellungen*'—all that he denies to faculties, giving them the power to act and react on each other in such a variety of ways, and with independent energies, as to explain all the varying psychical phenomena. While he contends most earnestly that the soul is one—a monad without relations to space—he makes it the theatre, literally the 'show-place,' of all manner of active and antagonistic agents, which are evolved from its own being by the objects that excite them.

The associational and cerebral psychologists reject the doctrine of faculties as commonly received, and resolve all the operations and products of the soul into the single power of association between its ideas, this being in their view the single function either of the soul or its ideas, and that into which all its remaining powers and activities may be resolved.

Faculty, power, capacity.

§ 33. We call these endowments of the soul *faculties, powers, capacities*, with some difference of meaning and application for each.

Faculty is properly limited to the endowments which are natural to man and universal with the race. We also limit the term, by a sense of natural propriety, to those endowments which are especially spiritual, and which manifest the independent and higher energy of the soul.

The word *power* is applied to the active properties of material objects, as well as to those which pertain to spirit. Originally it was employed by Aristotle in contradistinction to act. Hence, power and action are always contrasted, and beings are always contemplated by him as ἐν δυνάμει and ἐν ἐνεργίᾳ. *Force* is quite as frequently used as *power*, of material objects and agents, and in the collective sense, *the forces of nature* are more frequently spoken of than its powers. When power is applied to the soul, it is used in a larger signification than faculty; for by it we designate the capacities which are acquired, as well as those which are original. All men are said to be endowed with the *faculty* of memory. A few are said to have, or to have attained to, the *power* of remembering with surprising reach and accuracy. All men have the *faculty* of sense-perception, but seamen gain the *power* of seeing objects at a very great distance.

Capacity signifies greater passiveness or receptivity than either of the others. Hence it is more usually applied to that in the soul by which it does or can suffer, or to dormant and inert possibilities of being aroused to exertions of strength or skill, or of making striking advances through education and habit.

Function, state, phenomenon.

§ 34. The normal operations of each of these faculties are called its *functions*. The term is taken from the action of the bodily organs. From these it is transferred to organs in the metaphorical sense, as the 'organs of government,' and the 'functions' which they perform. In both these applications it has come to mean, first, the appropriate operations of each, and then the activities to which they are appointed, or destined. This signification appears when the term is applied to the activities of the powers of the soul. In this use it is assumed that there are activities for which the soul is *designed*—modes of operations which are adapted, or conduce to, the end of its being. Hence the *normal activities* of these powers are called functions.

States of the soul are often spoken of. The phrase has passed into current if not into technical use. Strictly interpreted, it would designate the more permanent or enduring, as contrasted with the more transient phenomena. It has come, however, to mean any conditions of the soul whatever.

Phenomenon is used as properly of spiritual as of material beings or agents. Literally, it means that which appears to, or is known directly by the senses: next that which is known as a fact by the mind. In science, it signifies more precisely that which is known as a fact, in distinction from its explanation by a force, principle, or law. Whether this explanation has or has not yet been attained, makes no difference. Whatever is or is not yet explained, when viewed solely as a fact, is called a phenomenon. The English word *appearance* carries with it the meaning, or at least the suggestion, of unreality. It often means and is understood as a mere appearance, a possible illusion. No such signification belongs to phenomenon, as a technical term that has become established in psychical as well as in material science, to signify an observed fact or event.

IV.

IS PSYCHOLOGY A SCIENCE, AND WHAT ARE ITS PRINCIPLES AND METHODS?

Materials of psychology; It is an inductive science, and the science of induction.

§ 35. In the preceding chapters we have impliedly answered these questions. In the subsequent examination of consciousness they will be discussed more fully, and also the nature and authority of psychological science.

Our own theory may be briefly stated, thus: The facts or materials with which psychology has to do are derived primarily from consciousness. These materials psychology seeks to arrange in a scientific method, and to explain by scientific principles. At the same time physiology comes to the aid of consciousness, by furnishing a knowledge of the functions and states of the body which prepare the objects of the sense-perceptions, and are the essential conditions of the development and the activities of the soul. Both these classes of facts must be considered in conjunction, must be observed with attention, must be analyzed into their ultimate elements, must be compared, classed, and interpreted according to the methods which are common to all the inductive sciences.

So far it would seem that psychology is truly an inductive science. It is distinguished however by two striking peculiarities. First. Its subject-matter is attested by consciousness to be *sui generis*, consisting of phenomena which cannot be resolved into material entities or agents, and cannot always be subjected to or judged by analogies furnished by material agents, phenomena, or laws. Second. This subject-matter is in part the function of knowledge itself, the very agency by which all scientific knowledge is produced, whether of matter or of the mind. This special and fundamental function, psychology must examine, in its various processes, and their products. By this peculiar feature, the science of the human soul involves the scientific study of the principles and laws of all knowledge whatsoever, and of each one of the sciences. In every other feature except this, psychology takes rank with the other inductive sciences, and is co-ordinate with them in its subjection to a common method. But by this last feature it becomes in a sense the arbiter of them all, as it tries and tests the methods and principles common to them all, itself included. While, then, psychology is an inductive science, with a subject-matter of its own, it is also in a certain sense, the *science of induction* itself. It requires us to find, and in a sense to justify the fundamental principles of all the sciences, by showing that such principles exist, and demand verification. So far as psychology concerns itself with the explanation of these principles, it is the science of sciences, the *Prima Philosophia*.

These views are very generally received in respect to the nature of psychology as a science, in answer to the question whether such a science is possible. The opinions of those who dissent from them may be classed as follows:

§ 36. A very large number of persons deny that psychology can ever become a science, because of the vagueness and uncertainty of its subject-matter. Science, they allege, knows nothing of powers, either in matter or in spirit. It does not concern itself with the constituents of things, or with the essence and ultimate properties of matter or spirit. It has to do with phenomena only, and it seeks to learn the order and laws of their occurrence by definite statements concerning

Some hold psychology too vague to be a science.

their mathematical relations. Force is measured by number; so is the quantity of matter; so are pressure, motion, attraction, and repulsion, in short, every thing with which science, as such, has to do. The range of science proper, they contend, is limited within the domain where mathematical relations apply, and cannot include the facts of psychology to any effective or valuable result.

It is sufficient to say in reply, that this view of scientific knowledge, would exclude the science of life in all its forms as truly as the science of the soul. It proves too much, and therefore cannot be true. Science does inquire after the powers, the conditions, and causes of phenomena, as truly as it concerns itself with the mathematical relations of either. Besides, it is always pertinent to observe, that the power by which we are impelled to seek, and by which we attain scientific knowledge, is the only authority for our confidence in science itself. To distrust the possibility of exact and determinate knowledge of the conditions and laws of *this power*, is to distrust the authority of science. If the soul, as the agent of science, cannot itself be known in its processes and their results, then the processes have no value, and the products no binding force.

The materialistic view of psychology.

§ 37. The materialists of every sort contend that a science of the soul is possible and real, because the substance of the soul is material, and its phenomena can therefore all be explained by the laws and relations of matter. Their cardinal axiom is: there is nothing substantially existent in the universe except what has extension and sensible properties. The phenomena of the soul must therefore be the manifestations or actings of an existence of this kind, and can be resolved by scientific methods just so far as they can be referred to changes in the constitution or the actings of an extended and material substratum. We pass over the grosser and cruder theories of the ancient schools, which resolved the soul into some form of refined but unorganized matter, as now universally outgrown and rejected, and observe and notice only that form of modern materialism which passes current with not a few scientific men. This theory makes the brain and nervous system the proper substance of the soul, and explains its phenomena by the peculiar activity of this highly organized material substance. It has this in common with the materialism of the grosser sort, that it holds it to be impossible that there should be any agent of psychical phenomena except matter.

The cerebralist theory.

§ 38. The materialists of the present day are properly called Cerebral Psychologists, and plant themselves on the more recent discoveries of physiology in respect to the brain and nervous system. These discoveries are those of the reflex nervous action by the agency of the *afferent* and *efferent* nerves, made by Sir Charles Bell; the discovery of the independent activity of the several systems of nerves, made by Marshall Hall; of the capacity for increased nervous energy, and the flow of a more effective nervous stimulus, which is induced by the repeated action of any organ, whether internal or external, whether muscle or brain; of the changes in the substance of the brain attendant upon a high mental development—a change in bulk and complexity; and, last of all, the discovery of the provision for the consentient or consilient action of different organs of the body, by the coördinating agency of the great nerve centres, which tendency can be greatly augmented and modified by culture and habit. These physiological facts, combined with the doctrine of the association of ideas, which is resolved by many into the physical coaction and coalescence of brain movements and brain cells, are the data or materials out of which the Cerebral Psychologists construct their science of the human soul. Some cerebralists venture to avail themselves of the as yet partially established doctrine of the correlation of physical forces, in support of the conclusion that mind, or soul-energy, is but the spiritual correlate or metamorphose of so much brain or nervous energy. Many of these views are ably represented in the works of Professor Alexander Bain, of Aberdeen, entitled *The Senses and the Intellect*, and *The Emotions and the Will*, also, *Mental and Moral Science*, etc.

The facts and phenomena recognized by the cerebralists are true and important. The most of them should be treated of in anthropology, or the science which treats of the relations of the soul to the body. We may even admit that they all deserve to be considered among the conditions of the purely psychical activities. But they are only the invariable antecedents or the essential conditions of these phenomena. There is no evidence that they produce these phenomena; they do not appear among the constituent elements of any psychical state or act; they cannot be found in them by analysis; they do not explain in the least the original capacity to produce them; they do not account for the dependence of one of these classes of states upon another, as of memory upon perception, or of reasoning upon both. These cerebral conditions might be supposed to exist, without the occurrence of any of the phenomena in question, without perception, memory, or reasoning.

Moreover, these professed explanations have neither meaning nor application except as they suppose the mind already to possess a knowledge of psychical phenomena as known by consciousness, and as connected by certain scientific relations which are purely psychical in their origin and authority. The cerebralist talks, like every other man, of perceiving, of being conscious, of remembering, of induction, and of reasoning. He proposes, as problems to be explained, these phenomena as dependent on and connected with one another in the experience of human consciousness. Of these facts of consciousness he continually avails himself, to give meaning and significance to his cerebral analysis. In short, he supposes a science of the mind's inner experiences which he proposes to supplement by facts or laws of sense-observation, using the facts to be explained to interpret the facts which explain them. Should he attempt to use the nomenclature of his own science in place of that given by the science founded on consciousness, he would fail to be understood. The one cannot be a substitute or an equivalent for the other. A science supposes a knowing agent, and a knowing agent is something other than a throbbing brain: and to know even the functions of the brain, especially after a scientific method, must surely be something more than for the brain to exercise a function in respect to itself and its own functions. Such a conception is more incredible and inconceivable than the conception, which is so often stigmatized, of the soul as conscious of its own operations. A soul that is self-conscious would not be so singular as a brain functionizing about itself.

The phrenological theory.

§ 39. The so-called phrenologists constitute a distinct branch of the cerebral school, if, indeed, their doctrines have not been superseded by the more exact and comprehensive knowledge of the brain on which the cerebralists build. To the claims of the phrenologists to have established a science of the soul, the following objections may be urged: 1. They have not proved that the protuberances of the brain, or the cranium, on which their science is founded, correspond to the psychical powers or functions which it is claimed they decisively indicate. 2. The classification of these very psychical powers which they adopt is illogical, inasmuch as it is chargeable with not a few cross divisions. 3. The classifications and arrangements of the whole science rest for their verification on the knowledge of the soul which is given by consciousness. It even requires this knowledge to supplement its observations of the cranium. It is consciousness which furnishes all the facts which are to be explained, and which is the test of the correctness of the classifications. Were phrenology established, it would not be a science of psychical facts: it would serve only as a guide in the use of certain external indications as explaining the psychical characteristics of individuals.

The question may here properly be raised, whether the brain is not the organ of the soul. We reply, that there is an important difference between asserting that the brain is the substance of which psychical processes are the functions, and the very general statement that the brain is the organ of the soul. This last would seem of itself to imply that the brain is one substance and the soul is another, each having proper features and functions of its own. To say that the soul, so long as it exists with its present corporeal environments, uses and depends upon the brain as its organ of communication with the material world, and sympathizes with the physical condition of the brain in its capacity to act with effect, is to say no more than the truth. This dependence and sympathy may hereafter be established in a multitude of particulars which have not yet been discovered. The brain might itself be subdivided into special organs, and for each of these a separate and as yet unknown function might be ascertained. The relations of these organs and their functions to the powers and acts of the soul might be traced out with surprising minuteness, and still the brain would not be proved to be identical with the soul itself.

The Associationalist theory.

§ 40. The Associational Psychology represents still another theory of the science of the soul. It is founded, as its name imports, upon the fact or law recognized by all psychologists, that the ideas or acts of the soul which are often united tend to recall one another more readily. This law is applied by this school to take the place of every other law or condition of psychical activity, and to exclude every other power or capacity. It is made to stand in the place of the so-called faculties, and even to explain the origin of all necessary and intuitive truths. The school numbers many adherents, among whom are conspicuous Hobbes, Hume, Hartley, Bonnet, James Mill, John Stuart Mill, Bain, and Herbert Spencer. Some of these are more consistent and extreme in their conclusions than others, but all may be fairly said to adopt the associationalist theory in its principal features. These common features are the following. They hold, 1. That a psychical state is analogous to a change or effect in a material object as being a simple impression, or changed condition which is simple—not complex, as is claimed by those who find in every such state a conscious relation to the *ego*. They also, hold, that it is necessarily produced by its cause, condition, or object. They deny, distinctly or impliedly, the truth that every state of the soul must be performed by the conscious *ego*, and that in many of these states this *ego* is active, and in no sense passive. 2. They teach that every such state thus necessarily produced and passively experienced, tends to be reproduced with its attendants. 3. A reproduced state, unless in some way reinforced, as by similar conditions, of itself tends to be and is reproduced with an energy that is weaker than that of the original. (Cf. Hume, Bain, and Spencer.) 4. If it is often reproduced and is reinforced in every act, its energy is greatly increased. This increased energy is manifested subjectively by its stronger tendency to recur again, and objectively by the greater vividness with which the object is represented. Herbert Spencer insists that the facility thus acquired becomes literally mechanical, and that the acts in question pass entirely out of the domain of consciousness, and are taken up by the passive energies, first of the associational faculty, and then of the brain and nerve-cells. In this way they become the material for propagation, through transformations of the nervous substance which are transmitted from one generation to another. A few physiologists, who are not so extreme, account for the phenomena in question by what they call processes of 'unconscious cerebration.' Every activity of the mind not occasioned by some new or original impression, is the action or product of this tendency to recurrent action, thus weakened or strengthened in whole or in part. Imagination is a weakened impression. An act of memory is a somewhat stronger and recurring activity, bringing up a more perfect reproduction of the past. Generalization is a more vigorous revival of some part of many original impressions, which is capable of being suggested by each of these originals or their parts, and made common to them all. Judgment and induction are similar experiences of partial elements of more widely ramified impressions. All these processes are reduced to the more vivid experiences which result from many similar impressions; never to the discernment and affirmation of similarity in the parts of each of the objects to which they belong. Similarity itself, as the ground and motive to the classification and interpretation of nature, is only the result of two or more passive impressions, and never an intelligent cognition or judgment. It is not an objective fact of relation knowable by the intellect, but a subjective sensation or impression more or less frequently recurring.

The belief of necessary truths or fundamental relations, is the result of the frequent conjunction of similar experiences made inseparable by repetition. Thus, the relation of causation is resolved by Hume into the customary connection of ideas or objects. Thus, J. Stuart Mill resolves the belief in any necessary truths, even the simplest mathematical postulates or axioms, into "inseparable association," and gravely suggests that their opposites might be and appear just as axiomatic to a community trained under different associations. Thus, Herbert Spencer, in his *Principles of Psychology*, resolves our *a priori* convictions concerning the reality of space and time, and the relations which they involve (for the necessity of which, as realities, he contends, against Kant and Hamilton), into the invariable conjunctions which first created a persistent tendency to recurrence, which tendency has been fixed by being propagated through countless generations of human beings.

It is necessarily implied in this theory that it dispenses with what it calls the scholastic doctrine of separate faculties of the soul. This, indeed, is its pride and boast, that it makes these several faculties to be but varied results of the single tendency or law of association.

The fundamental defect of the associational school, consists in this, that it does not distinguish between those activities of the soul by which, so to speak, objects are prepared for and presented to the soul for its varied activities, preeminently that of knowledge, and the activities which the soul performs with respect to them when so prepared and presented. An impression on the sensorium, even when responded to by reflex nervous activity, is not the act of knowledge by which the mind distinguishes the object from itself and from other objects; nor does the tendency thereby created to its repetition explain the act of imagination or memory with respect to it when represented a second time. A similar impression, in whole or in part, is a very different thing from that apprehension of a whole or part as similar which is essential to generalization and reasoning as acts of knowledge. The constant conjunction of two ideas, in consequence of which the one will always suggest the other, does not explain the relation under which the mind connects them in an act of judgment; least of all the relation by which it joins them in those beliefs which are necessary and intuitive, as are those which concern the relations of space, time, causation, and design.

It is worthy of notice, that though the associational school is plausibly successful in its explanations of the lower activities and products of the intellect, they fail most signally in explaining the higher operations. J. S. Mill supplements the functions of the associational power in his theory of reasoning and induction by resorting to an 'expectation concerning the uniformity of nature,' which neither association nor induction can account for. Bain resorts to the emotional nature to explain belief, and Herbert Spencer must fall back upon the growth of two nerve-cells into one, propagated indefinitely through successive generations, to account for *a priori* and necessary beliefs.

The associational school can only explain the higher processes and products of the mind by explaining them away—by causing them, under the pressure of its theory, to become something else than what they are. Its theories and explanations are plausible, because the single principle on which they rest is so nearly allied to the pervasive law of attraction, which is so potent in mechanical and chemical philosophy. The extensive and ready favor with which they are received as the only truly scientific theory of the mind, is but a single example of the power of materialistic analogies and prepossessions in the judgments of spiritual facts and relations.

Usually materialistic.

The associational theory, though in its fundamental principle not necessarily materialistic, has been uniformly received by the cerebralists, especially by the cerebralists of the modern school. The doctrine that every mental process is the result of the association and blending of ideas, when united with the principle which explains association by the conjunction of nerve-cells into nerve-growths, and the consilience of nerve activities by the increased energy of nervous stimuli, commends itself as demonstrable, reasonable, and true to all those who find in the movements and growths of the brain the scientific explanation of psychical processes. Bonnet, Hartley, Bain, and Herbert Spencer impliedly, are eminent examples of the union of both cerebralism and associationalism in the same scientific theory.

Metaphysical or *a priori* Psychology.

§ 41. The Metaphysical, or, as it is called by some, the Constructive theory of the science, remains to be noticed. This assumes that psychology can become a science only as it is expounded in the spirit of a system of speculative philosophy which is first assumed or proved to be true, and which must be established as true, before the study of the mind can be made truly scientific, or even before it can begin. There is a truth in the assumption, that every special science is only so far scientific as it rests upon true metaphysics. But there is an important difference between the correct and adjusted statement of this underlying philosophy in a perfected system, and the investigation of these truths in their concrete applications without the aid of such a system. In psychological studies the temptation is particularly strong to view the facts in the light of some preconceived and half-learned philosophy; but it ought for this very reason to be more vigorously resisted. It is in the order of nature that the study of metaphysics should follow after the study of the mind, inasmuch as it is by the analysis of the power to know, that we are supposed first to discover what it is to know, and especially what are the objects and relations which are essential to science; in other words, what conceptions and relations are philosophically valid as the axioms and postulates of scientific knowledge.

To pursue the reversed order, is to weaken the certainty of knowledge, as well as to confuse and embarrass the mind of the student. Such an error of method is certain to be revenged on speculative philosophy itself. It opens the way for fantastic dogmatism on the part of the teacher; for, as soon as he is emancipated from the necessity of justifying his speculative system to the consciousness of his learners by the facts of inner experience, he will be tempted to be positive when he is not certain, and to be fantastic when he is neither logical nor clear. It breeds haziness and pretension on the part of the student. In attempting to follow a guide who deviates from the order of nature, his steps cease to be confident and firm. The want of clear insight he will supply by pretension and conceit, which are both parent and offspring of credulity and dependence.

No maxim deserves to be recorded by the student of philosophy in letters more clear and bright than this: 'The man who seeks to enter the temple of philosophy by any other approach than the vestibule of psychology, can never penetrate into its inner sanctuary; for psychology alone leads to and evolves philosophical truth, even though it is itself subordinate to philosophy.' The investigator who attempts to construct psychology by the aid and under the direction of a metaphysical system, contradicts the order by which both psychology and philosophy are developed and acquired.

THE HUMAN INTELLECT:

ITS FUNCTION, DEVELOPMENT, AND FACULTIES.

A PRELIMINARY CHAPTER.

Knowledge defined. What is it to know?

§ 42. We have considered the soul as capable of various functions or operations, which are manifested to consciousness as psychical facts or phenomena. *The intellect* has been defined:—the soul as endowed with and exercising the power to know. We now proceed to make the intellect the special object of our study, that is, we enter upon that special division of psychology which is concerned with the capacities, operations, and laws of the human intellect.

The distinctive function of the intellect being to know, we at once inquire, 'What is it, for the soul to know?' The fact that we exercise the function of knowing is attested by consciousness and also that it differs from feeling and willing. For this conscious experience there can be no substitute. All definitions and descriptions presuppose that the person to whom they are addressed can understand their import and verify their truth by referring to his own conscious acts.

What consciousness apprehends and distinguishes may be more exactly defined as follows:

1. To know, is an operation of the soul acting as the intellect —an operation in which it is preëminently active. In knowing, we are not so much recipients as actors. We do not merely submit to the impressions made upon the senses from without. Nor are we the passive subjects of the mechanical operations of ideas already acquired, acting upon us by an independent force and movement of their own. But in all states of knowledge the soul knows itself to be active.

2. The intellect exercises its capacity to know under certain

conditions. Like every other agent in nature, it is limited in respect to the mode, energy, and results of its action, by the occasions and circumstances under which it acts.

Thus the intellect cannot perceive a color, a taste, a tree, a house, unless these objects are presented to the mind, for it to act concerning or upon. So, too, it cannot remember unless an event has occurred which it may proceed to recall and recognize. Nor can it imagine or believe, without certain materials or data, by means of which it creates or infers.

These conditions are objective only. There are also conditions which are subjective, as the mind's capacity to know, which is always implied; its disposition for present activity, its bodily conditions of health and reason; also certain favoring circumstances, as absence of preoccupation; and, last of all, the direction and fixing of the attention to the so-called objects.

3. The objects which condition the acts of the intellect are diverse in their character. Some are presented from the world without: as the objects of sense, for the existence and nature of which, the soul itself may be in no way responsible. Others are presented from within, as the operations of the soul itself, in the various forms and the endless variety of the states of knowledge, feeling, and will, all of which are apprehended as objects by consciousness.

Others still are the products or results of precedent acts or energies of the soul—*residua* from objects once perceived, waiting to be re-awakened—the so-called images or pictures once present, now absent, yet capable of being revived.

It is manifest from this enumeration that the word *object* is used in two widely divergent senses—either as the external or material object, the *object-object*, as it is often called, and which may be explained as the object eminently objective; or as the *subject-object, i. e.*, the mental object, or the object created by the mind's own energy. The adjectives *objective* and *subjective*, also, follow the import of the nouns. *Objective* is applied to whatever the mind contemplates as an object, whether it be a *subject-object* or an *object-object*. Every relation which such an object holds is called *objective*. On the other hand, *subjective* is applied to the knowing mind, whether it is conceived as apprehending a *subject-object* or an *object-object*. *Subjective* is also applied to all the psychical experiences and acts; to the feeling and willing, as well as the knowing soul.

The process which prepares objects of knowledge.

§ 43. 4. If the soul can create objects for itself to know—as in the cases already referred to of consciousness and memory,—we ought carefully to distinguish those of its activities by which objects are, so to speak, *prepared* for the mind's cognition, from the special activity of the intellect in knowing these objects when prepared or presented for its apprehension. For example, the energy of the soul in what is called the *association of ideas*—by which, on occasion of the presence of an object known, another object presents itself in order to be known—is clearly distinguishable from the act of the intellect in apprehending that object when presented. In like manner, all the antecedent preparation by which material things are made ready to be known through the joint action of body and spirit in the sensorium, is plainly diverse. and ought to be distinguished from the act of the mind in perceiving the object when thus made ready.

We observe also, that these acts or functions of preparation, are generally not conscious acts, in the sense in which the acts of knowledge are. Some of them may be wholly removed from consciousness, as is the activity by which the soul preserves and suggests objects once known, even though this very activity largely depends on previous conscious operations. Some of these may be entirely removed from consciousness, as the physiological or *psycho-physical* operations which conditionate sense-perception. Others may be entirely within the range of conscious observation, though performed with rapid,spontaneous and uncontrolled exertion.

They are all properly psychical acts, and are appropriately treated in connection with those activities with which consciousness has to do. We cannot understand the one class of activities without constant reference to the other.

To know, implies the certainty of being.

§ 44. 5. To know—the conditions of knowledge being fulfilled—is *to be certain* that something *is*. Knowledge and being are correlative to one another. There must be *being*, in order that there may be knowledge. But it belongs to the very essence of knowledge to apprehend or cognize its object to be. *Subjectively* viewed, to know, involves *certainty; objectively* it requires *reality*.

We distinguish different kinds of objects and different kinds of reality. Objects may be psychical or material. Their reality may be mental and internal, or material and external, but in either case it is equally a reality. The spectrum which the camera paints on the screen, the reddened landscape seen through a colored lens; the illusion that crosses the brain of the lunatic, the vision that frightens the ghost-seer; the thought that darts into the fancy and is gone as soon, each as really exists as does the matter of the solid earth, or the external forces of the cosmical system. It is true, one kind of existence and reality is not as important to us as is the other; we dignify one class as real, and call the other unreal. We name some of these objects realities, and others shadows and unreal; but, philosophically speaking, and so far as the act of knowledge is concerned, they are alike real and are alike known to be.

The word *being* is sometimes contrasted with *phenomenon*. It is obvious that in that case being is not used in the sense in which we have defined it; *i. e.*, as equivalent to *a knowable object*. When used in such a contrast, we oppose permanent, or independent being, to transient, or dependent being.

We often *err* in making one kind of reality indicate another. We do not *err* in not *knowing* that something *is*, but in *mistaking* it for something which it is not. We do not err as to *that* the being is, but as *to what* it is. We do not err as to its beingness or entity, but as to its relations.

This leads us to observe:

Also the reality of relations.

§ 45. 6. In knowing, we apprehend not only that objects exist, but also that they exist in certain *relations*. It is essential to the definition of knowledge, not only that we know objects as existing, but that we know them as *related*. We cannot even know two thought-objects as existing without also knowing that the one is not the other. We cannot notice two leaves, without knowing that they are alike or unlike in form, surface, or color. We cannot observe two occurrences without referring them to the same or different causes, etc., etc.

It may be objected that, although it may be true that whenever two objects are known by a single act, they must be known in relation, yet it is not so when the object is single. To this we reply, that it is impossible that an object should be known singly and apart from every other. A single object must be known by some agent, and it cannot be known by that agent unless the object is distinguished from the agent, and from his act in knowing: but to be distinguished is to be known in the relation of diversity. The attention may not be strongly fixed on the relation—it may seem to be engrossed by either of the two objects; but their diversity cannot be unknown.

But there is scarcely such a thing supposable as a single object. No single object actually exists in the world of matter or of mind. Every so-called object or event in nature, every single state of mind, will readily resolve itself before the attentive eye into many separable elements existing in relations to each other, and held together as one thing by the cementing force of these bonds. An apple, an orange, a pebble, nay, even a grain of sand, consists of parts not a few, united into one perceived whole. A mental state, however simple, is in its essential nature complex, to say nothing of the special relations of time and quality which distinguish it from every other.

This prepares us to assert that to know, always involves two comprehensive acts, each of which corresponds to the other—the act of separation, or resolving objects as wholes into their parts or distinguishable elements, and the act of uniting or combining these parts into their wholes. These acts are technically termed *analysis* and *synthesis*, and they are present in every form and variety of knowledge. In *sense-perception* the different parts of material objects and the objects themselves, are first distinguished and then united under relations of space and time. In *consciousness* they are connected as coexistent, successive, or produced by the active *ego*. In *imagination* they are again separated and reunited. In *thought* or *intelligence*, they are again divided, to be re-combined as constituents of general notions or concepts, of judgments, arguments, inferences, and systems. Thought, indeed, tends to bring all knowledge into the unity of common properties, powers, laws, and ends.

When is the process of knowledge complete?

§ 46. 7. The process or act of knowledge is complete when it is matured into a *product* and this product itself becomes an object to the mind's future knowing. At one time the whole of a mental state becomes such an object; at another, some one element of a single mental state is detached from the act that produced it, and becomes endowed, so to speak, with a separate life. This product, so far as it exists, exists as a mental transcript or representation of the original, whether that original were a *subject-object* or an *object-object*. It is also capable of being recalled, and of itself recalling its original.

The power of producing such permanent and reproducible results is essential to the perfection and the utility of the act of knowing. It is so essential, that upon it depend the simplest acts of the memory and the imagination, without which the mind would be limited to the present, and could neither gather instruction from the past, nor apply wisdom to the future. The higher processes by which man explains the powers and laws of nature would otherwise be impossible, and the capacity to use these powers and to apply these laws in any practical service would be excluded altogether.

The knowledge which is thus separated from the original activity is called *representative knowledge*, with reference to the original act of acquiring, and *mediate or represented knowledge*, with reference to the original objects known. The products thus preserved are called *acquired or positive knowledge.*

The act of knowing is diverse in its energy. Attention.

§ 47. 8. The same act of knowledge, with similar objective conditions, may be performed with greater or less energy. This greater or less energy in the operation of knowing is called *attention;* which word, as its etymology suggests, is another term for tension or effort, and was doubtless first transferred to the spiritual operation from the strained condition of the part or whole of the bodily organism, which accompanies or follows such effort. This effort is manifested in the more or less exclusive and complete occupation of the knowing power by the object or relation that is apprehended. This greater or less effort of attention is followed by the greater or less distinctness, vividness, and completeness in the objects apprehended, and in the objects retained among the mind's permanent possessions, as also by a greater or less facility in exercising a similar activity a second time.

Some of these beings and relations are discerned by the mind with far greater ease than others. To hold the mind to certain classes of objects and relations, is comparatively easy, requires little or no exertion, and is accomplished with spontaneous facility. To know so as to master an unfamiliar object, always involves effort at the first; and a ready facility can only be attained by frequent repetition. Why or how this is so, we need not here explain. The fact is attested by universal observation. It is natural and soon becomes easy to all men to attend to material objects, up to a certain degree of minuteness. It is comparatively difficult and unnatural to consider closely the experiences and processes of the soul. It is easy to decide upon the comparative length and breath of two corporeal objects. It is not so easy to apprehend the parts and relations of a mathematical theorem or of a logical argument. The easier and more natural processes are performed by all men. The more difficult and less natural are reserved for the few. For facility in the one, that education which nature furnishes to all, is amply sufficient. For skill and readiness in the other, special discipline and culture,—literally great pains-taking,—are requisite.

The easier and spontaneous processes are first performed, and are therefore the earliest perfected and matured. The more difficult and artificial are exercised next in order; and readiness and skill in using them is reached at a later period. The powers of sense and outward observation are first developed, next those of memory and imagination, and last of all, those of reflection, thought, and reason.

As it is with the intellectual processes, so is it with their products. We have seen how the products are related to the processes; that as the mental processes are employed and perfected with energetic attention, so the mental products are evolved in completed perfection, as naturally and as certainly as the ripe fruit or perfected seed drops from the plant or tree which has rightly elaborated its organic processes.

The psychological and logical relation of processes and products.

§ 48. 9. In this way there comes to be an organic connection among the products of the intellect, corresponding to the organic relations of the several processes out of which they grow. This relation, as it depends on the development of the soul itself, is called *psychological;* as it implies antecedence and subsequence of time, it is called *chronological.*

Besides the psychological or chronological relation of the powers and products to one another, there is still another, which is more important and fundamental, and that is their *philosophical* or *logical* relation.

We use one of these kinds of knowing to supplement the other, and often not only to supplement, but even to correct its operations and results. Thus we reason to conclusions which we cannot observe by the senses or experience in consciousness. We infer results which we cannot try by experiment, and we predict them before it is time for them to occur. We correct rash conclusions, by looking at principles and laws. We deny assertions, however confident, by employing arguments. We question so-called facts because they do not square with an established theory.

Corresponding to the relation between these *processes* of knowing, there is the relation of *logical dependence* or of *rational connection* between their *products.* One conception is subordinate to another, as a species to a genus ; or one is a property or attribute of another, as a quality of a substance ; or one is contained in another, as an element in its definition ; or is given as a reason for another, as a proof for an assertion, a premise for a conclusion, a datum for an induction, or a means to an end. Many conceptions and truths are also capable of being united in mutual relations of classification and explanation, as constituents of a system. All these are examples of logical relations in mental products.

The logical relations of the products grow out of the philosophical dependence of the processes from which the products are evolved. But inasmuch as the products are expressed in language, and are made objective to the mind, their logical and objective relations are more striking and prominent than the subordination of the acts of knowledge to one another when psychologically considered. The rational faculty asserts for itself intellectual authority over the lower powers, by asserting for its products, the place of criteria, rules, reasons, and principles in respect to the products of the lower. Hence the objective or logical relations are more conspicuous than the psychological and subjective.

We therefore set up a broad distinction between two kinds of knowledge, calling the one *empirical* and the other *philosophical,* the one, knowledge by observation, and the other, knowledge by principles or reasons. We should remember, when we make this distinction, that the same mind uses two ways or processes of knowing, and that these supplement one another. There must, then, be a relation of dependence between the two. The one must be subject to the other, in the mind's own judgment, and according to the ordinances of the mind's own constitution. The mind that *observes* and *acquires*, knows that, by *thinking*, it can correct and aid its own observing, and that the one method of knowing has a certain authority over the other.

Thus, when we analyze a substance, we determine the qualities that are common to its class, and so are enabled to define a general conception, by resolving it into its constituent or necessary elements. We account for or explain a phenomenon which we observe, or a fact of which we hear, by referring to the causes or forces by which it was produced; and these very causes or forces we interpret still further by the laws according to which they act; or we round off and complete the explanation by stating the adaptations to an end or assumed design.

The *psychological* and *logical* relations of knowledge do not always coincide. The order of intellectual growth and of psychological development does not agree with the order of logical dependence and of philosophical arrangement. That which is last in actual attainment, is first in logical importance. The truths and relations which the mind is the latest and the slowest to develop and assent to, may be those which are fundamental to its rational knowledge. It may even be taken as a maxim, that what is psychologically last, is first in logic and in reason.

The critical, or speculative, application of knowledge.

Another and still higher activity of the intellect is the *critical* or *speculative*. It reaches this when having attained the command of its higher faculties, and developed the familiar principles and rules which they involve, it applies them in judging the mind itself, and preëminently its higher powers, for the purpose of testing their trustworthiness and examining their authority. After questioning every other agent in the universe, and judging of its workings, it turns its scrutiny in upon itself, to test the processes by which it knows, and even the very rules and principles which it imposes upon every thing besides; itself included.

Order of intellectual development, growth and studies.

§ 49. The consideration of these facts and relations, enables us to trace the growth of the mind through the stages of its normal development. This development begins with the beginnings of attention. Before this, its activities are, as it were, rudimental only. From this condition the mind awakes when some object attracts and holds its attention. The infant's power to know begins to be developed when it begins to attend. As soon as the infant begins to notice, its vacant countenance for the first time assumes the expression of intelligence, and is lighted with the dawn of intellectual activity. Attention gives discrimination, and discrimination implies objects discriminated. The first objects distinguished are objects of sense. The sensible objects that are first mastered are those which relate to its wants, and generally, so far only as they are related to these wants; first to its appetites, then to its affections and desires. With the discernment of these objects, in their relations to these sensibilities and desires, begins also the direction of the active powers by intelligence.

But though the attention is at first chiefly occupied with sensible objects, and these prominently in their relations to the sensibilities and the practical wants, it is not wholly neglectful of the psychical operations and the psychical self. At a very early period the body is distinguished from the material world of which it forms a part. The soul also begins to be apprehended as diverse from the body, as soon as the purely psychical emotions, as the love of power and sympathy, or the irascible passions, are vividly experienced.

As fast as the attention masters distinct objects, it must separate them into separable ideas or images, which are henceforth at the service of the *imagination and the memory*. These reappear in the occasional dream-life that begins to disturb what was hitherto the animal sleep of the infant. Memory begins to recall past experiences of knowledge and feeling. Recognition finds old and familiar acquaintances in the objects seen a second time. At a later period, imagination begins to imitate the actions and occupations of older persons, and furnishes endless and varied playwork for childhood in the busy constructions of the never-wearied fancy; while it irradiates the emotional life with perpetual and inextinguishable sunshine.

Slowly, the rudiments of *thinking, or the rational processes*, begin to be learned and practised. The attention not only discriminates, but compares. As it compares, it discerns likenesses and differences in qualities and relations. These it thinks, apart from the individual objects to which they pertain. It groups and arranges, under the general conceptions thus formed, the individuals and species to which they belong. To these activities language furnishes its stimulus and lends its aid. Inasmuch as there can be but a limited language without generalization, the infant or child is forced to think, by the multitude of words which catch its ear and force themselves upon its attention; each representing the previous thinking of other men, and even of other generations.

With classifying, are intimately allied the higher acts of tracing effects to causes and illustrating causes by effects. Then, inductions are made by interpreting similar qualities and causes, as exhibited in experience and elicited by experiments. The mind becomes possessed of principles and rules, which it applies in deductions both to prove and explain. The powers and forces of matter and spirit begin to be discerned, as the result of induction and deduction combined. The relations of these powers to their conditions, and to one another, as well as to motion, time, and space, begin to be fixed and definitely stated and the laws of matter and of spirit are ascertained in a wider or more limited range and application. Science arranges all beings and all events into the order of completed systems, by means of the processes of thought; the world of nature is recast into a new spiritual structure, under the relations by which thought decomposes and recombines its individual beings and events, as presented to observation under the relations of space and time. Adaptation and design shoot golden threads of light and order through that otherwise pale and lifeless system of nature, which science reconstructs out of blind forces and fixed mechanical laws. The originating and intelligent intellect of the Eternal Creator and Designer is reached, as the first assumption and the last result of scientific thought.

Last of all, thought turns back upon itself, and *critically* analyzes all its knowledge, and its very power to know. It inquires into and scrutinizes its acquisitions and its assumptions, and challenges its own confidence in its most familiar processes and beliefs. It seeks to justify to itself its acquired knowledge, its science, and its faith, by retracing, under the guidance of logical relations, every step it has taken, and every stage through which it has passed in its development and growth. It lays bare the necessary assumptions, the primary and universal relations, which are acknowledged and acted upon in all observation, in all science, and in all faith. It returns again from the course of its speculative criticism, to confide a second time in this knowledge and the faith which it could not but acquire and apply in its progressive synthesis, and which it now has learned to vindicate by its retrogressive analysis.

These critical and speculative processes of thought are reserved for but a few of the race to prosecute. They are, however, the normal and the necessary consummation of the completed growth of the fully developed man.

The consideration of the development and growth of the intellect furnishes the principles by which to regulate the *culture of the intellect,* and to arrange the *order of its studies.*

The studies which should be first pursued are those which require and discipline the powers of observation and acquisition, and which involve imagination and memory, in contrast with those which demand severe efforts and trained habits of thought. In early life, objective and material studies should have almost the exclusive precedence. The capacity of exact and discriminating perception, and of clear and retentive memory, should be developed as largely as possible. The imagination, in all its forms, should be directed and elevated—we do not say stimulated, because, in the case of most children, its activity is never-tiring, whether they be at study, work or play.

We do not say, cultivate perception, memory, and fancy, to the exclusion or repression of thought, for this is impossible. These powers, if exercised by human beings, must be interpenetrated by thought. If wisely cultivated by studies properly arranged, they will necessarily involve discrimination, comparison, and explanation. To teach pure observation, or the mastery of objects or words, without classification and interpretation, is to commit the error of simple stupidity. But, on the other hand, to stimulate the thought-processes to unnatural and prematurely painful efforts, is to do violence to the laws which nature has written in the constitution of the intellect. Even thought and reflection teach us that, before the processes of thought can be applied, materials must be gathered in large abundance; and to provide for these, nature has made acquisition and memory easy and spontaneous for childhood, and reasoning and science difficult and unnatural.

The study of language should be prosecuted in childhood, as it is, in fact, in the acquisition of the mother-tongue. In the acquisition of other languages the methods by which the vernacular is learned should be followed so far as is possible. Grammar, so far as it is required, should be simple, plain, and practical. Its theories should be kept in the background, its terminology and principles should be the reverse of the abstract. The contrasts and comparisons involved between the strange and the familiar, will stimulate and guide to the first beginnings of reflective grammar. The memory for words should be exercised and stimulated. Choice tales and poems—narrative and lyric,—should be learned for recitation. Natural history in all its branches, as contrasted with the sciences of nature or scientific physics, should be pursued with the objects before the eye—flowers, minerals, shells, birds, and beasts. These studies should all be mastered in the spring-time of life, when the tastes are simple, the heart is fresh, and the eye is sharp and clear. The facts of history and geography should be fixed by repetition and stored away in order.

But science of every kind, whether of language, of nature, of the soul, or of God, *as science*, should not be prematurely taught. For the consequence is, either disgust and hostility to all study on the one hand, or, on the other, superficial thinking, presumptuous conceit, and, worst of all, sated curiosity.

The law of intellectual progress involves effort and discipline severely imposed and constantly maintained, but the effort and discipline should follow the guidance of nature.

§ 50. The consideration of the nature and the development of knowledge teaches on what principles we may divide and classify the powers of the intellect.

Principles of classifying the powers of the intellect.

In assigning different faculties to the intellect, we do not divide it into separable parts or organs. When we say that the intellect has faculties, we mean only that the soul, as the intellect, acts under certain conditions in clearly distinguishable operations which terminate in definite and determinable results or products. The consideration of the soul's development gives the conditions of these faculties. The consideration of the logical relations of the products assigns to these faculties their relative authority and importance.

In tracing the development of the intellectual powers in their succession, we do not exclude the co-action of the other so-called faculties of the soul, as of feeling and will. Their presence and agency have already been recognized with sufficient prominence. Nor do we deny or overlook the truth, that the several powers of the intellect act together in the earlier stages of its growth, and in all the periods of its history aid and direct one another. The action of a single power of the intellect does not exclude the co-action of the other powers. On the other hand, it is to be remembered, that as the energy of the whole soul is so far limited that one psychical state is preëminently a state of feeling, another intellectual, and another voluntary, so, of the intellectual activities, one is likely to be predominantly an act of sense rather than of memory, and another an act of the imagination rather than of intelligence.

When it is said that one power, as defined, is, in the order of time and growth, developed sooner than another, it is not intended that each lower power is completely matured before the other and higher is used at all, or that distinctly traced boundary lines mark off the several stages of the mind's development. This would involve the absurdity of teaching that the child perceives with the senses for a long time before it begins to remember, and that it remembers and imagines for another long period, before it generalizes and explains. What is asserted is, that sense must begin before memory and thought are possible, and that, as a power, it is perfected before thought has reached its consummation. Conversely, it will be found to be true in fact, that many acts which we call acts of sense-perception are largely intermingled with acts of representation and thought; also that acts of memory recall past objects under the laws of association which thought makes possible; while imagination, in which thought is not largely conspicuous, is scarcely worthy the name.

These cautions being premised, we observe that the powers of the intellect are clearly distinguishable by *the order of their development and application, as manifested in the character and relations of their products.*

The leading faculties of the intellect are three: THE PRESENTATIVE, OR OBSERVING FACULTY; THE REPRESENTATIVE, OR CREATIVE FACULTY; THE THINKING, OR THE GENERALIZING FACULTY; or, more briefly, the FACULTY OF EXPERIENCE, the FACULTY OF REPRESENTATION, and the FACULTY OF INTELLIGENCE. Each of these has its place in the order of intellectual growth and development. Each has its appropriate products or objects. Each acts under certain conditions or laws. Each of these leading faculties is subdivided into subordinate powers, which are distinguishable from one another in like manner with their primaries.

The presentative faculty.

§ 51. I. The *presentative faculty*, or the faculty of acquisition and experience, is subdivided into *sense-perception* and *consciousness;* or, as they are sometimes called, the outer and the inner sense.

In the order of the mind's development these are exercised *first and earliest of all.* The intellect begins its activity with observing objects of sense. Closely connected with this is the consciousness of the soul's inner experiences, prominent among which are its sensations of pleasure and pain. Not only does this order actually occur, but it is impossible for us to conceive of any other as possible. The mind must observe before it remembers; unless it had previously observed and acquired, it would have nothing to remember or imagine.

The objects or products with which this power is concerned, or which it evolves, are *individual* objects. In this respect they are distinguished from the objects of thought, which are always *general.* But this feature they share with those of memory and imagination, which are also individual. From these last they are still further distinguished by being presented for the first time; hence the epithet *presentative* is applied to the faculty by which they are known. This feature is made still more precise by their individual *relations in space and in time.* The objects of sense are known in space, as being *here,* and the objects of consciousness are known as *now* in time. These two relations they share with the objects of no other power. They are also mutually related to one another, the one being an individualized *non-ego,* the other being a determinate state of the *ego.*

The conditions of the acts of sense-knowledge are the existence of the living body in connection with a sentient spirit, and the excitement of the same by material agents. Some of these are bodily, some are psychical. Some of these are known to physiology, others are wholly unknown, but so far as they are knowable, they are appropriately considered in explaining the power of sense-knowledge.

The condition which furnishes or constitutes the object for the *act* of *consciousness,* is that the soul should in fact act or suffer in a present and individual state. Consciousness takes heed of the fact, *i. e.*, of the operation, and cognizes that it is. Whence or how it is that the soul furnishes this material, or how the soul is able to act in these varied forms, it can do little to explain. These operations lie out of the range of consciousness; they are presupposed by it.

But these *objective* conditions are not alone. There are also subjective conditions of the presentative power in both consciousness and perception. Let the external world and the quick sensibility both conjoin to furnish ample material through eye and ear; let the active and eager soul exercise the most varied forms of act or affection; if the perceiving or conscious spirit does not attend, it will fail to notice, and of course will fail to know.

The representative faculty.

§ 52. II. Next to the presentative comes *the faculty of representation.* That this is developed second in order of growth and of time to the soul's power to acquire and observe, is obvious.

The objects or products of this power are individual objects, like the objects of sense and of consciousness. They differ from them in this, that they are representative of them. They are therefore not real, but mental objects. They are wrought or created by the mind itself, but always with respect to some real object actually experienced. This is their common characteristic, that they represent observed and experienced objects. They are *representative; i. e.*, they present a second time, and so take the place of objects previously known.

In representing these objects, the mind acts in two ways—as the *memory;* and as the *imagination* or *phantasy:* and hence the representative power is divided into these two. In memory it knows that the mental object represents an object previously known. In imagination it changes the representative object into another, such as it has never actually experienced. According as it changes the object in more or fewer particulars, and with special applications, does the imagination receive different names.

The conditions of the representing power, are, that the soul should retain and reproduce past objects for the memory to recognize and the imagination to modify. If the soul refuses to furnish these appropriate objects, neither the memory nor the imagination can know their objects. For this reason, the *power* of the soul to *retain* and *recall* is essential to the power to know these mental objects when represented. Concerning the actings of the capacity of the soul to retain and reproduce we know little directly, but indirectly we know very much: that is, we know how we can affect its actings by our own conscious energies in acquiring. The relations and laws by which acquired objects can be reproduced are more obvious and better established than almost any other psychological truths. These are all comprehended under the familiar title of the *association of ideas*, and they very properly enter largely into the consideration of the representative power.

§ 53. III. The power of *thought* is developed last of all in the order of the soul's evolution or growth. It is also called the *intelligence*, and *the rational faculty*.

Thought, or intelligence.

This power requires for its possible exercise some range of observation, some acquisitions of memory, and some creative activity of imagination. For its effective energy and its actual application it must be preceded by many separate exercises of all these functions. To the thorough and persistent use and the complete development of this power, the soul is most of all disinclined; and therefore it is perfected and developed later in the order of time.

But though this power is last and reluctantly developed, it surpasses all the others in dignity and importance. It explains facts and events by powers and laws. It enforces conclusions by premises. It accounts for inferences by data. It lifts observation up to the dignity of science, and establishes it on the firm foundation of principles. It enables us to interpret the past and to predict the future.

The *products* or *objects* of this power are always *generalized* objects. They are *universals*, as contrasted with individuals. This difference distinguishes this power of the intellect widely from the two others. These products are known by various names, as the *concept*, the *class*, the *judgment*, the *argument*, the *induction*, the *interpretation*, and the *system*.

In accordance with these distinguishable products, the intellect is said to perform all the acts which require the several powers or faculties of generalizing, classifying, judging, reasoning, inferring, explaining, and methodizing the individual objects given by experience. Hence the intellect is sometimes said to be endowed with as many subordinate faculties.

The most obvious aid or instrument provided by Nature for furthering these processes and retaining their products, is *language.* For this reason the existence of language is regarded as a necessary result of the power of thought, and the use of language is regarded as the indication of its presence and exercise.

The *conditions* of thought, as distinguished from the materials or occasions of thought which experience furnishes, are certain relations discerned and generalized by the power of thought itself. The reality of these relations is an assumed condition of these peculiar operations; and when the mind comes to apprehend them, it must proceed upon the belief that they are universally present and incontestably valid. In this sense the mind itself prepares for itself these objects of its own apprehension. For the service of thought, all individual objects must be believed to be connected or bound together under *universal and necessary relations* or *categories.* Such are the relations of *substance and attribute, cause* and *effect, means and end.* Thus the relation of *substance* and *attribute* is assumed as real in order to the possibility and truth of the acts of generalizing and of judgment. The relation of *cause* and *effect* must be presupposed to give meaning and force to acts of reasoning and explanation. The relations of *design* are the prefatory conditions of acts of induction. But universal or generalized objects presuppose the existence of individual concepts and their relations. To individual beings and events, space and time relations are presupposed. Therefore, in order to the products of thought, the intuitions of *space* and *time* are presupposed. These relations are said to be *a priori,* for the reason that they are presupposed in these processes. They are called *intuitions, categories, primitive cognitions,* etc., etc. They are said to be *universal,* because applicable to every individual object in the way explained. They are *necessary notions,* because they are necessarily applied by the mind in all its thought-activities, and to all thought-objects.

They are, however, no more necesary to thought than they are to presentation and representation. We imply and suppose them as truly, though not as conspicuously, in perception and consciousness, in memory and imagination, as we do in classification and reasoning.

But it is by means of thought that we discern and define these categories. It is only as we use thought-processes *critically* —*i. e.*, as we generalize and analyze our own mental processes— that we discover these relations as everywhere and necessarily present. Though they are actually present as the conditions and elements of all our knowing, it is only by thought that we discover and demonstrate their presence and their application, as the conditions of all knowledge.

Two aspects or forms of thought.

In view of the two methods in which the thought power is employed, the power itself has been subdivided by many writers into two: the *elaborative faculty*, as performing the processes, and the *regulative*, as furnishing the rules—or more properly as prescribing the sphere and possibility—of thought. These are named also the *dianoetic* and the *noetic* faculty. By some writers they are distinguished as the *understanding* and *reason*, in a usage suggested by Kant, but deviating materially from his own. Milton and others call them the *discursive* and *intuitive reason.*

We prefer to say that the analysis of the thinking power involves two heads of inquiry:

(1.) What are the several processes of thought of which the intellect is capable in the order of their development, the manner of their action, their conditions, and their products? So far as psychology prosecutes these inquiries, it considers them subjectively as processes of the soul. When we go further, and proceed to define their products as expressed in language, to derive rules to direct the knowing processes or to test what is known, psychology passes over into the service of *logic.*

(2.) What are the ultimate relations or categories which thought brings to light, and, which, all knowledge presupposes? What is their authority and trustworthiness? What is their relation to special acts of knowledge? What application can be made of them to the discovery of truth and the detection of error? Last of all, how can they be applied to vindicate man's confidence in his own knowledge, and in his very power to know?

All these questions when prosecuted with reference to the subjective power of the soul to evolve and apply these intuitions, belong legitimately and necessarily to psychology.

So far as the intuitions themselves, objectively considered, are made the subjects of analysis and discussion; so far as their relations to one another, and the structure of human knowledge, are examined: so far, in short, as they are made the subject of critical or speculative discussion, they lead us within the field of *metaphysics*, ontology, or speculative philosophy, for which, as has been already explained, psychology is the direct and necessary preparation.

We divide therefore, our treatise into FOUR parts, with the following titles: I. PRESENTATION; II. REPRESENTATION; III. THOUGHT; IV. INTUITION; the last two being devoted to *Thought proper* and *Thought critically applied* to the analysis of knowledge and the discovery of the categories or ultimate relations which are the conditions of its processes and products. For the explanation and justification of this division we must refer to the foregoing remarks, and the subsequent treatment of the topics themselves.

THE HUMAN INTELLECT.

PART FIRST.

PRESENTATION AND PRESENTATIVE KNOWLEDGE.

CHAPTER I.

CONSCIOUSNESS—NATURAL CONSCIOUSNESS.

Consciousness defined. Variously applied.

§ 54. WE begin with PRESENTATIVE KNOWLEDGE. Of objects presented to the mind there are two classes; objects of matter, and objects of spirit. Corresponding to these two classes of objects, two powers or faculties are distinguished, viz., CONSCIOUSNESS and SENSE-PERCEPTION. We shall first treat of *consciousness.* It is briefly defined as *the power by which the soul knows its own acts and states.* The soul is aware of the fleeting and transitory *acts* which it performs; as when it perceives, remembers, feels, and decides. It also knows its own *states;* as when it is conscious of a continued condition of intellectual activity, a gay or melancholy mood of feeling, or a fixed and enduring preference. Whether the state is in such cases in fact prolonged, or only repeated by successive renewals, we need not here inquire; it is sufficient that states of the soul are distinguished from acts by their seeming continuance.

Again, the terms *conscious* and *consciousness* are often applied to any act whatever of direct cognition, whether its object be internal or external. In other words, they are used as equivalent to knowing, perceiving, etc., or to knowledge in any form. Thus we say, 'I was not conscious that you were in the room;' or, 'I was not conscious that he was speaking;' as well as, 'I was not conscious of being angry.' In cases like these the terms designate an act of simple perception and knowledge. The reason why they come to do so is, that every act of knowledge, whatever be its nature or object, is attended by consciousness. The phrase, 'I was not conscious that you were in the room,' is explained as meaning, 'I was not conscious of seeing you in the room.'

Consciousness is also employed as a *collective term for all the psychical states.* In the words of Sir William Hamilton, "it is a comprehensive term for the complement of our cognitive energies." Every such state or energy is attended by consciousness; it is an act or state of which we are conscious, or, as we sometimes say, it is a conscious act or state. The sum-total of all such acts is therefore expressively described as the consciousness of an individual. It is equally true that we are conscious of our *states of feeling,* and these may also be designated by the same general and comprehensive term, though with somewhat less propriety.

Consciousness is often figuratively described as the '*witness*' of the states of the soul, as though it were an observer separate from the soul itself, inspecting and beholding its processes. It is called the 'inner light,' 'an inner illumination,' as though a sudden flash or steady radiance could be thrown within the spirit, revealing objects that would otherwise be indistinct, or causing those to appear which would otherwise not be seen at all. Appellations like these are so obviously figurative, that it is surprising that any philosopher should use them for scientific purposes, or should reason upon, or apply them with scientific rigor.

The terms *conscious* and *consciousness* explain their own meaning, and confirm the truth of the assumption and belief that the fact is true which this language implies. They describe a *knowing with,* or within the knowing agent, and they imply that the states of the human soul may be known by the soul to which they pertain.

The power of the soul thus to know itself is often called *the internal,* or *the inner sense.* This term is suggested by analogy. As the soul, by the external sense or senses, apprehends the properties and qualities of matter, so it is said to know its own states and powers by another, *i. e.,* an inner sense.

Consciousness has also, for the same reason, been called by many philosophers, as Leibnitz, *ad-* or *ap-perception,* by which term the same fact is recognized that the word consciousness implies, viz., a perception of the mind's own activities, in addition to the perception of the objects of those activities.

The term *Bewusstseyn,* and its cognates in the Teutonic languages, recognizes the distinct—rather than the accompanying—knowledge which consciousness always involves. It describes a *be-*, rather than a *con*-knowing; *i. e.,* the clear and completed knowledge which the mind usually attains by a second and more attentive look. Hence it is with eminent propriety applied to that knowledge which the soul has of its inner states, as this, to be of any service, must be earnest and attentive. The word in German, however, is not so closely limited to this internal knowledge, as is consciousness, in English. It is supplemented by self-consciousness—*Selbst-bewusstseyn.* Hence sometimes, when we should use consciousness only, the Germans would say self-consciousness. Their more usual technical appellation for the power is the inner or internal sense.

Reflection is the appellation used by Locke for this power; or, more exactly, it is under this appellation that he discusses its nature and authority. Hence, among many English writers reflection is freely used as the exact equivalent of consciousness. It is the great and distinctive merit of Locke to have called attention to this as a separate source of knowledge, and to have claimed for the knowledge which it furnishes equal authority and certainty with that which is received through the senses. We quote a passage memorable in the history of psychology.

"The other fountain from which experience furnisheth the understanding with ideas, is the perception of the operations of our own minds within us, as it is employed about the ideas which it has got; which operations, when the soul comes to reflect on and consider, do furnish the understanding with another set of ideas, which could not be had from things without; and such are *perception, thinking, doubting, believing, reasoning, knowing, willing,* and all the different actings of our own minds; which we, being conscious of, and observing in ourselves, do from these receive into our understandings as distinct ideas as we do from bodies affecting our senses. This source of ideas every man has wholly in himself; and though it be not sense, as having nothing to do with external objects, yet it is very like it, and might properly enough be called internal sense. But as I call the other, sensation, so I call this reflection, the ideas it affords being such only as the mind gets by reflecting on its own operations within itself."—*Essay,* Book ii. chap. i. § 4.

§ 55. Consciousness is exercised in two forms, or species of activity, viz., *the natural* or *spontaneous* and *the artificial* or *reflective.* They are also called by some writers *the primary* and *the secondary consciousness.* The one form is employed by all men; the other is attained by few.

Two forms of consciousness.

The first is a gift of nature and the product of spontaneous growth; the second is an accomplishment of art and the reward of special discipline. The natural precedes the reflective in the order of time and of actual development. But it does not differ from it in kind, only in an accidental element, which brings its results within our reach and retains them for our service. This is the general conception which we form of both, as preliminary to the special consideration of each.

The capacity to attend to the psychical states in the lowest appreciable degree—*i. e.*, with that energy which leaves any permanent product or result for the memory or imagination—is matured by the slow education of infancy and childhood (§ 64). During this period, even under the most favorable circumstances, the growth and development of consciousness is steady, but slow.

Where consciousness is energized by attention, and applied to psychical phenomena for scientific purposes in the interest of psychological science, it is called the secondary, the artificial, the philosophical or reflective consciousness, or simply, reflection. As such, it is distinguished from and contrasted with the primary, the natural, the common, the unreflecting consciousness, or simply, consciousness. The division indicated by these contrasted terms is convenient and important. It should always be remembered, however, that the two so-called species of consciousness do not differ from one another in kind, but in degree, and that there is no well-defined and sharp line of distinction that divides off the one from the other.

Natural consciousness *as an act.*

§ 56. We notice first *the natural, or primary consciousness.* Natural consciousness is the power which the mind naturally and necessarily possesses of knowing its own acts and states. It may be further described by considering it in its operation and its objects, or as consciousness *the act,* and consciousness *the object.*

We begin with *consciousness the act.* As an act, it is a necessary and essential constituent of many active conditions of the soul. The soul cannot know, without knowing that it knows. It cannot feel, without knowing that it feels; nor can it desire, will, and act, without knowing that it desires, wills, and acts.

Consciousness is an act of *knowledge*, and is therefore an act purely and simply intellectual. The states observed may be psychical, in any form, *i. e.*, states of intellect, sensibility, or will —but the act by which they are known is intellectual only. It is an act of *direct or intuitive* knowledge. To attain it, neither memory nor reasoning are required, nor any indirect process or succession of acts, but the soul immediately knows its present condition or act. It confronts it face to face. It knows it as now existing. It is eminently presentative knowledge.

Consciousness, as an act of knowledge, is matured into, or results in a peculiar product. When it is complete, it furnishes for the mind's recall an idea of the object known. This is a purely intellectual result. What the mind is conscious of may be a state of knowledge, feeling, or choice, but the feeling and choice which we reproduce in memory is not a feeling or choice, but our idea or image of a feeling or choice, and this is purely intellectual. As an act of knowledge, it involves the discernment of relations (§ 45). We know the state to be our own; *i. e.*, we discern its relation to the ego. We know that the present is not the past state of the soul; *i. e.*, we know the two under the relations of contrast and of time. Again, the knowing agent distinguishes itself as the conscious observer from itself and its own states as the object observed. Like every act of knowledge it is at once an act of analytic separation and synthetic union.

The act of consciousness is a *peculiar* intellectual act—an act that is preëminently *sui generis*. Especially is it peculiar in the conditions of its exercise. To most of the other acts of knowledge it is required that their objects should exist before they are known. But in this peculiar process the object and act are blended in one. Thus, the landscape on which I gaze is a permanent object, to which I can bring and from which I can withdraw my mind. The thought or feeling which I remember must have been experienced in order that it may be known a second time. It is rashly concluded by many that this is a necessary and universal condition of all knowledge. What is asserted of consciousness violates, as is objected, the first and essential requirement, that something should *have* existed, in order to be known. 'How can I know that I know,' it is urged, 'unless I have first known, in order to furnish an object for me to know?' Or it is concluded that consciousness is, at best, but a kind of memory, an act that immediately follows the act or state of which we are said to be conscious. "No one," says Herbert Spencer, "is conscious of what he *is*, but of what he *was* a moment before. That which thinks, can never be the object of direct contemplation; seeing that, to be this, it must become that which is thought of, not that which thinks. It is impossible to be at the same time that which regards and that which is regarded." (*Principles of Psychology, Part i. chap. i.*) The force of this objection lies in the assumption, that every thing which is known must have already existed. But this assumption is unauthorized. It is founded on a supposed analogy between this and other acts of knowledge. It by no means follows, because the landscape must have existed before we see it, or the mental state must have occurred before we remember it, that a perception or feeling must be past before we can be conscious of it. Besides, how can one remember that which he did not know at the time when it occurred? How can one recall the state in which he was a moment before, and know that he had been in that state, if he was not conscious of it at the precise instant in which it occurred? Those that resolve acts of consciousness into acts of memory, make an act of memory itself impossible. The remembering act necessarily follows the act which is remembered however closely. We cannot recall the act itself, nor that it was our own act, unless we knew both, when the act occurred.

Consciousness the object.

§ 57. From the consideration of consciousness the act, we pass to *consciousness the object*. The object of consciousness has already been defined to be an act or state of the soul; more exactly, the soul acting and suffering in an individual state. That such an object should be peculiar and unlike any other, we are prepared to believe, by what we have already noticed under consciousness as an act. Other peculiarities will reveal themselves to a closer inspection.

We observe, in general, that objects of consciousness are unlike the phenomena of matter in this, that they are given to observation as essentially *complex* even in their greatest simplicity. Every state or condition of the spirit in actual experience and as known by the soul, is complex, even in its extremest simplicity. The object is *threefold* in its elements, every one of which must be recognized by the conscious spirit. The elements are, the identical *ego,* either agent or patient according as the case may be; the *object* with respect to which it acts or suffers; and the present *state or action* in which it exists or acts. Every psychical state of which we are conscious implies an acting or existing *ego,* to which the state pertains. A condition of the soul without an individual person acting or feeling, is impossible as a conception, and is never experienced as a fact. Again, this *ego* is known to be in a definite form or condition of action or suffering. The states are transient, the agent remains. The states are as fleeting and as transitory as the flying moments; indeed, they come and go more swiftly than any instants which we can count; the individual self remains unchanged, referring all these changes to itself. Again, the *ego,* in its acting and suffering, is concerned with some object. It must have some object to be employed upon, either material or mental. One state is as often distinguished from another by its object, as by any thing besides. These are the elements which make up that complex whole which we call the object of consciousness.

Relation of consciousness to each of the elements of a psychical state.

§ 58. It is a natural question, What is the relation of consciousness to each of these essential constituents, either as combined together in a general view, or as each calls forth special and separate attention? To this question we give this general preliminary answer: The soul, in consciousness, is directly cognizant of all these elements, as entering into every one of its states. It knows them as distinguishable from one another, and yet as, in their union, constituting a single whole.

Here we observe that, in an act of direct or intuitive knowledge like consciousness, it is as essential that the connecting relations should be apprehended, as the parts which they bind or connect. In logical analysis, the parts are considered separately, and to each we assign a separate word or phrase; but in the synthesis of real knowledge the parts are viewed together. The verbal expression of a mental state is not a single word, as *I, perceive* [or] *love, this apple,* each apprehended apart, and then somehow aggregated into a phrase or proposition; but it is a finished proposition, in all its parts and relations, as, I perceive [or love] this apple. In other words, we can analyze or separate only what is given as united in the concrete or real. If the parts and connecting relations are not discerned together by an intuitive act, they can neither be separated nor united by any other act or process. The objects known by consciousness are intuitively known. All the materials which mediate or abstract knowledge *evolves* from these objects, the objects must be known already to *involve.*

But though these elements are always recognized in every object of which we are conscious—*i. e.*, in every conscious mental state—they are not regarded with equal attention. According as one or other of these elements receives the chief attention and is most absorbing, so is each state of consciousness definitely and peculiarly marked. We will consider the predominance of each of these elements singly and apart.

The activity may be chiefly noticed.

§ 59. First let the *soul's own activity* be the special object of its own conscious observation.

The states come and go, they rise and fall, they are varying and restless as the waves of the ocean, each pushing forward the one that went before. Moreover, these states are the products of the soul's own energy, or the sufferings or joyful experiences of its own sensibility. What can it be conscious of, if it knows not these? For these reasons no one has ever doubted that the operation or state of the soul is the appropriate object of consciousness—is the central element, the element *par éminence,* if the object is believed to be complex; the sole object, if the object is conceded to be simple.

Consciousness of the *ego.*

§ 60. Second. Of the *ego* itself we are also directly conscious. Not only are we conscious of the varying states and conditions, but we know them to be *our own states; i. e.*, each individual observer knows his changing individual states to belong to his individual self, or to himself, the individual. The states we know as varying and transitory. The self we know as unchanged and permanent.

It is of the very nature and essence of a psychical state to be the act or experience of an individual *ego.* We are not first conscious of the state or operation, and then forced to look around for a something to which it is to be referred, or to which it may belong. A mental state which is not produced or felt by an individual self, is as inconceivable as a triangle without three angles, or a square without four sides. This relation of the act to the self is not inferred, but is directly known.

The fact of memory proves this beyond dispute. In every act of memory we know or believe that the object now recalled was formerly before the mind; in other words, I, the person remembering, did previously know or experience that which I now recall. But how could this be possible, if the first act or state was not known, when it occurred, to belong to the same *ego* which now recalls it? This truth has been extensively overlooked or denied. Thus Hume says: "For my part, when I enter most intimately into what I call *myself* I always stumble on some particular perception or other, of heat or cold, light or shade, love or hatred, pain or pleasure. I can never catch myself at any time without a perception, and never can observe anything *but* the perception." "If any one, upon serious and unprejudiced reflection, thinks he has a different notion of himself, I must confess I can no longer reason with him. . . . He may, perhaps, perceive something simple and continued, which he calls *himself*, though I am certain there is no such principle in me."—*Human Nature, Part* iv. *sec.* 2. Dr. Thomas Reid says: "I am conscious of perception, but not of the object I perceive; I am conscious of memory, but not of the object I remember." But he guards himself against the conclusion drawn by Hume from their common assumption, by insisting that, though consciousness does not give us the intuition of self, yet we have a firm belief of the reality of the self, through a native and necessary suggestion, for "our sensations and thoughts do also suggest the notion of a mind, and the belief of its existence and of its relation to our thoughts."—*Inquiry, chap.* ii, § 7. Dugald Stewart says: "We are conscious of sensation, thought, desire, volition, but we are not conscious of the existence of the mind itself. This is made known to us by a suggestion of the understanding consequent on the sensation, but so intimately connected with it that it is not surprising that our belief of both should be generally referred to the same origin."—*Phil. Essays,* p. i, c. i. Dr. Thomas Brown says of a special sensation, as of fragrance: "There will be, in the first momentary state, no separation of *self* and the *sensation,* no little proposition formed in the mind—I feel, or I am conscious of a feeling, but the feeling and the sentient *I*, will for the moment be the same. If the remembrance of the former feeling arise, and the two different feelings be considered by the mind at once, it will now, by that irresistible law of our nature which impresses us with the conviction of our identity, conceive the two sensations which it recognizes as different in themselves, to have belonged to the same human being—that being to which, when it has the use of language, it gives the name of self, and in relation to which it speaks as often as it uses the pronoun I."—*Lecture* xi. Hamilton says: "On the other hand, as there exists no intuitive or immediate knowledge of self as the absolute subject of thought, feeling and desire, but, on the contrary, there is only possible a deduced, relative and secondary knowledge of self as the permanent basis of these transient modifications of which we are directly conscious, it follows," etc.—*Notes on Reid, (H.,)* p. 29, *b.*—*Cf. Met. Lec.* 19, *on Mental Unity.* Mansel dissents from Hamilton on this point. (*Prolegom. Log.*, c. v.) "I am immediately conscious of myself, seeing and hearing, willing and thinking." James Mill agrees with Brown, etc.: "To say that I am conscious of a feeling, is merely to say that I feel it. To have a feeling is to be conscious, and to be conscious is to have a feeling. To be conscious of the prick of a pin, is merely to have the sensation."—(*Analysis of the Human Mind, Chap.* v.) But he corrects himself in another passage, as follows: "The consciousness of the present moment is not absolutely simple, for whether I have a sensation or an idea, the idea of what I call myself is always inseparably combined with it. The consciousness, then, of the second of the two moments in the case supposed, [the case of remembering a preceding state,] is the sensation combined with the idea of myself, which compound I call 'myself sentient,'" etc.—(*Id. Chap.* x.) John Stuart Mill says, in the same strain: "My mind is but a series of feelings," and *defines it* as, "a thread of consciousness," "a series of feelings with a background of possibilities of feeling."—(*Exam. of the Phil. of Hamilton*, c. 12; *cf. McCosh, Fundamental Truth*, etc., c. 5.)

It will be found, moreover, that all those writers who deny or doubt this, do yet incidentally betray their faith in the reality which they by words or reasonings oppose. Dr. Brown, who is so earnest in opposing it, cannot thread together the several experiences of the soul's life, without resorting to "the irresistible law of our nature which impresses us with the conviction of our identity," and James Mill himself is forced in one sentence to confess what he stoutly denies in another, "for whether I have a sensation or an idea, the idea of what I call myself, is always inseparably combined with it." These are more or less distinct acknowledgments of that direct knowledge of the *ego* which enters as an essential constituent into every conscious state of the soul.

The relation of consciousness to the objects of psychical activity.

§ 61. Third, we inquire still further, What are the relations of consciousness to the *objects* of the psychical acts and states? Is the soul conscious of the objects as truly as it is of the states themselves? When I gaze upon a landscape, and am delighted, am I conscious of the landscape which I see, as truly as I am conscious of the act of seeing and of the delight which it gives? It is maintained that it is a gross impropriety to say that we are conscious of the landscape, except in the general sense in which we use *conscious* as the equivalent of *knowing*. Thus Reid says in the words already cited: "I am conscious of perception but not of the object I perceive, I am conscious of memory, but not of the object I remember."

The truth is, that we are conscious of the object somewhat as we are conscious of the *ego*. The state or operation is the central object of apprehension; but as the state can neither occur nor be known except as having a relation to the unchanging *ego*, so each separate state is distinguished in part by its object. This is especially true if it is preëminently a state of knowledge. We distinguish one such state from another by what we know; *e. g.*, in one moment I perceive a tree, in another a house, etc., *i. e.*, I cannot be conscious that I perceive a house or a tree, unless I notice the relation of the act itself to the house or tree.

We do not say that two states of knowledge cannot be distinguished subjectively as well as by their objects. We know that an act of knowledge never can occur by itself without some feeling, desire and will. So far as it is a state of feeling and will it is purely subjective. These subjective elements attract the notice of consciousness preëminently, and these mark and individualize the state to the soul's memory. But when such states are described in language or recalled to the thoughts by an explicit statement, they are described by their objects. Even a state of the most absorbed feeling is indicated by the object or event which excited the emotion. We cannot conceive it possible that we should know that we know, enjoy, or choose, without knowing *what* we know, enjoy or choose. In other words, in being conscious of an act or state, we must be conscious of the state or act in relation to, and as therefore including the object.

We recapitulate thus: The object of consciousness is a state or act of the soul; this state or act must occur or exist in order that it may be known; but it does not exist before it is known in the order of time, but only in the order of dependence, or of logical necessity. So far as the order of time is concerned, it exists while it is known. What is known of this object must depend on the nature of the matter to be known, and also on the reach or capacity of consciousness to observe it.

A psychical act or state, as we have seen, is in its nature complex, consisting of three elements in intimate relation to each other: the *ego;* the object; the acting or suffering of the passing moment. But the act or suffering is inconceivable, except as belonging to the *ego* and defined by the object. Of this double relation consciousness must take notice. It must, therefore, also take notice of the terms or elements which are related.

The object of consciousness is a state of being.

§ 62. We observe still further, that consciousness the object, as contradistinguished from consciousness the act, is a state or condition of *being,* as contrasted with an act of knowledge. Knowledge of every kind as has been shown, supposes and requires being as its objective correlate. The being, known by consciousness, is a spiritual being, a permanent identical agent or producer of the states and acts which are known; *i. e.,* a being in the eminent and higher sense, substantial or real being. This the mind knows to be, or to exist, by a direct or immediate act of its own. In every act of consciousness, knowledge is directly confronted with actual being, and the being which is known is affirmed to be identical with the being which knows.

Special sense of *cogito, ergo sum.*

The saying of Descartes, *Cogito, ergo sum,* has preëminent propriety and obvious truth when applied to the act of consciousness. It means more than, I find myself a thinking being, and therefore I, the thinking being, exist; but it means *conscius sum*, that is, I know directly and positively myself as a being. It has been said with eminent truth that absolute skepticism is incompatible with *the act* of consciousness; because, if I doubt or question any reality, or whatever reality I doubt or question, I cannot doubt or question that I myself doubt or question. The same truth is confirmed by the view already taken, that to consciousness as the act, an object must be present and known; and this object is an existing being, which is known or affirmed by the very act of consciousness to exist.

The validity of relations is also established.

§ 63. Not only is the reality and validity of *being* thus established, because involved in the act and object of consciousness, but the *relations* of being are as necessarily affirmed. The several states of the soul are not only discriminated as diverse from one another, but they are known to be like and unlike. They are also known to be produced by the soul which is conscious that they exist; that is, they are known under the relation of causation.

In view of these facts, we need not wonder that even the ancient philosophers counted the human soul, thus known by and to itself, to be a microcosm or epitome of the great universe. In the spirit of man, and in the exercise of the simplest and the most essential of its powers, thought and being are both conjoined; the one is confronted with the other, the one is essential to the other. Thought is perpetually springing out of being, and apprehending being to exist—not only simple being, but being in all its forms of activity and the relations which they involve.

Nor should we be surprised to find that all the conceptions which are necessary to scientific knowledge—those categories which cannot be proved, but which must be assumed—those prime relations and first truths on which all our higher intelligence of matter or spirit depends, are affirmed of spiritual being in the act of consciousness itself. It is natural to man to make himself the measure of the universe—*i. e.*, to take the little universe of being which he knows so directly and so well, with the relations involved, to be the analogon of the greater universe which lies beyond, and which is more indirectly known. This is the process by which many explain our belief in the authority and universality of the categories or first truths.

The development and growth of consciousness.

§ 64. It has been already stated that consciousness, though natural and necessary to every human soul whose powers are normally developed, is not exercised at the beginning of its existence, but only after certain conditions and stages of growth have been attained, and the power to apply them has been matured. The order of this development and maturity may be sketched as follows:

The first activities are those of simple life. These, whether they pertain to the body or the soul, are unconscious. All forms of reflex nerve-action, all the purely instinctive movements of either body or soul, or of both combined, are known to be unattended by conscious apprehension. But all these activities are exercised in great number and for a long time before the experience of sensations.

As soon as a sensation occurs, whether painful or pleasant, it must be felt. It is essential to its very nature to be experienced by a sentient being, and to be felt as painful or pleasant. This experience, whether in man or animal, involves some sort of possible apprehension of self as the subject of its pain or pleasure. This is not consciousness, as we use the term, but only consciousness in its lowest and most rudimentary form. By some it is called the *feeling* as distinguished from the *knowledge* of self, or *self-feeling*. As long as the sensations are confused together and are not discriminated, whether they are weak or strong the soul remains in this elementary condition of comparative unconsciousness. This is the condition of the infant. It is also the condition into which the developed man relapses in swooning, distraction, intoxication, or approaching sleep. In the infant such a condition cannot be remembered, for reasons which we will give in their place. The man can recall it but dimly, and only as he measures and imagines the state, by contrast with those of which he is distinctly conscious, and which he can clearly recall.

But when the several sensations are discriminated from one another, the soul reaches a higher stage. But even this does not involve consciousness, unless the sensations are also discriminated from the self to which they pertain. Observation attests that the one is possible without the other. Even the external objects that occasion the sensations, may be distinguished from one another and from the sensations which attend them, before the soul distinctly recognizes these sensations as its own. No fact is more patent to universal observation, than that, in infancy and childhood, man is occupied with the objective, with very infrequent cognition of self as contrasted with his sensations or their objects, or with the impulse that carries the feelings and actions without.

As soon as feelings of another character are experienced—emotions proper, and not sensations, emotions which are perhaps antagonistic to sensations and their impulses—the opportunity is presented for the soul to distinguish its own agency and itself as an actor or sufferer, as contrasted with itself as purely sentient; *i. e.*, as carried out of itself by its sensations and appetites. The soul furnishes in itself the condition for that reflex act which we call the conscious discrimination of its states as its own. It can know itself as an actor and sufferer, while the act of consciousness is not explained by its conditions, and is not developed from nor produced by these conditions. We concede that it does not occur before these conditions are furnished, and these conditions do not exist till the soul has reached a stage of development that is somewhat advanced, and has had ample experience of the world without as well as the world within.

> The baby, new to earth and sky,
> What time his tender palm is pressed
> Against the circle of the breast,
> Has never thought that this is I.
>
> But as he grows, he gathers much,
> And learns the use of I and me,
> And finds I am not what I see,
> And other than the things I touch;
>
> So rounds he to a separate mind,
> From whence clear memory may begin,
> As thro' the frame that binds him in,
> His isolation grows defined.
>
> TENNYSON.—*In Memoriam.*

The object discerned by the act of consciousness is not, as we have already observed, the soul itself, as a substance or subject, with all its capacities and powers; for, besides those capacities which consciousness apprehends, there are others which it does not reach. Even the cause or source of many which it does discern are beyond its direct cognition. In all of these operations the sentient power acts out of sight, receiving or rejecting those objects for which nature has or has not adapted its action. Even after the soul acts and appears as the *ego*, and, as such is the conscious subject of its higher acts, it also acts as the unconscious subject of many others. As the subject of many similar acts and states objectively known to the conscious *ego*, it is called the self; as the agent which is actor, and also conscious of individual acts, it is called the *ego*, or I. Pre-eminently it is the *ego*, or I, when it makes itself manifest in an act of will, as the regulator or controller of the blind impulses and desires.

The act of conscious self-apprehension may also be more or less frequently exercised by different men, after the capacity for it has been reached. The conditions may be more or less favorable for its exercise, after the power has been matured. First, the objective conditions may be more ample and energetic in one man than in another. The corporeal nature of one may so hold the spirit by obtrusive and engrossing sensations as to preclude the possibility of that discrimination which is the first condition of conscious knowledge. Thus the body of the idiot or the half-witted may so preoccupy the energies as almost to detain it in the animalized state. Moral obliquity, especially in early life, may almost literally brutify or sensualize its condition. Various morbid conditions of the body may come in at an early period of the soul's development to arrest its natural progress, by filling up its experience with continued sensations of weakness and pain. Even a low energy of vital force may give to consciousness only feeble sensational activity and inert impelling forces, which are too unobtrusive to elicit discriminating cognition. The occupations, cares and interests may be so material and sordid, as to fill up the life with activities that are solely objective. The psychical nature of one person may also be far richer and more varied in its capacities than that of another, furnishing the material for conscious observation that is comparatively copious and inviting. Second, the subjective capacity of conscious activity differs in degrees in different persons. The natural powers, the acquired facility, and the inclination to look inward, are stronger in some than in others; and hence in some men that is a passion which in others is rarely and ineffectually performed. Nature, habit, and art exhibit surprising diversities and contrasts in this respect.

On the other hand, the capacity for consciousness is not the product of accidental conditions or circumstances, nor is it the result of any development from any lower existence, but is provided in the nature of man and the designs of his Creator. The brute is not self-conscious under the most favorable circumstances, nor can he become so as the result of any development whatever. He may be like man in the lower stages of being, in the experience of what we call bodily sensations and animal appetites; but he never discriminates one sensation from another by a self-conscious act, simply because he has not the capacity. Much less does he distinguish the self from its states, because there is no self and no states to be thus distinguished. Hence he can, in the proper sense of the word, neither remember, nor generalize, nor reason, nor judge, so far as these involve the reference of acts or objects to himself by appropriate acts and products. He cannot purpose or choose, for a similar reason. Neither the objective conditions of these acts are furnished in his own nature, nor is the subjective capacity to discern them.

Latent modifications of consciousness.

§ 65. The question has been discussed of late among English psychologists, whether there can be any *latent modifications of consciousness.* The phrase is infelicitous, because apparently self-contradictory—a *latent* modification of that which, in its very essence, is an act or an object of *knowledge*, being apparently, both in word and thought, impossible. The truth which the phrase was designed to describe is, however, real and important, and deserves to be clearly stated. That the soul may act without being conscious of what it does, or even that it acts at all, has been already established. That these unconscious acts affect those acts of which it is conscious, and their objects, is equally evident. We have already distinguished between those processes by which the soul, so to speak, prepares objects for its conscious apprehension, and the acts of knowing these objects when thus prepared. All effects of this kind are accomplished by modifications of the soul which are latent—*i. e.*, unknown to the direct inspection of consciousness.

Many of the instances cited as examples of latent modifications of consciousness are only examples of objects observed *with less attention*—objects comparatively unheeded, which may be afterward revived with greater distinctness. For example, I write hastily, to-day, a word or a phrase which is incorrect or ungrammatical. I do not notice the error, but I recall it to-morrow, and notice the mistake by an act of memory. Or, I see a person, and, at the time, do not notice some article of his dress or some peculiarity in his look or language, but recall either distinctly on reflection. Or some part of a total perception, as of a crowded and active company, or a varied landscape, apparently escapes my notice. It is a mere accessory, a subordinate, nearly overlooked in comparison with the central figures or objects ; and yet it may serve as a link in the restoration of a train of connected objects. These objects are not latent, though very little attended to. Leibnitz (*Nouveaux Essais*, ii. c. i.) cites the case of the sound of the sea as an example. A single wave does not affect the ear, but only many, when combined. And yet each wave must contribute its share in affecting the conscious mind, or the whole could not be heard. A distinction is to be made in this instance between the impulse of a single wave upon the organ of hearing, and the experience of the sensation. The action of many waves together may be required to bring the organ into the condition necessary for the sensation in question, or any other. To the total effect upon the organ each wave may contribute its part, without moving the consciousness in the least, even latently.

The general truth cannot, however, be controverted, that the unconscious and conscious processes of the soul act and react on one another continually, and that neither should be overlooked in the science which explains its phenomena. Consciousness, though the most important, is, therefore, not the only source of our knowledge of the soul, and its powers and laws.

CHAPTER II.

THE REFLECTIVE, OR PHILOSOPHICAL CONSCIOUSNESS.

The reflective consciousness defined.

§ 66. HITHERTO we have considered consciousness as the common endowment and universal characteristic of the human race. Every human being is capable of being conscious of his psychical states. Every man who is normally developed is actually conscious of these states at a very early period of his existence.

We have, however, distinguished and defined another species of consciousness. This is the artificial, or secondary consciousness, and it is attained by comparatively few. Though all men can understand and appreciate the descriptions and appeals of the dramatist and the orator, there are but few who can originate and enforce them. The consciousness which discovers and teaches is properly called the *philosophical* and *reflective* consciousness. We proceed to consider more particularly, "What is the reflective consciousness? and, What are its relations to the natural consciousness?"

The reflective consciousness is the natural consciousness exercised with earnest and persistent *attention*. It has already been shown that every intellectual power may be used with a greater or less degree of energy. We have also seen that the development of the natural consciousness through its successive stages is but the development of an increase of attention. When the habit is carried to a still higher degree of energy, and the subjective states and activities become familiar and frequent objects of contemplation, the natural or spontaneous becomes the artificial or reflective consciousness.

It may help us still further to accept the possibility and to understand the nature of consciousness as modified by attention, to consider it in the two forms of *the abnormal* and *the ethical self-consciousness.*

The abnormal consciousness in children and adults.

The abnormal or the morbid self-consciousness is distinguished by any degree of attention to one's own psychical state which interferes with the normal use and development of the powers. Children are appointed by nature to an objective, and, in one sense, an animal life. But now and then a child, through an unfortunate bias, or some ill-judged training, has learned to look inward upon itself with unnatural precocity. As a consequence, the subjective predominates over the objective, the tendency to reflect hinders the power to acquire; and that easy and spontaneous play of observation, memory, imagination, wit, and invention, which is the strength and the charm of childhood, is excluded or hindered.

Among adults frequent examples occur of a morbid or unnatural attention to the inner life. Hypochondriacs, who are haunted by disturbing sensations which proceed from bodily disease, till their attention is so absorbed in watching these sensations that it cannot respond to the objects that are fitted to amuse and incite to action, furnish one example. Men who have inherited or indulged a sensitive nature till it has become their tyrant; who watch their feelings with a selfish exclusiveness, or who pamper them with a dainty fastidiousness, like Rousseau, may become half insane through brooding over their own exaggerated sufferings and wrongs.

Another type of the abnormal consciousness is that which results from an egoistic thoughtfulness of one's appearance, manners, words, looks, actions or achievements, which shows itself in the countless forms of affectation that are displayed in manners, art, or literature. So common has this become in the artificial society of modern times, that it has given a new sense to the words *conscious* and *consciousness*, with and without *self* as a prefix.

The ethical consciousness.

The ethical type is that attention to one's inner states which is applied in view of a moral standard, for the purposes of self-correction and self-improvement. That this is not abnormal is obvious from the fact that the word *reflection*, which originally signified any reflex action of the soul, has acquired the secondary signification of its use and application for ethical purposes. This kind of reflective consciousness always brings with it some intellectual discipline. Christianity has trained the intellect of the human race to this activity, and hence has been so efficient in educating and elevating the masses of men, even when it has furnished little formal intellectual culture.

§ 67. The type of the reflective consciousness with which we are specially concerned is that which is properly called *philosophical,* because it is used for scientific ends. In common with the types already referred to, it involves attention. But if the attention is to yield important scientific results, it must be employed in a peculiar way, with distinct reference to peculiar ends, and with the aid of special appliances. Its characteristics are the following:

The scientific reflective consciousness. Characterized by persistent attention.

First: It is *persistent* in its observations. It not only attends to the phenomena of the soul as inclination or duty may decide, but it attends continuously, in order that it may carefully observe and accurately remember. But how can the mind attend continuously to the same mental state? Of material objects many of the phenomena are permanent; they retain an unchanging identity. We can observe them again and again, till we are certain that we have attained a definite impression, and can bring away a satisfying recollection. But the mental phenomenon is but for an instant. If we look *for it,* in order that we may look *at it* the second time, it is not there. It existed only so long as, by our own act, we gave it being; and when that activity is intermitted, the object which we would fain examine by a second look is no *longer* and *nowhere* to be found. The only resource which we have, is to prolong the state by continually renewing or repeating it. To this act or effort of prolongation Locke gives the name of *retention,* and this he describes as a peculiar mental act (*Essay,* B. ii. c. x. § 1). But can we prolong a single state beyond its assigned period of time? Is not a single state limited to a definite period of duration? The question is trivial, and it is of no consequence how it is answered. Whether we can prolong a state or not, we can certainly repeat it again and again, allowing no other activity to intervene. What we fail to notice at one view, we observe in another. What we only faintly apprehend at the first sight, we fix and confirm at the second. What we observe incorrectly or partially in one act, we discern truly and completely in the act which follows. The uninterrupted repetition of similar psychical states is the substitute for literal continuity in the object observed, and hence is a distinguishing characteristic of the philosophic consciousness. It is because the mind, as it were, turns thus in upon itself, that this effort of consciousness is termed *reflection—i. e.,* the bending back or retortion of the soul on itself. It is because this repetition of the object, and retortion in the act, are found to be practically necessary, in order to any accurate and successful observation of consciousness, that consciousness, the act, has been supposed to be a remembrance, a sort of second thought, and the power has been resolved into memory.

Other advantages are secured by this repetition of the mind's activity, and one especially, that it is capable of being viewed more coolly. If I am absorbed by the beauty of a splendid picture, or a glorious sunset, I shall not be likely, when these objects first break upon my sight, to give much attention to the act or process by which I view them in order to ascertain its exact nature, or to the emotion with which I am literally rapt or carried out of myself, to discover whether delight prevails over wonder. But when my curiosity is satisfied, and my feelings are calmer, then I have some energy to withdraw from the act of seeing and the feeling of admiration, which I can employ in reflex attention to the act and the emotion.

It attends to all the psychical phenomena.

§ 68. Second: The philosophical consciousness is *comprehensive* in its observations. It brings within its field of view all the phenomena of the soul. Its object being to know all its powers, it must of course consider and attend to all its phenomena. The philosopher may not, like the man of morbid or abnormal tendencies, give an exclusive and one-sided regard to certain feelings, or to a few species of intellectual acts; but he must regard all the variety of experiences of which his being is capable, omitting none, being partial to none, doing full justice to each and to all. This principle is accepted as a cardinal maxim of the inductive method. To whatever object-matter this method is applied, it is essential that all the facts should be fairly considered. Nature is an honest witness, and stands pledged to tell not only the truth, but the whole truth. Those who examine the witness are equally bound to *hear the whole truth,* and to open their minds attentively to consider it.

Compares and classifies them.

§ 69. Third: The philosophical consciousness attends to psychical phenomena, *in order that it may compare them;* and it compares these phenomena, in order that it may unite those which are alike, and distinguish those which are unlike. Its aim is scientific knowledge; and science is knowledge that is comparative and discriminating. In other words, it is classified and arranged knowledge.

The power to discern relations sharply, surely, and quickly, may to a certain extent be a special endowment or gift of nature. Its successful exercise or application, however, is the result of attentive comparison. The observer must bring the facts together, placing them side by side. He must then consider them in their connections, leaving the various relations to suggest themselves.

Interprets and explains them by powers and laws.

§ 70. Fourth: The philosophical consciousness *interprets* the phenomena which it unites and discriminates. In other words, it explains them by a reference to powers and laws. But the classification of phenomena is a condition of science, rather than science itself. It is science begun, but not science completed. The object of science is to ascertain what is familiarly called the nature, essence, or constitution, whether of the material or the spiritual beings with which it has to do. It may not be easy to define what is intended by these terms. It is obvious, however, that something more is meant than a bundle of classified phenomena. The phenomena are supposed to indicate or reveal some power which the being possesses. They are to the power as an effect is to its cause. The power is conceived as a capacity to cause some result or phenomenon. Hence science is said to be the investigation of causes, principles, or powers. The scientific consciousness, therefore, *reflects*, that it may refer phenomena to their *causes or powers* in the soul.

But again; The powers or agents of nature act according to *laws*. These laws are fixed methods or rules according to which phenomena occur, when the conditions of their presence are furnished. The laws of the soul are, therefore, to be discovered and established, in order that the science of the soul may be complete, and the aims of the philosophical consciousness may be accomplished.

Relations of the philosophical to the natural consciousness.

§ 71. Our second inquiry respected the relations of the natural to the philosophical consciousness. These relations need to be carefully considered. Neither the natural, nor the reflective consciousness creates these facts; each only observes them; the one cursorily and to little scientific purpose, the other patiently and with comprehensive and sagacious comparisons. Psychology does not add newly-created phenomena to our stock of knowledge, nor even in one sense newly-discovered facts. It has to do with old and in one sense well-known facts, only carefully and comprehensively observed and exhibited in new relations. The facts, and many of the relations of the facts, are as obvious, and in one sense as truly known, to the peasant as to the philosopher. When the philosopher teaches the peasant, he does not impart new knowledge concerning the soul, by mere testimony, on the authority of his own observations and experiments, or those of others; he simply teaches him to attend to the phenomena of his own inner self. He says to him, Look, and you will find this or that. In so far, he only teaches him what in one sense he knew before.

But does not the reflective consciousness discover and impart *new* knowledge? Most certainly. It by no means follows, because the natural furnishes to the reflective consciousness all its facts, and the reflective must go to the natural consciousness for all its materials, that the philosophic consciousness makes no important additions to the stock of human knowledge. The same starry heavens are pictured on the eye of the stupid or superstitious savage, as upon that of the scientific astronomer; but how much more does the one see in them than the other! A simple child and a skilful engineer look upon a steam-engine, both in one sense seeing the same objects; but how much more does the one perceive in the engine than the other, of the powers, the laws and the uses of each separate part, and of their action with respect to the whole. The same natural consciousness is the common possession of the race; but how great is the store of important scientific truth which reflective thought has superinduced upon, and discovered in it. The reflective consciousness imparts new knowledge as it fixes the attention upon phenomena which the natural consciousness fails to observe, and as it places these phenomena in novel relations by comparison, classification and explanation.

The difference between the knowledge given by the natural and that acquired through the philosophical consciousness, is well illustrated by the individual conception of the *ego*, which is common to all, and the generalized conception of the *self* which is the product of reflection. In every act and condition of the natural consciousness there is necessarily present, the recognition of the *ego*, as the unchanging subject of the changing psychical states. It is plain that neither reflection nor memory can create or evolve this knowledge; for both reflection and memory pre-suppose and require it as their essential condition. It must be given to the mind by the intuition of the natural consciousness, or it is not given at all. But the intuition is of the *individual ego*—the one single being to which, and to which alone, belong the various and changing states which are its experiences and its doings, or rather into which it is constantly passing by suffering and by action.

The conception of the *self*, which is expressed in language and defined by its constituent elements or characteristics, is the generalized product of the philosophical consciousness. *A self* is one of the individual agents or *egos*, which, so to speak, is like every other, in those common characteristics or powers which make them alike. It is, however, an *ego* stripped of its individuality by the process of abstraction, and considered only in those attributes and qualities which it has in common with others. The natural consciousness must begin with the apprehension of the *ego*, as the condition of knowing a single mental state. It cannot connect one state with another except by means of this identical *ego*. We begin with the natural consciousness of the individual *ego*, and end with the philosophical *concept* of the *self*; *i. e.*, with its nature and capacities as developed to the reflective consciousness.

Office of language in respect to each.

§ 72. The relations of the natural to the philosophic consciousness cannot be fully appreciated, unless we advert to the *office of language* with respect to each. Language is of essential aid in giving precision and permanence to the observations and results of the reflective consciousness. The subject-matter, as we have seen, is fleeting. It endures but for an instant. The state which we observe and record no sooner appears, than it is gone. But we can give it outward form and definite shape by embodying it in words and expressing it in speech. The frequent use of the word, makes familiar the state and the discerned relations of which it is both the symbol and the record. The thought, however evanescent, is held before the mind for the purposes of comparison and philosophy, when the word is often sounded to the ear or pictured before the eye. Within the sharply-cut outlines of language, psychical objects are so presented that we can avoid a crowded, feeble, or bewildered gaze, when we would summon our energies to compare, classify, and explain.

But language neither creates phenomena nor furnishes observations. It simply records both, and directs and stimulates others to repeat like efforts of thought, each for himself. To attempt to observe without language, is to reject the aid which nature furnishes to our hand, and to the use of which it prompts us by an impulse which we cannot resist if we would. But we should ever remember that language is only an aid, and that the ready use of it either by ourselves or others cannot release us from the obligation to think and observe, to consider attentively and reflectively judge the states of our own souls, to reproduce and study which the words of others simply direct and aid us. We ought especially to guard ourselves against the liability to be imposed on by the use of a refined and technical terminology, or the exhibition of a well-rounded and carefully-adjusted system. Technical language is essential to the use of the reflective consciousness, but it is not nearly so certain to exhibit the facts just as they are, with the beliefs and relations which they involve as the language of the natural consciousness or the utterances of common life.

The language of common life sometimes the most trustworthy.

Indeed, as an expression of psychological facts and a touchstone of psychological theories, the language of common life is far more worthy to be trusted than the language of the schools. It is the outspeaking of those beliefs and feelings, of which man is naturally conscious and which he therefore spontaneously expresses. It is the unconstrained embodiment of all the experiences of his inner self; the subtle robe which the spirit is continually weaving for its inner processes. Each fold and adjustment is a natural and necessary product. Not one is assumed for a purpose. The language of the people is free from all those biassing influences which are incident to speculation, by reason of preconceived theories, whether these are fondly cherished by their originator, or traditionally accepted from revered teachers; whether adopted or defended through pride of opinion, the tenacity of consistency, or the heat of controversy. It is expressed in too great a variety of forms, and under circumstances too dissimilar to admit the supposition of a common prejudice or a common interest. For these reasons we accept the common discourse of men as expressing the unbiassed convictions of those who are competent to discern and decide upon the truth.

"But are uncultivated men competent to understand and decide upon such truths as are in question among philosophers? Let it be granted that their language expresses their own judgments, and that these judgments are worthy to be trusted as far as they go. But do they reach the questions and distinctions of the schools? Can common men understand these questions and distinctions? And if they cannot understand their import, how can they decide upon their validity or their truth?" These inquiries are often urged, in the way of exception and reply to this view of the importance and authority of the language of common life. The answer is obvious, and ought, as it seems to us, to be decisive. The facts which the philosopher seeks to discover are the facts or phenomena which are common to all men, and of which all men are actually conscious. They are not the phenomena which are experienced exclusively by philosophers, but those which are co-extensive with the experience of the human race. What all men experience when they know or feel, they will be likely to express in language; for they cannot know or feel, without knowing that they know and feel. So far, then, as they attend to these processes, and express in language what they discern, they are likely to express the real facts which consciousness discerns; and these are the very facts which the philosopher desires to know.

To detect and correct the mistakes of philosophy, the unbiassed and unreflecting language of common life is often one of the most efficient instrumentalities. The questions are often grave and difficult. What are the elementary facts of human experience? What does analysis show to be the real and the ultimate elements in our knowing and feeling? To answer questions like these, there is no readier and surer expedient than to ask, How do men express themselves all the world over, when they have no theory to maintain and no points to carry? What are the unthinking utterances of common men? Language we say is thought made visible. But thought is belief that something is true. The language of common life is, then, the beliefs of unbiassed men made visible, concerning points in regard to which we simply desire to ascertain the testimony of their unbiassed consciousness.

The actions of men also an important test of truth.

§ 73. *The actions* of men are also of great importance in ascertaining what are the real beliefs of men. Their actions speak louder than their words. When the actions of men can only be explained on the supposition that they are conscious of certain knowledges or believe certain facts which they may deny in their philosophical speculations, we conclude that their philosophy is defective or wrong. We appeal from the propositions and reasonings of the reflective consciousness, to those actual beliefs of the natural consciousness which their actions demonstrate that they hold. When men act persistently and habitually as if they believed certain facts were true, we cannot doubt that they do believe them, however they may seek to persuade themselves or others to the contrary.

These thoughts suggest the truth, which ought ever to be kept in mind and applied, that the teacher of psychology must appeal for the truth of his assertions to the consciousness of the learner. He can communicate nothing upon authority. His duty is to ascertain and classify and interpret the phenomena of his own soul, and to set forth the processes and the results in a manner so clear and so self-evidencing that his pupils will be enabled to consult their own consciousness as he proceeds, and to find in it a confirmation of all which he propounds. Whatever is asserted by the teacher or guide, should be constantly met with the inquiry, Is this confirmed by my experience, or rendered probable by the analogous facts which this experience furnishes? The testimony of others, and the authority of their opinions, should influence us greatly, not to change our opinions against the evidence of consciousness, but to revise these opinions with care, and often to suspect the exactness or the candor of our own observations, whenever the weight of authority is against our convictions. But in psychology, simple authority has no weight against the final decision of consciousness itself.

Conditions of reaching the decisions of consciousness.

§ 74. To reach this decision, two conditions are necessary: *First*, that we fully understand the questions which we are to decide, in their entire import and all the relations which they involve; and *second*, that we patiently and candidly use all the appliances and tests which are at hand to determine the answer. The greatest practical difficulty in settling questions in psychology arises from the circumstance that the student does not, first and foremost, make himself familiarly acquainted with the questions which are to be decided. He too often assumes that he fully understands what he has only imperfectly mastered. Or if he apprehends the point in question for a moment, he fails to make it so familiar as is necessary in order to view it in all its relations, and to decide with a full and distinct appreciation of its entire import. Men are reluctant to bestow this preliminary reflection, because they think that they are already fully acquainted with the question in discussion, and the terms and distinctions which it involves.

All men know something about their own souls, and are able to pronounce with confidence upon many questions that are in controversy. They therefore conclude that they understand every question as soon as it is propounded, and are often in haste to decide, before they have fairly ascertained what the question is. Hence the misunderstandings and disputes between men who are apparently in earnest to discover the truth; hence the warmth with which each disputant maintains his opinion, and the obstinacy with which he defends it against attack. Each man is quite certain that what he has in mind is true; but is he equally sure that his antagonist and himself have the same thing in mind? or that either has all and no more in mind than is properly expressed by the terms? All men know something about psychology, therefore many men decide upon any question which comes before them before they have been careful to learn what its import is. All men are theologians and metaphysicians by nature; therefore they conclude that there is no question in theology or philosophy which they are not at once competent to decide. They hastily and confidently pronounce upon the problem before they are fully possessed of the terms, the data, or the means of solving it.

§ 75. These considerations explain in part the apparent paradox which is presented in the claim, on the one side, that the facts of consciousness are the most certain of all facts, and in the notorious fact, on the other, that many of the simplest and most fundamental principles in psychology are yet undecided, while its philosophical theories are endless themes for never-settled controversy.

Uncertainty and slow progress of psychology explained.

The claim is a just one. The facts of consciousness are the most certain of all facts. The objects which consciousness presents are, if possible, more real and better attested than the objects of sense. We can question whether the eye and the ear do not deceive us; but we cannot doubt whether we perform the acts of seeing and hearing. We may question whether these objects are what they seem to be, but not whether certain psychical acts are in reality performed. We may doubt whether this or that object be a reality or a phantasm, but we cannot doubt that we doubt. Nothing in the universe is so certain, and deserves so well to be trusted, as the psychical phenomena of which each man is conscious.

On the other hand, the fact adduced in objection cannot be disputed. Psychology is unsettled, and every treatise which professes to give the facts of the soul in a scientific form, abounds in criticisms of theories that are still adhered to, and that are maintained by eminent writers. How can this fact be reconciled with the claims to superior clearness and certainty that are asserted for the facts of consciousness?

The positions which we have laid down in respect to the relations of the natural to the reflective consciousness, enable us to reconcile this apparent inconsistency. First, the truth deserves attention, that there is as much vagueness and dispute in respect to the less obvious conceptions and relations of material objects, as in respect to the more recondite relations of psychical phenomena. The obvious facts and relations of matter are accepted without controversy, and are described in popular language. Those which are less obvious, or which involve nice observation, careful discrimination, or some speculative inference, are quite as much in controversy as are the obvious phenomena of the soul when these are subjected to philosophical elaboration. The metaphysics of mathematics, of physics, of chemistry, are as unsettled as the metaphysics of psychical facts. It is because psychology always resolves itself into metaphysics, that psychology always rushes into controversy.

Moreover, it not only concerns itself with its own metaphysics—those which are appropriate to its own facts—but it shoulders the metaphysics of all the material sciences, and transfers to its own arena the smoke and dust that properly belong to the doubtful questions in other fields, and therefore incurs the special reproach to which we have alluded. One reason why psychology is always vague and unsettled, is that it attempts more than the physical sciences, going more deeply than they into the philosophy of its appropriate facts. It is also true that it is not so easy to shape our philosophy to our facts, nor to test our philosophy by our facts, in the psychical as in the physical sciences. This leads us to notice the peculiar difficulties which the student of psychology must expect to encounter.

They are the following:

Peculiar difficulties in the study of the soul.

§ 76. First: The objects of contemplation are not, as in the material world, permanent objects, to which the mind can come and go, so as to bestow repeated observations, till every feature and relation has been carefully and minutely examined. In the science of the soul, the objects—*i. e.*, the phenomena, cease to be, while consciousness surveys them. The soul has at its command only a given quantity of energy, which it must divide between each direct activity and the consciousness which accompanies it. The energy employed in knowing or feeling, *i. e.*, in producing the material for the inspection of consciousness, must consequently be withdrawn from the activity of inspection; any special effort to attend to our processes involves a corresponding weakening of the activity to which we summon ourselves to attend. Material objects become more vivid and distinct the more keenly the attention is fixed upon them; but the objects of consciousness are dissipated before the concentrated gaze which would master their secrets. The repeated creation of a similar object for the subsequent application of consciousness is an imperfect substitute for the continued examination of the same object.

Second: Two observers, and, if need be, twenty, or twenty thousand, can examine and reëxamine the same material object. But the objects of the soul can be surveyed by a single observer for a single instant only. If many observers agree to examine in order to analyze an object which they conceive to be the same, it is sometimes difficult for them to be entirely sure that the objects before their minds are identical in fact.

Third: The testimony or report which one observer brings from his own examination, avails little as a substitute for personal inspection by the student himself. Should the latter even confide entirely in the competence and the candor of another party, he needs to observe for himself in order to be sure of the identity of the object concerning which he accepts the testimony of another witness than himself.

Fourth: Objects of sense are clearly distinguished from and set over against the soul that observes them. In the very act of observation the soul separates them from itself. Objects of the soul are known not to be severed in fact from the soul which observes. For the soul attentively to view its own states as objects to itself, there is required a special and constrained effort. "The understanding," says Locke, "like the eye, while it makes us see and perceive all other things, takes no notice of itself; and it requires art and pains to set it at a distance, and make it its own object."

Fifth: The act of reflection, or second-thinking, for the sole purpose of examining the nature of any act or state already experienced, is especially artificial, and against nature, for the reason that men usually act for some direct motive of use, enjoyment, or duty, and, in thus acting, their look must necessarily be outward and objective. It is necessary, if men would act with interest and energy, that their feelings be strongly aroused by some existing object. But to reproduce the act a second time, or its pale reflection, for the sole purpose of seeing of what sort or nature it is, is not natural, because most men are not greatly interested thoroughly and scientifically to know what their actions are. Or, if they are interested in this as as an end, yet the reproduction, and the continuation through successive reproductions of an act or state, for the mere object of examining its nature, is embarrassed by the difficulty of reproducing it without the excitement of its original motive. We perceive, remember, and imagine,—we hope and fear, choose and reject, naturally and readily enough, when the objects arouse and excite us; but to perceive and re-perceive, to hope and fear again and again, simply that we may know more exactly how it seems or what it is to perform or experience these states, are, at best, forced and unnatural efforts. Nothing but the deepest convictions of the dignity and value of the results, in the acquisition of intellectual discipline and the advancement of psychological science, can impel to the earnest undertaking of such efforts, and the patient prosecution of them to a successful issue.

Sixth: Material objects invite to an analysis by their obtrusive likenesses and differences. The phenomena of the soul do not present such obvious occasions for discernment. Material objects as it were, indicate by dividing lines, by intersecting seams, by salient and projecting points, the sections into which they readily divide themselves under the eye of analysis. Indeed, Nature herself is continually separating and combining these objects before our eyes, changing color and form, disintegrating and throwing apart materials mechanically united, as when the frost breaks up and rolls out the different ingredients of a rock; or she decomposes the ingredients chemically united, as when, by fermentation or solvents, gases and precipitates are evolved. The so-called five senses so soon as they are applied together or in succession to any object, at once suggest five sets of qualities or attributes, to say nothing of the ever recurring relations of extension and number.

To the analysis of the phenomena of the soul there are no such forward promptings of nature. A psychical state, when viewed by consciousness, does not suggest diverse attributes or relations. To bring these to light, it must be brought into comparison with states like and unlike itself. These must be recalled by memory, and vividly reproduced to the imagination. One state must be artificially confronted with another, for the sake of evolving some common points of likeness or contrast.

All these circumstances combined explain the inherent difficulties of philosophical self-observation, and the slow progress and uncertain conquests of the science of the soul in contrast with the rapid advances and the certain results of the sciences of matter. The history of psychology, attests that its progress though slow is real, and that its acquisitions, though often disputed, are more and more assured.

CHAPTER III.

SENSE-PERCEPTION: THE CONDITIONS AND THE PROCESS.

Sense-perception defined and distinguished.

§ 77. From consciousness, the faculty or form of presentative knowledge which is concerned with the objects of spirit and their relations, we proceed to the second, which is employed upon the objects and relations of matter. We define *sense-perception as that power of the intellect by which it gains the knowledge of material objects.* It is also called sensible perception, or simply, perception. We apply these terms to the power, the act, and even to the object. Thus we say, Man is endowed with perception; *i. e.*, with the power to perceive. We say, My perception of the color or sound was clear and vivid—describing the act of perceiving. We also ask, Do you recall certain perceptions, as of color or form?—emphasizing the object.

The terms to *perceive* and *perception*, are applied freely to other acts and objects of knowledge besides those which require the agency of the senses. We are said to perceive mathematical distinctions, the drift and force of reasoning, the design of a machine, and the purpose of an antagonist. But perception, in the technical sense, is appropriated to the knowledge of material objects. This knowledge is acquired by means of the senses, and hence, we call it *sensible perception*, or, more briefly, *sense-perception*.

Is developed earliest of all the powers.

Sense-perception is called into activity *first of all the powers of the intellect*. It is educated and fully developed in our earliest years, at a period and by processes which we cannot distinctly recall to memory.

Seems to be the most familiar. Is not the most easily understood.

But though this power is developed so early and exercised so constantly, and, at first view, seems so easy to be understood; it is far from easy to analyze its elements, or to explain its processes. To understand sense-perception, we must study the body as well as the mind; we must trace out, and, as it were, unravel the subtle connections by which the two are united; we must show how far the one is dependent on the other; what each furnishes towards the result, and what are the separable acts or processes in the action of each. For these and other reasons, it naturally receives the earliest attention in the study of the intellectual powers. The processes of sense-perception seem to most men to be the most familiar and the best understood of all their intellectual acts. Some of the senses are all the while in action. Sense-perceptions are present in our loftiest speculations and our most refined reasonings. The world of sense holds man to its realities in the most ethereal of his flights, and never ceases to be the dark or radiant background to the most vivid pictures of his fancy. Sensations visit man in sleep. They disturb or soothe his repose. They haunt him in his very dreams. With sensations and sense-perceptions man begins and ends his earthly existence.

Distinguished from other mental acts.

The first requisite to a correct theory of perception is to separate the act from every other with which it is likely to be confounded. It is not unnatural to suppose that much, if not all, of the knowledge we have of material objects, is gained by this process alone. A more careful examination shows that we gain very much of our knowledge of these objects by the exercise of the other and higher intellectual powers.

Knowledge of matter not gained by sense-perception.

For example, we take an orange: and inquire first what acts of knowledge in respect to it are not acts of perception; and second, what knowledge is properly ascribed to this power. We first look at the orange, and immediately supply the half which we do not see—the portion of the sphere which is hidden. We know, or believe, the orange to be spherical. The part which we supply we do not perceive by the eye of the body; we only image it to the "mind's eye." This is an act of *imagination* or *representation*, but not an act of perception. We can separate its form, as spherical, from all material reality, and can construct the abstract or mathematical sphere for the mind to consider and analyze. We can reflect on its properties and its relations to the circle by the revolution of which it is conceived to be produced. The discernment of the *mathematical forms, properties and relations*, which may be applied to the orange is not perception. We know, or believe, that its sensible qualities, as of taste, color, feeling, smell, are inherent in or belong to the something which we call their substance. The knowledge of the orange as *substance and qualities* is not necessarily involved in perception. We observe that other objects possess qualities like some of those which belong to the orange—as yellow, round, etc.—and are therefore properly classed with and receive the same appellation. But *classification and naming* are not included in perception. We can know that this fruit has been produced by *the powers* and under *the laws* of vegetable life; knowledge of this sort is not essential to perception. We can know, by *reasoning*, that it will produce certain effects if eaten, or used in illness; but this we do not know by simple perception. We can go still further, and know, or certainly believe, that it is adapted to and was designed for certain uses or ends; as to minister comfort and afford nutriment to man. The knowledge of the *uses and designs* of the orange is not included in sense-perception.

What are acts of sense-perception?

It is evident that all these acts of knowledge may be performed with respect to the orange, and that none of them are acts of simple sense-perception. It is equally clear that they presuppose such acts as their preliminary conditions; so that, if we did not already know something of the orange by certain antecedent acts, we could never know the orange by these higher methods. This preliminary knowledge remains to be considered, after these higher processes have been eliminated.

Knowledge that is gained by sense-perception.

What is the knowledge gained by these preliminary acts? We answer at once, It is the knowledge which is necessarily involved in the use of the organs of sense. Let us try these organs upon the orange, one by one; and first the sense of smell, suspending the action of every other. We perceive a grateful odor, and that is all we know by this means. Were we limited to the agency of smell, this is all the knowledge that the orange would ever give us. We open the ear, and the orange falls, or is struck. We hear the sound from the fall or the stroke, and this is all that we know by the ear. We taste the orange. At once two kinds of knowledge are given, as two senses awake to action—the senses of taste and of touch. Could we separate the touch from the taste, we should by taste perceive only the flavor of the orange.

We grasp it with the hand, first lightly, so as only to be aware of its presence, then with greater force of pressure, so as to encounter resistance. We pass the hand over the surface, and perceive that it is smooth or rough. We come to its limits; for the hand is in contact with another something. Through the hand we can perceive the object as impinging and resisting, as smooth or rough, as having extension and form.

Last of all, we open the eye. A surface of color presents itself, separated from other shaded and colored surfaces by an encircling ring. The color is shaded by the most delicate transitions, deepening here, almost vanishing there. As the orange is near or remote, the limiting or bounding circle widens or is contracted, and the colors are feeble or bright. The eye gives colored extension, form, contrasts, and relative size. Were we all eye, we should perceive nothing more.

In connection with the use of these organs, we perceive or are aware of certain changing affections that attend upon the varying condition of the muscles which direct and move the sense-organs. We know the muscles as tense and as relaxed: we apprehend the affection that accompanies the grasp that is firm and that which is relaxed; the sensation that attends the stretching forth and the withdrawment of the hand. Certain vital and muscular affections are known in connection with the sense-perceptions.

These various knowledges, or *percepts*, obtained by these several means, we combine into one separate and single object, occupying a limited portion of space. The process of perception is not complete till we have attained the knowledge of single objects, made up by the mind of separate parts corresponding to the several senses, and having definite relations of form and magnitude. Such an object we call a *material thing*. When we have gained such a knowledge of the object as enables us to recall and otherwise use it as a mental representation or image, we have completed all that is essential to the process.

Much of our knowledge of sense-objects is acquired indirectly.

We make the knowledge received by one sense a substitute for that which we might receive by another. Thus, by the color of the orange we know its taste; by its appearance to the eye, its feeling to the hand—whether it is hard or soft, whether it is green or ripe. We know an object to be near, by the distinctness or sharpness of its outline and the vividness of its color. We know it is remote by the dimness of the line and the dulness of the color. We determine its distance by its size, and its size by its distance. Knowledge obtained by such processes is called *acquired perception.* The knowledge of sense-objects under the relations of *substance and qualities* involves the application of still higher powers and relations.

Results of analysis. Eight topics proposed.

This general outline or preliminary analysis of sense-perception has shown that it is dependent on corporeal organs or instruments; that it is attended by special sensations, each differing in quality and intensity according to the constitution and condition of its appropriate organ; that in connection with each of these sensations we gain a positive knowledge of material objects; that we unite these knowledges, so as to gain and retain perceptions of separate material things, and that we gain this knowledge of things both by direct observation and indirect inference. It opens for us the following distinct topics of inquiry:

I. *The Conditions or Media of Sense-Perception.*—II. *The Process of Sense-Perception, in its two elements of Sensation and Perception.*—III. *The Classes of Sense-Perceptions.*—IV. *The Acquired Sense-Perceptions.*—V. *The Development and growth of Sense-Perception.*—VI. *The Products of Sense-Perception.*—VII. *Activity of the Soul in Sense-Perception.*—VIII. *Theories of Sense-Perception.*

I. *The conditions or media of sense-perception.*

The conditions enumerated. The first condition.

§ 78. We perceive by means of certain bodily organs, and on the condition that these organs are excited by their appropriate objects or *stimuli,* and that the nervous system with which these organs are connected, shares in this excitation. These conditions of sense-perception are purely physiological, and are discovered by the senses. Prominent among them is *the existence of a material, nervous, and sensorial organism.*

5

The human body is material in its composition; *i. e.*, it consists of particles of matter which are endowed with the properties, and subject to the laws which belong to matter in general, and which are united into bones, viscera, etc. It is also an organism which differs from a machine, in that each of its separate portions performs certain functions, as digestion, secretion, circulation, respiration, each of which is peculiar, and appropriate to no other organ. This function is essential to the existence and action and to the performance of the special function of every other organ: while all must act together in order to further or render possible the special action of each. If digestion is weakened or arrested, the blood ceases to move and the lungs to expand, or both these functions are irregularly and imperfectly performed. Death may ensue, *i. e.*, the once living organism, may be decomposed into particles of unorganized matter.

The nervous system. The sensorium.

In this living organism is present a system of organs, consisting of the brain, the ganglia, and the nerves. The nerves are filaments which terminate on every surface and at every extremity of the body, and penetrate every portion, even the hardest bones. They are interlaced with one another, and their substance is occasionally expanded into large knots or masses. These expansions are called *ganglia*, and serve as independent centres of nervous activity and force. The nerves increase in size as they approach the ganglia, the spinal marrow, and the brain. By means of the ganglia and the spinal marrow, they are all connected with the brain, which is itself a larger ganglion, or system of ganglia. This system of nerves performs several distinct functions, and for each of these functions there is a distinct set of nerves. If the nerves are diseased, single organs fail, or the entire body perishes. If the spinal marrow is injured by disease or violence, the limbs are wholly or in part disabled. If the brain is shocked by concussion, life is suspended or returns no more.

The function of the nervous system with which we are specially concerned, relates to sensation. To fit the nerves for this function, they are connected with various organs, the most noticeable of which are the eye, the ear, the nostril, and the hand. These organs with the nerves attached as capable of the sentient functions when acting in a living organism, are known by the collective term, the *sensorium*, or *sensory*. The term is technical, and is appropriate to *those organs and nerves, which bear some part in the process of perception, and so far only as their functions are concerned in this process.*

The reflex action of the nerves.

We must notice another function of the nervous system which is intimately connected with perception, viz., *their capacity for reflex action.* The nervous filaments which proceed from the external and other organs run side by side in pairs, two being united within the same covering or sheath, and connected by interwoven fibres. If any part where they terminate is irritated, or excited in any way, one of these filaments conveys the notice to the brain or ganglion, and the other conveys the stimulus back to the place where the impression or sensation occurred. We say the sensation *or* impression, for it is by no means essential that the soul should feel pleasure or pain, or in any way be aware of any object. Whatever the excitement may be, the companion nerve responds to the call of its associate, and contracts, convulses, or appropriately moves the muscle or the organ which is aroused. A message of invitation or warning flashes inward along one of these mysterious filaments, the *afferent.* An answer is sent at once outward by the *efferent* to the place from which it came, and the answer is obeyed. This may be done without the intervention or the knowledge of the soul. The nerves arranged for this special service of the senses and of motion are called the *senso-motor*, and the general action which we have described is called their *reflex action.*

The nerves, it will be observed, are the subjects of *diverse affections or phenomena.* First, they are subject to mechanical action and change. Like other filaments, they can be bruised, rent, or cut. Second, their constituent elements suffer chemical changes. Third, they minister to the healthy or unhealthy action of all the vital and sense-organs. Fourth, they are capable of various reflex actions, both occasional, in response to casual excitements, and regular, as when they sustain the involuntary action of the heart, lungs. and other organs. Fifth, the highest of all, when a sentient soul makes this organism living, they are capable of a special affection or excitement, which is the condition of sensation and sense-perception.

The first and essential requisite to sense-perception is the existence of the sensorium as thus defined.

The second condition is an object or excitant.

§ 78 *a*. The second requisite to sense-perception is *the existence and the presence of appropriate objects.* We say in general, there must be visible objects in order to vision: audible objects in order to hearing: tangible objects in order to touch. In other language we say, objects, to be perceived, must be luminous, sonorous, resisting; or, more abstractly, there must be *light, sound,* and *hardness*, or there cannot be *vision, hearing,* or *touch.*

One apparent exception to this principle occurs in the case of the so-called *subjective sensations* which are excited by stimulating the nerves by peculiar agents. Thus the optic nerve, under electrical applications, may be so excited as to occasion flashes of light. Sparks are perceived from a blow or contusion. Slight sensations of smell and of taste, also a ringing or whizzing in the ears, are occasioned by electrical action. Experiments of this kind prove that the sensation depends entirely on the excitement of a part of the sensory to a given species of activity, and that this excitement is idiopathic, or limited to the nerve or nerves concerned; *e. g.*, the optic nerve alone emits light; the acoustic nerve, sound, etc., etc.

The third condition. Its action on the sensorium.

§ 79. The third condition of sense-perception is *the action of the object upon the sensorium.* In order to receive this action, the external organs must be in a normal condition—*e. g.*, the eye, the ear, the palate, and the skin. If any lesion or disease occurs, the perception is irregular or impossible. In like manner, if the nerves are diseased or destroyed, the perceptions are disturbed or prevented. Let the optic nerve be injured, and the vision is dimmed, clouded, or extinguished. So it is with hearing, with touch, with smell, and with taste.

It may be asked, how do we know that these *three* requisites must be present? We reply, Only indirectly. We learn it by inference. If the sensorium no longer exists, there is no perception. If the object is withdrawn, as the luminous or sonorous matter, there can be no perception. If the organ or the nerve is destroyed, the soul does not perceive. We conclude that all these are its essential conditions. But that these conditions are not the acts themselves, will be still more manifest from the analysis of these acts. We proceed next to:

II. *The process of sense-perception.*

The process of sense-perception in the simplest form; what is it?

§ 80. The simplest form in which sense-perception is experienced is in its connection with a single organ of sense. The states or acts which we ordinarily call sense-perceptions, by which we apprehend the most familiar objects, as a table, a chair, a horse, or a dog, are made up of too many elements to allow us to discern the precise character of the elements or the steps of the process itself. It is only when we consider a single act, as of seeing and hearing, and of the simplest object, as a single color or sound, that we are in a condition to determine the essential nature and elements of the act itself.

It is psychical, not physiological.

The most general assertion which we make is, that sense-perception is clearly and distinctively a psychical and not a physiological phenomenon. We are prepared, by our previous analysis, to distinguish perception from the organic instruments and conditions that are essential to it. Neither the eye nor the optic nerve, nor the image formed on the retina, nor the nervous response to the image—none of these, nor all of them together, constitute vision. The picture may be formed, the nerve may be stimulated to reflex activity, so as to contract the iris or let fall the eyelid, and yet there may be no sight. If a hot iron is applied to the flesh, and the soul does not feel and apprehend, there is no sense-perception. It may disorganize and destroy the flesh, consuming it to the bone, and yet, if the soul does not respond, the phenomenon which we seek for does not occur. In order to this, an energy must be aroused from the soul itself. Its presence and its nature are known by consciousness. Its physical conditions are observed by the senses and traced out by physiological analysis. The anatomist separates and follows the one class of phenomena by his dissecting knife, interpreting the functions which he does not observe. Consciousness watches the other, notes their similarities and differences, refers them to their agent and records their relations and laws.

It is complex; of two elements.

Let us, then, leave these physical or physiological conditions, and consult consciousness alone. We inquire of consciousness, What is the psychical act or state? She replies, It is a process complex in its nature, but instantaneous in time. It is complex, because the soul, in its single act, distinguishes two objects—its own condition and some material reality: one of these is subjective, and hence is called a *subject-object;* the other is objective, and is denominated an *object-object.* One element is called *sensation,* or *sensation proper;* the other is called *perception,* or *perception proper.* The one of these is an element involving feeling; the other is intellectual, being an act of knowledge. Each requires the other. Each is the attendant of the other. There can be no perception without sensation, nor can sensation occur without perception.

The elements unequal in energy; in the same, and the different senses.

But though these two elements coexist, it is with unequal energy. The one activity is always at the expense of the other. If sensation is intense, perception is feeble. If perception is energetic and absorbing, sensation is weak and scarcely observed. The operation of this law is seen in the several senses, and in the differing states or energies of single and separate senses. In vision, as compared with smell and hearing, perception prevails; while in both the latter, sensation is in excess. In the perception of bright and stimulating color, as contrasted with the discernment of form and outlines, sensation is conspicuous in the one, and perception in the other. If we look at the unclouded sun at midday, we cannot perceive distinctly, by reason of the blinding and painful sensations; if its disc is overcast, or a darkened glass is interposed, the perception is more distinct and easy, by the repression of the sensations.

Sensation proper pertains to the soul.

§ 81. *Sensation proper,* or the sensational element, comes first in order. This does not occur alone or apart. Pure sensation is simply an ideal or imaginary experience. Though sensation always occurs with perception, it may be clearly distinguished from it. Sensation, thus considered, is

A subjective experience of the soul as animating an extended sensorium, usually more or less pleasurable or painful, and always occasioned by some excitement of the organism. This definition implies:—

First of all, that sensation pertains properly *to the soul,* as contra-distinguished from material things or corporeal agents. The sensation of touch is not in the orange, the sensation of heat is not in the burning flame, but both are experienced by the sentient soul. The sensation of sweetness is not in the sugar, that of sourness is not in the vinegar. There can be no music when orchestra and audience are both stone-deaf. As all sensations pertain to the soul which experiences them, they are properly said to be subjective.

Yet experienced by the soul connected with an organism.

§ 82. Second, the sensations, though subjective in the sense already defined, are yet experienced by the soul *as connected with a corporeal organism*, and are directly distinguished in this from emotions proper, on the one hand, and from perceptions proper, on the other. The soul has a subjective experience of heat, hardness, sweetness, sourness, etc., but it has this experience as an agent connected with and animating an extended sensorium. The several sensations, though like the purely spiritual emotions in being agreeable, or the opposite, are unlike them in being felt by the soul as existing in a peculiar form of being and activity, viz., that of corporeal sensibility. That which feels is not the soul as pure spirit, but spirit animating an organism.

It is but a part of the truth which Reid utters, when he says: "This sensation [of smell] can be nothing else than it is felt to be. Its very essence consists in being felt; and when it is not felt, it is not. There is no difference between the sensation, and the feeling of it; they are one and the same thing." "As to the sensations and feelings that are agreeable or disagreeable, they differ much, not only in degree, but in kind and dignity. Some belong to the animal part of our nature, and are common to us with the brutes; others belong to the rational and moral part. The first are more properly called *sensations*, the last, *feelings*." (*Essays, Intell. Powers*, ii. c. 16.)

Berkeley, *Theory of Vision*, says to the same effect: "The objects intromitted by sight would seem to him [a man born blind], as indeed they are, no other than a new set of thoughts or sensations, each whereof is as near to him as the perceptions of pain and pleasure, or the most inward passions of the soul."

Reid certainly would not say that the pain, or the painful sensation, which is occasioned by a burn, a cut, or a blow, is precisely like the pain which is occasioned by the death of a friend, the loss of fortune, or the failure of a darling project. Both these classes of states, when not felt, have no existence; they both pertain to the soul, and to the soul only, as distinguished from the objects which occasion them. Both are alike subjective. Both are alike in being disagreeable, hence both are called painful. But one is experienced by the soul as connected with an organism, while the other is felt in the soul without reference to the sensorium at all.

Hamilton on the other hand asserts, "It may appear, not a paradox merely, but a contradiction, to say, that the organism is at once within and without the mind; is at once subjective and objective; is at once *ego* and *non-ego*. But so it is, and so we must admit it to be, unless, on the one hand, as materialists, we identify mind with matter, or, on the other, as idealists, we identify matter with mind. The organism, as animated, as sentient, is necessarily ours; and its affections are only felt as affections of the indivisible *ego*. In this respect, and to this extent, our organs are not external to ourselves. But our organism is not merely a sentient subject, it is at the same time an extended, figured, divisible, in a word, a material, subject; and the same sensations which are reduced to unity in the indivisibility of consciousness are in the divisible organism recognized as plural and reciprocally external, and, therefore, as extended, figured, and divided. Such is the fact: but how the immaterial can be united with matter, how the unextended can apprehend extension, how the indivisible can measure the divided,—this is the mystery of mysteries to man."—*Works of Reid*, Note D* 18 and *foot-note*, p. 880 (Cf. 35, 38, 39).

The sensations localized.

§ 83. It is implied, in what has been said, that all sensations are experienced with *a more or less distinct and definite relation of place in the sensorium.* This relation of place is at first very indefinitely apprehended; indeed, it may not be attended to at all; but there must be furnished the means of discerning such a relation, provided the attention is directed to the sensation. It is impossible to believe that a pain in the teeth or a pain in the head should not be known as apart in place from a pain in the foot; that a burn in the foot and a wound in the arm should not give directly to the mind the apprehension of a different place for each.

When it is asserted that every sensation gives or might give a relation of place, it is not intended that the relations of place involved in and given by the direct experience of an original sensation are or could be apprehended so completely and so definitely, as they are by the aid of experience and the acquired perceptions; but only that some knowledge, or the materials for such knowledge, must be furnished in every original sensation.

The different sensations differ in respect to the greater or less definiteness of the part or place of the sensorium which is affected. Thus a sound or a smell is far less distinctly defined in any relations of place than a sight or a touch. But more of this in another place.

Differ from one another in quality and definiteness.

§ 84. Fourth: The different sensations, as subjective experiences of the soul, *differ greatly from one another in respect to quality and intensity;* in other words, they differ in kind and degree. Each of the leading classes of sensations differs from each of the other classes, as the sensations of sight from the sensations of touch. Under each of these broadly distinguished classes or kinds, special sensations differ from one another; as the different tastes, feelings, smells, colors, etc., etc. What are called the same sensations, differ also in energy, strength, or intensity; as one shade of the same color, as red, is deeper or more intense than another shade; one odor is more pungent than another.

We come next to *perception* or *perception proper.*

Perception proper, an act of pure knowledge.

§ 85. This, as has already been explained, is no separate act or state of the soul; it is only a separable or distinguishable element of a single complex act. Perception, as such, is,

First: an act of *knowledge and of knowledge only.* The sensational element is an element of feeling, attended, indeed, with the knowledge that the soul which feels animates an extended organism; but in the perceptional act the soul knows, and only knows.

Its object a non-*ego.* What kind of a non-*ego.*

§ 86. Second: This knowledge is *objective—i. e.,* the soul not only knows the object to be, but it knows it is not itself. What it knows is a non-*ego*, a not-me, a not-self. But from what self, or *ego*, does it distinguish the object? or what kind of non-*ego* does the perceiving soul distinguish? Is it what is usually called a material object, distinguished from the organism or the body which the soul animates and moves? or is it the organism itself which the soul distinguishes from itself, though it animates and moves it? It should be carefully kept in mind, that, as there are three *non-egos*—viz., the not body as distinguished from the body and soul united; the body as distinguished from the soul; and the senso-

5*

rium as distinguished from the soul as pure spirit—so there are three *egos*, viz.: the soul as united with the body sensed and perceived, *i. e.*, the living body as a whole; the soul as animating or connecting with the sensorium; and the soul as distinguishable from both sensorium and body.

Our present inquiry is, Which of these objects is apprehended in perception proper? Which is known, or might be known, in connection with every sensation, or in every act of sense-perception? We answer, The bodily organism itself, or rather that part of the sensorium which is excited to action. What the soul directly perceives—*i. e.*, distinguishes from itself—is its own sensitive organism, so far as it is excited to sensation. This is that which it knows to be not itself, even though it knows that in sensation it is intimately connected with it. The immediate object of perception proper is the *sensorium in some form of excited action.* (§ 98)

It is not intended that, in the order of time, the infant does, in the earliest development of the reflective consciousness, apply the pronoun I to the soul as distinguished from the sensorium or the body. It is most evident that at first, and for a very long period often, this appellation is applied to the soul and the body as a complex whole, and this *ego* is distinguished from what is usually called a material thing.

An extended *non-ego.*

§ 87. Third: The object in perception proper is not only known as a *non-ego*, but it is known as *extended.* Even in sensation proper the soul knows itself as united with the extended sensorium; much more when the soul, by an act of intelligence, distinguishes this sensorium from itself as a purely psychical agent, must it know that object to be extended which it as it were sets over against itself. We do not here ask what extension is, or how it is possible that the unextended spirit can know extended matter; nor do we ask what are the relations of extension to space, either in the order of knowledge or of being. These questions are reserved for future discussion. We record only what the mind actually perceives, as attested by our experience of the act of perception.

Perception attends all the sensations.

§ 88. We ask, fourth: In the exercise of which of the senses does the mind distinguish this non-egoistic and extended object—in the exercise of *one or two*, *or of each and all?* The views which we have proposed concerning sensation involve the necessary consequence that perception proper occurs in connection with each of the senses. If every sensation involves the apprehension of the extended sensorium with which the soul is connected, then it follows that it is possible to perceive this sensorium, to whatever sensation it is excited, and that every sense gives the knowledge of an extended *non-ego*. Some of these senses do this with greater indefiniteness than others, it is true—as the sense of smell compared with the sense of touch, but all with equal reality; if, indeed, it is true that no sensation can in fact occur without perception.

Those psychologists who make sensation to be a purely *spiritual* or *subjective* experience of merely *intensive* quality, and make perception to be the apprehension of the cause of these so-called feelings, either limit perception to the sensations of touch and sight, excluding it from smell, taste, and hearing—as does Reid—or confine it to touch only, as Dugald Stewart and Dr. Thomas Brown.

The extension and externality of all objects not given with equal clearness.

But while each and all of the senses alike give us an extended and external object, they do not give it with equal distinctness and clearness. As we have already observed, the senses of smell and hearing are far inferior in this respect to the senses of sight and touch; and so far inferior, that they seem to many not to give it at all. The muscular sensations are also more conspicuously present in the movement and direction of certain organs than in the management and experience of others.

The varying relations of sensation and perception proper.

§ 89. We pass, fifth: to the varying relation of the sensational and perceptional element in different states of sense-perception. The general law is, that *these elements vary inversely—i. e., as the sensation is stronger, the perception is weaker, and vice-versâ.* The operation of this law is illustrated in the different sensations of the same sense as compared with one another, and also in the different senses.

In different sensations of the same sense.

Of different sensations of the same sense we observe, that in some the attention is occupied more with the sensation, while in others it is fixed upon the object which the sensation reveals. This is true of tastes, smells, sounds, touches and sights. If any of these are very agreeable or disagreeable, the subjective pain or pleasure which they give, solicits and absorbs the soul's energy, almost or entirely to the exclusion of all apprehension of the organism, or of any thing external. If they are what we call indifferent or unexciting, there is opportunity for the mind to attend to the relations of diverse quality, of place, form, outline, which the particular sense admits of. It has passed into a proverb, that certain sensations are absorbing, transporting, ravishing, enrapturing, and ecstatic; all of which terms indicate the complete occupation of the soul's energy in subjective enjoyment, or, as the case may be, in pain or agony. We freely remark of others, that in them we are cool, unexcited, not carried away, self-controlled; which epithets imply the possibility of any intellectual activity which may be required, the energy of simple perception being, of course, included.

In the different senses.

In vision, the apprehensions of color are more sensuous; those of form and outline are more perceptional and intellectual. In gazing upon rich and gorgeous coloring, as of a splendid sunset, of autumn foliage, or a glowing painting, the enjoyment is more intense and the excitement is akin to pure emotion. In the apprehension and comparison of form, outline, and grouping, color is less conspicuous, the perceptional element predominates, and approaches the purely intellectual. But just in this proportion does the sensuous and passionate give way.

In touch, if we take a burning or frosted implement, we are so occupied with the pain, that we do not notice its form, surface, weight, and many other peculiarities which a nicer handling would reveal, which delicate handling is rendered impossible by the absorption of the soul with its sensations. On the other hand, the delicate intellectual touch, which apprehends minute constituents, slightly varying surfaces, gentle outlines, fine edges, etc., requires as an essential condition that the sensations be not at all obtrusive. He that passes his finger over the edge of a razor in order to judge of its fineness, must be careful that no painful sensations, as from a cut, or pleasant sensations as of titillation, disturb or distract the delicacy of his perceptive touch. In all these examples it is to be noticed, that so far as we exercise sensation proper we are occupied with our subjective condition as pleasant or painful; while in perception proper we apprehend an extended non-*ego*.

The illustration of the varying energy of the sensational and perceptional elements in the different senses will be given in the following chapter.

CHAPTER IV.

CLASSES OF SENSE-PERCEPTIONS.

Three classes of sense-perceptions. The muscular.

§ 90. The sense-perceptions may be divided into three leading classes: the *muscular*, the *organic*, and the *special sense-perceptions*. This division is in part directed by the character of the sensations themselves, and in part by their bodily conditions.

The *muscular* sensations, or sense-perceptions, comprehend all those which arise from the varying conditions of the muscles, whether in action or at rest. The muscles constitute a very large portion of the substance or structure of the body. They also pervade or are closely connected with those parts and organs which are not muscular. The affections appropriately called muscular sense-perceptions are those which depend on the contraction and relaxation of the muscular fibres, or the varying relative position of the muscles. As we slowly stretch or violently jerk out the arm or the finger, as we rotate the wrist, as we tread or kick with the foot, as we strain the whole body to lift a heavy weight or to push against a resisting obstacle, or as we exert a part or the whole of the body in manifold conceivable motions or efforts, we experience as great a variety of muscular sensations. Scarcely one of these is distinguished by a separate name; and the greater part of them escape common observation.

They are ranked lowest in the scale of the sense-perceptions, because they are least definitely placed in the sensorium, because they cannot be distinctly recalled to the memory, and because they are usually the least positive in the pleasure and pain which they occasion. They serve most important uses, however, as we shall see, in enabling us so to direct and regulate the bodily motions as to distinguish the individual body from the rest of the material universe, and to defend it against serious or fatal injuries. It is contended by many that we derive our first knowledge of extended matter from the muscular sensations, as through their varying movements the infant first explores every part of the organism within, and from the sensorium thus explored derives the standard by which it measures the material world without. (Cf. § 98.)

The organic.

§ 91. The *organic* sensations are those which depend *on the healthful or diseased condition of the vital organs;* such as the stomach, the lungs, the heart, the other viscera, and the nerves. When these organs are entirely healthy, and their functions are normally performed, they are attended with no very positive or distinctly noticed sensations. When they are injured or diseased, the sensations which attend these conditions are always unpleasant, often distressing, and invariably most readily distinguished and recognized. The healthy man does not know that he has a stomach. The dyspeptic scarcely knows that he has anything besides; he is so absorbed by the uncomfortable or painful sensations that are occasioned by the diseased organ. The same is true of a man whose lungs, heart, or nerves are diseased. This class of sensations are more readily distinguished and recalled than the muscular, because they are more definite and positive.

The organic sensations are often blended with the muscular. The vital organs are in part muscular, or intertwined with muscular fibre, as the heart, the stomach, etc. Their special affections are therefore experienced in constant connection with normal or abnormal muscular sensations, and both are assigned to the same parts of the sentient organism.

The special sense-perceptions. Smell; its organ, conditions, and objects.

§ 92. The *special sense-perceptions* constitute the remaining and the most important class. All these are distinguished by this marked peculiarity, that *they are experienced through organs specially constructed for the sole function of sense-perception.*

They are the so-called five senses: *Smell, taste, hearing, touch,* and *sight.* Each of these is clearly distinguished from every other, and to each of them is assigned its own organ or organs.

The organ of *smell* is the nostrils, which open into the two nasal fossæ, the plates of which are overlaid by a mucous membrane called the pituitary membrane. The passages between these plates are somewhat tortuous, giving extent of surface for the expanse of membrane, and the ramifications of the olfactory nerve.

This organ is in immediate contiguity with the organs of taste, with which it acts in ready sympathy. Offensive smells occasion nausea and disinclination to food. Savory odors, on the other hand, stimulate the appetite.

It is generally believed that smell is excited only by the contact of the interior surface of the organ with minute portions of matter, or gases diffused through the atmosphere. But whatever uncertainty there may be in respect to the occasions of these sensations, with the sensations themselves we are all familiar. Their varieties are almost endless. The odors from flowers, from food, from perfumes, from woods, from earths, from metals, and from many other objects, are too numerous to be classed or named except in a very general way. We *class* them in a few general and obvious groups, as quickening, refreshing, depressing, sickening, aromatic, spicy, etc., etc. We *name* them usually from the objects which excite them, as the odor of the violet and the lilac, of the rose and the tuberose, of the peach and the apple, of cedar and camphor-wood.

It is to be remembered that the so-called sensations are in truth sense-perceptions—*i. e.*, they involve apprehended relations of externality and extension. The experience of every odor, according to the explanation already given, must be referred to some part of the sensorium. These sensations are, however, very undefined in their places and limits, and hence it has been supposed they are purely psychical. They cannot be distinctly recalled in the imagination or memory. Hence, in our actual perceptions of objects, they are referred directly to the object as seen or handled. That is, the object seen or touched occupies the attention and engrosses the memory, and not the object smelled.

The language and terms taken from this sense are transferred to supersensual objects, especially to the moral and the religious. The odor of incense, the offense that is rank, and smells to heaven, and the like, are examples of such an application.

Taste: organs and objects. Variety of the sensations.

§ 93. The organs of *taste* are the tongue, the palate, and a portion of the pharynx. These are also truly, though imperfectly, organs of touch; but they are coated with a membrane which is organized in such a manner as to yield a variety of special sensations called tastes. The tasting organ, so far as it can be traced, consists of minute papillæ, which cover the upper surface of the tongue and the inner cavity of the mouth.

Sapid substances to be prepared for tasting, must be made liquid. Those which are hard and compact, must be broken by mastication and dissolved in the saliva. The harder the substance and the slower the process of dissolving, the longer does the taste continue.

The sensations of taste are various in kind and almost countless in number. They are capable of being so combined as to produce singular modifications and striking contrasts. They can thus, to some extent, be changed by custom and formed by art. Tastes that are at first positively disagreeable, become pleasant by being connected with a stimulant effect upon the nervous system—as the pungent and fiery taste of strong liquors, and the nauseating taste of tobacco. Or the sense-organ itself becomes less sensitive in its energy, and of course less offended by the sensations which were at first more intense, and therefore positively disagreeable.

Tastes, like smells, are designated by a few general epithets, as pungent, bitter, sweet, spicy, acrid, sharp; more precisely by the objects which occasion them, as the taste of pepper or alum, of the peach or the plum, of different vegetables and meats. Of this language or vocabulary of taste we may say in general, that it is taken originally from the sense of touch, as the obvious meaning of some of the terms, and the less obvious roots of others, both indicate. The reason is obvious. The organ of taste is also an organ of touch. The tongue touches as well as tastes. Certain tastes are attended with certain touches.

It ought not to escape our notice in this connection, that the sense of the beautiful and the sublime in nature, art, and literature, and the capacity for judging rightly of its occasions or sources, is called *taste* in many languages; a singular transfer of a term from one of the grossest of the animal capacities to one of the highest of the psychical endowments. It is explained by the fact that the corporeal sense of taste is susceptible of fine and delicate discriminations.

The question is never mooted, whether the sensations of taste are purely subjective, or independent of all relations of externality and extension. Taste, as a sensation, is inconceivable except as an affection of that part of the sensorium which pervades the surface of the tongue and palate.

Hearing; its organ and objects.

§ 94. The sense of *hearing* comes next in order. Its organ is a complicated and convoluted bony tube or chamber, resembling somewhat the interior of a snail-shell, and furnished externally with an expanded appendage, the surface of which is corrugated very much after the manner of the bony passage within. The object of the external ear (which with the internal constitutes the organ), is to receive, convey, and quicken the vibratory action of the air till it reaches the tympanum. This is a parchment-like substance, which, by the aid of a chain of bones, bears upon a liquid within. The arrangement of this entire structure, when judged by mechanical principles, is obviously adapted and designed to carry and increase vibratory action. But the vibrating tympanum is not itself hearing. Though we seek for the spirit of sound in all these narrow and winding chambers, we cannot find it there; but it flees from our search like a shadow or a mocking spirit. It is the soul which lives in the sensorium that hears. When the tympanum is made to vibrate with the requisite intensity and rapidity, and the nervous apparatus is unharmed, the soul, if attent, experiences the sense-perceptions which we call the sensations of sound.

Every body which emits or conveys sound is susceptible of vibration. The sonorous body with which we are most familiar, is the atmosphere, which, by being everywhere present, is the constant and the pervading medium of sound. Many solid bodies are, however, capable of more delicate vibrations, and hence are more perfect conductors of sound; or perhaps they owe their effect on the sensorium in part to the vibrations which touch conveys through the bony structure. A stick of timber will convey to the ear in contact with it, a whisper or the scratch of a pin, scores or hundreds of feet. If the ear is brought into contact with a musical instrument, either directly or through the medium of some intervening substance, the intensity of the sound is greatly increased.

The sensations various. In what respects distinguishable.

Of these sensations there is a great variety. What deserves especial notice is, that each one of this endless variety is readily distinguished from every other, and very many of them can be recalled and recognized. A single human voice is capable of emitting a great variety in respect to quality, tone, and pitch. The voice of each individual has its distinguishable characteristic in each of these particulars. The wind sighs and whistles and groans in the forest, or beats and rolls among the clouds like resounding waves. Almost every substance has a sound of its own when it strikes or falls upon another, and this sound can be varied in quantity and quality.

Single sensations of sound are distinguished by quality, by intensity or loudness, and by volume or quantity. Besides these obvious differences, there are others less discernible to common apprehension, which are observed and named by elocutionists and musicians. The epithets which we commonly hear are such as low and high, feeble and loud, soft and harsh, smooth and rough—sweet, gentle, clear, piercing, light, heavy, etc., etc. All these epithets were originally appropriated to the other senses, especially to those of touch. Some few are derived from taste and sight. To a limited extent, sounds are named from the objects which excite them: as the bell and glass, like the wooden, the metallic, etc., etc.

Sounds in succession and combination. Melody and harmony.

Besides these distinguishing differences in single sensations of sound, there are others which belong to sounds when in succession and combination. Sounds of almost any quality become pleasing when uttered in any regular succession; especially when a series is made to repeat and to return upon itself, and its measures or intervals are marked by accent or beat. Examples of these are the beating of a drum to a tune, the rhythmical measure of well-sounding prose, or the more regular and marked repetitions of poetic verse. If the sounds possess musical quality, these repetitions constitute melody, giving exquisite sensuous pleasure to the ear, and, by expression, speaking movingly to the soul. To this is superadded the more refined attribute of harmony, when sounds of different musical quality are given in concord, greatly enlarging, enriching, and elevating both the sensuous and expressional resources of music. Melody and harmony combined, when added to what culture has done for the voice, and art for the improvement of instruments, are the grounds of the elevated enjoyment that music affords.

The condition of oral language. Expressive of feeling.

The sensations of sound are invested with even a higher interest, and applied to a still more elevated use. Without the sense of hearing, vocal utterances do not become sounds; and without vocal utterances as heard, there could be no language. As addressed to and affecting the senses, sounds are pleasing or displeasing, musical and melodious or the contrary, harmonious or discordant; as significant of human thought and feeling, they are endowed with a wondrous and almost a sublime power. When we listen to a foreign language of which we are ignorant, or when we cannot catch the sense of our mother-tongue, it is to our ears a jargon or a chatter, or, at best, but a pleasing flow of insignificant sense-perceptions. But as soon as these sounds are understood, they become the audible expressions of thought, in its most subtle distinctions and its most complicated connections.

Not only are sounds significant of thought; they also express feeling. Even simple and inarticulate tones do this, especially if the tones are musical, or partake of musical quality. The whine of the beggar, the command of the master, and the threat of the enraged, are expressive as tones, even when no words are uttered, or when the uttered words fail to be understood. A plaintive or a triumphant strain of music is easily interpreted, though no thoughts are uttered in words. But when thought and feeling are both conveyed, the one by clear and well-chosen words, and the other by an expressive elocution, and the soul is enraptured and elevated by eloquent speech, then the resources of sound and the importance of hearing begin to be appreciated. When, again, poetry and music lend both grace and expression to thought and feeling, we have a still higher example of the dignity of a single sense, and the wondrous uses to which it may be applied in the service of the soul.

In view of these relations, the sense of hearing has been

ranked higher than any other. It effects a connection between one soul and another; it enables the spirit to breathe out feelings which even articulate speech cannot utter. Its dignity and worth are especially illustrated in the case of the blind. It is to them the subtle interpreter of those emotions, which are expressed to others by the eye, the countenance, the attitude, and the gesture all combined. To the blind the voice softens in tenderness, thrills with love, is harsh from anger, and lingers in entreaty. To them every tone breathes some shade of emotion. An intelligent and educated blind man once remarked with great energy, "The human voice is to me the divinest endowment of man."

The sense of touch. Its organ.

§ 95. The *sense of touch* comes next in order. The organ of this sense is the skin. The skin is the external covering of the body, and the lining of certain internal cavities, as the mouth. Its sensations depend on the action of certain minute *papillæ*, which are placed beneath the external cuticle, each one of which encloses the termination of some nerve, or nervous branch or branchlet. Different portions of the skin are more or less sensitive, and the perceptions which are gained through them are more or less delicate, according to the number of the nerves and the fineness and frequency of the nervous terminations. The thickness or thinness of the external covering or cuticle is also an important circumstance. In general, those portions of the body in which the perceptions are least acute and discriminating are the most scantily supplied with nerves, and their branches extend over a very large surface—in some cases over several square inches. In the more sensitive parts of the body, on the other hand, there are very many distinct nerves and nervous branches and branchlets.

The distinguished physiologist, E. H. Weber, was the first who instituted a series of careful experiments, in order definitely to ascertain the different degrees of sensitiveness in touch of different parts of the body. He applied for this purpose the points of a pair of dividers, which were separated more or less widely. He ascertained that in some parts of the body these points could not be perceived as separate, unless the dividers were opened as widely as three inches; while in others the extremities needed to be only the thirty-sixth of an inch apart in order to be distinctly perceived. Similar experiments have been made by other physiologists. The tip of the tongue, the lips, and the ends of the fingers, are the most sensitive and discriminating portions. The human hand, inasmuch as it is lined with a sensitive covering, and—through its connection with the arm and shoulder, and its division into thumb and fingers—is provided with an apparatus especially adapted to regulate and direct the application of touch and pressure, is preëminently the organ of touch.

Essential condition of touch.

It is an essential condition of a sense-perception of touch, that the object should be actually applied to or brought in contact with the organ—*i. e.*, with some portion of the surface of the body. According as this application is made with greater or less force, the sensation varies in intensity and the perception in distinctness, and sometimes the quality of the sensation is changed. A light pressure or gentle touch, is usually favorable to distinct or delicate perception. If the pressure is increased, the sensation may become excessive and unpleasant, and even positively painful; while the perception is less acute, owing, probably, to the compression of the nerve or nerves. In some cases, the very slightest contact that is possible, with a careful avoidance of pressure, as in the touch of a feather, is attended with the greatest sensibility and the acutest discernment. But the force of the application of the organ to the object of touch depends usually on muscular effort. It scarcely ever can happen that muscular effort is not called into requisition, either in positive and direct pressure, as of the hand or finger, or in withholding from pressure beyond a certain degree, or in resisting pressure when it is imposed from without.

Variety of sensations involved in touch. Sensations of gentle touch.

§ 96. Hence it is that *the muscular sensations* always attend and often seem to be blended with the perceptions that are appropriate to touch. In the acquired or complex perceptions of touch, these muscular sensations play a conspicuous part. In the classifications of common life and in those of the earlier philosophers,—both psychologists and physiologists,—the muscular sensations were assigned to the sense of touch. So were the sensations of temperature, many of which arise from contact with a body warmer or colder than the touching organ, and hence were referred to touch proper. Inasmuch as these various classes of sensations are all concerned in many of the perceptions of touch, it is necessary to consider each apart.

The first class are the sensations of *gentle touch*, or of *touch proper*. These sensations are occasioned more frequently by feeling an extended surface, but they may, and often do, arise from gentle contact with the extremity of a pointed body. Sensations thus arising are neither pleasurable nor painful, and one is scarcely distinguishable from another. Hence none of them can be readily reproduced in the memory. Pressure against a surface, or motion over it, each involving muscular sensations, seems to be required as the condition of sensations sufficiently positive and energetic to enable us to distinguish the objects themselves, and to recall to memory the sensations which they occasion.

Sensations involving violence or injury.

The second class are the *acute* and *often painful* sensations that come from any affection that does violence to the organ, as the prick of a pointed substance, the cut of a knife, the stroke of a whip, the bruise from a stick. These sensations are all distinct and energetic, and occasion a shock to the nervous system which is more or less violent. They are more definitely localized than the sensations of touch proper, and more distinctly revived and recalled. The sensitiveness of the skin to affections of this kind is not proportioned to the sensitiveness of its touch. It has been proved by the experiments of Weber and others, that those parts of the surface of the body which are furnished with the fewest and the most sparsely ramified nerves and branches of nerves, and are the most incapable of sensations of touch proper, are none the less susceptible to exquisite sensations of this sort. These sensations are not confined to the surface of the body, its interior portions being capable of exquisite suffering from pricking, cutting, and laceration. Sensations of this class seem to be more nearly allied to those which we have called organic, and which are most conspicuous when an organ is injured or diseased.

Sensations of temperature.

The third class are sensations of *temperature*. These arise usually from contact of the body with some material object differing in temperature from itself. They are also experienced, by what is called radiation, from an object not in contact with the body. In such cases the body may be said to be in direct communication or contact with the heated atmosphere, or the vibrating medium of heat. The sensations of temperature are, in many particulars, like the painful sensations which we have just described. They are like them in not being confined to the surface. In case of scalding from water or steam, or of a severe burn from fire, or of violent internal inflammation, or of febrile excitement, their causes are purely internal, and the affections are organic. The sensitiveness of the body to heat and cold is not proportioned to its susceptibility to touch.

Sensations of pressure and weight.

The fourth class are the sensations of *pressure* or *weight*. These, so far as they are definite and peculiar, are the slightly benumbed and painful feeling which a weight occasions when laid upon the hand or arm, when there is no muscular effort to sustain or resist the pressure. In such a case slight additions may be made to the bulk of the body imposed, without being perceived. If the same experiments are made upon the parts of the body which are more mobile—as upon the lips, when resistance and muscular effort is provoked and made necessary—minute differences will be perceived and appreciated. Accurate experiments of this kind were made by Weber, eliciting surprising results. Hence the so-called sensations of weight are very largely complex in their nature, consisting largely of muscular sensations.

The muscular sensations.

The fifth class are the *muscular sensations*, which have been already sufficiently characterized. Not only do they enter very largely into the sensations of weight, but into all those sensations which require motion upon, and application to, the surface of the body which is touched. The sensations of the rough and smooth, of the adhesive and slippery, of the elastic and non-elastic, are of this character. According to the nicety with which these sensations are distinguished, is the delicacy of perception. Success in any manual art depends upon this sort of delicacy. Skill in sewing, engraving, and drawing, in the handling of tools, in driving, rowing, and playing on musical instruments, depends on the natural capacity for and the nice attention to these muscular sensations. They are equally, if not more important, to our judgments of form, size, distance, and the various relations of extension, as we shall see in considering the acquired perceptions.

Sensations localized.

One feature all these sensations share in common. Though sufficiently alike to be classed together as tactual, muscular, etc., etc., yet they differ in quality according to the part of the body which is their seat. The tactual sensations on the palm are different from those on the back of the hand; those on the hand are different from those on the different parts of the arm, and so on through every portion of the surface of the body. The same is true of the different muscular sensations. The muscular sensations which attend the opening and closing of one finger, differ from those which are experienced in opening and shutting the hand. Those which we feel in managing the arm differ from those which are used in controlling the position of the head. The same is true of the other classes of sensations which are appropriate to the interior of the trunk or the vital organs. This fact is of great importance in the explanation of the acquired perceptions.

Perceptions proper of touch.

§ 97. From considering the sensational element in touch, we pass to *the perceptional.* By perception proper, in touch, as in the other senses, we apprehend objects as *extended and external.* To touch has been assigned especial superiority in these discriminations. Many limit them exclusively to touch, making it the only agent through which we perceive, and assigning to all the other senses the sensational function only. Others, as we have already said, limit perception proper to touch and sight. We have given our reasons for holding that through every sensation, and of course in connection with every one of the senses, we perceive—*i. e.*, we apprehend objects as extended and external. The perceptions of touch, however, differ from those of the other senses not only in being more definite and minute, in consequence of the greater energy of the sensations, but also (with the exception of sight) in their immeasurably superior variety. For this reason they deserve special consideration.

Extension perceived by touch. Not explicable by extension in the organism.

Let it be observed as a preliminary, that we do not, by touch alone, know mathematical extension, nor mathematical qualities, nor the relations of pure mathematical quantities to one another, nor to the pure or abstract space or time which we conceive to exist. We simply perceive extended and external somethings.

It is contended by many that the reason why we perceive extension by touch, either exclusively, or in common with sight, is, that the organism itself is extended. We find, they say, that in those parts of the skin in which our perception of extension is the most definite and acute, the nerves and the nervous endings are most frequent; while in those portions in which its dimensions are most vaguely perceived, these are more sparse. Hence it is concluded that two nervous terminations at least are required for the apprehension of superficial extension. Moreover, it is urged that, as the remaining organs, except those of sight and touch, are each furnished with a single nerve only, or, at most, with a single pair, that is the sufficient reason why, by means of these, we have no perception of extension. In touch and sight, it is said, the soul being affected by sensations through nerves placed side by side, must necessarily perceive objects as extended. This view is held chiefly by physiologists, and, among them, by the distinguished John Müller, with whom many others agree.

Of this theory we observe, that it overlooks entirely the differences between the physical conditions of perception and the act of perception. It may be, and probably is, a necessary condition to the perception of extension by touch and sight, that many nerves should terminate side by side or be spread over an extended expanse in the organs. But it is one thing for the nervous apparatus to occupy an extended organ, and entirely another for the mind, by means, or on occasion of the sensations which follow the excitement of these nerves, to perceive an extended object. The soul is not aware that it has nerves at all, or that one or more are called into action. Nor is it aware that separate parts of the skin, or other organs, are thus affected. It knows neither nerves nor extended organs as organs. The spatial arrangement of the nervous endings may be a physiological fact, but this fact does not in the least explain the apprehension of extension as a psychical process. Moreover, this theory, and many others adopted by physiologists, involve the absurdity of making the soul first to know extension physiologically, in order to know extension psychologically—*i. e.*, they require it to know the nerves as side by side, in order to know that very property which is essential to knowing one object as side by side with another.

The correct analysis of the psychical process is that as the tactual and muscular and other more subjective sensations, are called into action, they are known to pertain to the soul, as connected with an extended sensorium. This sensorium is known to the soul not as a collection of nerve-endings or nerve-expansions, but as found in various conditions of activity, involving the soul's own active sympathy of either suffering or enjoyment. All these sensations involve some relation of extension and place, very vague at first, but sure to be more positive and definite as soon as the soul fixes its attention upon each. The soul, as it were, occupies and pervades the sensorium as extended in all directions. Its attention is first fixed upon certain of the sensations that are most positive or energetic, both the muscular and the tactual. Then the local diversities and likenesses are noticed, and the relations of place within and upon the surface of the body become fixed. Differences in direction, form, size, etc., are known, by processes which we shall explain under the acquired perceptions. But in order to any one of these discriminations it must be assumed that in the original perceptions of touch, extension, the sensorium as extended in three dimensions, is directly perceived. Unless such knowledge is gained directly in connection with touch, it cannot afterwards be acquired. But tangible objects are not only known as extended; they are also known as external. This brings us to our next division:

The perception of externality by touch.

§ 98. *Externality*, or outness, is involved in the extension which is known by the sensations of touch. Externality differs from simple diversity, or difference. Diversity may pertain to objects that are purely spiritual, as a series of mental activities or mental entities.

Two meanings of externality.

But externality as apprehended in perception, as has already been explained, is the diversity or distinguishability of an extended object from the spirit as non-spatial and non-extended; and again, it is the separateness or separableness of the surrounding material universe from the animated body. Both these relations are apprehended in sense-perception, and pre-eminently by the sense of touch. It is not only important, but essential, that these two meanings be not confounded.

It is also important to observe, that the externality which we perceive, is, like the extension, not abstract, but concrete; or, in more familiar terms, an external object, or an object as external.

Externality in the first signification.

We will consider the two senses of externality in their order. As to the first, we ask, How does the soul, in touch, perceive its own body to be external to itself? We answer,—as we have done already, (§ 86),—Precisely as through the other senses, by an immediate and inexplicable act of its own. It perceives directly its own body as a not-self or a *non-ego;* originally its own sensorium excited to sensation. We open this question a second time in connection with the sense of touch, because it has been often urged that its sensations are peculiar in revealing *outness,* or externality.

Some—as Reid—contend that the simple sense of resistance or hardness, or that affection of the sensorium which every solid body occasions, directly *suggests* outness.

Dr. Thomas Brown teaches that all proper tactual sensations, like other sensations proper, are purely subjective and spiritual, without the suggestions of externality and extension, and that it is only through the muscular sensations that the knowledge of the *non-ego* is gained. "We open the hand or the arm, as we have done in a score of previous instances, without striking against an object. All that we experience is a succession of purely subjective affections—affections simply and solely spiritual. But we strike against a wall, or other resisting medium, and we ask, What has caused this new sensation? We answer, it is not myself, for I have previously had, or rather produced, only a succession of spiritual states, in a series of muscular sensations. But here is a change. I have a sensation uncaused by myself, but caused by a being different from myself. There exists, therefore, a being not myself, and so I reach the *non-ego,* or externality." To this solution or explanation there is this fatal objection, that allowing that the order of sensations has been previously the same, and that the order is for the first time changed by some resisting object, the change would consist simply in a new subjective experience. The resisting object would give a novel sensation, but it would still be subjective. However unusual this may be, it is only subjective and psychical, and, according to Brown's theory, can give no relation of extension, and therefore no relation of externality. Even if in the way supposed, a cause other than the agent could be reached, it might be purely spiritual, and not necessarily spatial.

All these, and every other theory of the sort, have one common weakness—that they require us, by some arrangement or series or combination of sensations purely subjective, to account for or develop an objective, *i. e.*, an external *non-ego*. But it is obvious that it is not the greater or less positiveness of a subjective sensation, nor any change in the order of such sensations, which would elicit a *non-ego*, unless this were immediately discerned by the mind itself.

Objection answered.

But what! it may be asked, when I grasp a pebble, or an ivory ball, or a stick, is that which I perceive as external to myself simply the sensorium excited by the object grasped? Is this the *non-ego* which I perceive, and this only? We reply, that this is the only *non-ego*, which we reach by direct and original perception. The question is not what is in fact first noticed in the order of time, but what is first and ultimate in the analysis of thought. But do we not perceive also the object which produces these sensations? Do we not directly perceive the surface of the pebble, the ball, or the stick, as diverse from the sensorium, and the body which it pervades? Not by immediate perception. If we did, it would involve the inference that we perceive a *non-ego*, viz., the surface of the pebble as touched, and producing a sensation, viz., the felt sensation, which is also *non-ego*. That is, we should have immediate perception of two *non-egos*—the sensorium excited, and the object exciting it to a sensation. This is possible, but it must be shown to be necessary. We shall show in its place (§ 113), that externality in the *second* sense—*i. e.*, the distinction of the not-body from the body—is discerned not by an original, but by an acquired perception. If this is true, it is the result, not of a single act, but of a series of processes.

Sense of touch the leading sense.

§ 99. The sense of touch is the most positive of all the senses in the character of its sensations. In many respects it is worthy to be called the leading sense. The sense-perceptions which it gives, and those which are called into action in connection with it, are felt on every part of the surface, and throughout the interior of the body and all its members. The sensations themselves are the most energetic of any that we experience.

Moreover, the organ of every other sense is also an organ of touch, and, as such, is more or less sensitive. We touch the food which we taste, and unless we touch it, we cannot taste it. Though the eye does not literally touch the undulating light—*i. e.*, in response to the touch of light, gives no tactual sensations—yet when the surface of the eye is pressed by the finger, or strikes against any solid object, it feels and is pained. It is also acutely sensitive at times as a touching organ. The inner surfaces of the nostril and of the ear, like the outer surface of the body, are susceptible of tactual sensations. All of these organs are more or less completely provided with a muscular apparatus, by which they are moved, directed, accommodated, and made more ready for and subservient to their appropriate sensations. Hence the tactual and muscular sensations are very intimately connected with seeing, hearing, smelling, and tasting. In view of these considerations, it was said long ago by Democritus, that 'all the senses are modifications of the sense of touch.'

In view of these facts, touch has been called, by some physiologists, *general sensibility*, or the power of *general sensibility;* and the four remaining senses have been called the *special* senses.

Furnishes intellectual terms.

It ought not to surprise us to learn that the sense of touch furnishes most of the terms for the intellectual acts and states. Sight itself is indebted to touch for many of its terms. We *take* or *apprehend* a meaning; we *hold* an opinion; we *comprehend* or *grasp* a train of thought or a course of reasoning; we *accept* a proposition. Especially does touch furnish the words for those acts of the intellect in which the feelings and the will have a share. The reason is obvious. We touch and handle objects in order familiarly to understand their properties and laws. The objects which we touch, and the ways we touch or handle them, are determined very largely by our feelings, whether of curiosity or indifference, of love or dislike, of caution or boldness. All these feelings are expressed through acts appropriate to the sense of touch, or by the modes of using its principal organs. Hence the spiritual acts or states generally, are expressed by terms and phrases primarily applied to this class of bodily activities.

Sight; its organ and the conditions of vision.

§ 100. The *sense of sight* is the last which we are to consider. The *organ of vision* is the eye. The eye is in a structure like an optical instrument, and adapted to the refraction of light by a combination of lenses, and to the production, by this means, of a distinct miniature image of the objects seen upon the retina, *i. e.*, the dark network of nerves which lines the inner chamber. This image can be seen in the eye of some animals if separated carefully from its socket, and divested of the sclerotic coating behind. The surface of the eye is small compared with that of the organ of touch, but it is susceptible of the readiest and most rapid motions, and of adjustments of position and direction with little muscular effort, and as little muscular sensation as is sufficient for the discrimination and regulation of its motions. This susceptibility of easy and swift motion and adjustment is one of its most remarkable physical features, and is the condition of its marvellous superiority.

The *conditions of distinct vision* are a proper quantity of light, and the formation of a well-refracted image upon the retina. If the light is deficient or excessive in quantity or intensity, there can be no distinct vision. There is a particular distance for every eye, at which the most perfect vision of a near object can be attained. This distance varies considerably, from that of the so-called near-sighted, to that of the far-sighted. This variety is occasioned by a difference in the degree of the convexity in the lenses of the eyes of different persons, requiring a different distance of the object in order to bring the rays to a focus upon the retina. There is in every case, however, a certain range within which distinct vision may be had by a more or less constrained adjustment of the retina and one or both lenses, through certain muscles provided for the purpose. The muscular sensations experienced by the adjustments of the eye in order to discern objects distinctly, are important *media* in forming and applying the acquired perceptions. In order that the vision by both eyes may be single—and it must be single to be distinct—the two axes must be steadily fixed upon the same point; and in order that they may be fixed, they must be inclined together. The muscular sensations, varying with the different adjustments of the two axes are important in the acquired perceptions or judgments of vision.

Function of the image on the retina.

These conditions are completed or finished when a distinct picture on the retina is formed. This leads us to consider *the function of the image on the retina*, or its relations to the act and the object of vision. Concerning this there is confusion and error of opinion. The mind can not see the image on the retina. If it could, it must see it by means of another image, and so on *ad infinitum*. Nor does it perceive the image by any direct act, knowing it to be an image on the retina. It does not know that there is a retina, till the anatomist or the optician brings this fact to its notice, nor does it know of nerves, or nerve endings, or nerve expansions, in the act of seeing, nor can it in any other way be aware of the image as an image. That its formation is essential to the act of vision, we know by physiological researches, but not in psychical experience. Physiologically, we know that the one is necessary to the other. Psychically, we are not only not conscious of using it as a known means of the act of seeing, but we are conscious that we do not employ it as such an aid or means. If this fact were kept in mind, serious difficulties in the explanation of the process of vision would be set aside. For example, it has been often asked, How can we see objects upright, of which the images on the retina are inverted? How can we see objects as single, whose images are double? The answer to questions like these, and the difficulties which they involve, is, that the mind neither knows nor uses the image in the psychical act. It is by a purely physiological analysis that it subsequently discovers such an image as the last member or link in the series of physical conditions.

The act of vision as a sense-perception includes two elements, the sensational and the perceptional.

Sensations proper of vision.

The sensations proper from light and colors are scarcely marked in our conscious experience as pleasurable or painful. Hence they are feebly obtrusive. They rarely if ever attract the attention except when they are painful through disease, or an excess of energy which induces abnormal action. Some colors, however, seem to give a positive sensuous pleasure, as rich violet or purple; and a series of such colors, finely blended, occasions extreme satisfaction. So far as this is æsthetic, it is not sensuous at all. The pleasure from form and outline, as distinguished from color, is still less sensuous. These facts explain why it is that the sensations of vision are less definitely located in the sensorium, and why, when the eye is known as their subject, the percepts are so readily detached from the eye and projected before it. The equally unobtrusive character of the muscular sensations which are experienced in using the eye contributes to the same result.

Perception proper in vision. The object of vision.

§ 101. Vision as *perception proper* apprehends illuminated, shaded, and colored *visibilia*. When we call them objects, we do not intend that they are objects in the sense that they can be felt or handled, but that they are illuminated and colored *percepts*, set over against the soul by itself, and distinguished from itself by its own act of perception. The spectrum, as of a color refracted by the prism, or of a flame depicted on a screen, is a real object of vision. So is the image that seems to lurk behind a mirror, or to lie in the depth of a glassy pool. The colored network that is projected when the eyes are closed is an object. The visible percept is always colored. When we say it is colored, we include, under color, light and shade. Darkness even, is discerned by the eye only as the intensest and gravest of positive colors.

Is always extended.

This visual object is always extended. The colored percept is an extended object, and it cannot be apprehended as colored without being perceived as extended also. Brown (*Lectures*, 28, 9) insists most earnestly that the extension is not originally given in the sense-perception of color, and that we connect the two only because of an oft-experienced and inveterate association from touch. Dugald Stewart (*Elements*) sanctions this view. James Mill, and all the associationalists, must of necessity adopt this solution. The following suppositions refute the doctrine: If two or more bands of color are beheld by the infant which has never exercised touch, it must see them both at once; and, if it sees them both, it must see them as expanded or extended; otherwise it could not see them at all, nor the lines of transition or separation between them. Or if a disc of red were presented in the midst of, and surrounded by, a field of yellow or blue, or if a bright band of red were painted so as to return as a circle upon itself on a field of black, the band could not be traced by the eye without requiring that the eye should contemplate as an extended percept the included surface or disc of red.

Visible extension superficial only.

The object of vision is, however, an extended *superficies* only. By vision only, a sphere is perceived simply as a delicately-shaded circular disc. A cube is a flat surface with abruptly-shaded portions, bounded by converging lines. If we draw or paint from nature, we do it on a surface perfectly flat or even. In order to do this with truth, we must first see the object as without obtruding or receding portions. "The whole technical power of painting, says Ruskin, depends on our recovery of what may be called the *innocence of the eye;* that is to say, of a sort of a childish perception of these flat stains of color merely as such, without consciousness of what they signify, as a blind man would see them if suddenly gifted with sight." (*Elements of Drawing.*)

Indeed, in some visible objects certain of these original aspects are apparent and obtrusive, and we cannot substitute the reality for the appearance. When, for example, we stand at the end of a long street, the lines of houses, or of trees, or posts, approach one another till they nearly meet in a point. But they do not converge in fact; they are exactly parallel.

It has been insisted by some that the eye perceives more than superficial extension—that we discern by vision, *depth,* or the *third* dimension; that the eye, as it were, sees around a sphere, or along the receding sides of a cube. An appeal is confidently made to Wheatstone's discoveries in respect to binocular vision, and the application of the same in the stereoscope. The conclusion very far outruns the data from which it is derived. The objects seen through the stereoscope are not in relief, but are in a superficies or plane. No third dimension exists, but the usual signs of its presence are so striking, that the mind leaps for the instant to the conclusion that it is there in fact. The experiment of the stereoscope is so far from confirming the view that the third dimension is actually seen, that it shows most decisively that it cannot be, by effecting an illusion, which is well-nigh perfect, by means of objects drawn and actually seen upon a plane.

A single object seen with two eyes.

The question has been very frequently and very earnestly discussed, "How is it possible that the mind should apprehend but a single object by means of two eyes?" The question has been variously answered by physiologists. Some have insisted that one eye only is in fact used in the act of vision, the office of the second being to strengthen or reinforce the nervous or physiological action of the first. Others teach that the mind beholds two objects in fact, but passes so readily from the one to the other, as in effect to apprehend only one. Others have sought to solve the problem by tracing the impressions made upon the corresponding parts of each retina, through the corresponding nerves of each, to a common blending or meeting-place in the organism, where the two are fused into one. So far as these facts are purely physiological, if they are to throw any light on the psychical act or object, they must assume that the mind performs the act by a conscious recognition of the retina, or the nervous apparatus, which cannot be admitted as true.

The psychical act is occupied with a visible object, which, as has been explained, is colored extension. It sometimes happens that, in consequence of a diseased or abnormal condition of the eye or its nervous apparatus, the mind perceives two objects, when it ought to perceive but one. How is this to be explained, and what light does the fact shed upon the relation of vision with one eye, to vision with two? We answer: In double vision the mind beholds two similar objects in two situations. In single vision two percepts are perceived in the same part of the field of view. They must necessarily coincide. If the one overlaps the other, the one must strengthen the other.

Original place of the visible percept.

The question also suggests itself, *Where*, in relation to the retina or the eye, is the visible object [*i. e.*, the variously-colored plane or disc first apprehended] placed in the original act of vision: is it in the retina itself, or in the front of the eye? or is it projected in space—say at the proper focal distance before the eye? The question, in all its forms, supposes a more extensive or a more matured knowledge of space, distance, and position than the mind can possess when it begins to see.

Position, or place, as applied to perceived objects, is relative. It supposes some objects to be fixed as starting-points, and others as standards of measuring or estimating distance from them. None such can be definitely fixed and familiar before the *not-body* is distinguished from the body, and before the hand, the eye, and the parts of the external body have been fixed in their relative positions. The vague knowledge of extended matter which the sensorium gives must first be made definite by a bounding outline; and the most familiar extra-organic objects must first be placed apart from one another, before the eye or the retina can be known as the instrument of vision, or either can be distinguished as the place or the seat of the sense-percept. Long before these cognitions are attained, the sense-percept seen by the eye will have been carried by the hand into the space without the body, and irrecoverably connected with its correspondent touch-percepts, in the way hereafter to be described.

Superiority and dignity of the eye.

§ 102. The superiority of the eye to the other senses is owing in part to the unobtrusive delicacy of its sensations. They do not occupy the attention and detain it from the object itself and its relations. The force and tension of the soul's activity are given to these. Vision is capable of far finer discriminations than touch. A hair of the diameter of .002 of an inch can be distinctly seen.

The eye can also pass from one object to another with a swiftness which none of the other organs can imitate. In so doing, it can place data at the service of the intellect as quickly as the intellect can use them, however rapid may be its movements. By its swift and wide-reaching motions it can imitate the slower and limited motions of the hand, drawing outlines, constructing figures, measuring distances, combining groups and elements, with surprising rapidity and precision. The cultivated eye sweeps across a landscape, and in an instant the mind computes the size and distance of its principal objects, and unites them together within a frame-work of mathematical relations. The minuteness of the observed distinctions, the vividness of the contrasts, the cheerfulness of the colors, the stimulus of the light, the sharpness of the outlines, enable the mind to hold fast its perceptions, to recall them vividly and at will, and to employ them for science, art, or practical life. The eye has always been estimated as the noblest of the senses; and many of the words which describe the actions of the pure intellect, as to *see*, to *perceive*, to *discern*, are taken apparently from this sense, though perhaps all are finally to be traced to the sense of touch.

CHAPTER V.

THE ACQUIRED SENSE-PERCEPTIONS.

The sense-perceptions, as original and acquired.

§ 103. Thus far in our inquiries we have considered each of the senses singly. We have seen that by each of these we gain peculiar knowledge. We perceive sights only by the eye, and sounds only by the ear. In connection with these diverse objects, we apprehend certain relations common to all, viz., externality and extension. In other words, by each of the organs we experience a determinate sensation, and apprehend an object that is both extended, and also distinguishable from the sentient and perceiving mind.

But the range of our sense-perceptions is far wider than this. We early learn to use one sense in place of another, or of several, and to apply the knowledge which is given by one, in place of that which belongs to one or more of those which are unused. Thus, if I go into a darkened room and perceive a peculiar fragrance, I know and say there is a rose or a tuberose in the apartment—though I can see or handle neither. If I hear a sound, I know it is from a piano, a guitar, or the human voice, and I know the direction from which it comes, and from how great a distance. If I look at an iron that is at glowing white heat, I say, It *looks* hot; though heat is properly felt.

The two classes of sense-perceptions thus characterized are *the original* and *the acquired.* They are thus defined: *An original perception is one that is gained by a single sense, when exercised alone;* of an object, or in respect to its relations. *An acquired perception is gained by using the knowledge given directly by one sense, as the sign or evidence of the knowledge which we might gain by another.*

Importance and time of gaining the acquired perceptions.

The importance of the acquired perceptions is manifest from the greater frequency with which we bring them into use, and the confidence with which we rely on them, as well as from their greater convenience. Thus, a man strikes with a hammer upon the head of a barrel, and knows in an instant whether it is full or empty, without the trouble of opening it. A surgeon applies his ear to the breast of his patient, and determines whether the lungs or heart are diseased, where, and how far. An architect, by a glance of the eye, sees whether the framing of a bridge or roof is safe; or he measures off the dimensions of its parts by the eye as accurately as he could by his hand, or an instrument.

The time when many of the acquired perceptions are gained, is very early. The most important, and those which are universally applied, are made in infancy, at a period earlier than the memory can recall, and by processes which the memory cannot untwine, nor any subtle analysis easily resolve. Others, which are commenced in infancy, are perfected in youth and early manhood. Many are not complete till the senses through age begin to fail, and the attention becomes less energetic and agile. We begin the education of the senses in the earliest moments of infancy. The artist, the mechanic, the musician, and the observer of nature, never finish it till the organs refuse to aid and to serve the observing mind.

Many of these acquisitions are made so early, that they cannot be distinguished from the original teachings of nature. In very many, the process is performed so rapidly that it is difficult for us to believe that the mind goes through any process at all, the knowledge comes so simply and directly.

It is more convenient to begin with those which have been made within our memory, of which the stages and the means are within our view and at our command. We may afterward venture to unravel the more delicate tissues that have been wrought by the finer and more dexterous arts of infancy, in that early yet mysterious period when Heaven lies close about us, and seems to direct the movements of the soul.

The acquired perceptions of smell and hearing.

§ 104. The acquired perceptions of *smell* and of *hearing* invite our first attention, because they can be most readily explained. Our first examples are of odors. We experience the sensations of smell, as from a lily or tuberose, from camphor or musk. We ascribe them to certain objects of given appearance and structure, without the use of the sight or the touch by which the appearance or structure is directly discerned. The ground of this confident knowledge is experience. There is no reason *a priori*, why the fragrance of the tuberose should not proceed from the lily, and the fragrance of the lily from the tuberose; no known cause why camphor and musk should not interchange their odors. We have simply learned by experience, that in all cases where the sensation is experienced, a certain object is present.

We do the same with *sounds*. We hear a sound, and believe that it comes from a bell. We hear another, and know it is from a drum; another still, and say, There goes a cart, or a coach. Each of these sounds we ascribe to its appropriate object with positive certainty, on the ground of simple experience.

We not only learn in this way the objects which occasion smells and sounds, but we learn the place and direction of both. This is especially true of sounds. We know whether a ringing bell is on our right, or on our left; whether it is high, or low: whether a military band is far, or near; whether it approaches or recedes. That knowledge of this kind is founded on experience only, is obvious from the fact, that when the usual or the assumed conditions or occasions of our knowledge are changed, we are mistaken in respect to the place, direction, and distance of a sound, and that mistakes in respect to these lead to error in regard to the object which occasions it. The beating of our own hearts may be mistaken for a knocking at the door; the trampling of horses in a neighboring stable, and the cutting of wood in a neighboring cellar, may be thought to be within our own dwelling. The rattling of a cart on a bridge may be mistaken for distant thunder; the humming of a mosquito, for a distant cry of alarm, or the sound of a trumpet.

Acquired perceptions of sight. Distance judged by size.

§ 105. The *acquired perceptions of sight* are still more numerous and interesting. These divide themselves into several classes. The *first* of these are the *judgments of distance by size*. If we know the real magnitude of an object, we judge how far distant it is by means of its apparent magnitude. If we hold any familiar object, as a globe two feet in diameter, near the eye, and then remove it slowly, it will dwindle away first to an inconsiderable ball, and then to a mere speck. If we know its real size, we judge by its apparent magnitude how far it is actually removed. So true is this, that from a magnitude that is falsely assumed, we mistake as to the real distance, and are as confident and as prompt in our mistaken perception as though the data and the inference were both correct.

Let a person look over the coping of a wall, or the ridge of an intervening building, and see only the spire of a miniature church—say of a bird-house—and believe it to be attached to a real church, and he will at once see it as a very distant spire.

Judgments of magnitude by distance.

Second: We judge of *magnitude by the assumed distance.* When we have a correct impression of the distance of objects, we perceive them in full size. We every day see men and other objects at long distances greatly diminished and dwarfed, and yet we do not perceive or judge them to be smaller than they really are. A lofty building viewed at a very great distance, or a tall ship far off at sea, will even seem loftier than when viewed from a position very near, from which the beholder looks upward, without distance and other aids by which to judge of their height. The most impressive judgments of the height of the loftiest mountains and edifices are gained by seeing them at a great distance over an intervening plain.

Judgments of distance by color, outline, clearness, etc.

Third: If the magnitude is unknown, or not considered, we judge of *distance* by means of *the intensity of the color*, the *sharpness of the outline*, and the relative clearness or confusion of the distinguishable parts. For example, should we view, through a tube, several trees of the same species, as the elm, the maple, or the oak, removed at different distances from one another, the nearest would be known by its brighter green, its more sharply defined outline, and its more clearly distinguished leaves and branches. By these circumstances, designated technically as "*atmosphere*," painters produce the effect of nearness or distance, with accessories, of relative magnitude and of more or fewer intervening objects.

The traveler in Italy, especially when he goes directly from England, judges the mountains to be far nearer than they are in fact. The atmosphere is so much more transparent than that to which he is accustomed, as to reveal the outlines and face of the mountains so distinctly that he cannot believe them to be as distant as they are.

Judgments of size by other equidistant objects.

Fourth: We judge also of *the size of objects,* by *comparing them* with *other* objects which are or seem to be at equal distance from ourselves. If the size or distance of our standard of comparison is incorrectly taken, we misjudge altogether. Dr. Abercrombie (*Intellectual Powers*) tells us that, on going up Ludgate Hill toward the great door of St. Paul's which was open, he took several persons who were standing under the opening to be children, whom he found, on coming up to them, to be full-grown men. The reason was, that he assumed the height of the door to be less than it really was, and, by this false standard, he misjudged the size of the persons who stood under it.

Influence of intermediate objects.

Fifth: Our *judgments of distance* vary according as there are more *or fewer intermediate* objects. Objects seen across the land seem further than objects at the same distance seen across the water. A given expanse of the sea is greatly enlarged to the eye when a score or two of vessels are anchored at different distances along its surface. A level meadow or prairie, with copses, trees, and dwellings interspersed, seems far more extended than without them. A salt marsh, when dotted with haystacks, seems wider than at the season when they are removed.

Sixth: Intermediate objects, by affecting our judgments of distance, affect our *judgments of size.* The sun and moon appear larger when near the horizon than when toward the zenith. Through the influence of intervening objects and the dimming influence of the atmosphere, they are removed to a greater distance, and then judged to be larger. The sky itself, for this reason, is not the half of a sphere, but a section of which the height is shorter than half the base.

When the ordinary standards of judgment are withdrawn, and our accustomed processes cannot be applied, we are either greatly embarrassed, and even bewildered, or we fall into serious and amusing errors. Captain Parry says: "We had frequent occasion, in our walks on shore, to remark the deception which takes place in estimating the distance and magnitude of objects over an unvaried surface of snow. It was not uncommon for us to direct our steps toward what we took to be a large mass of stone at the distance of half a mile from us, but which we were able to take up in our hands after one minute's walk. This was more particularly the case when ascending the brow of the hill."

§ 106. By means of sight we acquire perceptions *appropriate to the touch.* When we look at a sphere, we see by the eye only a circular disc on which the transitions of color or of light and shade blend so finely with one another, that we know if we grasp it with our hands we shall feel it to be spherical in form. A sphere may be so skilfully painted in fresco on a flat surface, that we actually take it to be a sphere in fact. We often seem to see projecting statues, graduated mouldings, depressed panels, receding corridors, vaulted domes; and yet as we approach, we find only a plane surface.

Judgments of form, etc., by sight.

When the blind from birth are restored to sight, they come into a new world, of the *percepts* of which, and their relations to the *percepts* already familiar to their touch, they have had no previous knowledge. They must therefore go through a special discipline in order to connect the well known objects of touch with the newly acquired experiences of the eye. Thus the blind boy whose sight was restored by Cheselden could not call the cat and dog by their right names, or could not tell which was the cat and which was the dog. He could not avoid distinguishing them by the eye, but he had not learned to connect the dog and cat as handled—to the appropriate forms of which he had attached the names—with the dog and cat which he saw, so as to be able *to feel* them by means of his eyes. Finding himself one day at fault, he carefully felt of the cat with his hands, his eyes being shut, and set her down, exclaiming, "So, puss, I shall know you another time." The question has been often asked (cf. Locke, *Essay*, B. ii. c. ix. § 8), whether a blind man, on being restored to sight, would know a cube from a sphere. It is obvious that, so far as mere vision is concerned, he could not but distinguish the two objects as soon as he attended to them with the eye. What he would need to acquire would be the capacity readily to connect the visible with the tangible cube and sphere.

In the examples which have been cited, we translate the *perceptions given by sight* into those which *are derived from touch.* The proposition is sometimes broadly and positively laid down, that from the touch is derived all perception whatever of form, distance, and magnitude; inasmuch as in all cases, we must come back to the touch as furnishing the ultimate standard. The position is sometimes stated thus: All visible extension must be reduced to that which is tangible. These propositions need to be somewhat qualified, if we hold that we can perceive superficial extension by the sight. They are true to the letter of all those perceptions which involve the relation of depth, or the third dimension of space; but to all judgments of superficial form and dimensions they cannot literally apply. To the blind, however, touch furnishes the only possible standard of definite form, distance and size.

The blind man applies his finger, his hand, or his arm, to every object which he encounters, and measures its size by any of these standards. But those who see, perceive objects extended superficially. Why, then, may they also not apply any of these objects as units of measurement, and as standards by which to judge of form and size? We reply, they may, and would do so always, if what is called the apparent magnitude of the standard, and of the object to which it is applied, did not constantly change as the two are near or remote. A yard-stick or a foot-rule may be so far removed from the eye, as to measure to the eye no more than a foot or an inch respectively. Even though the standard is unaltered in its position, the object measured may, by being itself carried near or far, measure a foot, a yard, or a rod. We can only be satisfied that the standard and its objects coincide, when we bring the standard in actual contact with the object by the hand. But even then we use the eye, in order to be certain that the two coincide. The hand of the blind, however surprising may be its delicacy of touch, can never attain the fineness of the eye in discerning exact adjustments. Give the practiced eye an assurance that its distances are correctly taken, and it will measure and judge with marvellous accuracy. It is a circumstance which is worthy of attention, and certainly ought not in this connection to be overlooked, that the point of distance from the eye at which vision is usually most satisfactory, coincides with that at which the hand can most conveniently handle and hold an object.

§ 107. It is by the *acquired perceptions* that we definitely assign the places of *our sensations to the different parts* of the body.

Acquired sense-perceptions of place, motion and expression within the body.

All the sense-perceptions must be known to have some place in the sensorium, though the limits of the place may not be definitely drawn, and the relative positions of each perception may not be exactly fixed. Whatever is involved in such a perception taken singly, is an original perception. Whatever is added or superinduced by combining several perceptions, is acquired by experience. For example: an adult person has a pain in one of his teeth, he does not know which—or a cut in a part of his arm, he does not know exactly where. If he touches the tooth with his tongue, or if he discovers in a mirror which tooth is defective, he ascertains which is the one affected; he learns as we say, where the pain is.

That much of this knowledge is acquired, is evident from some cases of lesion in different parts of the body, and of the loss of a limb by amputation. A man who has no foot, will feel pain in the foot. Why? Because he experiences precisely the same sensations which he suffered when he had the foot, and knew it was the seat of the pain. But if he had never had a foot, he would never have assigned pain to it; for he would never have had the means by eye or hand or muscular sensations, of connecting these sense-perceptions with it.

It is also by the acquired perceptions that we *learn to regulate and control the movements of the body.* Man was made to move. When the soul, so to speak, finds the body, it finds it in motion. Not only is this true, but the body is, by its very structure, adapted to certain specific motions, as of walking, speaking, and singing, all having definite relation either to its present or its future wants or enjoyments. These bodily capacities the soul acquires the power to use in definite ways for special ends. The motions to which nature prompts, the intellect learns to control and regulate, so as to bring to pass determinate results. A more particular consideration of this subject presents two separate questions: *What does nature provide?* and *How does the intellect apply these provisions of nature?*

We ask, first: *What does nature provide?*

The provisions of nature for these ends.

We have already adverted to the fact, that with the *sentient nerves* which conditionate sensation, there are provided the *reflex motor* which impel to motion. In obedience to the stimulus furnished by the one, there is awakened in the other an unbidden and often an uncontrollable tendency to motion. Consciousness need not, and often does, not intervene. Thus, we wink in response to the stimulus of light; the flesh quivers and withdraws itself from the knife; the muscles knit themselves into convulsions and cramps. Under the same law, the excitements being diverse, the heart beats, the lungs expand, and other involuntary motions are performed. These functions and operations relate to the body, and their effects terminate in its well-being.

There are other movements that are connatural and at first involuntary, which the intellect has the power to apprehend and the will to control. Such are the muscular efforts that are involved in speaking, singing, and walking, and in feats of skill or dexterity. Many of these relate to the soul as well as to the body, in the way of use or enjoyment. Some of them are made ready for the spirit against the time when it shall be sufficiently developed to apply them with intelligence and design. To all these movements the stimulant comes not from without, but from within. When the infant weeps from pain, and laughs and shouts from delight, it is under an excitement proceeding directly from the soul, that the muscles are moved to laughter and to tears. In the same way, every emotion seeks and finds expression by attitudes, looks, and gestures.

In the same way man is prompted to speech: first to inarticulate cries expressing emotion only, and then to articulate language and words significant of definite thought. Nature provides for all this, by making man capable of a limited range of vocal sounds, through the action of those muscles that move the larynx; and nature prompts to the use of these muscles in various ways, according to the varying excitements of feeling and thought. To very many, if not to all of these effects, the consentient action of many muscles is required. For this nature provides, by so arranging the structure of the nerves through which these consentient muscles are excited, that, under the stimulus of feeling or thought, those needed, and those alone, shall be aroused to the united activities which conspire to the single effect which is required.

Not only does nature provide for the conspiring action of several muscles to one effect, but she even arranges for and prompts to the combined action of different parts of the body in obedience to a single impulse. In order to make progress by walking, each leg must alternately advance and wait for the other. To these alternate motions there is an original impulse. These are movements which the infant makes long before it begins to walk. The arms, on the other hand, tend to move together. So do the fingers. It is difficult, and sometimes impossible, by any effort to bring certain of the fingers to a separate action. But it is in the eyes that this tendency to joint action is most conspicuous. The eyes will persistently move together in the same direction. They cannot be forced to act apart. One eye cannot by any violence be made to look upward while the other is directed downward. Nor will one tend to the right, and the other to the left.

Even more than this is true. There seems to be, so to speak, a natural aptitude for the joint action of organs that are not paired together, but which yet are fitted to aid one another in important uses. This is preëminently true of the eye and the hand. The eye must lead the hand, and the hand follow the eye, in a multitude of actions. When we would touch or grasp a small object at the first trial, the eye must guide. When we would strike it with a stick which we hold, or with a projectile, the eye must conspire with a fixed and earnest gaze. There must be some physiological reason for this concurrent action of nerves and muscles connected with two organs, though it has not yet been discovered.

We ask, second: *How does the intellect apply what nature provides.*

The control by the intellect of these arrangements.

The intellect finds itself furnished with this corporeal instrument, and actually using it under the promptings of nature; it finds it laughing, or weeping, speaking, and walking, under the promptings of nature, and it acquires the power of directing these activities in particular methods and to certain definite results, and of doing this so readily, that it does not notice its own processes, or advert to the elements of which these processes consist. First, it observes the muscular sensations which are employed when certain effects occur, and the effects it observes by the appropriate sense-perceptions. It experiments upon these, and notices how the sensations which are connected with the varying use of its muscles are connected with varying effects. Then it tentatively and designedly repeats the effect which it has chanced to produce, or it seeks to imitate the effect which another has accomplished; *e. g.*, to utter a sound, to refrain from laughter or from weeping, to walk slowly or rapidly, or with a particular gait. By repetition of the effort, the effect is produced with little attention to the means, till at last the effect seems to occur without the use of these means at all. When the mind would accomplish an object, as utter a sound, hold a book, or let it fall, walk, run, or leap, it thinks only of the effect, and wills it, and it is accomplished.

How we learn to talk and to walk.

In learning the unfamiliar sounds or combinations of a foreign language, we try one experiment after another, till at last we succeed. When the ear is satisfied that the result is reached, we repeat the muscular effort required, guided by the muscular sensations, till our command over the organs is complete, and we can produce at will the sounds which we seek for. The infant pursues the same method in learning to talk. It is awakened from its purposeless lispings by the desire to produce a sound, as to pronounce a word, or a brief sentence. It succeeds imperfectly at first, but well enough to guide its efforts in the direction toward complete success. It triumphs at last, and it attentively observes the sensations which are connected with the word which it has learned to speak. Guided by these sensations, it can repeat the word or sentence a second time.

The deaf-mute cannot learn to speak, not because he is mute by reason of any defect in the organs of speech, but because he is deaf, and cannot regulate these organs. He has the vocal apparatus in complete perfection and he can make all the varieties of vocal utterances which are required in speech, but not having the ear by which to direct his efforts, he can neither form his own efforts to definite results, nor can he retain the acquisitions which he has made. In a few cases, the deaf and dumb have been taught to articulate by a discipline specially directed to the management of the vocal apparatus; but the articulation is imperfect, and easily lost.

The infant learns to walk as it learns to talk. It notices the sensations which attend those adjustments of the muscles which are necessary to quick or slow progress, to rising or sitting, to running or leaping. In all these effects we are usually guided by the eye. But sometimes we have not the eye to guide us. We ascend a flight of stairs to which we are accustomed, by a vague remembrance of the height and width of the steps. The blind depend on the direction of others, both in their first essays and in many of the subsequent uses which they make of their limbs.

Feats of dexterity. Expressional effects.

By similar processes, facility is acquired in those uses of the limbs which are required in feats of dexterity, as in sleight of hand, or in playing on a musical instrument. It is to be observed, however, that whatever movements nature fails to provide for, she gracefully accepts as a second or an acquired endowment. The effort to constrain the organs or limbs to an unnatural position or adjustment, may at first be painful, and it may cost constant and severe application. But if it is persevered in, and especially if the intervals in which it is remitted are short, these new adjustments of the muscles are secured, and they even shape themselves to new forms. While the mind is renewing its efforts at brief intervals for a succcession of months or years, the substance of the body, in obedience to the laws of life, is continually changing; and as it changes in its material, it is also changed in form, under the moulding pressure of psychical tension.

In infancy and early childhood the merely physical capacity of receiving directions and impressions from within is incomparably more ready and quick than in later years. In early life, every single distinct effort in the use of any bodily organ seems to initiate a definite physical predisposition toward a permanent physical effect, either in the force or direction of the nervous stimulus, or in a new combination of muscles, or in fixing some form or attitude. A few repetitions, a brief perseverance, and the body is permanently moulded or fixed to the special service of the soul, in some new aptitude or habit. Hence it is that the bodily habits acquired in early life are so readily contracted and so inveterately retained. But whether the law acts with greater or less efficiency at an early or a later period, the principle is the same.

The errors of the senses explained.

§ 108. What are called *the errors of the senses lie wholly within the sphere of the acquired perceptions.* A person needs only to fall into a few such mistakes to be convinced that they are mistakes of judgment only, and that, as in the cases when he judges correctly, the process is a processs of judgment or induction. When a man sees, as he says, a bent stick in the water, he judges that it is bent by what he sees; or, in other words, he judges by what he sees, that, if the stick is handled or otherwise tested by the sense of touch, it will be found to be crooked. And yet he seems to perceive by the eye that it is bent. So, when he looks into a kaleidoscope, and sees scores of brilliant objects arranged in symmetrical groups, he perceives them all by the eye, and can count their number, and does not doubt that he can grasp them all by the hand. It is common in such cases for a person to say that his senses deceive him. But the senses are not treacherous: they cannot deceive. It is the man who is deceived in the judgments which he pronounces on the evidence which the senses furnish. He is simply hasty and premature in judging by the eye. He rashly connects, with what he sees by the eye, something which he believes with his mind. The bent stick is perceived when out of the water just as is a bent stick in the water; in either case a judgment is pronounced—in the one case a judgment which is right, in the other a judgment which is wrong.

The muscular sensations of the fingers may also be disturbed. We cross the fingers, and at the points of both a single pea is felt as two. The reason is that the convex surfaces, which as they are usually touched are interpreted as looking inward forming a single sphere, seem to look outward, and by the imagination are interpreted as requiring two to complete them.

This class of the so-called errors and deceptions of the senses ought to be sharply distinguished from another, which is caused by *the physical conditions of the sensations themselves.* Some men, for example, are color-blind—*i. e.*, they see every object in one uniform, dingy hue, instead of under the bright and diversified colors which are granted to the majority of men. Some men, through a disease of the stomach or liver, see every object tinged with yellow. It occasionally happens that a man is afflicted with double vision—seeing two objects where other men see only one. Others see spectra, or visible images which have no tangible reality, and no reality at all except to the individual who beholds them. Others hear sounds, as of ringing in the ears, when there is no sonorous body, and no vibration of the atmosphere. Cases of this kind are never deceptions of the senses, for the objects perceived are the natural and legitimate product of the physical conditions that are present; these conditions being the physical excitants or stimuli and the sensorium excited, whether to normal or abnormal activity.

The acquired perceptions as forms of knowledge.

§ 109. The acquired perceptions differ from the original as *forms of knowledge.* Acts of original perception are acts of *direct or immediate knowledge.* In such acts the objects are present to the intellect, and the intellect knows directly that they are, and that they exist in certain relations. Acts of acquired perception are *acts of mediate knowledge.* In such acts it is by the medium of another act of original perception, that the object is said to be perceived. Thus, when I know the place of an object, the size or distance of an object seen, I use a direct or immediate perception as the medium through which I reach what I know indirectly.

Again: an act of acquired perception requires for its fulfilment the *representative power,* in the form of phantasy or memory. When the mind, on occasion of a direct perception, supplies that which it does not directly feel, or see, or measure, it must reproduce its object from something previously experienced, either in the form of a perception precisely like what is reproduced, or else similar or analogous. But the original perception apprehends its object directly.

Again: if the act of acquired perception rests upon the representing power or agency, it must involve the action of *the associative* power. At the experience of one odor, we think of a lily; at the experience of another, of a tuberose. At the sight of a distant moving object, no larger than a mote, we think of a man or a horse. What brings the form of a rose or a tuberose, the picture of a man or a horse, before my mind's eye on occasion of these direct perceptions? We must anticipate

7

our knowledge of the laws which govern the representative power, in order to answer—The laws of association.

They involve induction.

Every act of acquired perception is an act of *induction*. The mind does more than represent some picture or remembrance out of the stores of its past experience; it *believes* there is a real object corresponding to this picture. In so doing, it performs a process of induction. It judges, by the signs or indications which the original perceptions furnish, that there are existing objects which the other senses would find to exist should they make the trial. The process by which this belief is attained is variously named *inference, induction, judgment, interpretation,* etc. It is peculiar in this, that it knows by *media* or signs. It also assumes that these signs always indicate the same accompaniments, and that the laws and operations of nature are uniform in respect to the connections which are indicated.

It may surprise many to learn that the processes employed in the acquired perceptions are processes of induction. Induction is usually conceived and described as a process which is appropriated to philosophical discovery, which requires wide generalization and profound reflection, and issues only in comprehensive principles and laws. A little reflection will satisfy any one, however, that the act of mind is the same with that performed in every one of the acquired perceptions. The difference between the two kinds of induction is not in the process, but in the materials upon which the mind performs them. But the acts, the fundamental assumptions, and the liabilities to error in both, are essentially the same.

But it cannot be possible, it will be urged, that the perceptions which the infant so rapidly acquires, and which the most ignorant and unreflecting so skilfully apply, are in their nature similar to those profound and daring acts by which the astronomer scales the heavens, and the naturalist penetrates and resolves the mysteries of the universe. The difficulties and objections which are expressed in this language can be most effectually set aside, if we notice the differences in the circumstances and conditions of the acts performed by the infant and the philosopher.

We notice 1. that the infant employs its perceptions upon a very limited number of objects.

2. The few objects which the infant mind distinguishes are constantly recurring to view.

3. All the objects and parts of objects with which the infant has to do—in other words, all its sense-perceptions—have an immediate relation to its appetites and desires.

4. When any experiment has been successfully made in the way of connecting the known and the untried, the gratification at success will stimulate to repetition: and this again holds the attention to every element and step in the process, till the whole is fixed in the memory. The infant repeats all its lessons as fast as it learns them, because it rejoices over its acquisitions.

5. The associating power unites what observation notices. So few are the combinations which it has made as yet, and so closely were they connected by the original acts which first bound them together, that the one cannot be perceived or thought of without its companion.

6. The resemblances which the infant apprehends are few, and discerned with little effort. It might better be said that similar objects are at first recognized as the *same,* rather than discerned as similar. Hence the inductions of the infant are at first simple acts of spontaneous memory, rather than beliefs founded on similar instances.

In induction proper, the similarities are remote—not obvious, not directly discerned, but indirectly surmised; the data themselves are the results of previous research and reflection, instead of being forced upon the attention.

7. The infant cares for the result, and, in its eagerness to reach it, slights or disregards the means. What it finds to be true, occupies its attention, and not the evidence or data by which it has discovered it.

8. The freshness and energy of the activity of the human soul in the earliest periods of its life continually surprise and astonish us. The activity of the intellect, the freshness of interest, the energy of will, the eagerness of the desires, the variety of the experiments upon itself, upon nature, and man, are ceaseless occasions of interest and surprise to older persons whose powers are torpid or overwrought, and whose curiosity is partially sated.

Whatever objections may be urged against the possibility that acquisitions like these should be made in infancy and early life, are satisfactorily met by the unquestioned fact, that the infant is constantly making experiments and falling into errors in this very sphere of induction and acquired knowledge. It makes awkward attempts to grasp, to reach, to stand, and to walk; it misjudges in respect to the distance, form, size, and nature of the objects beyond its reach; it is taught by experience, and it applies the lessons which experience imparts, whether painful or pleasant. It is never so busy as in the earliest years of its life. All this time it is chiefly occupied with experiments upon the material world and its own bodily powers, its energy being employed in the very directions, and being busied with the very objects, with which the acquired perceptions are concerned.

It ought also to be remembered that, during the same period, it makes the surprising acquisition of language; always of the mother-tongue, and, if circumstances favor, of one or two languages more. To acquire a new language so as to speak it well, costs an adult whose powers are well disciplined many months, if not years of labor. With how much greater ease, rapidity, and perfection, is the same task achieved by the infant! Surely it is not surprising that at an age as early, or even earlier, it should master the acquired perceptions.

Objections from the cases of animals.

It might be urged in objection still further, that there is no evidence that animals have what are properly acquired perceptions. On the contrary, observation shows decisively that they perceive directly the distance, size, and properties of the objects with which they are concerned. The chicken, with the young of certain birds, strikes its beak with precision and success at the food brought within its reach, even before it is released from the shell. The young of the partridge and the grouse run swiftly through the stubble, avoiding projecting objects as if with practiced skill. The young of quadrupeds run and leap with little previous discipline or training. In view of these facts, it is confidently urged that, if these animals are taught by instinct to perceive correctly, it is not to be supposed that man would be left to the slow and uncertain processes of feeling his way along to certain beliefs. Surely nature would do as much for its noblest work, as for the inferior species.

To this objection is to be opposed the indisputable fact that the human species *is* slowly disciplined to feel its way on to matured and trustworthy acquisitions. The reason why, is obvious. The animal has not the capacity to judge by signs, to that extent and with that discrimination which would qualify it to build up the power of perception. This deficiency is supplemented by instinct, about which we know but little, but enough to be certain that it effects by blind and unintelligent impulse what reason discerns and performs with discriminating judgment.

Some facts are observed in infants which are supposed to be inconsistent with these conclusions, and to prove decisively that the infant, as well as the animal, has a so-called instinctive perception of distance. Thus, for example, Adam Smith reasons: "A child that is scarcely a month old, stretches out its hands to feel any little plaything that is presented toward it." It is more than possible that in infancy the eye cannot be excited by a visible object, especially if the object gives pleasure, without a consentient movement of the hands, and of both hands and eyes, in the same direction. That some provision should be made for such a conspiring movement or impulse to motion of two members of the body that perform many functions in common, may be received as probable, and believed to be true. But this would not prove that the eye, in the proper sense of the term, discerns distance. All the movements with both hand and eye show that this is judged or inferred by indications or signs.

Reasons why the perceptions of animals and of man should differ.

Important reasons suggest themselves, however, why the animal is taught and impelled by instinct to do at once, and with little exposure to failure, what man can only attain by slow and painful acquisition, and at the risk of many failures and sufferings. The discipline to which man is subjected has respect to his moral culture as well as to his intellectual discipline. He needs to learn patience, caution, foresight, and circumspection, as well as the higher virtues. All of these are furthered by the disciplinary processes through which he gains the acquired perceptions. It is by the adaptation of this discipline to high moral uses, that we explain the law of nature by which man is born the most ignorant and helpless of all the animals, and forced, as it were, to make his acquisitions by his own sagacity, as fast as he is impelled by his awakened appetites, desires, and affections.

We conclude, then, that the processes of the acquired perceptions are processes of induction, and that they involve the powers of representation, and judgment by indications. In other words, in the very act of perception, usually considered as the lowest and the most elementary of all the acts of the intellect, there is required the presence of the higher powers with the intuitions and relations which they involve. This is a striking instance of the principle already enounced, that no faculty of the intellect can act apart from the rest. For we have found that, in the very lowest of all, the rudimentary action of the very highest must be present, in order that its perceptions may be human and rational.

CHAPTER VI.

DEVELOPMENT AND GROWTH OF SENSE-PERCEPTION.

Nature, interest, and difficulty of the problem.

§ 110. We propose next to trace the growth and development of the sense-perceptions in earliest infancy. We take our guidance from what we have observed of those processes which we are certain that we acquire, and, going back to that period of which memory brings no report, we ask, From what beginnings, in what order, and by what steps does the infant mind develop and mature the power of sense-perception of which it finds itself in possession when it awakes to distinct and remembered consciousness?

The question is full of interest. It seems like a proposal to revive the experience of our earliest years, and restore, as it were, the forgotten past of our lives. There is a mystery about those months and years which we would fain unravel, while the difficulty and apparent insuperableness of the problem incite and challenge us to the effort.

The difficulty which attends the effort arises from the fact that it is impossible, by memory, to bring back a single fragment of our infant life. We cannot penetrate the darkness and obscurity which overhang the whole of this period of our existence.

We can not recall to the memory any single perception in which all visible objects were depicted on an extended plane, without distance or depth. Nor can we by imagination feign such an experience. The effort to do either must be fruitless. The new elements which we have incorporated into our constant habitudes of perception and knowledge we can never throw off. We can not lay off the new growth which has overgrown the original germ. But the problem, though difficult, is not *insolvable.* To the judgment only is it explicable, but not to the imagination. We can demonstrate what our infant life *must have been,* but we cannot imagine how this infant life *must have seemed.*

To attempt to retrace and thus to reconstruct the processes of the earliest perceptions of childhood, is not *irrational.* We have at our command the materials with which to prosecute our analysis and to construct our synthesis. These are the known facts of experience and observation within our conscious experience, the facts observed of infants and very young children, and the probable conclusions which analogy warrants us in deriving from both.

> Who can tell what a baby thinks?
> Who can follow the gossamer links
>   By which the manikin feels his way
> Out from the shore of the great unknown,
> Blind, and wailing, and alone,
>   Into the light of day?
>
> *   *   *   *   *
>
> What does he think of his mother's eyes?
>   What does he think of his mother's hair?
> What of the cradle-roof, that flies
>   Forward and backward through the air? etc.
>
> J. G. HOLLAND.—*Bitter-Sweet.*

All that we observe of the actions of infants and young children is entirely consistent with the theory, that they develop the power of perception by many experiments and many mistakes.

The known methods and laws of nature in the education of men and of animals give the strongest confirmation to these conclusions. We rely with confidence upon the view that, so far as it is possible to account for the acquired perceptions by the theory of intelligent activity rather than by that of blind instinct, so far we are bound to go. Where intelligent activity cannot be presumed or proved, there instinct and intuition must be assumed.

Synthesis or *combination*, however, cannot account for every process or solve every problem. There must be original elements with which to begin, or else there would be nothing with which to combine, or which could be added when it was sought for. There must be capacities or powers of original knowledge, beyond or behind which we cannot go in our analysis; which capacities, indeed, give the elements which we evolve by such analysis.

The condition of the intellect before sense-perception begins.

§ 111. These things being premised, we observe: The first condition in which the soul exists before the beginnings of conscious activity, is nearly allied to the state of sleep undisturbed by dreams, or of a dead fainting, in which the most indistinct and feeblest sensations possible are experienced without distinct perception. The undeveloped condition of man is not dreamlike in the sense of being confused, or bewildered; it is rather such a vague and low condition of sense-perception as would attend the activity of those muscular and vital sensations which belong to the processes of the animal life. These sensations, when closely attended to in later knowledge, are at best but vague and indefinite; and when they fill up the whole world of our conscious life, they must be obscure indeed.

The beginnings and development of attention.

From this condition the soul is aroused when it begins to attend either to a sensational excitement, or to the responsive perceptional act. The soul scarcely can be said to have sensations even, till it is conscious of some sharp or positive experience of pain or pleasure. Much less can it be said to perceive, till its attention is aroused, repeated, and fixed upon some single sensible *percept*.

We are not to suppose that the attention, in either of these directions, is developed at a single bound, or that its energy is attained by one spasm of effort; nor that the soul maintains itself always in the attent condition which at first it attains only now and then. All analogies from the states of our mature experience would lead us to believe that the soul now rises for a moment into fixed attention, and then sinks again into blank inanity.

Nor, again, are we to believe that the attention can only be aroused or occupied by a single sense at once, but rather that two or more of the senses may be exercised at the same time upon their appropriate objects, and thus the development of one of the senses may aid that of the others. This view is altogether consistent with nature and experience, and with the observations which we are able to make of the successive efforts which the infant makes to correct his mistakes and to perfect the training of his powers. As it is true with the adult, so is it with the infant; the several capacities are developed together and aid one another.

The order in which the perceptions are developed.

§ 112. The sense-perceptions which we should expect would be developed first are the *muscular and vital.* If, however, we perceive only so far as we attend, we ought not to call these sense-perceptions till they are connected with other perceptions which are more positive and objective, as the perceptions of sight and touch.

We should also suppose that the three senses of hearing, taste, and smell, would spring into activity next in order. Observation does not, however, confirm these anticipations. The sense of hearing is used, in some feeble degree, a few days after birth, scarcely in such a manner or degree as to be called attentive or discriminating. The sense of taste is still later. At first, the infant swallows medicine as readily as milk. It is not till some four weeks have elapsed that it distinguishes the one from the other. The sense of smell is exercised still later. Others say taste and smell are active from the first. Hearing, though feebly developed at first, remains the longest, as death comes on.

It is *with the eye and the hand* that the soul begins fixedly to attend, and of course, effectively to perceive. But with which does it first begin—with the eye, or with the hand? It is impossible to answer. Perhaps it were safer and more exact to say that it begins with neither alone, but with both, each aiding the other.

In our analysis we begin with the hand. Whatever may be true of the eye, we are certain that intelligent perception by touch must be acquired very early for those who can see.

The development of touch.

§ 113. We begin, then, with touch. Our problem is, to show by what steps of touch we acquire the perception of extension and of outness or externality —by which we mean separableness from the body—or the not-body. We assume that by original perception the *non-ego* proper is distinguished from the sentient *ego*, or the *ego* which animates the sensorium. We do not ask at what time this distinction is consciously developed; we only contend that it can not be acquired. Our present inquiry is by what process the knowledge of the *non-ego* as the *not-body*, is attained.

The *first* step is for the soul to know familiarly *its own* body as bounded by a *limiting surface*. This knowledge it *acquires* by contrasting the muscular and tactual perceptions. The muscular and tactual perceptions we suppose to be familiarly known. By means of the distinguished muscular sensations we perceive the interior of the body which the spirit inhabits and controls. Upon contact of the sensorium with what are afterwards discovered to be material objects, we have only certain affections upon its own surface. When an infant lays its hand on anything flat and smooth, it perceives a portion of its own body in a given state of activity. If this surface is triangular, a corresponding portion of the sensorium is similarly excited, and so on. As soon as the two classes of sense-perceptions are familiar by attention, the muscular sensations give us the knowledge of the interior space that the sensorium occupies, and the tactual sensations give the knowledge of its bounding or limiting enclosure. The infant is constantly made aware of this limit, by contact with the surrounding objects that excite it to sentient activity. In the warm surroundings of a bath, bed, or heated apartment, the surface of the body is defined by a gentle glow. If the temperature is cool, it is revealed by the rough and comfortless chill, that creeps over and pinches the sensitive wrapping.

The *second* step is to distinguish the two descriptions of tactual sense-perceptions which are experienced as the hand is applied to any part of the body as the arm, or to the non-sentient table. In the one case the surface that is touched, also gives the sense-perceptions of being touched; in the other it gives or so to speak *experiences* none. The absence of capacities for sensation distinguishes a certain class of objects as unlike all those which have them. This is the distinguishing mark of extra-corporeal objects. It is not, however, enough that objects are distinguished as extra-corporeal. They must be also known as separated in space—*i. e.*, they must be known as extended, and thereby involving a space which is beyond or without the body. This suggests the next acquisition.

Third, Objects corporeal and extra-corporeal can be grasped by the hand, and in this way can be known as *occupying space.* When a blind man grasps his own arm or wrist, he knows certain muscular sensations as extended through and posited in the space that lies within the surfaces that he touches. If his wrist is withdrawn from the enclosing grasp, and an extra-corporeal object is inserted in its place, the adjustments of the grasping hand are the same as before, and the dim knowledge of the space which these adjustments involve is also the same. All is the same, except the sensations located within the wrist. The wrist is known by direct perception as space-filling. The enclosing hand is a measure of the space enclosed. The same enclosing or grasping hand measures the surface of another body, whether it is applied to a sentient or a non-sentient object. The last is measured by the first, by means of the extension of the enclosing hand. It occupies, however, precisely the space which the other filled. It is known, therefore, as space-filling, and as filling other space than that occupied by any part of the body.

In this way it is possible for the mind, by touch alone to reach the extra-corporeal world, and to know that all its objects, like the body with which it is directly connected, occupy space.

These processes are all acquired, and that which is acquired in them all is the facility of using one *percept* as the sign of another, or of some relation which is indicated by the percept as its invariable attendant—*e. g.*, outness, extension, direction, distance, size, and the like.

Hamilton's theory of the perception of the extra organic.

The theory of sense-perception, taught in this volume, coincides with the theories of John Müller and Sir William Hamilton, so far as they agree, viz., that we have a direct or intuitive perception of the extended organism, and an indirect or acquired perception of extra-organic matter. Müller explains the last process, substantially as we have done, though with less detail. Hamilton explains it thus: "The existence of an extra-organic world is apprehended * * * in the consciousness that our locomotive energy is resisted, and not resisted by aught in our organism itself. For in the consciousness of being thus resisted is involved as a correlative, the consciousness of a resisting something." (*Appendix to Works of Reid,* Note D*, 28; cf. 20, 23, 24, 25, 26; cf. 864, Note D.)

This explanation of the process supposes the application of the relation of causation. For it represents the locomotive energy as a causative energy which, unresisted, would produce certain effects, which effects are overborne or set aside by an agent which is known to be neither the *ego* nor the organism with which the *ego* is connected. From the presence of this new and strange effect, the existence of an extra-organic agent is inferred. The theory is in *principle* the same with that of Dr. Thomas Brown which we have already noticed, with this difference, that Brown supposes the cause and its activities to be both spiritual and non-extended, while Hamilton supposes the locomotive energy to be known directly as extended. The distinction of body and not-body is better explained by the presence and absence of certain tactual and muscular sense-perceptions. When the reflective consciousness has been developed and the relation of causation is familiarly handled by the mind, this relation would confirm and make definite the belief in extra-organic beings and agents.

A more serious difficulty is involved in Hamilton's theory—the same, indeed, which in another way is fatal to that of Brown's, viz., it seems not to explain how with the necessity of finding for this effect an extra-organic cause, this "correlative" "resisting something" must also be proved to be *extended.* The agent, the *ego,* as a percipient and actor is not extended; then why may not the extra-organic agent and *non-ego* be non-extended, or why must it be extended? How is it shown to be correlative so far as to be extended, except it is taken to be the analogon of the extended organism, *i. e.,* like it in being spatial in many percepts, etc., etc., but unlike it in respect to other sense-percepts, as we have explained.

Development of vision.

§ 114. We consider next the development of the eye. Vision seems to begin at that early period when the bright and steady light attracts and holds the infant's eye, or when it carries the eye with itself wherever it leads. Certain objects that glisten with reflected rays, or that are brilliant with intense color, are soon separated from the background of undistinguished things against which they are projected, or athwart which they are moved. It is not easy to decide how much of intellectual perception attends this early moving and fixing of the eyes, and how much is an unconscious and reflex response of the nervous organism to the stimulating light. The eye is so constructed that only a single portion of the retina can give a perfect image of an object that comes within the field of view; so that when a bright object comes before the eye at all, it will hold or draw the eye to or after it, by the reflex action of the nerves which its brightness excites. Whenever the mind perceives such an object as a distinct and definite percept, then vision begins. Such a percept, as has already been explained, is known as a *non-ego*, and is known to be extended in two dimensions.

We have already given the reasons why, in the beginnings of vision, the percept should not be located in the eye (§ 101). It remains for us to show why it should be projected in space. With this projection of visible objects afront of the eye, begins the development, or education of the sense of vision, if the act of location is acquired, and not intuitive. It is not easy to explain the steps of the process, or the grounds why the percepts are carried forward into space, even if they are not located in the eye. Some contend that no explanation can be given, because none is required; that there is no problem, because there is no process, it being, in their view, by an ordinance of nature that the object seen should first be seen at the eye's focal distance forward, and thus here is fixed the original starting-point from which all the acquired judgments of distance proceed. They insist that all objects, as viewed by the act of original vision, are seen in a hollow sphere—forward, above, below, on this side and that—whose radius is this focal distance. Such must of necessity hold that the act of projection is original, and not in any sense acquired.

Those who hold that it is acquired, give various explanations of the process; in all of which they must call in the aid of the hand. The most plausible is the following: The eye, though, like the hand, it is moved by muscles which are directed by the aid of the appropriate sensations, does not, when in its normal or healthy state, give any tactual sensations by the felt contact of its surface with the objects which affect it, nor do the muscular sensations themselves attract the attention. We may assume that, in the way explained, space and spatial objects external to the body have become familiar through the sense of touch and the use of the hand. At the surface of the eye such tactual experiences are wanting, and of course no outer limits can be defined. So soon as the lids are raised and the experiences of color are made, the eye gropes after these strange objects, but cannot touch them. It reaches after them, as it were, but they are beyond its reach. But still they exist. If they draw near, while the eye regards them, they fill more of its field of view; if they withdraw, they occupy a less extensive plane. Meanwhile, as they draw near or remove, the eye is adjusted to perfect vision, and its adjustments and motions are known by changing sensations; but still the objects cannot be touched, nor can they be reached. By all these criteria, visible percepts are strikingly contrasted with those which are tangible—they exist, but they cannot be touched by the eye, nor can the eye reach them. They are in space somewhere without the body. This *somewhere* is definitely fixed as soon as the seen object is also touched. The *where* of the percept after which the eye inquires, is answered as soon as the hand touches the object seen. The limited distance which is measured by the sensations proper to the extended hand, becomes fixed and clear, and the object held by the hand and gazed at by the eye is distinctly projected in space. Henceforward the eye and the hand go together beyond the limited range which is at first allotted to them, into the unexplored infinitude that awaits their labors.

Then comes the power to set up a field of vision. This supposes some knowledge of place, of relative distance and size, in gaining which the eye is aided greatly by the hand. First, the mind must construct certain definite objects of vision out of the bewildering multitude of colors and outlines which present themselves to the unpracticed eye. Next, it must select a few of these objects for its observation at a single look. These it must place in a plane more or less distant, leaving out of distinct vision objects near and remote, estimating distance and judging size in the ways already explained. These acts and judgments of the quick and sensitive eye, aided by the slower and cooler hand, must be repeated again and again, till any required field of vision can be selected and constructed with ease and precision, so that we seem to see space, distance, and dimensions by the simple glance of the eye. These space relations, when once learned, are so few, so simple, so easily indicated, and so permanently established, that they seem never to have been learned at all. They become entwined in all our associations; they leap at once to the imagination; they preoccupy it so completely as to shut out the possibility of the opposite; their suggestions are accepted by the intellect with a rapidity that often leads to illusion and error.

Hence is it that all the so-called subjective sensations are at once projected into space. Hence, when the veins of the retina themselves become the objects of vision, they are seen afront of the eye, a dark arborescence projected on an illuminated background. Hence, when we look into a mirror, either natural or artificial, we see all its reflected objects in the depths of space. Hence the spectra of the imagination, the visions which haunt the phantasy of the diseased and insane, are all distributed in space.

Combination of touch and vision.

§ 115. We are next to show how the infant learns to combine the perceptions of touch with those of vision. We may do this by considering how the infant learns to connect the hands as seen with the hands as directly felt. Before this is possible, the hands as seen must become familiar as definite and separated objects, with forms that are easily recognized. The muscular sensations must also have become definite and distinct to the attentive intellect.

This knowledge being given, the mind must learn to connect the hands as seen, with the hands as moved and touched. To unite these two percepts is one of the first and most important of the acquired perceptions which the infant masters. How this can be effected, seems not difficult to explain. It should be considered, for the reason already given, that these two classes of objects are the only objects with which the infant is conversant. These occupy its chief attention. They constitute and complete its universe.

Let one hand lie upon another, or let the hand rest upon a material object that does not belong to its body. The eye watches the process, and as the hand holds the surface with its sentient touch, so the eye holds it with its gaze; it observes that what was still is now in motion; that what was seen is now covered, and by the interposing hand. Or, if the process be described in terms taken from the language of vision only, one patch of color or shade or light is obscured by another which moves before it and hides it from the view. Or, one is moved behind another, and is hidden from sight. In this way the two percepts coincide in place, and one is made the sign of the other; when one is seen, it is expected that the other will be felt; when one is felt, the mind expects that the other will be seen. As the mind proceeds and masters the other relations of form, place, size, and distance, etc., the import of either percept as a sign of the other becomes to the same extent enlarged. It is a sign not only of the other as a percept simply, but of all the relations which it signifies.

These acquisitions are in fact achieved by every person born blind, to whom sight is given in later years. In infancy, the eye performs a service similar to that which it renders to the blind who learn to see in mature life; with this difference, that the eye does not wait to furnish its aid till the hand has done all that can possibly be accomplished without it. When the eye and the hand are developed together, by their mutual aid they greatly shorten the processes of acquisition, and of making the results more sure. What each can do apart, we have already considered. It is fair to infer that in the processes by which infancy makes its acquisitions, whatever each can do best it will perform for the other. If the touch gives the first distinct knowledge of the third dimension of space, it places this knowledge at the service of the eye. The eye, if it can not directly discern distance, can yet observe and interpret the signs of distance. The hand can determine the relative distances of objects only within its reach; or the foot must measure off distance by counting the steps, carrying the body as it goes. But the eye can, by a glance, reach for rods and furlongs and miles, and measure with sufficient accuracy for the common occasions of life.

Observations upon infants.

That the eye and the hand must conspire in infancy, is not only fairly to be inferred, but it is evident from observation of the experiments which the infant is continually making with both. The infant *learns* to touch; by which we mean not merely that it learns to use its hands, but that it learns to use them with intelligence, and to interpret its touch-perceptions. It is equally evident that it learns by practice not only to use its eyes in seeing, and to judge what its sight-perceptions signify, but also to combine its sight and touch-perceptions together, and thus makes the one serve as the signs of the other.

As the eye of the infant rolls or rests in the socket, or is caught for an instant by the excitement of the stimulating light, so the hands and arms, at first, hang uselessly from the shoulders, or dangle hither and thither, resting on whatever may sustain them. They can neither grasp nor hold, much less can they be carried to a point on which desire fixes the eye; nor can they, in obedience to desire, hold and carry an object,—as food to the mouth,—or release it when brought to its destined place. All these uses of the hand must be learned by attention. That they are learned, is evident from the aimless use of the hands at first, from the many experiments and failures and final successes which follow, and from the gratification that is manifested at success.

The earliest objects which attract the persistent attention of the infant's eye are the hands. As these are to be the instruments of its activity and the arbiters of its earthly destiny, it is natural and appropriate that they should occupy the largest share of its earliest notice. It is impossible that it should be otherwise for two or three reasons. They are always before its eyes, ever flitting to and fro in aimless and convulsive movements, and challenging its notice as they are passing across its limited field of vision. As if to concentrate the whole energy of the attention upon the action of the hands, the infant is short-sighted, and, till it is four months old, observes only the nearest objects, and then objects somewhat more remote, till, by gradual advances, the whole spectacle of the universe is unveiled and opened to its view.

It is manifest that the explanation of the process by which the infant learns to connect and unite the visual and tactual percepts of its hands, applies equally well to those acts by which it learns to connect the percepts of all material objects, so as to view them as single things. That this power is acquired, and neither innate nor connate, is obvious. That it is acquired by observation and experiment, is equally clear. The world of the eye and the world of the hand are at first diverse and apart. How to bring them together, is the first problem of infancy. Upon this problem it tasks its earliest powers. When it is achieved these two worlds rush together, coinciding so completely that it seems inconceivable that they should ever have been perceived apart.

We need not pursue our synthesis further. We need not ask further how the infant builds up the rest of its knowledge, or acquires its infant skill. We need not ask how the infant learns to use its hands, to grasp, to hold, and to handle a spoon, a fork, or a knife, or how it learns to walk or talk ; for all these processes can be explained by analogous activities which occur within our recollection. Still less need we ask how it learns to connect the percepts of smell, of taste, and of sound, with their appropriate sight or touch objects. These problems present no difficulty and require no solution.

It is instructive to watch the timid yet adventurous experiments which an infant makes, especially with its hands. First, it strikes about in aimless efforts, or makes a play for its eyes with the half convulsive motions of its little fists. By a gradual progress it learns to reach after the few objects which the eye has separated from the background—the infinite unknown which lies beyond its reach and beyond its aims. Soon it endeavors to lay hold of objects which the eye rests upon, though quite beyond its reach. It clutches after the distant lamp, the fire-blaze, or the polished fire-iron. By slow but sure progress it masters the objects within its own apartment, and learns to apply its rude standards of size and distance to the world within its vision, the finite universe which its four walls enclose. All beyond is infinitude. During this time, as has been said, the infant is short-sighted, till many months of its life have elapsed, with the manifest design that it should be forced to master all near objects before it is tempted beyond.

If we would conceive how the world out of doors may appear to an infant brought to the window, after it is somewhat familiar with the form, size, and relative positions of the objects within, we may read what is told of Caspar Hauser, who is said to have been confined, till the age of seventeen, in a darkened apartment, without communication with nature by the senses, or with man by language. The story, whether true or false, meets the case. "I directed him," says his teacher, "to look out of the window, pointing to the wide and extensive prospect of a beautiful landscape that presented itself in all the glory of summer, and asked him whether what he saw was not very beautiful. He obeyed, but instantly drew back with visible horror, exclaiming, 'ugly, ugly!' and then pointing to the white wall of his chamber, he said, 'there not ugly.' Several years after, his friend asked him if he recalled the remembrance of the scene, and of his own feelings, and he said: 'What I then saw was very ugly; for when I looked at the window, it always appeared to me as if a window-shutter had been placed before my eyes, upon which a wall-painter had spattered the contents of his different brushes, filled with white, blue, green, yellow, and red paint, all mingled together. Single things, as I now see things, I could not at that time recognize and distinguish from each other. That what I then saw were fields, hills, and houses; that many things which at that time appeared much larger were in reality much smaller, while many other things which appeared smaller were in reality larger than other things, is a fact of which I was afterward convinced in the experience gained in my walks.' He also said, 'that in the beginning, he could not distinguish between what was really round and what was only painted as round or triangular. The men and horses represented on sheets of pictures appeared to be precisely as men and horses carved on wood.'" —*Caspar Hauser: An Account*, etc. (translated from the German), pp. 88, 89. 2d edition. Boston, 1833.

§ 116. The phenomena attendant upon the acquisition of sight by persons who had been blind from birth have already been referred to as illustrating and establishing some of these positions. They deserve a separate and more particular notice.

The blind from birth, upon the recovery of sight.

The cases which are most easily accessible to the English reader—which are, indeed, the most satisfactory and decisive of any on record—are those reported in the *Philosophical Transactions of the Royal Society of London* for the years respectively, 1728, 1801, 1807, 1826, and 1841. The persons operated upon differed greatly in respect to age, mental capacity, and the degree of their previous blindness. The observations and experiments with all of them may be accepted as having established the following facts and truths:—

The patients, as soon as they began to see, saw objects not only as colored, but as extended. Their experiences gave no countenance whatever to the views of Stuart and Brown, that color can be perceived without extension, and that the two are united by inseparable association. It is true that in almost every case the patients, previously to their recovery of sight, had some experience of light, and of course of light superficially extended or diffused. But this experience of light was so obviously dependent upon the affection of the retina, as to indicate, if not to prove, that any experience of light whatever involves the perception of extension.

The extension which they perceived by sight was in two dimensions only. This was made evident from a few experiments instituted with express reference to this point in the case of one of the most intelligent. A solid cube and a solid sphere were both taken by him to be simply discs or planes. A solid cube and a flat projection of the same were both taken to be flat and in every respect alike. A pyramid, when turned toward him so as to present one of its sides only, was called a triangle. When the pyramid was turned so as to expose a part of another side, he could not make out what it was.

As to distance from the eye, or the place where objects are located in original perception, the testimony is unanimous and decisive that objects at first seem very near—how near, could not be exactly known—and that the relative distance of each object beyond this indeterminate limit is learned by experience. Most of the patients were afraid to move, lest they should hit against objects that were comparatively remote. Two or three of the patients, in attempting to reach objects extended to them, clutched behind the objects when held near before them, and when more remote, only succeeded in grasping them after repeated efforts. Cheselden's boy said, at first, that all objects touched his eye. The boy reported by Sir Edward Home (1807) said the sun and the candle touched his eye, even before the cataracts were removed; and, just after the first operation, said the head of the surgeon did the same. But after a second operation, he said the sun and the candle did not touch his eye. It is probable that the objects which were said to touch the eyes, in these two cases, stimulated them so actively as to present some analogy to the muscular sensations accompanying the touch, with which, in every possible form, the patient was so familiar. Hence they interpreted and called these experiences perceptions of touch.

All these persons were forced to learn by experience to combine the percepts of sight with the familiar impressions of touch, so as to translate the one into the other. All experienced a difficulty similar to that of Cheselden's boy with the dog and cat. When they saw objects a second time, and were not certain that they could recall them, they reached for them with the hand, and could not be content till they handled them a second time. Their judgments of size and form all needed to be acquired. Visible mathematical figures, as a square, a circle, and rectangle, could not be recognized till the fingers were resorted to. One patient did make out one or two of these figures, by drawing the outline with her finger in the air, and, as it were, constructing the figure with the finger after the lines presented to the eye. Another could not understand how drawings of objects could represent the objects, till he revived the percepts of the objects by his fingers. Most of them were embarrassed by drawings and pictures, not being able to see likenesses or to understand perspective, or to perceive that light and shade represented form and distance. Their judgments of the comparative size of objects were embarrassing to them. Cheselden's boy knew that his own room was a part of the house, but could not easily believe the house was so much larger than the apartment.

The testimony is uniform, also, that, in learning to see objects as separate things, the constructive power is brought into play, requiring intelligent attention and constant memory on the part of the percipient, and that it is only slowly, at best, that the mind learns to separate material objects, to construct its field of vision, and to locate objects as near and remote by the various signs which it learns to interpret. In short, these observations and experiments confirm and illustrate all that has been said in this chapter in respect to the early development and growth of sense-perception.

CHAPTER VII.

THE PRODUCTS OF SENSE-PERCEPTION; OR, THE PERCEPTION OF MATERIAL THINGS.

Material things and sense-percepts.

§ 117. Thus far we have considered sense-perception as a process, and in its growth. We proceed next to discuss its products as the permanent possessions of the mind. We have already explained of knowledge in general, that, as an activity of the intellect, it is brought to its appropriate termination when its objects can, so to speak, be detached from the process by which they were so matured as afterward to be retained, recalled, and recognized. This is eminently true of sense-perception, which is only complete when it results in the knowledge of material things. *A material thing or object, as known by sense-perception, is a completed whole made up of separate percepts.* We distinguish the knowledge of things from the knowledge of percepts. A *percept*, as has been explained, is the appropriate object of the mind's knowledge through a single organ of sense. A *thing* is the product of the mind's knowledge in apprehending several percepts as united into a finished whole, with the relations which such a combination involves.

As an example of the difference, take an apple. The apple seen, touched, smelled, tasted, and heard, are separate percepts. The object perceived by the combination of all these percepts is the apple, as a material thing. The separate original perceptions give as many percepts. The original and acquired perceptions, when united, give material objects or things.

Two questions now present themselves for consideration: By what means, and under what relations, does the mind unite separate percepts into things or objects? Under what conditions does the mind so complete its knowledge of percepts and of things, as to be able to retain and recall them as permanent objects of knowledge?

We begin with the first of these questions: By what steps, and under what relations, does the mind unite percepts into things or material objects? We answer:—

Percepts are united into things by two successive steps or stages, to each of which there is an appropriate product. By the first the mind unites these percepts into a material thing or whole, under the *relations of space and time*. By the second, it connects the parts of the whole under the relation of *substance and attributed quality*. The several percepts united in both these relations constitute what is commonly known as *a material thing*.

It has already been shown how the percepts of sight and the percepts of touch are referred by the mind to the same portion of space. The seen hand and the touched hand are found to lie in the same direction, and to be at the same distance from any and every part of the body. In like manner the apple or the egg, the chair or the table, which is seen and that which is touched are found to coincide in the same portion of space. They are in the same place. By a similar process the sentient body itself must have been previously perceived to be one material thing.

The first stage of perception; limited to coincidence in space and time.

§ 117 *a*. This coincidence in place is the product of the *first constructive* or *synthetic act* by which the mind, in sense-perception, unites percepts into a thing. Such an act is complete when it gives a material object or whole, in this lower sense, viz., a combination of the percepts that are appropriate to different organs of sense, by means of the relations of space and time. The percepts of sight and touch are inseparably united in space, and this is the earliest combination made by the intellect which may properly be called a material thing. With these two are connected the percepts of taste, smell, and sound, at first under the relation of simultaneous occurrence in time.

It is obvious that the several percepts, when viewed as connected into a whole under these relations, have a very unequal relative importance. The percepts of sight and touch, to those who can see and feel, as they are defined in place and eminently objective, constitute the material object as it is usually conceived and named. The percepts of smell, sound and taste, are its invariable attendants in time, until they are connected with it by another relation. To those who see, even though they can also feel, the leading percepts are those of sight. The name of an object suggests its visible form and color, etc., rather than the object as touched; a certain and decisive evidence that the object as seen is that which is most prominent and attractive to the mind, and therefore is most readily recalled to the imagination. To the blind, on the other hand, it is the object as touched, or the tangible percept, which is suggested by the name, and is represented to his imagination as the thing perceived. The other percepts, of taste, smell, and sound, are connected with the combined percepts of touch and sight less readily, and by a looser bond. As at first experienced, they are referred to the sentient organism, and are less readily separated from it. They are more sensational and subjective, less perceptional and objective. As to the manner and the relations by which they are first connected with the percepts of sight and touch, philosophers are not agreed. It must at least be true, that whatever other relations unite them to material things, they must at the very earliest period be their constant attendants in place and time.

The conception of a material thing or whole, made up of extended parts or single percepts, is, however, very equivocal in its import and varied in its application. To an infant with limited experience, the greater part of an apartment may be perceived as a single object or thing; the only separable objects in it being the chair, table, and a few utensils, the position of which is often changed. To a child, a horse and vehicle, seen together for the first time, may be a whole, or a single object. The savage perceives a ship or steamer as one huge animal. Many observations and experiments, much information from others, repeated lessons inferred from words and names properly applied, are required to enable the child to distinguish things as wholes and parts; to hold apart objects that should not be united; and to unite objects that should not be divided. The point of view from which objects are observed, and the purpose or use to which they are to be applied, direct in the formation and application of names, and determine whether this or that object shall be regarded as a whole or part of a thing. A house with its grounds, the house alone, an apartment, a door, a window, the smallest perceived portion of either, each and all, are things or parts of things, according to the principle or use which regulates the application of the respective terms. But whether a perceived whole is greater or smaller in its spatial dimensions, it must have defined spatial dimensions and be capable of being perceived by one of the leading senses. Whatever the thing may be, the percepts of which it consists must at least be capable of being perceived as occupying the same space, and of occurring together in time.

The second stage: The relation of substance and attribute.

§ 118. By the *second* stage or step of the perceptive process, the several percepts or parts are connected with one another, or with the whole which they constitute, as *substance and attribute.* Thus the objects of the sense of touch are known as hard or soft, rough or smooth, elastic or non-elastic, etc., etc. Those of sight are red, yellow, orange, violet, and green; those of hearing are sharp, smooth, harsh, and sweet; those of smell are pungent, exhilarant, fetid; and all these qualities are ascribed to an object to which they belong, and of which they are affirmed to be attributes. Certain relations of time and extension, as long and short, square and round, are in like manner treated as properties or attributes. They are more than parts of the wholes which they help to constitute; they are connected with a being or agent, the nature of which they define, the presence of which they signify, and the powers of which they manifest.

It is not here in place to discuss the nature of this special relation which has occasioned so much speculation and dispute among metaphysicians (*P. iv. c. vii*). It is sufficient to say here, that as we have already shown that knowledge of every kind necessarily gives beings and relations, or beings as related, we are prepared to understand the definition of a *substance as a being that is capable of being distinguished by relations;* and of *attributes, qualities,* and *properties,* as relations used to distinguish and describe or define beings. That the objects of perception, both wholes and parts—*i. e.*, combined and single percepts—are in fact connected in this way, is too obvious to require illustration and proof.

The relations most frequently employed as attributes are the relations of *time, space, and causality.* As soon as beings are known as enduring for a longer or shorter period, or having this or that size or form, and these relations are used to designate or distinguish them from other beings, these relations are *attributed* to them as distinguishing characteristics. As soon as the sense-object is known—*i. e.*, thought of as the producer of sensations, as of smell, taste, or sound, it would be known as endowed with distinguishable capacities to produce these effects. The sensations would, in their turn, be referred to these beings as their causes or originators. No illustration is needed to prove that the sense-element in these three percepts is very early regarded as an effect. So far as the mind is passive in sensation, it must always be so regarded. The sensation is experienced when the object or being is near; it is felt less intensely when the object is remote; its quality or intensity vary with the varying conditions of the object. An object with a certain form, feel, or color, when brought into contact with the tongue or palate, causes a certain taste. Touched by the hand, no special sensation follows; but touched by the tongue and palate, there ensues the specific sensation of taste. The object touched might have been regarded simply as a being or thing; but the object tasted is known as also occasioning a sensation.

It is conceivable, as has been already suggested, that before these coexistent and successive percepts and sensations are known as substance and attribute, they should be known as constant attendants, and that, simply as conjoined, the presence or the thought of the one should, under the laws of association, suggest the thought of the other. Under this relation sense-objects are known to animals, which can not and do not distinguish the relation of conjunction from that of causation. If one sensation has been experienced in connection with another, the repetition of the one brings up the image of the other, and the pain and pleasure, the hope and fear which are appropriate to it. The dog connects with the whip in the hand of his master the thought of chastisement and pain; with the sight of his gun or his walking stick, the excitement of sport or of a ramble. It is not easy to assert when and how the two relations are distinguished by man; that they are distinguished, is obvious, for reasons which this is not the place to give.

This relation supposes reflex and indirect knowledge.

That it is not till the second or advanced stage of the perceptive process that percepts are connected under the relation of substance and attribute, is still further evident from the fact that the knowledge involved is indirect and reflex, as distinguished from that which is direct and objective. It supposes the objects related—the subject of sensations, and the object which occasions them—to be more or less familiar, and that both subject and object are projected in the view of the mind upon the same plane, so that both become objects to its thought. A thing cannot be known as capable of producing sensations as effects, unless the body or the soul, one or both, are known as the conditions or subjects of its action; and this requires that they should be placed afront the reflecting mind by a special effort, which involves a maturity of discipline which time alone can develop. Moreover, it supposes some progress in generalization, and some sort of induction. Many objects must have been touched and seen, before they are so far recognized as *similar*, as to be taken for *the same* in their causal efficiency. Many experiences must be had with the sensations of smell, taste, and sound, before these could be invariably referred to the same substances, because dependent on their properties or attributes.

In one sense it is true, that an act of sense-perception is not complete, and its product is not perfected, until the soul's higher energies are awakened, and the object of them has been viewed in the higher relations. The human being can scarcely be said truly to have perceived even a pebble as a man, till he has brought into action all the powers with which he is endowed as a man. The infant's eye may not glisten with the penetrating sharpness of the eye of the young eagle, and yet may wear the softer lustre which betokens the dawning intelligence. The soul leaps into no single form of activity, least of all into the full development of its higher powers.

Thus far we have conceived the substance as an object seen and touched, and its attributes as capacities to occasion the sensations of smell, taste, and sound.

We have connected a *percept* with a *percept* as substance and attribute—a leading percept, as of sight, with a sensational percept as of smell—and called the one a thing, and the other its quality. If we push our inquiries a step backward, and inquire, Which is the substance and which the attribute when the object consists solely of a percept of touch and a percept of sight conjoined? we answer, That sense-percept is made the substance, which is regarded in the relation of cause to the sense-element involved in the other. The object as touched and the object as seen, may respectively be substances, in their respective relations to the sensations of sight and of touch. We say, *it* is white—*i. e.,* the object which I touch; and again, *it* is hard—*i. e.,* the object I see—the touch-percept and sight-percept being each in their turn taken as beings.

We may narrow our view still more, and inquire which is the being or substance, and which the attribute or quality, when we have a single percept only, and view it in relation to the sentient mind? We reply, The object, perceived by sense *to be,* is known as a substance when considered as the producer of the sensation which is the condition of the perception. The tangible or visible object, as a being, is distinguishable as a space-occupying or extended something. As causing or producing the sensations of sight or touch, it is known as possessing the attributes of color or touch. The elements involved in every act of sense-perception provide for the possibility of this relation. But the relation is not, in fact, discerned until the mind projects and brings up the perceived *non-ego* and the sentient *ego* into the same field of vision, by a reflex and comparing act.

The *sensation*—*i. e.,* the effect—is not the *property* or quality which produces it, though the two are called by the same name. Sweetness means one thing when it is said to be in the sugar, and another when it is experienced by the sentient soul. The heat, in one sense, *is,* and in another *is not,* in the fire.

The conditions of complete perception.

§ 119. Our second question is, *Under what conditions* does the mind attain a definite, permanent knowledge of the objects of sense-perception, whether *percepts* or *things,* so that they can be readily recalled and recognized? It is only when they are placed so completely in the possession of the mind as to be at its disposal, that the process of perception can be said to be complete. When this is done, the object of perception is converted into an idea or image. The real object apprehended by the mind becomes an intellectual object, having a purely ideal or psychical existence. By some writers the special term *ideation* is appropriated to this process. Sense-perception is said to be complete *in the highest sense* when its object is *ideated,* or becomes *an idea.*

But as every perceived object is composed of parts, it follows that the perception of a thing can only be complete when the mind separates by distinct *analysis* the parts or percepts of which the thing is composed, and unites them by perfected *synthesis.* In other words, the mind must distinguish the constituent *per-*

cepts by completed acts of *original perception*, and combine these percepts into *things*, by finished acts of *acquired perception*. We are naturally led to consider the conditions of complete perception of the parts and relations of material things.

(1.) Objects are most easily distinguished which are apprehended with *special energy*—which are very strikingly contrasted with, or which are similar to other objects. A lively color, a loud sound, a positive taste, etc., are more readily apprehended than a color which is faint, a sound which is feeble, or a taste which is not positive. Things are more or less readily perceived with effect and permanence according as the percepts of which they are constituted are more or less readily known.

The definiteness with which objects are perceived depends in part also on their likeness or unlikeness to other objects in connection with which they are presented to the mind. Of two percepts and two things that are very similar, and of two that are very unlike, those are more likely to be perceived which are in striking contrast to each other, than those which closely resemble one another. Two colors, two sounds, etc., as well as two apples or two paintings, are each more readily perceived and retained if they are strikingly contrasted, than if they are very similar. The ground of the likeness or unlikeness, the resemblance or contrast, is in part objective,—pertaining solely to the object perceived. In part it is subjective, and arises from the natural or acquired capability of the individual to feel and know. Thus, one class of persons are physically incapable of distinguishing different colors, *i. e.*, those who are color-blind. Others, who can discern the colors which are commonly named, can with difficulty distinguish shades of color that are nearly allied. Some persons are very insensible to differences and similarities of sounds to which others are keenly alive. Even when the original sensibilities of the senses and aptitudes of the intellect present no diversity, there are the greatest possible differences of susceptibility, arising from differences of habit and attention.

(2.) *Motion* heightens the contrasts of perceived objects, and gives definiteness to the outline and limits, especially of visible percepts. To the infant's eye, moving objects are the first which, so to speak, are separated from the undistinguished mass of blended color, in which the world of matter is at first arrayed. From this extended surface of color certain objects are detached, as the moving lamp, the walking person, the portable furniture and utensils. They pass to and fro athwart the background upon which they are projected, and are brought into contrast with its unbroken surface, till they take their place in the memory, as the first distinct objects with which it is provided. By degrees this undistinguished mass of blended light and shade, of form and color, is broken up, as one and another separate percept and distinguished thing is detached by the mind's observation and is set apart in the mind's storehouse as a distinct idea. The influence of motion is not limited to visible objects. It is most important in giving distinct percepts to the sense of touch. The hand must move over the surface felt, or the surface must move over the hand, to leave distinct percepts of its limits and qualities.

(3.) *Repetition* is an efficient and often an indispensable condition to the completion of an act of perception. Even the simple percept, as a sound, a color, a taste, is more perfectly mastered by being apprehended in successive acts of attention. If several percepts are to be united as a single and separate thing, it is still more requisite that they be often apprehended by the same or continuously connected acts, in order that the object may be brought completely into possession and placed entirely at command. This is especially necessary if the percept, or object, by reason of its spatial extent or the complexity of its elements, is beyond the power of the mind to master in a single act. In some cases, repetition serves to make the impression more vivid and definite. In others, it is required in order that there be any impression at all.

(*a.*) Repetition often excites and gratifies *the interest* of the soul in the objects perceived, and thus arouses greater energy of attention.

This is illustrated by the example of many single percepts. A color or sound gives pleasure when once perceived. Let it solicit the mind's notice a second time, and the remembrance of the gratification which it gave, will arouse the mind to attend with increased energy to the object which had previously imparted so pleasant an experience. In the recollection of that experience, and with the hope of its renewal, it summons again all its energy of perception. The result is a definite remembrance of every thing which the man is competent or prepared to know in respect to it. When the attention is solicited again, the mind at once responds to the call, withdraws its divided or distracted activity, and, according to its sense of the value of the good to be enjoyed, responds with an energetic and attentive gaze.

(*b.*) Repetition is still more essential to enable the mind to *unite into a whole* the separate parts of objects which cannot be grasped by a single act of perception. The examples already cited, belong to those objects which require but a single act of attention in order to be completely possessed by the mind. There is a very large class of objects, however, which consist of too many parts to be known by a single effort of perception. These must be combined together into one, *by successive acts.* For example, if we perceive a mathematical figure with a very irregular and complicated outline, it is necessary that we view it in separate portions, in order to master the whole. Not only is this true, but we often need to review each portion which we have already perceived, in order to connect it with the part which was perceived previously. After we have followed the outline by repeated acts of observation, we need often to review the whole as a whole by a rapid succession of acts, or by a single glance of the eye to unite the several parts. If we look at a painting, we study its several parts, perhaps for hours together, in order to gain and carry away a distinct and satisfactory impression of the whole. If we look at the front of an edifice that is elaborately adorned, we follow the several features one by one in their order, often returning upon our course, that we may retain the perceptions which we have gained.

The first efforts of the eye upon such an object are like voyages of discovery or movements of military reconnoissance. They serve the same purpose as the use of the finding-glass of a telescope. The eye runs hither and there with a vague and quickly-shifting gaze. It finds one feature after another which excites its interest and attracts its attention, and thus learns in a general way what material is present for it to work upon. After this preliminary work, a second and still another look may be required, that the mind may determine which of these parts it is worth while to unite together into a continuous and connected whole, by successive acts of attentive perception. That this theory is correct, is manifest from the difference which we notice between observing a complex object when seen for the first time, and when it has become familiar by repeated acts of perception. If the object is new and strange, we must view it again and again in order to bring away any distinct perception. If it is familiar, or like a familiar object, a single and hasty look is often enough to secure a clear and permanent knowledge. In such a case we know beforehand what we expect to find, and to what points we need to direct the eye in order to assure ourselves.

When the object contains a greater number of parts than we can grasp at a single view, there is need of repetition for another reason. Let the outline of a mathematical figure be made up of many sides, or the face of an edifice consist of a very great number of salient features, and it is impossible—let either be ever so familiar—that they should be perceived distinctly by any single effort of perception. The eye must pass around the outline, or sweep across the face by successive acts, and master each portion in detail, in order to perceive the whole so as to recall it.

Here again we notice a striking difference between objects that are regular and uniform, and those which are irregular and multiform. Of two figures of fifty sides, let one be a regular and another an irregular polygon. Let the façade of a building be made up of similar parts combined after a uniform law of recurrence and symmetry; or let the parts have no relation of likeness, order, or correspondence. A few repetitions of attention enable us to master the one; very many are required to put us in possession of the other.

(4.) *Familiar objects* are readily and rapidly perceived. Novel or unfamiliar objects are slowly and painfully mastered. The fact is unquestioned. The explanation of it is furnished by the principles which have been already laid down.

To familiar shades of color, sounds, forms, touches, tastes, and smells, the mind is ready to attend, being guided by its remembrance of what it had perceived before, and incited to attention by remembered pleasure. If the combination is also familiar—*i. e.*, the union of the taste or smell with the color, or of the touch with the form—the same law holds good. In looking at an individual chair or table which I have often perceived, or the aspect of which is familiar, one percept prepares the way for another —the color for the form, the form for the weight; one part for another, as the leg of the chair or the table for the back of the one or the bed of the other; so that the mind is at once prepared for what it expects and readily apprehends what it is waiting for.

But let the object be unfamiliar, we are detained upon its parts in the way already explained, in order that we may discover what they are, so far as to decide which, if any, shall receive our attention. If a novel piece of furniture is seen, or a new implement, or an edifice singularly planned, or a work of art executed after peculiar principles, or if an animal or plant of an unfamiliar species or a dress of a new fashion is presented for our inspection, we find it necessary to look again and again at the object. We must feel our way step by step and part by part, to find the parts of which it consists, so that we can recall them.

The acts of repeated perception which are required in such cases, are not to be confounded with acts of *recognition*, or with acts of *comparison* for the purpose of discerning similarities or other relations.

Acts of recognition and of comparison do indeed usually accompany these efforts of perception. But though they often facilitate, they do not constitute the acts. This is manifest from the analysis of the acts. A single percept, or an object consisting of several percepts, must first be perceived in order to be recognized. It must be known the first time, or by a first act, in order to be known the second time, or by a subsequent act. So, two objects must be perceived, before they can be compared and discerned to be similar or alike.

Some psychologists distinguish perception from sensation thus: "a sensation, when recognized as similar to one previously experienced, becomes a perception." So Herbert Spencer: "As there can be no classification or recognition of objects without perception of them; so there can be no perception of them without classification or recognition." "A perception of it [an object] can arise only when the group of sensations is consciously coördinated, and their meaning understood." "The perception of any object therefore, is impossible, save under the form of recognition or classification." (*Principles of Psychology*, § 46.)

Morell says: "To *perceive* a thing, means, first of all, to *recognize* it;" and again: "When we come to perceive special objects, then it is implied that we not only *recognize*, but that we also begin to *classify* them."—(*Introduction to Mental Philosophy*, pp. 85, 86. London, 1862.) That this is really impossible and logically self-contradictory, is obvious from what has been said. Recognition and classification attend and assist perception, but they do not constitute the act. It is obvious that this definition would exclude from the act of perception-proper, all that is material to it, or by which it is distinguished from sensation-proper, viz., the apprehension of spatial relations and of externality. Neither of these are necessarily involved in the recognition or comparison of sensations. The view would shut us up to a purely idealistic theory.

(5.) To complete and successful perception, *some continuance of time* is necessary. This necessity for time is partly physical or organic, and partly mental or psychical.

The organic necessity lies in the unexplained and ultimate fact, that in order to a complete and definite physical impression upon the organ, there must be a continued action of its excitant or stimulus for a brief but appreciable period. The eye and the ear, and the other organs, with their connected nervous apparatus, must be occupied with that which excites them, in order to give a sensation of which the mind can avail itself to distinct perception. Indeed, after the stimulant has ceased to affect the organ, the impression, and with it the perception, remains; as is evident from the experiment by which we revolve a burning coal so swiftly as to perceive a circle of fire.

The psychical necessity is obvious from the fact that the mind can remit or increase the energy of the organ by its own voluntary agency, and that, to exert this energy also requires time, if for no other reason, because the mind acts through and under the laws of its physical organism. An increase of energy in a part or the whole of the organism is an affair of time, and is often a measure of its lapse.

Jugglers, prestidigitators, etc., perform many of their feats by having acquired a capacity of rapid movement which does not allow time enough for the sense-perceptions of lookers-on to respond to the objects. Often they do not furnish time enough for the requisite impressions to be made upon the sense-organs. Still more frequently they do not furnish time in which perception or intelligence may perceive the objects in their relations, so as to discriminate, construct, and interpret what the sense-organs respond to. Quickness of movement and quickness of thought are the prime requisites for a successful juggler. To this should be added the capacity to divert the attention by lively sallies, by sudden gestures, rapid speech, exciting tones, and a bold address, as well as skill in inventing the physical appliances of illusion. A man endowed by nature with aptitudes like these, who has learned to make them efficient by art, can almost cheat the eyes and ears of the soberest and most practiced observer.

Can we attend to more than one thing at a time?

§ 120. It is in place here to consider the doctrine which is insisted on so earnestly, particularly by Dugald Stewart (*Elements*, c. ii.), that the mind, in perception, can attend to but one object at a time. This position he endeavors to sustain and enforce by examples like the following: In viewing a mathematical figure, say of a thousand sides, we view each side by a separate effort of attentive regard, till we have passed around the outline by successive acts of perception. The eye and the mind do this so rapidly, that when the outline is not very complicated, they seem to grasp and master the whole by a single and instantaneous act. So, in listening to a concert of music, we think we hear—*i. e.*, attentively listen to—all the instruments and separate parts together, whereas we in fact can attend to but one. But when we seem to ourselves to listen to all, we in fact pass so rapidly from one to another as to think we attend to all together. A single object he defines as the *minimum visible* in connection with the eye—that is, the smallest extension of color or shaded light by which the eye can be affected—and would by a similar rule, assert that the *minimum audible*, or the simplest and shortest appreciable sound only, can be attended to at a single instant.

The theory of Stewart labors under the following difficulties: It excludes the possibility of comparing objects with one another. In order to compare objects so as to discern that they are alike or diverse, they must be considered *together*—that is, they must be attentively perceived in combination. In the cases supposed by Stewart of the several sides of a complicated outline, or the separate sounds of the instruments in an orchestra, the parts of the figure must be considered together, to be known to be adjoining, near, or remote: the separate notes or sounds also must be heard together, to be discerned to be alike or harmonious, to be known as higher or lower, or to be connected as before and after one another. If the mind could apprehend no more than a single object at once, it would be forever and entirely cut off from the most important part of its knowledge, viz., the knowledge *of relations;* or every description of knowledge by *synthesis.*

It might perhaps be said, that what Stewart intended to assert was this: that in *sense-perception* the mind can only attend to one object at the same indivisible instant; that in those cases in which it compares two objects, it connects an object perceived with an object represented, a percept with a representation. For example, in viewing a complex outline, or hearing the sounds of an orchestra, it sees at the present instant a single side or the smallest possible part of a side—the *minimum visible*—or hears a single sound or note, and, while seeing or hearing, compares with it the side just seen or the sound just heard before. But in order to do this, it must apprehend at the same undivided instant of time both the side which is seen and the side which is remembered, etc. The doctrine that the mind can apprehend or know but a single object at a single instant of time, must be abandoned as incompatible with all the higher functions and acquisitions of the soul, as well as with the most obvious facts within our experience.

To the knowledge of relations, the knowledge of at least *two related objects* is necessary. To successful or permanent knowledge, even of relations, attention is requisite. The mind must then be able to attend to more than a single object. Inasmuch, also, as by far the most important of our sense-perceptions are concerned with the union of percepts either of the same or different senses, it follows, that the mind can attentively perceive more than a single percept. That the mind, in any single act of perception, usually attends with unequal energy to each of the related percepts, is a point which might be urged with some show of reason. When we view two or more objects together for the purpose of comparing them, and strain the mind to its utmost energy, the excess of energy is directed now to one and now to another. Both are attended to, but not with the same intenseness. The mind regards one object with more attention than the other, in order that it may receive a vivid and distinct impression of it, and then compares or in some other way connects it with that received from the other. When this is done, the process of comparison or connection is complete. This fact has given occasion to the unwarranted inference, that the mind can attend to but a single object at the same indivisible instant.

CHAPTER VIII.

ACTIVITY OF THE SOUL IN SENSE-PERCEPTION.

Sense-perception held by many to be passive only.

§ 121. The impression is very common, that the soul, in its sense-perception, is simply receptive of material objects—that it passively receives whatever imprints are made from without, exerting no active agency of its own.

By many, this is stated as a positive doctrine, which is consistently carried out into all its logical inferences and applications. Thus Kant and his disciples, as well as many psychologists not of his school, assert that the soul, in sense-perception—as indeed in all the intuitions of consciousness—is simply receptive, while in the higher functions of thought it is self-active.

Psychologists of the materialistic school, and many who are not materialists, but are more or less influenced by forms of expression and habits of association that are borrowed from materialistic theories, not only assert that the mind is passive in its sense-perceptions, but even in the higher activities of imagination and thought. Locke often inadvertently expresses himself in language and by illustrations and analogies borrowed from the physics of his time. Condillac not only makes all sensations to be impressions imprinted upon the *tabula rasa,* but makes all ideas, or the intellectual copies of sensations, to be simply "transformed sensations." With him agree in principle the ideologists of the French school. The schools of Benecke and Herbart in Germany, as also of Herbert Spencer and his disciples in England and America, all formally accept and positively teach the same doctrine, or unconsciously assume it to be true in their theories and discussions.

Grounds on which the theory rests.

The grounds on which these theories and assumptions rest are the following: 1. The general misconception of the nature of the soul, and the powers and laws of its working, by which it is invested with material properties, and interpreted by material analogies. 2. The unquestioned fact, that the soul, in sense-perception, apprehends and acts by means of a material organism, and has to do solely with material objects. 3. The soul is known to be entirely dependent on matter for the objects which it perceives. It cannot perceive any material object when the object or stimulus does not exist. Moreover, the efficiency of the material organ or instrument which it employs, depends on the material conditions which are required for healthful and vigorous activity.

Evidence that the soul is active.

§ 122. We maintain that in sense-perception the intellect is active, and for the following reasons: The soul, in sense-perception, is known through consciousness to be active, and in a special sense to be self-active. To perceive by the senses, is only a special form of the soul's general capacity or power to know. To know, is not to receive or suffer an impression, but to be certain of a fact; whenever this function is exercised, the soul is self-active, whether the objects known are material or spiritual.

That the soul is active in sense-perception, is still further evident from the following facts, most of which have already been noticed. The power of the intellect to perceive any objects of sense is developed by degrees in the mind of the infant, and, after it is fully developed, is exercised at different times and by different persons with a greater or less degree of energy. The infant at first feels many sensations, but it can scarcely be said to know objects at all. In other words, it only perceives with the lowest activity possible of a power undeveloped by exercise. It is only when its attention is aroused and its power to know is acquired and fixed, that it is properly said to perceive. Its attention is first limited to the objects of a single sense. One after another, each of the senses is awakened to action, and, as each is aroused, the mind seems to bestow for the time the whole of its energy upon the world which a single sense unfolds before it. It studies light, it studies colors, it studies forms, it studies sounds, it studies touches. Soon, in connection with the movements of its body, it learns to apprehend the relations of space, viz., position, distance, and dimensions. It then gathers its percepts together, locates them together or apart, attaching them to their appropriate places or objects. Then it uses one class of percepts in place of another, or as signs of distance, size, etc., in all the varieties of acquired perception.

As real and as great a difference is to be observed in the perceptions of different men; also in those of the same men at different times. If we suppose the powers of perception to be developed in any number of persons, we cannot fail to notice important differences in the energy and effectiveness with which they are used. Two persons look out upon a landscape, but how much more does the one behold than the other! One sees countless objects which the other entirely overlooks—houses, trees, lawns, lines of beauty, contrasted and varying colors, artistic groupings, none of which are observed by the other. Numberless sounds await the notice of each. One hears, the other fails to hear the crowing cock, the sharp report of the rifle, the rattling and rumbling of distant vehicles, the cawing crow, the singing of birds. The same is true of the percepts of taste, smell, and touch, though in a manner and to a degree less striking.

Different modes of this activity.

§ 123. The methods in which the soul exerts its activity are various. First: The soul imparts special energy to single organs, so that they perform their functions with more than usual efficiency. It can determine an unusual flow or excitement of the nervous power to the eye, the ear, or the hand, thereby rendering each capable of more vivid sensations. The process and its effect are both called the *innervation* of the organs. This is accomplished, in all probability, by means of the *reflex* or *efferent nervous organism.* Whatever may be the physical or physiological medium by which the effect is produced, its cause is often purely psychical; the soul itself is the originating agent.

This innervation of a single organ or pair of organs is observed in cases like the following: The eye rests listlessly or wanders vaguely over a landscape or a crowd of men. In a moment it is fixed by some single object; perhaps through some physical stimulus, as a bright light or glaring color; perhaps by something attractive to the feelings only. The curiosity is aroused, and stimulates the organ to do its utmost. Under the innervation of the agent of vision, the picture which had before been painted dimly on the retina, is suddenly lighted up as though a new force of sunlight had poured upon the object a fresh illumination. In a similar way, the soul can awaken the ear to more distinct hearing, by summoning its physical capacities to do their utmost. "Did you hear that shriek?" says one man to another: The ears of both are made attent at once, and are physically excited to catch even the feeblest sound, as well as mentally to interpret its meaning.

That the soul possesses and uses this power, is evident still further from the fact, that, in order to increase the energy of single organs, the mind is often forced to suspend the action of the others. We close the eyes, that we may hear distinctly a doubtful call, or mark the faint ticking of the clock, or do full justice to the skill and power with which a superior singer manages delicately shaded sounds. We find it difficult, and sometimes impossible, to give full effect to two of the senses at the same time. We cannot at the same instant read the degrees from a measuring scale, and listen to a musical air.

Second: The mind exercises its activity in its sense-perceptions, by directing its attention to a limited number of sense-objects, and neglecting the remainder.

The mind, as we have seen, can, in a single act of apprehension, be occupied with only a few objects, whether they are objects of sense, or psychical creations. To do justice to those objects, so as to bring away distinct and vivid images of their nature and relations, requires that they be exclusively before the mind. If they are exclusively present, other objects must be withdrawn, unnoticed, or neglected. The fact is unquestioned, that the mind does both admit and shut out the objects of sense by its active efforts. If we notice and follow our own processes in sense-perception, we shall observe that we are constantly employing our energies in this twofold way. When, for example, we listen to a full orchestra, we may single out the fife, and follow its shrill piping, in spite of the crashing masses of sound that assail the ear from trumpet, trombone, and drum; or we trace the silver threading of the leading violin, or we combine into a single and almost exclusive impression the sounds which the stringed or wind instruments make together; or we give the ear to a single part as rendered by its appropriate agents, soaring with the air, or sustained by the animating tenor, or sympathizing with the bass, leaving, in each instance, all the other parts unheard. The power of the mind *not* to perceive or *not* to notice, is illustrated by examples like the following: The miller does not hear the sounds from his own mill, while the visitor can hear nothing else. The operative does not notice, and therefore is not disturbed by the whir of the spindles and the clash of the looms. He can speak and hear with entire freedom, while the bystander can do neither, from the distracting and deafening din.

Third: The activity of the mind in sense-perception is still further illustrated in the great variety of acts and processes which we are compelled to perform, in order to create percepts and images which we can carry away and retain. These acts and processes are acts of selective analysis and constructive synthesis, by which the soul chooses for itself the objects which it will separate and remember as distinct images or things.

When we are confronted with an object wholly strange and new, we often find ourselves making distinct efforts to study it part by part, adding one after another, till we have combined all its elements into a definite product. Even when the eye is introduced to a new landscape, it first runs with rapid glances along the horizon, resting here and there upon any point or feature which invites a prolonged or second look; then it sweeps hither and thither, crossing its path as often as need be, searching out whatever may attract its gaze. After having thus constructed the outline of the picture, it leisurely paints in the details one by one, till the whole is finished, and it can carry away the remembrance of it as a single object; or perhaps it divides it into separate portions, and treasures in the memory cabinet pictures of selected parts. But how much does the most careful and active observer overlook! How much is reserved for after-efforts! A recognition of the activity of the mind in perception is altogether essential to a right conception of the nature and conditions of acts of memory and imagination. The mind can recreate by the representative power only what it has first created by the power of perception. The memory and imagination can recall and reshape no more of the objects of sense than the perceptive power has shaped and fixed and carried away for the service of both. The acquisitions of the memory and the reach of the imagination do not depend so much upon the number of objects which we have perceived, as upon the manner in which we have perceived them.

Fourth: The activity of the mind in sense-perception is required in early life to separate the mass of perceived or perceivable material into the distinct objects which are apprehended and named by men of average intelligence.

We have already seen that the work of thus uniting different percepts into distinguishable wholes is performed to a great extent before the time when we can distinctly remember. To the infant's eye the whole world of perceivable matter, so far as it is perceived at all, is perceived as a single whole, or one undivided object. The apartment within which it tries its first experiments of activity is literally a universe; the walls, the ceiling, the table, the chairs, all blending together in a total impression. This whole is divided into parts by successive efforts. One mind does this with greater perfection than another. Its discriminations are more subtle, its combinations more exact, and its interpretations more sagacious, even upon such objects as apples, oranges, chairs, tables, horses, and dogs.

Fifth: The activity of the mind is conspicuous in the diversity of the sense-perceptions which are reached by different men as they advance in life, or differ in their employments and culture.

A single general example may illustrate the diversity of perception in which all these conditions exert their influence. Let two men together inspect a complicated machine or engine; let the one be a person of average knowledge and experience, and the other an accomplished engineer: how much more will the one perceive in the engine than the other! Before the practiced eye, each separate part takes its appropriate place, being sharply distinguished from every other, the dividing surfaces and connecting members being all discerned at a glance, and all these separate portions being united into a complete and symmetrical whole. To the eye of the uninstructed person, however keen may be his physical vision, there is neither whole nor parts, but a confused and bewildering impression. The difference cannot be accounted for by any physical defect or excellence in the organs of vision, but only by the previous intellectual training.

These intellectual conditions are the result of the mind's own energy, and that they are most significant is convincingly demonstrated by a multitude of similar cases. The sharp but uninstructed eye of the child or the savage looks out listlessly upon the stars; the reflecting eye of the astronomer groups them in figures, threads them upon lines, and arrays them in mystical curves. The mechanic perceives much that every other man overlooks, and the objects which each mechanic perceives, or, as we say, has an eye for, depend on the particular trade to which he has been trained. It is true that in such cases, some activity of phantasy and memory attends and often precedes the special activity of sense. But if the memory and the phantasy are first aroused, their action determines and decides what is perceived by the senses; it directs and holds the attention to their appropriate objects, and so enables the mind to master and retain them as permanent possessions.

It follows from these truths, by a necessary inference, that the mind's activity in perception, and its mastery over a greater or smaller number of objects, must depend very largely upon the *interest* which these objects excite. In other words, the feelings and the character affect the accuracy and the reach, and of course the permanence of the sense-perceptions.

The eye that is sharpened by the lust of gain, detects objects and qualities to which the less interested observer is totally blind. The ear that is quickened by expectation or terror can catch the sound of deliverance when all other ears are deaf. The hand that palpitates with hope or fear, can apprehend delicate monitions of good or evil, which the stranger would not notice. The living soul, as intellect, sensibility, and will, is present in the acts of every sense, and largely determines the report which each shall make of the material universe. What a man is, is exemplified in what he perceives,—his tastes, his desires, and even his moral habits and resolves.

Is elementary, and easily exercised.

The activity of sense-perception, though an activity of knowledge, is however the most elementary of all these activities, and the one which is most easily performed. In one aspect it is the lowest in the scale in respect to its dignity and disciplinary value. It is the least intellectual of all the intellectual acts. It is performed with great ease and with surprising perfection by the infant. All the manifold processes of combination and judgment which it involves are executed with the greatest rapidity, at the very earliest age, and by persons of the least cultivation in the higher discriminations of the intellect, and apparently of the very lowest capacity for such cultivation. The habits and aptitudes which are the results of these efforts seem to be more completely controlled by association, to displace and almost to defy reflection, more entirely than is true of the higher activities and applications of the intellect. That some activities and processes of the intellect are capable of being more readily performed than others, is an original fact of our being. It can only be accepted as a psychological fact, which, to our knowledge is ultimate and inexplicable. But though this fact cannot be resolved by any higher or more comprehensive psychical or physical law, it is readily explained by the still higher relations of adaptation and design.

SENSE-PERCEPTION: SUMMARY AND REVIEW.

§ 124. (1.) The processes involved in sense-perception, as our analysis has shown, are by no means simple. The product, when complete in a perceived material object, is in its constituent elements and relations more complex than is usually believed.

We will briefly review and recapitulate the several steps of the processes and the elements of the product.

(2.) Sense-perception is an act of knowledge by means of sensations and the sense-organs. As the term indicates, the act implies two elements, which are distinguished as sensation and perception; more exactly as sensation-proper and perception-proper. These are distinguished in thought, but not separable in fact. The act of consciousness by which we know the process, separates these elements by an analysis of thought, but connects them by a synthesis of time relations, as constituting a single and instantaneous psychical state. They are distinguished in the relation of dependence, but are united as instantaneous in time.

(3.) Sensation, or the sensation element, is known still further: First, *physiologically*, as dependent on the excitement of the sensorium, in whole or in part, by some physical excitant or object. The sensorium is a collective term for the nervous organism and the senso-organs conjoined. This organism, animated by the sentient soul, acts as the agent or instrument of the several sensations. How it is fitted thus to act, we do not know. What there is in its nature which renders it capable of responding, as it does, to the impressions or excitements which it suffers, we cannot explain. We know that each class or portion of the sentient nerves is capable of a special sensation, and so far is idiopathic. In order to produce it, the excitement or impression must usually be applied to the nerve-endings, in the sense-organs. A class of exceptions to this rule is found in the effect upon the nervous filaments, of electric and chemical action, of pressure, of certain morbid and abnormal bodily conditions, which occasion what are called the subjective sensations of light and sound, and perhaps of taste.

(4.) Second, *psychologically* considered, sensation is a more or less positively pleasant or painful experience of the soul, as *consciously animating and acting with an extended sensorium.* The sensations are in this respect sharply distinguished by the soul itself from the desires which attend them as well as from the purely spiritual emotions. When the soul is said to be conscious of its sensations, consciousness can not be used in the technical sense of a direct cognizance of purely spiritual acts or states, but as a direct or intuitive cognizance of this peculiar experience. It follows that the several sensations, inasmuch as they are experienced by the soul as connected with the extended sensorium, must be indefinitely but really separated from each other in distance and place.

(5.) Perception, as an act of the mind, is *subjective* and *objective;* as *subjective,* it is distinguished by several steps or processes. As *objective,* it apprehends some being. The result is a *product,* or *the object as known.*

Subjectively viewed, sense-perception is distinguished as *original* and *acquired,* or *simple* and *complex;* also as *direct* and *indirect.* In original or simple perception, the mind knows the single percepts which are appropriate to single organs of sense. In acquired or complex perception, it connects these with one another under a variety of relations. In direct perception, the relations used are those of extension and diversity; in indirect, those of likeness, causation, and design are also employed.

Objectively viewed, perception always knows a material *non-ego.* But the objects of simple and complex perception are unlike.

(6.) In *simple* or *original* perception, the *object* is a *simple percept—i. e.,* an extended *non-ego.* But the term *non-ego* is equivocal, being capable of three distinct meanings, corresponding to the three distinguishable *egos* with which they are contrasted. These are the following: (1.) The perceiving agent as a pure spirit; (2.) the percipient agent as a spirit animating an extended sensorium; (3.) the individual as spirit, sensorium, and body. The three *non-egos* contrasted with these are: (1.) The sensorium in excited action, distinguished by the soul from itself as a pure spirit; (2.) the body perceived as other than the sentient soul—*i. e.,* the soul as animating the sensorium; and (3.) the surrounding universe as distinguished from the soul, sensorium, and body—*i. e.,* from the man as soul and body united.

(7.) In original perception, *the object* directly apprehended is the sensorium as excited to some definite action. This is distinguished from the soul as percipient, by the soul's own act of discrimination. In other words, the *ego* and *non-ego* contrasted are the first named above. This *non-ego* is the *percept* appropriate to each of the sense organs.

Some contend that there are but two organs and two forms of direct perception —those of touch and sight; the senses of taste, smell, and hearing, giving sensations only.

(8.) *Indirect* or *acquired* perception first combines single percepts into material wholes or objects, by referring them to the same portion of space. The first experiment is made with the body itself, the perception of which the soul completes, knowing it within and without. This gives the *non-ego in the second sense.* Other percepts it proceeds to combine and construct into other bodies, by processes of comparison, measurement, and induction, after the analogon of the body which the soul inhabits. These are distinguished from the body itself, giving the *non-ego in the third sense,* the distances, forms, sizes, etc., being assigned by the various processes of judgment, which are usually called acts of acquired perception.

(9.) Later still, the intellect knows the percepts thus united as *substance* and *attribute*, when it connects the objects with the sensations which they excite under the relation of causality, or compares one object with another under the relations of form and dimension. To do the one, the material object must be contrasted with the sentient soul, by *an act of reflexive comparison*, both being projected into the mind's field of view. To do the other, motion, measurement, and analysis are required to separate length, breadth, size, and form, from the things to which they pertain. Recognition, generalization, and other acts of the higher intelligence greatly stimulate and aid this activity, but are not essential to it. Many, not to say all, of these acts of acquired or indirect perception are acts of natural and unconscious induction, which, like other such acts, must assume in the objects known adaptation to the mind that knows them; in other words must assume design and order in the universe.

When the material object is known in these elements and relations as a product familiar to the mind, the process of sense-perception is complete.

(10.) When, moreover, consciousness is so matured as to distinguish the soul's spiritual acts and emotions from its sensations and their objects, then the *non-ego* is distinguished from the *ego* in the *first sense* required, and all the relations of matter to the spirit, which are objects of common observation, are attained and made familiar to the intellect.

(11.) In the processes of sense-perception the state of the intellect is *active*, and *active* only. It is a form of that knowledge, by which beings and relations are cognized as real. This activity is intimately allied to the higher processes of which it is the essential condition, and like them is directed by the emotions and the will, which together with the intellect make up the endowments of the conscious *soul*.

CHAPTER IX.

THEORIES OF SENSE-PERCEPTION.

Interest of the theories and their history.

§ 125. All philosophers have undertaken to give some theory or explanation of the perceptions of sense. These perceptions are among the most striking and interesting of all phenomena, and would naturally attract the attention of all inquisitive minds. They vary in uniformity with the changing condition of the bodily organs, and of the objects and media with which these organs are concerned. For this reason, men of philosophic tastes would be prompted to devise some theory to explain how and why these perceptions so often change.

It is not strange that these explanations have usually been derived from the generally received opinions or philosophical theories concerning the forces and laws of nature, and the powers and laws of the human soul. As the sciences of nature and of the soul have been continually changing, one theory of sense-perception has given place to another.

On the other hand, erroneous theories of sense-perception have, by a reflex influence, affected to a very large extent the philosophy of the soul. The conditions and laws of sense-perception would readily be taken as the types of all the intellectual processes. Whatever theory was adopted in respect to the nature of sight and hearing, would be extended to memory and the imagination. It is not surprising, therefore, that these theories have exerted so powerful an influence upon psychology and speculative philosophy.

Theories of sense-perception are especially liable to be erroneous, from the circumstance that they involve so many elements. The processes are themselves most complicated, involving, as they do, corporeal and psychical agencies. In order fully to understand the processes of sense-perception, we must know their conditions or media; this involves a correct, if not a complete, knowledge of such agents as light and sound. A grossly erroneous theory of either might vitiate our theory of the psychological processes of sight and hearing. The scientific knowledge of these agents and their laws includes assumptions both mathematical and metaphysical, which may be correct and complete, or erroneous and defective.

The instruments of sense-perception are the bodily organs; and to understand these organs we must not only have a correct theory of the living organism, but also of its relations to the rational soul. The psychical element in perception is also complex. The consideration of perception as a special act or kind of knowledge, requires some just views of knowledge in general. A serious error in respect to this fundamental point would, by a logical necessity, involve mistake or defect in respect to knowledge by perception. The element of *feeling* is also present in sense-perception in what is called bodily sensibility, the correct theory of which involves just views of the nature of feeling in general, and of the relation of feeling to knowledge. Of the various theories of sense-perception which are so prominent in the history of philosophy, the errors and defects are to be traced to some false assumption or oversight in physics, physiology, or metaphysics, or in all these sciences combined.

Theories of sense-perception are, to a great extent, theories of vision. This is not surprising. The phenomena of vision are the most prominent in our experience, and the most attractive to our attention. The organs of vision are more complicated than those of any other sense, and at the same time more easily separated into their component parts. As might be expected, the theories of sense-perception which are recorded in the history of philosophy, are, for the most part, theories of vision, and the illustrations and examples of the power of sense-perception, its actings and its laws, are almost universally drawn from the power of seeing with the eye.

The early Greek philosophers.

§ 126. We begin with the theories of the earlier Greek philosophers. In these there is very little to interest or instruct us, except as they serve to illustrate the causes of error, and to show us the beginnings and germs of almost every one of the false theories which deform and mislead modern speculation. They are all alike, in not sharply distinguishing the soul from the body, and scarcely from inorganic matter, in respect either of essence or functions. The first effort of philosophy was to resolve all agents and all phenomena—beginning with those most obviously material and mechanical, and terminating with the most spiritual and free—into some single element, as original and all-pervading.

Empedocles of Agrigentum introduced the distinction between sensuous and divine knowledge—teaching that the impressions of sense must be corrected by the notions of reason. It was an axiom with him in explaining sensuous knowledge, that like can only be known by its like,—this assumption pervades the great majority of the theories of perception down to the present moment; and, as we have seen, it is with the greatest difficulty that the mind can rid itself of its influence. (Cf. Hamilton, *Works of Reid*, p. 300, note.) In conformity with this view, he seeks to show that sense-perception can only be explained by our knowledge of the composition of the body perceived, and of the forces which act upon it. The objects of sense send off certain effluxes, ἀποῤῥοαι, from their surface, which pass into the human body through pores [provided in the several organs].

Democritus was the first avowed materialist; resolving all the different kinds of being, with their phenomena, into combinations of atoms, differing in size and shape. He taught that the soul differs from the body, by being composed of finer particles. All sense-perceptions are occasioned by contact. In modern phrase, he resolved all the senses into the sense of touch. That which is brought into contact with the soul is not, however, the material object; but its εἴδωλον, or image, being detached from its surface, reaches the soul by passing through the pores of an organ of sense. The εἴδωλον and the ἀποῤῥοή were nearly the same, unless the ἀποῤῥοή was used to emphasize the material element, and the εἴδωλον that which is subjective and spiritual. The nature and signification of either do not seem to have been held with greater intelligence and precision in earlier times than the corresponding terms [as image, representation, species] and conceptions are employed and understood in modern philosophy. At one time they were used in a signification simply and grossly material; at another, as the product of the combined activity of the spiritual and material. (Cf. Ritter, vol. i. B. vi. c. ii., note.)

From Democritus, Epicurus borrowed the notion of *effluxes, simulacra rerum*, which he conceived in the grossest form—viz., that they "are like pellicles flying off from objects; and that these material likenesses, diffusing themselves everywhere" in the air, are propagated to the perceptive organs. In the words of Lucretius: "*Quæ, quasi membranæ, summo de corpore rerum Dereptæ volitant ultro citroque per auras.*"

The philosophers of the Socratic school [Plato and Aristotle] recognized the doctrines of their predecessors to some extent, either to expand or refute them. They also made important additions to the philosophy of the previous times in respect to the theory of sense-perception. The doctrines of Aristotle and Plato, and even the terms which they employed, can be traced among philosophers of every age since their time; and they still reappear and exert their influence among the most recent schools. Aristotle especially gave the law to the schoolmen, from whose teachings the modern theories have retained many traditions. Plato is still appealed to and quoted by his admirers for his eloquent and just psychological discriminations, even in respect to the theory of sense-perception.

Plato taught very distinctly and emphatically, especially in his Theatetus, that sensation [proper] is an effect jointly produced by the force, motion, or action (φορά) of the material object and the sentient agent, and that it varies, of course, with this joint activity; that the sensations of no two sentient beings need necessarily be the same, under the same material conditions at the same time; and that the sensations of the same being, from the same object at different times, need not be the same, but may vary very greatly. Sense-knowledge, αἴσθησις, is therefore untrustworthy, illusive, and, it may be, deceptive. With this he contrasts the higher kind of knowledge, ἡ ἐπιστήμη, viz., that which is rational and intellectual—the knowledge of ideas, or of objects in their ideas. This knowledge, in its subjective character, is certain and satisfactory; in its objects it is permanent and fixed.

Aristotle. We find in *Aristotle* also the beginnings of the attempt to consider apart and to distinguish the intellectual act of perceiving on the one hand, and the physical conditions or media by which objects are actually perceived.

In respect to vision, Aristotle made a great advance upon his predecessors, in teaching that visible objects do not act directly upon the eye of the percipient, but through a transparent agent or medium. He also taught a doctrine of the refraction of light. Of this refraction the transparent medium spoken of is susceptible when it appears as water and air. In respect to the construction of the eye, he made little advance upon his predecessors, and knew little or nothing of the discoveries made by modern anatomy and physiology. The other senses require a medium as truly as does vision. The medium is in every case set in motion or brought into action by the perceived object, and is thus made capable of acting upon the appropriate sense. In respect to the construction and offices of the remaining organs of sense, Aristotle taught little that is worth reciting. All perceivable objects are extended, but their essence, as perceivable, does not consist in their being extended, but in a certain relation or proportion which they bear to the percipient.

In respect to the intellectual element in sense-perception, the element which we have called the discernment of relations, Aristotle is not clear and explicit. Now, he asserts that in perception, neither truth nor error are possible, but that these can only pertain to the higher powers of the soul. Again, he calls the power a judging faculty. The phenomena and products of sense-perception, he shows most clearly, have an element which does not pertain to the purely and properly intellectual powers; but he does not explain the higher element which both have in common. In this he gave the example for the confusion and defect of clearness which have prevailed from his day to the present.

He held however that there is a common percipient or sensory, by which the several sensations are measured, judged, and united together. Each separate sense apprehends its own object, as the eye color, and the ear sound; and each apprehends or discerns this object correctly. That which is common to all objects are these five: motion, rest, number, size, and form. The seat of this common sensory or common percipient, is the heart. This power combines and separates the percepts appropriate to the several senses, and prepares them, so to speak, for the phantasy and the memory, both of which are activities of the common percipient.

The doctrine that objects are not themselves perceived, but their species or perceptible forms, was sanctioned by Aristotle. As the wax receives only the impression or image from the device on a seal-ring, and not its matter, it making no difference whether the ring is gold or iron, such is perception by each of the senses. What is received, is not the matter of the object perceived, but that which it effects in conjunction with or in relation to the percipient. This is its form—τό εἶδος, species. What was intended by this form, was variously interpreted by the Greek commentators, Simplicius and Themistius contending that the percipient is the bodily organ, which received a corporeal impression; and

Alexander Aphrodisiensis and John Philiponus that it was a mental power, which, by perceiving, gained a mental impression or form. The last were doubtless in the right. (Cf. Hamilton's very valuable Notes, *Works of Reid*, pp. 827, 881; *Metaphysics*, *Lec.* xxi. vol. ii. pp. 36, 37, 38; Am. ed., pp. 292, 293.)

The distinction between *matter* and *form*, or *species*, was transmitted through the successors of Aristotle to the schools of the Middle Ages, and became an hereditary text for controversies and discussions, not only in respect to the nature and validity of the sense-perceptions, but of the objects and processes of our higher knowledge. These controversies have not yet terminated, nor have the terms over which they were fought been wholly laid aside.

The schoolmen. Their doctrine of species.

§ 127. The most of the *Schoolmen* retained in substance the distinctions and the doctrines of Aristotle, making such advances upon them as were to be expected from active disputants and well trained dialecticians, who employed their energies almost exclusively in defining more precisely what they supposed their great master intended, or in devising new inferences from the materials and data which he furnished. The schoolmen were not exclusively the followers of Aristotle. They were influenced more or less by the doctrines and the terminology of Plato.

The doctrine of the necessity and agency of species in sense-perception was prominent in their theories, and their views may be summed up in the following propositions: Objects are not and cannot be directly and immediately perceived, but only their *species*. The reasons given were the following: The object often is plainly not in contact with the sentient organ. It is also in its nature unlike the sensitive soul, and therefore cannot affect it. Every thing known must be *in* the knowing agent; but it is impossible that this should be true of the object; it can only be true of its species. Experience, also, proves that the image or species only is perceived. When a stick is thrust into the water, it is seen to be bent or broken. A change in the medium changes the object perceived. Our perceptions of the same object also vary at different times.

But the species is not a material entity or efflux. At least, it was not so regarded by the more intelligent of the schoolmen. It was scarcely possible, however, that it should not be treated as a material entity, and so have prepared the way for the grosser doctrine of an intermediate representative image. The species is not perceived, but only the object, through or by means of the species. And yet the species so far forth represents the object, that when it acts upon the organ of sense, it moves or excites the percipient to discern, by its means, the object itself. Some of the schoolmen taught that these species have some spatial relations—that they exist in every part of space, bridging over, by a continuous series the interval between, or binding together, the object and the sentient.

A few among the schoolmen rejected the doctrine of sensible and of intelligible species. Among the most conspicuous was William of Occam, who was led, by the boldness with which he urged the doctrines of the Nominalists, to reject also the doctrine of sensible species.

Descartes, Malebranche, and Arnauld.

§ 128. *Descartes*, made a permanent inroad upon the philosophy of the scholastics, and introduced the modern science of psychology. He prepared the way for the distinctions and discussions in respect to sense-perception which have played so important a part in modern speculation. The doctrines of Descartes which we need to notice are the following:

1. *Descartes* drew a sharply-defined line between spirit and matter in respect to both essence and phenomena, and of course distinguished clearly between the soul and the body.

2. All the affections of the body, being phenomena of matter (of which the essence is extension), must be resolved into positions and motions of its parts in space. Hence all those changes in the organs of sense by which we perceive must be changes in the relative positions of their constituent parts.

3. The medium by which they are conveyed to the brain was held to be the *animal spirits*. These serve as the instrument of sensation, by producing in the brain [conveying] changes corresponding to those occasioned in the organs of sense by the action of the object perceived.

But the soul does not, by a second or internal sense-perception, apprehend the last of these series of mechanical changes wrought in the brain, as though the soul were endowed with another interior apparatus of sense. How it becomes aware of these changes in the brain is not explained by Descartes; nor how, when these changes are made known to it, these serve as indications or signs of qualities in material objects. Descartes never asserted, as did some of his disciples, that these changes can act as representative ideas—that in vision, the image on the retina, or its reflex on the brain, appears as a copy or reflected picture, which is compared with the object itself. On the other hand, he held to the doctrine of a representative idea, in the sense that, on occasion of the apprehension of these changes, the mind has sense-perception of objects. As the schoolmen held that *by or through* the several species, the soul perceives objects, so he held that through or on occasion of these mechanical changes, excited and propagated through the corporeal machine, the soul apprehends the objects of which these are the indications or signs.

We see one object with two eyes, just as we touch one object with two sticks; the similar apprehended motions in the brain, (corresponding to the double muscular sensations with which we hold the two sticks), make the two sticks feel one object. But it is not explained how the soul is capable of knowing the last movements of the machine, or how it interprets the index in the brain. It is true, Descartes supposed the seat of the soul to be a small gland in the midst of a small cavity at the centre of the brain. To the plexus of tubes and interstices which constitute the walls of this cavity, the animal spirits bring the last changes which correspond to each sense-perception of material objects, and by means of the changes effected in these walls they transmit the orders of the soul.

4. All sensations are purely spiritual affections, being, in his language, "modes of thinking," or of thought, which, in its nature, has no relation whatever to extension. The sensation of pain which we refer to the foot, is simply in the mind. The sensation of color which we refer to an external object, is in the mind only; it is neither in the eye nor in the picture to which we ascribe it.

5. The soul, in its sensations, is purely and simply passive; even in its inclinations and desires, which are functions of the will, it is passive.

6. The diversity in the qualities of the sensations is owing to the diverse motions of the bodies which occasion them.

7. Material objects are known as external to the soul by the following process: The soul finds itself affected with certain sensations, or modes of thought. They are known not to be caused by the soul's own agency. Under the axiom that

every phenomenon must be referred to a cause, the mind believes in the existence of material objects as the external causes of its own sensations.

8. We confide in the indications of the senses, because we believe that God is too good a being to allow us to be deceived, or to bring objects before our senses in such a way as to make deception possible. That God is good, we know with innate certainty. Hence we confide in the truth that the ideas of sense correspond to the reality of things.

Malebranche developed a complete theory of sense-perception with far greater distinctness and detail than any of his predecessors, and did more to give direction and form to the modern theories than even Locke himself. The distinctions which he introduced are the following:

1. He distinguished, in sense-perception, the element of sensation from the element of judgment. Of the four different elements (which he says occur in almost every sensation, and are confounded by most persons, but which it is most important to distinguish) the *third* and *fourth* are the following: the *sensation*, or subjective state of the soul, as of warmth; and the *judgment* which the soul makes that this warmth is in the hand or in the fire. "This judgment is natural, or rather, it is only a compound or complex sensation"—"*ou plûtot ce n'est qu'une sensation composée.*" This natural judgment is usually followed by another (*i. e.* an acquired) judgment which the soul, through the force of habit, makes with the utmost rapidity.

2. Malebranche accepts the doctrine, that it is only through ideas that we can apprehend material objects, and thereby denies that we can know such objects as they are. He gives various reasons to show that these intermediate ideas are necessary. They are mostly drawn from the phenomena of vision. While he rejects the doctrine of species and effluxes, and every form of material representation, he as earnestly supports the doctrine of immaterial representatives, and holds that these are changing, uncertain, deceitful, and confused, when contrasted with the pure ideas which are attained in God. His favorite and peculiar doctrine was that "the soul sees all things in God."

Antony Arnauld maintained the following positions against Malebranche:

1. It is a false assumption that the soul cannot perceive except by means of representative ideas. What the soul perceives, is not the idea as distinguished from, and representative of, the material object, but it is the object itself. The idea is nothing else than the perception itself. To say that the soul has an idea, is the same as to say that the soul has a perception.

2. The soul, to perceive a material object, does not need to come into contact with the object perceived.

3. The soul is not passive in perception, but active. It is endowed directly by the Creator with the power to perceive.

4. We must be able to perceive material objects directly. Otherwise, we should not know that the representative ideas represent them.

§ 129. The speculations of *Locke* have exerted a powerful influence upon the course of modern philosophy, and incidentally upon the theories of sense-perception. John Locke.

His opinions, in respect to sense-perception, may be divided as follows:

1. Of the *media* or *physical conditions* of sense-perception he teaches little that is positive, and nothing that was new.

2. Of the *faculty*, he says only that it is a distinct source of knowledge, and

that from this we derive all that we know of material qualities—*i. e.*, of the separable elements given by each of the senses.

3. The *objects* apprehended by the faculty of sense are the *qualities* of matter. Of these there are two classes: the *primary* and the *secondary*. The *primary* are solidity, extension, figure, motion, rest, and number. The *secondary* are the so-called sensible qualities, as color, taste, smell, etc. The last are the capacities in material objects to produce certain impressions or affections of the soul by variations in the size, figure, position, and motions of the primary qualities.

These two classes of qualities make up all that we know of material objects, after we have added to them the "obscure idea" of *substance*, as that in which they inhere.

4. What *knowledge* is, or what it is for the mind to know, Locke teaches by the following definition:

"The mind knows not things immediately, but only by the intervention of the ideas it has of them. Our knowledge, therefore, is real only so far as there is a conformity between our ideas and the reality of things" (Essay, B. iv. c. iv. § 3).

Of the relation of these "ideas" to their correspondent qualities or objects, he says: "The ideas of primary qualities of bodies are resemblances of them, and their patterns do really exist in the bodies themselves; but the ideas produced in us by their secondary qualities have no resemblance of them at all." He expressly defines knowledge of every kind to be the discernment of an *agreement* or *disagreement* between two entities; in the case of sense-knowledge, between the representative idea and its counterpart.

The language of Locke in these passages, if strictly construed, would seem to declare that it is by the intervention of representative ideas that we perceive sensible objects, and that we can only know them so far as we discern that they "resemble" or "agree with" their objects. Hence it has been charged upon him that he taught the doctrine of perception by means of intervening images or ideas. It becomes a question of great interest therefore, what he actually intended by this careless and confused language. It is obvious that any such theory of knowledge, when applied to sense-perception, would involve a positive self-contradiction, or else an idle and useless expedient. If we can only know a material object by means of the intervening idea, which "represents" or agrees with it, then we can never reach or know the object at all; for we may go on by a succession of processes *ad infinitum*, and, when we have done, we shall only have reached a representative idea, but shall never have grasped the object itself. On the other hand, if it be conceded that we can and do perceive material objects, and, in perceiving them, discern that the idea is "conformed to," "agrees with" or "represents," its object, then we must be able to compare the two together—the material object and its idea. But in order to be able to compare the object with its idea, we must know the two terms which we compare—*i. e.*, the object itself as well as the idea. But if we know the object already, of what use is it, or how is it possible, to acquire knowledge of it by the idea? This would make it impossible to know the secondary qualities by any means whatever, for Locke expressly asserts that no similarity exists between the ideas of secondary qualities and the qualities themselves—as the smell, etc., of the violet, and the qualities in objects which produce them.

These consequences, so fatal to the representative theory, supposing Locke to have held it, would lead us to question whether he intended by "idea," in every or in any case, an intervening representative image; and by the words "to resemble," "to be conformed to," "to agree with," any relation discerned by the process of comparison.

But whatever doubt there may be in respect to the doctrines which Locke actually taught in respect to perception, there can be no question at all in respect to the construction which other writers gave them, or to the inferences which they derived from the principles which they imputed to Locke. (Cf. § 145.)

Bishop George Berkeley. David Hume.

§ 130. George Berkeley (*Principles of Human Knowledge*, § 18 *sqq.*), assuming that ideas only are the direct objects of the mind's knowledge in sense-perception, concludes that it is impossible that the mind should know that the material or external world exists at all. It is impossible that the mind should know the objects which the ideas are said to resemble. For, in the first place, one idea can only be like an idea, and can never be like an object; and second, if the idea was like the object, we could never know this likeness except by knowing both the idea and its object. All that the mind can know are its own sensations or modifications. The distinction between primary and secondary qualities is not well-founded. All we know is that on occasion of the ideas of extension, motion, and figure, we have the sensations of color, taste, and sound. Ideas exist only so far as they are perceived. The laws which we conceive to govern material things, only govern the combinations of our ideas. Real objects, as we call them, are only combinations of ideas; the only difference between them and the so-called imaginary ideas consists entirely in this, that the first are not dependent on our will to produce them, but are always present to our minds, whether we will or no. Imaginary ideas, on the other hand, come and go according as we will. Real ideas are also more lively and distinct, while those of the imagination are faint and confused. The knowledge of spirit is strikingly contrasted with that which we have of matter. We know ourselves and our own states or modifications directly. We know our thoughts, feelings, etc., not their ideas. That the universe is permanent in its objects—viz., ideas—and also in its laws, is to be explained by the fact, that the Eternal Spirit constantly sustains and presents these ideas for the contemplation of created spirits. By means of these, the attributes and government of God are made known. All the things that we perceive, are the ideas of God.

Berkeley's *Essay toward a New Theory of Vision*, 1709, was the most important contribution which he made to the theory of sense-perception. This was followed by *The Theory of Vision Vindicated and Explained*, 1733. In these essays Berkeley gave greater precision and fullness to the doctrine of the acquired perceptions. The fact that some of our perceptions are acquired was familiarly known and generally accepted before the time of Berkeley. It was generally held, however, that the acquired judgments were formed by means of the properties of light, as taught in the science of optics. This doctrine Berkeley sets aside, and clearly establishes the truth that it is by sensations attending the varied use of the eyes, by the confusion and clearness of the vision, etc., etc., that these judgments of distance and magnitude are formed, and that these judgments are wholly matters of experience concerning the ordinary course of nature.

David Hume was not content to apply the ideal theory to the world of matter, but he maintained that it was as true of the world of spirits, rejecting the distinction made in favor of the latter by Berkeley, and urging that we know nothing of the mind except only the ideas which we experience, thus resolving all real existences into mere collections of ideas.

T. Reid, Dugald Stewart, Dr. T. Brown.

§ 131. Dr. *Thomas Reid*, the father of the so-called Scottish philosophy, being startled by the consequences which Berkeley and Hume derived from their construction of Locke's theory of sense-perception, was led to review not only the doctrine of representative perception, but also some other principles which Locke was understood to advocate in respect to the origin and elements of knowledge. The features of his system are as follows:

1. He successfully exposed the groundlessness, inconsistency, and contradictions of the ancient and modern theories of representative perception, and cleared the way for a theory more accordant with experience and common sense.

2. Reid vindicated the general principle, that no theory of perception is entitled to confidence as truly philosophical, which contradicts the universal convictions and the common sense of mankind, when they apply their understandings to the judgment of truths which they are competent to decide upon. This was a special inference from the general axioms of Reid's philosophy.

3. Reid insisted that the mind is active in sense-perception; and did this with an earnestness rare among philosophers, not only of the English, but of any school whatever. The ancients, and the moderns before him, did indeed assert that the mind is active in its higher functions; but they as distinctly denied that it is active in the lower. It has been nearly the uniform doctrine of all the schools that, in sense-perception, objects act upon the mind so as to impress ideas, and that, in the reception of these ideas, the mind is chiefly or wholly passive. Against this doctrine Reid occasionally protests, in language like the following: "An object, in being perceived, does not act at all. I perceive the walls of the room where I sit: but they are perfectly inactive, and therefore act not upon the mind. To be perceived is what logicians call an external denomination, which implies neither action nor quality in the object perceived. Nor could men have ever gone into this notion that perception is owing to some action of the object upon the mind, were it not that we are so prone to form our notions of the mind from some similitude we conceive between it and body."

4. As intimately connected with the preceding, Reid asserts that the faculty and act of judgment are present in connection with the perceptions of sense.

5. Reid recognized and enforced the distinction between sensation and perception; and thus prepared the way for the correct and complete determination of these two elements in the process of sense-perception.

Dugald Stewart, the successor of Reid in the school of Scotch philosophers, followed closely and almost timidly in the footsteps of his predecessor, whom he greatly admired and revered.

1. He discriminated more carefully between sensation and perception than Reid. He limited perception to the act of apprehending the objects appropriate to each separate sense, and escaped the confusion and ambiguity which Reid committed, of confounding the original with the acquired perceptions.

Of three of the senses—smell, taste, and hearing—he denied perception altogether, in fact though not in form. He expressly asserted that these, by themselves, give no information of external objects (*Outlines of Moral Philosophy*, § 15). He asserts that the sensation of color, even as given in vision, can reside in the mind only, and is purely subjective; giving no relation of extension, and in our early experience clearly separable from it.

2. Stewart apprehended, far more clearly than Reid, the true character of what

he calls the mathematical affections of matter, and the relation of these affections to space and to our belief in space as a necessary existence. These mathematical affections are extension and figure, and are distinguished from the other primary qualities, such as hardness or solidity, and are thus characterized:—1. They presuppose the existence of our external senses. 2. The notion of them involves an irresistible conviction of the external existence of their objects—viz., of space. 3. This conviction is neither the result of reasoning, nor of experience, but must be considered as an ultimate and essential law of human thought. (*Phil. Essays.*)

3. Stewart adds to the doctrine of Reid, that we believe in the existence of the material world, by a necessary *suggestion*. The explanation of our belief in its permanence, he finds in our more comprehensive belief in the permanence of the laws of nature.

Dr. *Thomas Brown* followed in the same school with Reid and Stewart. The analysis which he has given of the processes and the products of the sense-perceptions, is one of the boldest and the most subtle which is to be found in the whole compass of English psychology.

1. Dr. Brown attached great importance to the muscular sensations. He was one of the earliest of English psychologists to recognize and to distinguish them from the sensations as usually accepted. This distinction is now almost universally adopted. Dr. Brown made so much of these sensations, as to derive from them the notions of extension and of externality.

2. He scarcely recognizes the distinction adopted by Reid between sensation and perception. So far as the original perceptions are concerned, he rejects it altogether. The only acts of perception which he acknowledges or describes are acts of acquired perception.

He refers our belief in the external and material world to the principle of causation. We know our sensations as subjective states of the soul. We believe that they must be produced by a cause. We know that they are not caused by ourselves. There must be causes other than ourselves. These causes are material *non-egos*. The existence of these *non-egos* is not suggested directly, as Reid teaches, but it is inferred. "Perception, then, even in that class of feelings by which we learn to consider ourselves as surrounded by substances extended and resisting, is only another name, as I have said, for the result of certain associations and inferences that flow from other more general principles of the mind." (Lec. 26.) Cf. § 40.

3. It is equally clear that Brown, to be consistent, would reject nearly or altogether the distinction between the primary and the secondary qualities of matter as explained by Reid, and in part adopted by Stewart.

§ 132. Sir W. Hamilton, the deservedly eminent Professor of Logic and Metaphysics in the University of Edinburrgh, was one of the greatest philosophers of Great Britain. He devoted his researches to two leading topics: Formal Logic, and the Theories of Sense-perception. He had studied the history of these theories with greater care than any one of his own time, and had gathered from his historical researches the most valuable results in the way of observation and analysis. His contributions are important in respect to all the points which have been noticed.

Sir William Hamilton.

1. Sensation and perception were more carefully discriminated by him, as to their nature and material relations, than by any philosopher before his time.

They are viewed by him as inseparable elements of a single mental state, and are called sensation and perception proper.

2. Hamilton asserts that sense-perception involves the action of the intelligence in the form of judgment, or the discrimination of relations. It follows of necessity that, in perception, man is active, and not simply receptive or passive. These important truths Hamilton enforces on every occasion.

3. In respect to *extension* and *space*, Hamilton teaches, with Kant and others, that while the spatial relations of every material body are known by sense-perception, yet space itself is pre-supposed by the intuition of the intellect, in order that it may be possible for any of these relations to be perceived as actual. Space must be known *a priori*, in order that extension may be known *a posteriori*.

4. In respect to *externality*, Hamilton teaches positively, though not with so great clearness as is desirable, that the term is used in two senses: (1) as denoting the diversity of the sentient organism from the perceiving intellect; and (2) the diversity of material objects from the material organism which the soul animates, and by which it apprehends.

In respect to the first of these relations, he asserts that it is directly apprehended in every act of sense-perception.

In respect to the second, he teaches that it is gained by the exercise of the locomotive power in the form of muscular effort. This effort is resisted, and with the resistance is gained the correlative of a resisting something, external to the body or sentient organism. "When I am conscious of the exertion of an enorganic volition to move, and aware that the muscles are obedient to my will, but at the same time aware that my limb is arrested in its motion by some external impediment, in this case I cannot be conscious of myself as the resisted relative, without at the same time being conscious, being immediately percipient of a not-self as the resisting correlative."

5. The qualities of material objects are treated by Hamilton as though, *as qualities*, they were the direct object of immediate sense-perception. This view is certainly implied in the whole of his doctrine, and his history of the sensible qualities of matter. This is a consequence of his failure occasionally to discriminate between *sense-perception* as direct and reflex. He does not always distinctly hold to the fact that if in original sense-perception, we can in any sense apprehend the qualities of matter, we can only apprehend those which pertain to the animated organism. We hold that the qualities of matter are only known by acquired perception in reflex action.

6. Hamilton sometimes confounds the conditions of perception with perception itself.

He falls into this error in applying the doctrines of latent modifications of the mind to the phenomena of vision and hearing. He argues that, because two portions of extension, or two parts of an extended substance, each of which by itself is invisible, become visible when annexed so as to form one continuum, that therefore each of them, by itself, must obscurely affect the sensorium or the mind. So, two separate sounds, each one of which might be too feeble to be heard alone, when uttered together, cannot fail to be heard. In both these cases the distinction is overlooked between the action of physical or physiological stimuli upon the sensorium, and their effect on the sensorium as the appropriate and indeed the only condition of the responses of conscious sentiency or perception.

7. Hamilton attaches too great importance to the subjective sensations, or the idiopathic affections of the nervous system, which are excited by electrical action, indigestion, or a blow. The sparks which are elicited by a blow over the eyes, the light, the sound, the taste, the ringing of the ears which electric or other agencies occasion, are doubtless owing to a special stimulus of the sensorium, and to this only.

8. Hamilton's theory of perception is vitiated still further by the metaphysical assumption that we know directly only phenomena, whether of matter or of mind. *We* hold that neither phenomena nor qualities, as such, are perceived, but objects, percepts, or beings; and that it is by an after-thought, or reflex process, that these are connected as qualities, and are referred to substances.

9. The most eminent service which Hamilton has rendered to the theory of sense-perception, is his criticism of all the possible forms of the doctrine of representative or mediate perception, and his demonstration that every such theory is untenable.

We give the substance of his criticism in our own language, for the sake of brevity, interposing such qualifications and explanations as may serve to illustrate and explain it.

In respect to the act of sense-perception, one of two positions may be taken: the mind is endowed with the power of perceiving material objects by a direct and intuitive energy, without the intervention of any intermediate object; or, the mind can perceive material objects only through the medium of some intervening object.

It will here be observed, that the alternative does not relate to the conditions of such perception whether material or physiological. It is simply a question whether there are or are not intermediate objects to the psychological act.

If the first position be taken, then the only obligation which rests upon the philosopher, is to state the conditions which are essential to the act, and to analyze the act into its elementary constituents, as given in, or inferred from, our conscious experience and careful observation.

The person who takes the second position is bound to show why this hypothesis is necessary. The natural and universal belief of mankind is, that objects are perceived directly. He who asserts that this is impossible, ought to give some reason for deviating from this belief. The several reasons that are to be found in the whole history of philosophy, are by Hamilton reduced to five groups, underlying each of which is a single fundamental principle. The first of them is, that an act of cognition is an act of the mind; and to suppose that the mind should know that which is not itself, is to suppose that it can go out of itself. To this it is replied: 1. That if we cannot explain how it is possible that the mind should act on that which is not itself, it does not follow that it cannot be a fact. The fact may be ultimate, and for this reason inexplicable. 2. The principle proves too much, for it would involve the inference that the mind cannot act upon matter, as it manifestly does in volition. 3. Moreover, it would carry with itself the consequence that matter cannot act out of itself upon the mind, and of course cannot produce a representative image of its object.

The second reason is, that mind and matter are substances not only of a different, but of the most opposite nature. That which knows immediately, must be of a nature corresponding or analogous to that which is known; the mind cannot, therefore, know matter directly; an intermediate something must be interposed.

This *assumption* is of the widest prevalence, and underlies almost every theory of representative perception. It accounts for the variety of the views of the nature of the interposed media held by both ancients and moderns. When this medium was conceived akin to the mind, it gave the *intentional species* of the schoolmen, or *the ideas* of Malebranche and Berkeley. When it was supposed to be identical with the mind, it gave the *gnostic reasons* of the Platonists, the *pre-existing species* of Avicenna, the *ideas* of Descartes, Arnauld, Leibnitz, Buffon, and Condillac, the *phenomena* of Kant, the *external states* of Dr. Brown. To the influence of this assumption, are to be traced the systems of the absolute identity of mind and matter in the opposite theories of exclusive materialism and of spiritual idealism.

This grand assumption should be rejected as arbitrary, unphilosophical, and contradictory to our plain experience.

The third reason for this hypothesis is, that the mind can only know that to which it is immediately present. External objects can hence be brought within reach of the mind only by means of some intermediate representative. The proper answer to this reason is, that the mind is present in every part of the body so far as to act and to be acted upon, and that the real object of immediate perception is some part of the body as excited to a specific sensation. The corrected view of the relation of the soul to the body, and of what is the real object of the mind's external perception, sets aside this third reason.

Reid and Stewart attempt to set this aside by a failure to conceive these points rightly, and they require some agency of the Deity, and an inexplicable connection between the sensation and perception, which is unphilosophical and unsatisfactory.

The fourth ground is stated by Hume, that the same object, as a table, at different distances changes its dimensions, but the object itself does not change; therefore the object must be apprehended by an intermediate and changing representation. To this it is answered, that the same table is not perceived, so far as vision is concerned, when near and remote, but a different object in each case is the immediate object of sense-perception.

The fifth reason is stated by the elder Fichte, that, as the will must act in view of intelligent objects, these must be within the mind; so far then as it acts in respect to material objects, these must be represented in the mind.

To this it may be replied, that the act of intelligence is in the mind, and this is all that is required as the condition of the act of will. Besides, the act of the will respects future results, which must necessarily be mediately represented. It is not denied that the mind is capable of mediate knowledge. The question at issue is, whether the act of sense-perception is an act of this kind.

After having shown that this hypothesis of a representative perception is unnecessary, Hamilton shows at length that it does not stand the tests by which every legitimate hypothesis may properly be tried. Those conditions are: (1.) That it be necessary, and be more intelligible than the fact which it explains. (2.) That it shall not subvert that which it proposes to explain, or the ground on which it rests. (3.) That the facts in explanation of which it is devised really exist, and are not themselves hypothetical. (4.) That it does not subvert the phenomena which it seeks to account for. (5.) That it works naturally and simply. The hypothesis of representative perception fails to answer to any of these conditions, and must therefore be rejected by every true philosopher. (*Met., Lec.* xxv. and xxvi.)

Immanuel Kant, and the German school.

§ 133. *Immanuel Kant*, the great metaphysician of Germany, has treated of sense-perception only indirectly. He has given no formal theory of its processes, but has metaphysically analyzed its results, and thus has indirectly taught a partial theory of the power itself and its functions. First of all, he implies that the soul, in its sense-perceptions, is passive or receptive only. He contrasts the receptivity of the soul in sense with its activity or spontaneity in the understanding. He indirectly teaches, by the assumptions that underlie his whole system, that the process of sense-perception is not complete until the understanding, by the judging power, conceives under some of its forms the matter given by sense. Had he distinguished between the natural judgments which concern individual things and their relations, and the secondary judgments that contemplate general conceptions, there could be little to object to in his theory; but this omission is fatal to its completeness and its truth. Sense stands on the one side as a purely passive receptivity of individual objects, and the understanding, on the other, as active, but as concerned solely with generalized concepts alone.

Of the relation of sensation to perception, Kant teaches that sensation gives the matter, and perception—*i. e.*,—intuition—furnishes the form. The form essential to any and every act of external intuition is space. All material objects, so far as they are perceived at all, are perceived in some relation to space—that is, they are perceived as extended objects. Kant recognizes this as a fact of actual experience. But the fact he subjects to no farther analysis, least of all does he examine farther the process by which the product is reached. Instead of studying the fact in its conditions and elements, he seeks to account for its possibility and the trustworthiness of its results, on grounds of speculative philosophy. For this reason, his discussion of space has an intimate relation to his theory of sense-perception, and the conclusion which he reached has explained the discussions of all physiologists and psychologists since his time. This conclusion was, that space and time must be assumed as the necessary conditions of our subjective experience in both consciousness and perception, but we are not thereby authorized to believe in their objective reality. We cannot, indeed, perceive any material object by means of the senses without involving necessary relations to space directly, and indirectly to time. It does not, however, follow that space is a reality. It is supposable, though not to us conceivable, that to minds constituted differently from our own, these forms, with the relations which they involve, should *not* be necessarily assumed. (*Kritik der reinen Vernunft. El. Lehre*, ii Th., 1 Abth., ii Buch, 2 Hauptst. 3 Abschn.)

In respect to the *reality* of external objects, Kant recognizes the fact in our psychical experience, that material objects are not only perceived as extended and spatial, but also as external; or in other words, as *non-egos*. In sense-perception this distinction is necessarily involved. Indeed, it is included as an essential element in the process and its result. But it does not follow, because the mind makes this distinction, that there is a reality corresponding to this *non-ego*. For (1.) The *non-ego*, as a being, is transcendental to all phenomena. (2.) It is posited in space, which is necessary as a form of sense, but which may be only an illusion. Kant would however demonstrate, on the ground of speculative necessity, that this is impossible. He contends that we must assume that there is something permanent and real without, in order to account for the changing modifications within. Of the existence of an external world, we can be rationally assured, but of it, we can have no direct perception. Even the self, or *ego*, is not apprehended as a permanent something. It is only concluded to exist as the thought-conception of a spiritual substance with capacities for spiritual acts. All that we are conscious of, are our changing modifications in time. But these can only be rationally explained by a permanent reality which causes them.

The theory of sense-perception was discussed by the successors of Kant chiefly in its purely metaphysical relations. In the writings of Fichte, Schelling, and Hegel, still less attention is given to psychological analysis, metaphysical principles and relations being almost exclusively considered.

Herbart, J. F.

J. F. Herbart's theory of sense-perception may be briefly stated as follows:

The soul, though a simple substance, is capable of being excited by the action of various material stimuli to various reactions of its own. Certain classes of these, when experienced, are sensations. A sensation is the soul's reception of, or its reaction against, this material stimulus. The sensations differ from one another in quality or kind on the one hand and in energy or intensity on the other.

As the several sensations are experienced, each continues to exist in the soul, with a force or tendency to be reproduced. As soon as favoring conditions present themselves, past sensations reappear in the order of the soul's original experience of them. When such a series is *viewed* [experienced?] from one sensation regarded as fixed, it has time-relations; and by means of the mutual struggles or tendencies of several series of experienced sensations to gain possession a second time of the soul without success, there is generated the idea of pure or simple time.

The apprehension of time prepares the body for that of space. Sensations experienced and recalled in the time series, are disputed by other sensations and series of sensations that struggle to occupy the soul. To provide for the possibility of these mutual struggles, and under the experience of the pressure which they create, the mind constructs a conception of space first as occupied, and then as empty or void.

Thus, time and space result to the mind as the effects of mutually blended or mutually repelling series of sensations.

When space and time are produced, that which is next developed is the apprehension of the difference between bodily affections and material objects. This results from an experience of certain positive sensations, particularly those of touch joined with those of the muscular sense. A certain portion of space within the body is measured in every direction by various time-series of sensations, terminated by those appropriate to superficial touch. Other sensations we project beyond the surface of the body, at greater or less distances, all of which are measured by successive time-series of sensations, in experience or imagination.

Sensations which do not occur within the space of the body, nor on its surface as explained, are projected beyond—*i. e.*, are apprehended as not within its space. This constitutes perception in the lowest, or the elementary stage. Afterwards are developed apperception, or the knowledge of mental states by a secondary act of knowledge; then the knowledge of substance and its attributes; then a knowledge of material things, or of material substances with material attributes and space-relations.

Schleiermacher, the distinguished philosopher and theologian, deserves also to be named for the very important contributions which he made to the theory of sense-perception. These were partly indirect, as he opposed so decidedly the current of the great leaders of metaphysical speculation in Germany, by rejecting many of the assumptions which are fundamental to their systems. In part, also, they were direct, in the positive doctrines which he taught in respect to the conditions and nature of sense-perception as a process. The relations of space, time, substance, and cause, he held, as against Kant, to be real forms of things, and not merely the forms of our apprehension of things. The reality of time and space must be assumed without misgivings or questionings. Being is directly apprehended, as well as phenomena and relations. To all the combinations and constructions which we make in knowledge, we attribute actual reality. Thought which, in Hegel, is the all in all, the originator of the relations and products of knowledge, according to Schleiermacher, is psychologically dependent upon sense-perception. In sense-perception there are two essential elements: the receptive, styled by Schleiermacher "*the organic function,*" and the *a priori* or spontaneous, called "*the intellectual function.*" This last is an act of knowing by relations, and, as so defined, is an important improvement upon Kant, and Reid, and even upon Hamilton.

Schleiermacher, moreover, teaches that the two elements, the organic and intellectual, are present in different proportions in the different faculties and acts of sense-perception, anticipating in this the law of Hamilton respecting the inverse proportion of sensation and perception proper. Important contributions have been made to the physiological and psychological theories of sense-perception by many distinguished German and English writers, whom it is not important that we should notice.

PART SECOND.

REPRESENTATION AND REPRESENTATIVE KNOWLEDGE.

CHAPTER I.

THE REPRESENTATIVE POWER DEFINED AND EXPLAINED.

Representation defined and illustrated.

§ 134. REPRESENTATION *or the representative power* is defined in general, *as the power to recall, represent, and reknow objects which have been previously known or experienced.* More briefly, it is the power to represent objects previously presented to the mind. Thus, I gaze upon a tree, a house, or a mountain. The object perceived is the tree, the house, or mountain, before my eyes. I close my eyes, and "my mind makes pictures when my eyes are shut." I at once represent or see with "my mind's eye" that which I saw just before with the eyes of the body.

> My eyes make pictures when they are shut.
> I see a fountain, large and fair,
> A willow, and a ruined hut. COLERIDGE.

> *Hamlet.*—My father—methinks I see my father!
> *Horatio.*—Oh, where, my lord?
> *Hamlet.*—In my *mind's eye*, Horatio. SHAKSPEARE.

In like manner we hear a sound, either singly, as the solitary note of the pigeon, or several sounds in succession, as the *caw, caw,* of the crow, the roll of a drum, or the notes of a musical air. Let the sounds cease. We can still distinctly recall them, and seem to hear them again with the mind, though the mind makes for itself all the sounds which it seems to hear. In a similar way we can represent the percepts that are appropriate to the senses of touch, of taste and of smell; reviving the touch, taste, and smell *by* and *for* the mind alone.

> Music, when soft voices die,
> Vibrates in the memory.
> Odors, when sweet violets sicken,
> Live within the sense they quicken.—SHELLEY.

We are not limited to sensible objects, or to sense-percepts, in the exercise of this power. We can as truly represent *the acts and the affections of the soul itself.* Not only can we with the mind's eye behold the tree and the mountain previously seen, but we can represent the act of the mind by which we beheld it, as also the delight which the sight occasioned. We not only hear a musical air the second time, but we revive again the idea of the accompanying pleasure. So is it with *the relations* in which the objects were presented at first. The objects themselves can not only be recalled as objects, but they can be recalled as related, or as totals made up of the objects connected by the several relations under which they were originally known. Whether these are relations of space or time, of self or not-self; whether necessary and permanent, or casual and changing; whether intellectual or emotional—whether objective or subjective;—whatever we apprehend in presentation, can be recalled in representation.

But the activity of the mind in this general function is not limited to the power of representing objects previously present. It can so far modify the objects of the past experience, as to transform them into new creations. It becomes in this way, in an eminent sense, a *creative* power. The mind not only can depict a man, a tree, or a mountain as actually witnessed, but it can alter the form, the dimensions, and the appendages or accidents of each, taking parts from the one and attaching them to parts belonging to the other. So, also, it can create or imagine a Lilliputian, a Centaur, a Parnassus, an Abdiel. The representative power in this higher form is called, the fancy or the imagination.

Appellations for the power.

§ 135. The power thus to act is called *the representative,* in distinction from, and in contrast with the *presentative* power. In sense-perception and consciousness, the mind *presents* to itself for the first time the objects of its direct and original knowledge. In representation, it presents these objects a second time, or *represents* them.

It is also called *reproduction,* or the *reproductive* power, because the mind, by its own energy, under appropriate circumstances and in obedience to certain laws, reproduces objects previously known.

It also involves the power to *retain* and *conserve,* in a certain sense, that which has been acquired by the mind. To this capacity the name of *retention* has been given, or the *retentive* power.

To these three distinguishable relations of this power, Hamilton has not only assigned separate appellations, but has treated them as separate faculties, viz., the conservative, reproductive, and representative faculties (*Met., Lec.* xx). But inasmuch as it is implied in the power to *represent*, that there is a power to *reproduce;* and in the power to *reproduce,* that the mind can *retain* or *conserve,* it seems more philosophical to consider and treat retention and reproduction as the essential conditions of representation, rather than as distinct faculties.

It is also called the *creative* power, the *constructive or productive* imagination, when it evolves new products. This exercise of the representative power has rarely received a technical appellation.

Objects of the representative power.

§ 136. The *objects of the representative power are,* as has already been implied, *mental objects.* They are not real things or real percepts, but the mind's creations after real things. They are spiritual or psychical, not material entities, although in many cases they concern material things, being psychical transcripts of them, either as believed to be real or as conceived to be possible. When they concern the soul only, they are not the real soul, or its present acts, but psychical transcripts of the real soul in a past or possible condition of action. They are in no sense *object-objects,* but are preëminently *subject-objects.* As *objects,* they are distinguished from the acts of the mind which apprehend them: as *subject*-objects, they are created by that very mind, and exist only for that mind. As *represented* subject-objects, they always indicate another reality, whether spiritual or mental.

But though the object of the representative power is a mental object, it is an *individual object.* By this characteristic it is distinguished from a *thought-object,* or an object of the intelligence. Thought-objects are both mental objects and subject objects, and, in an important sense, representative-objects: but they are also generalized objects or universals. Objects of representation are like them in that they are purely mental objects, yet are unlike them in being individual. Whether we recall these objects, or create them—whether we copy, as exactly as we can, from an original in nature, or create constructions the most fantastic, grotesque, or unnatural, they are all individual. Falstaff, Hamlet, Ivanhoe, Jeannie Deans, Don Quixote, Tam O'Shanter, the Eden of Milton, the Faëry Land of Spenser, were all individual beings in the imagination that originated, and are such in the imagination that reconstructs them as delineated by their originators.

These objects involve relations.

§ 137. The presented object was known by the mind not only as a being, but in its relations, as of diversity, space, time, etc.; so *the object as represented, may be known again in all these relations,* with all those in addition which are implied in its being represented. It should be remembered, however, that a relation as such—*i. e.*, a relation as separate from an object—as it cannot be apprehended by sense-perception or consciousness, so it cannot be recalled by representation. A relation, *as such,* cannot become an image to the representative power, but the object in its relations can be imaged.

The representative power, not only by its representative act recalls the object in the relations in which it was originally known, but the existence and exercise of this power *involves relations that are peculiar to itself.* Thus, in recalling a tree or a horse previously perceived, or a mental act of knowledge or state of feeling, I not only bring back the tree or horse as extended and external, and the psychical state as subjective and in time, but, in recalling it, I must know it as a *subject-object,* and as *having been previously perceived or experienced by myself.* These relations are necessary and peculiar to the representative power.

For the objects of this power we have no appropriate technical name. The words *image* and *picture* might be properly applied to the represented percepts of vision; but to speak of the image of a sound, smell, or touch, would be incongruous, if not offensive. Still less tolerable would it be to speak of the image of an act of knowledge or feeling. *Conception* cannot be accepted, as was proposed by Stewart, for it is too frequently applied to other and very different objects. *Idea* would be more significant, if it could be forced back to its original and etymological import; but idea has, since the time of Locke, been compelled to do all manner of service. In the earlier days of the English language the representative power was called *imagination,* or *phantasy,* and *images* and *phantasms* were appropriately and literally applied to its objects.

Conditions and laws of representation considered.

§ 138. The *conditions and laws* of the representing power should next be considered. The mind, in representation, as in the exercise of all its powers, acts under limitations and according to laws. In representation, man does not, like the great Originator, create by his own fiat, his world of mental objects. What he reproduces or constructs anew, is in some way dependent upon what he has previously experienced. Not only must every thing which is represented be reproduced from, or by the means of some past experience, but what is represented at any moment depends upon what was present the instant before.

The fact that one object or image brings up another to the mind, is called the *association of ideas*. The conditions or laws under which the mind recalls one object by means of another, are usually called the *laws of association*. The term is open to exception, because both percepts and experiences are connected with images, as truly as images (or ideas) with images. The phrase is, however, too firmly established in general acceptance and use to be set aside.

Representation divided into several varieties.

§ 139. The representative power, though marked by common characteristics and obeying common laws, is divided into several *varieties*, or species. These are distinguished by the completeness or incompleteness of the pictures which they make of the objects once presented; by the fidelity with which they adhere to, or the liberty with which they deviate from their originals; by the laws of association which predominate in each variety; and by the ends for which the power is exercised, and the uses to which it is applied.

The most perfect exemplification of the exercise of the representative power is an act of *perfect memory*. Such an act is always complex, involving the object, the action, and the agent, united by their mutual relations into one indivisible state. If the object is material, it involves certain relations of space; the action, being one of a continuous series, involves relations of time; the agent, being the body and soul united, must exist in every act under relations of both space and time. When a single act of presentative knowledge is recalled in all these elements of object and relation, the representation is complete, and the act is an act of perfect memory. For example, yesterday I took a walk to the top of a neighboring eminence. To-day I recall distinctly the landscape which I saw, in its minutest features—re-creating, as I do, a distinct and vivid picture of the scene; and not only of the scene, but of myself as beholding it, with the actions before and after, with my feelings also in viewing it, and the very accidents of the place where I sat or stood during the view. This is an act of perfect memory; it includes every element of the original.

As time goes on, it is possible that one or another of these elements should be recalled less distinctly, or should be omitted altogether. It is possible that I should be able to bring back the landscape only as an object, and be certain, as I see or think of it, only that I once saw it before; but how or when, or with what feelings or from what point, I do not recall. Or possibly the object may be lost, and the subjective feelings may alone be revived and recognized as having been before experienced. Relations of time and accessories of place may both be lost. Thus, when I see the face of a person in a crowd, I know that I have seen it before; but when, or where, or with what feelings I cannot recall. I remember a familiar passage of prose or poetry; I know that I have read or heard it; but when, or with what feelings or attendant circumstances, I cannot tell. All these are acts of what may be called *imperfect memory.*

But memory, whether perfect or imperfect, is clearly distinguishable from *phantasy*, or the imaging power. This is representation without the recognition that the objects recalled have ever been perceived or experienced before. Examples of this are such as the following: I look distinctly at the front of a dwelling, the form of a horse, or the outline of a tree, each of which I wish to retain and make wholly my own. I close my eyes and picture each distinctly to my mind. The undivided force of my attention is expended upon the object, and so successfully, that it becomes a permanent possession as an object, with few or no accessories of either place or time. In all cases of disturbed fancy, often called *phantasy*, visions of objects seen before, but not remembered or recognized, throng in upon the soul. There may be no recognition, no knowledge that the object is familiar or has been seen or felt before. These acts are more likely to occur in those conditions of the soul in which the action of the reason is nearly suspended, or permanently set aside, as in reverie, dreaming, monomania, and partial or complete insanity.

But the mind can do more than simply represent the past with greater or less perfection, with or without the act of recognition. It can recombine or construct anew the materials which the past furnishes for it to work with or upon. In such acts it becomes the *creative imagination.* Of imagination as thus defined, there are several forms or varieties.

1. The mind may neglect or leave out of view all things existing in space, and all events occurring in time, and form to itself pictures of void space, and of time more or less extended or limited. Within these voids it can construct geometrical figures, and arrange series of numbered objects, and thus provide for itself the materials of mathematical science. This is the *mathematical imagination.*

2. It can separate and unite the parts and attributes of objects and existences, both spiritual and material, in divisions and combinations which never actually occur, but are grotesque and irrational. These separations and unions may be made in obedience only to the more obvious and the lower laws of association. Thus, the chimney of a house can be set upon the hump of a camel, and the ears or head of a donkey upon the body of a man. Or horses may be colored red or yellow. This is *phantasy proper;* the products of which are simply *grotesque,* or as we say, *fantastic.*

3. Objects may be recalled in wholes or in parts, and recombined and reconstructed under the obvious and more natural laws of association, for the ends of wit, humor, or amusement. This is *fancy proper,* which, as exemplified in literature and some of the fine arts has been thus distinguished from the higher imagination.

4. When the higher objects of nature and spirit are recalled, recombined, and created, with the aid of the nobler laws of association, for the higher ends of ideal elevation and improvement—when the more elevated feelings are addressed and excited, and the nobler capacities of man are called into action, then the power becomes poetic imagination. The sphere of this power is not poetry alone, but eloquence, music, painting, sculpture, archi-

tecture, and landscape gardening; inasmuch as all afford opportunities for these higher sentiments and suggestions. This is *imagination* as contrasted with *fancy*.

5. When the combinations and creations are effected for the purposes of research, invention, and instruction, and under laws of association which are grounded on scientific or thought-relations, and directed to some definite result or product, we have the *philosophic imagination*.

When the philosophic or the poetic imagination are employed in the service of ethical improvement and religious incitement, they constitute an important element in *ethical ideality* and *religious faith*.

Interest and importance of the representative power.

§ 140. The *interest* and the *importance* of the representative power is enforced by the following considerations:

1. First of all, the exercise of this power ministers pleasure of a high order and in great variety, which is independent of the accidents of fortune and circumstances. Whether these acts are exercised by the infant in its endless combinations of play and sport, as in the simple story which it constructs out of two or three incidents, or whether they are employed by the novelist or poet in the fiction on which he lavishes all the resources of culture, the pleasure of creating is the same.

2. Man often flies to the unreal world of the fancy, to find rest and relief from the highly-wrought excitements of the too earnest and engrossing real world. Ideal objects and conditions furnish associations more pleasing and emotions more satisfying than any which the experience of reality can awaken. The sick man forgets for a brief moment his actual weariness and pain in the scenes of health and action which he imagines. The prisoner is enlarged from his cell. The oppressed forgets his wrong. The homeless dwells under the shelter of his own roof.

3. This power is the necessary condition of the higher functions of the intellect, and of every description of intellectual achievement and progress. The truth is common-place, that memory is the servant of thought and the conservator of our acquisitions. It was not in idle fancy that *Mnemosyne* was called by the ancients the mother of the Muses. Were the mind limited to the objects and the activities of the present, it could make little

progress of any kind. Thought would be almost impossible. Generalization, by which many objects are viewed as one, would be restricted to the few present objects that could be brought within the range of a single act of comparison. When such an act was finished, its product would be lost forever. It could never be reapplied to a new object, or be enlarged in its sphere. The new individual objects of sense and of consciousness would also be isolated. They could not even be named, for each would stand apart in the loneliness of its own individuality. Language would be impossible.

The induction of principles and of laws would be excluded, for, however surely the mind might infer that a common law controlled the objects perceived at a single gaze, neither the objects nor the principles learned through them, could present themselves a second time, the one to be exemplified or the other to be explained. There could be neither invention nor discovery. Even in mathematical science both would be impossible. The creations of art would be excluded. The inventor in mechanics, the composer in poetry or music, the thinker in morals, philosophy, and letters, the deviser of beneficent schemes for human well-being, are each and all dependent on the resources of the imagination for every possible conjunction of cause and effect, of tendency and result. No more manifest or more serious error can be committed, than for the philosopher to decry the imagination as injurious to, or inconsistent with, eminent scientific activity and achievement.

The practical uses of the imagination are not to be overlooked. It creates ideals of what we might be and do, which are far higher and nobler than any thing which we are, or which we perform. It lifts us above ourselves and the examples we observe in real life, furnishing loftier standards toward which we may aspire. A pure and elevated imagination is in many ways allied to a noble ethical nature, and favors an ardent and a sustained religious faith.

CHAPTER II.

THE REPRESENTATIVE OBJECT—ITS NATURE AND IMPORTANCE.

§ 141. The product of the representative power, or the object which the mind creates and apprehends in memory and imagination, has been the occasion of much confusion of thought, and not a little controversy. Scarcely any single topic has been more vexed in ancient or mediæval philosophy, than the nature of representative images. In the discussion of this topic, three topics or heads of inquiry present themselves: I. *The nature and mode of existence of the object which the mind remembers and imagines.* II. *Its relation to the original from which it is derived and to which it is referred.* III. *The special service which it renders in thought and action.*

Why the *object* of representation needs special discussion.

I. *The nature and mode of existence of the representative object.*

§ 142. These objects or products, as has already been stated (§ 136), are psychical existences. They exist in and for the soul only. They are at once the products of the mind which brings them into being, and objects for the same mind to cognize or contemplate. Whether they are transcribed from real beings and real acts, or whether they are created out of the materials or upon suggestions which real objects furnish, they are in all cases purely psychical and spiritual. It makes no difference whether the original is material, or spiritual; the idea or image of each is simply psychical.

It is a psychical object.

§ 143. The mental object is as transient and evanescent as the act by which it is brought into being. In this respect the mental object is strikingly contrasted with objects that are real. The acts by which we know both psychical and actual objects, are for a moment. They cease to *be* at the instant in which they begin. So is it with the psychical as contrasted with the real object. The real object alone is fixed and permanent. To it we can come and from it we can go, and find it still the same. But the psychical transcript or creation is as short-lived and evanescent as the act by which we behold it.

It is a transient and short-lived object.

These psychical objects of the representative power are to be distinguished from those *spectra* or *hallucinations* which result from an abnormal or morbid condition of the sensorium or the nervous organism. The first are psychical, the second are psycho-physical. The first are spiritual in their nature, the second are dependent upon the soul as connected with the sensorium. The hallucinations or spectra, are intimately related to those subjective sensations, which, as we have seen, are caused by any excitement of the sensorium by means of subjective agencies as distinguished from material objects (cf. § 78). They are not properly representative images or ideas, which are purely psychical creations and objects, being created by a psychical power under psychical conditions, and having only a psychical existence.

It is an intellectual object.

§ 144. These representative objects are not only psychical, but they are *intellectual* objects. It has been held by some that when memory recalls past psychical experiences of feeling and of will it recalls the experiences themselves, and not our ideas of them. "It is not ideas, notions, cognitions only, but feelings and conations, which are held fast, and which can, therefore, be again awakened." "Memory does not belong alone to the cognitive faculties, but the law extends in like manner over all the three primary classes of the mental phenomena." (*Ham. Met., Lec.* xxx). This opinion of H. Schmid is apparently sanctioned by Hamilton. It is a logical inference from one of the doctrines which he seems to advance concerning consciousness. But if consciousness is an act of knowledge, and knowledge, when matured, gives, as its products, intellectual objects which we can recall; then, as when we feel we know that we feel, so, when we remember that we have felt, we remember our past feeling as an object known—*i. e.*, we recall our idea of it (§ 56). The pleasure which I enjoy is not the original pleasure revived, but a fresh pleasure from the object recalled by the intellect, and perhaps a reflex pleasure from the fact that it is revived. But whatever it be which excites the pleasure, whether the exciting object or the pleasure excited, it is the object, or the pleasure *as remembered*—that is, it is an intellectual object which it apprehended by the mind. The representative object is not only a *psychical*, but it is also an *intellectual* object.

II. *The relation of the representative idea to its original.*

§ 145. The relation which the represented object holds to the real or presented object, is *sui generis*, and can neither be resolved into, nor explained by any other. It is important to distinguish it from those relations with which it is so often confounded, and thus to clear away many errors into which philosophy has often been betrayed.

The relation can be compared to no other.

§ 146. In doing so, we observe: (1.) That the ideas which we acquire by consciousness or perception cannot possibly *resemble* their *originals*, either as parts to parts or as wholes to wholes. Neither the single features nor the combined wholes of any mental transcripts can by any possibility resemble the single features or united wholes of any material or spiritual being or act. A mental object is wholly incapable of being confronted or compared with an existing reality. One material thing can be like another material thing as a whole and as a part; one spiritual being, or a single spiritual act, can be like another spiritual being or act; one tree can be like another tree; one mental state can be like another; one act of perception can be like another act; but the mental image of a tree cannot be like a tree, nor can the mental remembrance of a mental experience resemble or be like the original act or state.

Representative ideas of objects of consciousness and sense-perception do not resemble them. Memory.

It is true, one of these may be loosely and vaguely said to resemble or be like the other; but that this language is only employed in the way of analogy, is evident from the contradictions and absurdities into which those philosophers have involved themselves who have understood it literally.

We have seen (§ 129) to what contradictory and impossible conclusions Locke's definitions of knowledge, as the discernment of a conformity or resemblance of ideas with their objects, exposed himself, and actually conducted Berkeley and Hume. The representative idea is not known to consciousness as resembling any original.

We observe still further: (2.) When we *remember* or *recognize* objects which we have previously known, we do not discern any proper resemblance between the original and its mental transcript. For example, we look upon an object, as a house, a tree, a portrait, the page of a book; or we hear a sound, we perform some mental act, or experience some feeling; and when the object is removed, we recall it in our memory. It were simply absurd to say that we recall the material object by its mental object, or that we remember the object by its likeness to the mental picture which we revive to our minds. A discerned resemblance supposes two objects between which the likeness is seen; but in an act of simple memory it is plain that only one object is before the mind. It is therefore clearly impossible that any resemblance should be discerned; for that two objects would be necessarily required. In recalling or remembering a past object, event, or mental experience, we simply picture it as having been before discerned or experienced in fact, and we do this by a direct act of knowledge.

When it is said that this mental image is transcribed from the original, or represents it, the language describes an act and objects which are emphatically *sui generis*, and incomparable with any other.

The relation of these mental transcripts to their originals can only be understood by considering the acts of the mind by which we acquire and recall them. The nature of mental products can only be understood by the mental acts which give them birth. To understand the relation of a transcript to its original, we must consider the nature of the act by which we acquire it, as related to the act by which we recall and revive the same.

To bring these acts together, in order to compare them, let them be employed alternately upon the same object. As the eye opens and shuts upon the landscape seen and the landscape imaged, the real landscape is alternately remembered and perceived. When the eye is shut, it is remembered as *having been seen*. When it is recognized, it is recognized as *the same that we saw before*, and which we had remembered during the interval; but in neither case is any resemblance discerned. It is involved in the act of memory, that the object perceived should be recreated by the mind and recalled as real, and also that, when the object is remembered, it should be recognized as the same which was perceived. Moreover, there is also involved the knowledge that the object as perceived was real, and that the object as reproduced in memory is mental only.

§ 147. The nature of any product or object is determined by the mind's capacity to originate it; and the authority of the mind to trust it and accept the objects which its own activities involve, is to be found in the fact that it finds itself, so to speak, spontaneously exercising the power. Concerning this peculiar object and its relation to its original, we affirm *positively*:—(1.) The mental picture affects the sensibilities less powerfully than the perception or experience of the reality. By the supposition, if the original be a sense or material object, it must move or excite the senses; and this class of experiences are in their essential nature absorbing and vivid. If the experience be of a mental act or state, no recollection or transcript can match the reality in its power to interest and excite the soul.

Positive characteristics of mental pictures.

Different persons differ greatly in the power vividly to reproduce and make real the past, and as greatly in the capacity to be moved by it in their sensibilities. Some persons cannot revive a scene of pleasure or pain without ecstasy or horror; the very picture or remembrance of any thing which they have enjoyed or suffered seems to revive much of the delight or pain which the original experience occasioned. But even the sensibility of such persons to the pictures which their memory revives, is usually in direct ratio to their susceptibility to the present and the real. That the real object excites more feeling than the same object remembered, is assented to by common experience and confirmed by universal testimony.

> *Segnius irritant animos demissa per aurem*
> *Quam quæ sunt oculis subjecta fidelibus, et quæ*
> *Ipse sibi tradit spectator.*—HOR. *De. Art. Poet.*

> O, who can hold a fire in his hand,
> By thinking on the frosty Caucasus?
> Or cloy the hungry edge of appetite,
> By bare imagination of a feast?—SHAKSPEARE, *Rich. II.*

(2.) The mental picture consists of fewer elements than the original. It is but a scanty outline, as contrasted with its fullness—a skeleton as compared with its roundness and life. We look at a real tree, and in the background there is the confused or vague perception of the as yet undistinguished mass of form and color, while from it is projected in bold relief a few prominent parts that attract and hold the attention. If we test by the reality the best picture that we can frame in the fancy, we are surprised at the poverty of the one and the richness of the other.

(3.) The mental picture is recalled in parts under the laws by which one suggests another, and is constructed with comparative slowness. The reality displays its wealth of detail as coëxistent and at a single view. Or, if we study its details with attentive analysis, we do this with inconceivable rapidity, under the guidance and suggestion of the object itself. The object, when re-created in memory, is re-created in the several parts of which it is composed: if a material object, in the several sense-percepts which make it a thing or whole. If it is extended in space, or manifold or irregular in outline, the parts of the surface and outline must be recovered one by one, under the laws of association, and by acts that are successive to one another in time.

To illustrate these contrasted features, we need select but a single example. It is a precipice up which we gaze. First it impresses us as a whole, diversified by its varied features. Foremost are the broad faces of perpendicular or impending rock. These are buttressed by slopes strewn with accumulated fragments. Here and there are bushy crags and scattered boulders. The whole cuts against the sky with a notched outline, fringed here and there with nodding herbage, or broken by some daring tree, that, stayed upon its uncertain footing, reaches out and up toward heaven. If all this is apprehended by sense-perception, the quick eye first surveys the whole with a rapid sweep, then runs hither and thither, as it is caught and led by some salient feature, the rock itself bringing out new material faster than the mind can appropriate it, impressing the feelings with new emotions of wonder the longer we strive to master its wealth.

Let us seek to image that rock in the mind, at evening, when we are just returned from a fresh gaze upon its front. In place of the exhaustless confusion of the vaguely-seen whole to guide and excite the eye, there is slowly revived the scanty framework of the few parts which can be recalled by the mind. These parts are recovered one by one, as the mind resting upon what is already present brings back in fragments, and by repeated efforts, that which each present object suggests. However exciting the effort to recall and reconstruct, and however pleasing the picture that is recalled, the impressiveness and exciting power of the reality are wholly wanting.

The objects which the creative fancy or imagination in any way combines or constructs do not differ greatly from those which the memory transcribes, in their relation to the real existences of matter or spirit. The only material difference between the two can be expressed in a word—the one represents real, the other possible existences: the originals of the one in fact exist, and have in fact been perceived or experienced; realities corresponding to the other might exist. In every other respect the two classes of objects coincide.

III. *The usefulness of ideas in thought and action.*

In thought, we prefer ideas to realities.

§ 148. The special service of the products of the representative power for thought and action remain to be considered. It has already been observed (§§ 46, 56), that the process of perception, or consciousness, is normal and complete when it results in an idea or image—*i. e.*, when a transcript of the individual object is prepared for future recall. The usefulness of these acquired facts and of these ideas of possibilities of nature will be accepted by every one. That they are absolutely indispensable to secure the past, and to give range and reach to invention, is obvious to every mind. But it is not clearly, certainly it is not generally acknowledged, that, for the purposes of thought, remembrances are often better than percepts, and that the pale and scanty images which the mind creates are often superior to the fresh experiences which life presents. We often even prefer to employ mental images, when we might avail ourselves of actual observations. We often turn a fact into a mental picture or recollection, even while our eyes, our ears, and our attent consciousness seem to be occupied with a present reality.

The reason is, that the image, (supposed to be correct) presents to the mind fewer elements than the reality, and therefore does not distract, but aids the attention in the activities of thought. Moreover, the elements which it includes are usually the very elements or features with which thought concerns itself. For this reason recollection often guides thinking, and aids it in its work.

When we change our perceptions into *ideas*, or *ideate* our perceptions, we retain only what we attend to; hence the image presents *fewer points* or elements than the original. We are likely to attend to what is most important, especially if we bring to our observations an eye instructed by the previous training of thought, or the experiences of scientific inquiry. A disciplined mind will of necessity direct the observations of things to those features with which thought is concerned; and these points will remain recorded in the memory for thought to classify, or be recombined in the imagination for thought to invent and to explain.

In a certain sense, representation *abstracts* while it revives; as it omits much of what it perceives or feels, and retains only what it cares for.

Hence, in observations of things which are accompanied with any comparative analysis of judgment, we close and open the senses by alternate acts. We close the sense, that we may with undistracted thought think or judge of the image which it gives. We open and use it again, that we may correct or fix the image by or upon which we think.

Ideas especially useful in comparison and generalization.

§ 149. As the mind widens its range of materials for thought, and rises to higher generalizations, its images of things will need to consist of still fewer features—viz., those only which it needs to use in classification or reasoning. So far as it brings before its view concrete realities or individual examples, these need only contain those parts or elements which come into use in generalization, induction, or argument. The plastic power of representation here comes into play, which can readily omit all that is not necessary to be considered and can easily supply every thing that illustration or discovery may need.

Representation can go so far in its abstractions as to leave but a meagre outline, a mere skeleton of a concrete thing or group of objects. Such a skeleton has been called *a schema*. Such a *schema* or outline-image has been held not only to be the necessary condition for the formation and use of concepts, but it has been also contended that it is like the concept in being general and equally applicable to every individual thing to which the concept can be referred. For example, when we speak or think of such terms or things as *horse*, *dog*, or *flower*, it is urged that the mind frames a *schema*, or outline-image of the form or other relations of each subject, which is equally suitable to every individual horse, dog, or flower. This *schema*, it is urged, differs from the concept, in that it is not divided or severed into constituent elements, each one of which is regarded as an attribute of a substance, but it remains as an extremely abstracted whole, which may be applied to every individual horse, dog, or flower. This view contradicts the doctrine which we have laid down, that the object in representation is always individual, and never general. The image of a horse or dog need not be general because it is very scanty or meagre in its constituent elements, having to do only with a few that are characteristic, as the form, the head, the limbs, etc.; but so far as the object is imaged at all it must be individual. The reason why it seems to be general is, that being a creation of the imagination, it can readily be changed by addition or omission, so as to conform to the horse or dog before us. It is more exact to say that the *schema* is *conformable* rather than general; *i. e.*, it is capable of being readily adjusted to every object of its class, and hence its preëminent utility. Whatever form or features the individual image may take which we happen to construct, it can be easily shaped and adjusted to the individual example before us.

The nature of the outline image, or *schema*, and its relation to the concept, will be still further considered under the concept. (§ 107.)

We observe, however, in passing, that it is more than a mere conceit to say, that, as we rise from perception to thought, we interpose the image or idea as an intermediate object which is less gross and entangling than matter, and yet more substantial, definite, and concrete than thought. *The image* directs and aids the concept, standing, as it does, midway between it and the percept. On the other hand, the idea, especially when directed by thought, reacts upon perception itself, making it more intelligent and productive, as it directs the senses to what features it should attend, and often anticipates what they will find. In this way aimless efforts are spared, fruitless voyages of discovery are avoided, and the energies of the mind are expended upon productive objects.

Images prepare for and aid to action.

§ 150. Not only do images assist in perception and thought, but they prepare for and so prompt to action. If we recall an object which formerly moved us to excited feeling and impelled us to prompt and energetic action, the thought of the same object is fitted to excite us again in a similar manner, in real or mimic activity, in body and in soul. If an action is soon to be performed—if we are to sling a stone, or point a rifle, or throw a quoit, the image of the act and object held before the mind brings all the muscles into position, and makes ready for the act required, the instant the act is called for. Hence, in any discipline for feats of bodily dexterity, a vivid and concentrated fancy, a strong and kindling imagination, are of essential service, as they bring the powers into that position which effective activity requires. The same is true of discipline to mental exertion, so far as any purely spiritual activity depends on the distinct conception of its object. The thought of an enemy to be assailed, or of a wrong to be avenged, knits the muscles, braces the limbs, and convulses the features. The sensitive idealist is convulsed with horror at the pictures which his imagination draws of the scenes of cruelty which he reads of or conceives. He acts over, in fancy, the part which he himself would be ready to take in any depicted scene.

When men are to act in concert; as to row, or pull, or shout, in unison, or to repel an assault, or to storm a battery, or in any way to use their united strength, their imagination must be brought into active service in anticipating beforehand the objects which will soon present themselves, or the kind of activities in which they are to engage. The ideal is far better than the real scene for the purposes of discipline and anticipation. The real object may distract and bewilder as well as arouse and hold the attention. It may over-excite, and so unman. It may bring up unexpected objects, as well as those which are looked and hoped for. The reality, as compared with the idea, may hinder action, as it hinders thought. While, then, the idea cannot take the place of the reality, and discipline by means of the idea is of little avail unless it actually prepares for action, it is essential to such preparation. Nature has provided for this discipline by the strong impulse which she awakens toward it: she secures great deeds by first awakening grand pictures in the excited fancy.

CHAPTER III.

THE CONDITIONS AND LAWS OF REPRESENTATION—THE ASSOCIATION OF IDEAS.

Association of ideas. Importance and interest of the subject.

§ 151. We have noticed already that the soul, in representation, as in all its acts or functions, is limited to fixed conditions, and acts according to established laws. What is recalled at any moment, though recalled by the soul's proper activity, is always recalled by means of the cognitions and feelings which the soul possessed the moment previous. The general fact or truth that ideas are represented by means of ideas now present, is usually designated under the general title or phrase; "*the association of ideas.*"

The term *suggestion* has, by some writers, been preferred to *association.* They prefer to say, one idea suggests another idea, rather than, one idea is associated with another. This preference is partly a matter of taste in words, and in part is grounded on the philosophical theory which one of these terms is supposed to designate better than the other. Some object to the phrase, The suggestion or association of *ideas,* because ideas are not the only objects or elements that are concerned; real or existing objects or phenomena being as truly capable of exciting representations as the ideas or remembrances of things. But, the phrase is too well established in general use to be easily set aside, even though the reasons for so doing were vastly stronger than they are found to be in fact.

To seek to determine what are the conditions and laws of representation, is to propose an inquiry to which we are impelled by the intrinsic interest and even mystery with which the power itself and its actings are invested to all thoughtful minds. Hamilton observes (*Met., Lec.* xxxi.), that "the scholastic psychologists seem to have regarded the succession in the train of thought, or, as they called it, the excitation of *the species,* with peculiar wonder, as one of the most inscrutable mysteries of Nature." "The younger Scaliger says: 'My father declared that of the causes of three things in particular he was wholly ignorant—of the intervals of fevers, of the ebb and flow of the sea, and of reminiscence.'" "The excitation of species is declared by Poncius 'to be one of the most difficult secrets of Nature (*ex difficilioribus naturæ arcanis*);' and Oviedo, a Jesuit schoolman, says, 'Therein lies the very greatest mystery of all philosophy (*maximum totius philosophiæ sacramentum*).'" This impression of mystery and the wonder which it excites are not at all surprising. Thoughts and images come and go with the apparent caprice and lawlessness of wizards and fairies—now obtruding themselves when they are not wanted, and then hiding themselves most provokingly, notwithstanding the most earnest desires and the loudest calls for their return. To explain these phenomena by certain definite principles is an essential prerequisite to an enlightened theory of each of the special forms of this power, as the memory, the fancy, and the imagination, in all their varieties. All these so-called powers of the soul are, as has been explained, but special forms of the general power mentally to represent the actual past, and they must all depend upon common conditions, and obey common laws. A just and well founded theory of the association of ideas lies at the foundation of any satisfactory theory of all these several powers. Representations are also always employed in the actings of the other leading powers, viz., sense-perception and thought; and for this reason the consideration of the laws which regulate their presence or absence is essential to a complete elucidation of the powers with which, at first, they seem to have little concern. On the other hand, when the movements of representation are explained, this explanation is taken to explain almost every thing beside; so largely do the coming and going of represented objects enter into the other phenomena of the soul. A very considerable number of psychologists, as we have already remarked, have accordingly resolved all the psychical powers into the operation of the laws of association—viz., reasoning, induction, the belief in causality and adaptation, and even in time and space. The association of ideas has played a most conspicuous *rôle* in the modern theories of the soul and its operations, and its influence upon such theories was perhaps never so great as at present. Next to false or inadequate theories of sense-perception, have incorrect theories of the association of ideas exercised the most mischievous influence upon the scientific views of the soul, and indirectly on philosophical, ethical, and theological theories (cf. § 40).

§ 152. To form a correct theory, it is necessary, as in similar cases, to state at some length the defective or erroneous theories which have been accepted to explain these operations and laws. This will enable us to pronounce a critical judgment upon their error, as well as to recognize the truth which they include, and will prepare us to develop a theory that is true and satisfactory.

It will be observed, that the laws of association pertain to what Hamilton calls the reproductive, as distinguished from the representative power; in other words, to those operations of the soul which prepare objects for the soul's apprehension, as distinguished from the soul's acts in cognizing them when prepared and presented (§ 43). In representation in all its forms, these functions must necessarily be very prominent and important. In representation, the soul prepares and furnishes its own objects of cognition. The capacity to do this, and the laws under which the operation is performed, are analogous to the psycho-physiological capacities and acts of the soul by which sense-objects are prepared for the soul's sense-perceptions. (§ 135.)

The laws of association have been divided into two leading classes, the *primary* and *secondary*, which otherwise may be denominated *general* and *special*. They are distinguished thus: *the primary or general are those which act or tend to act at all times and in all persons*, while *the secondary and special are those which determine the associations of different persons or of the same persons at different times.*

The theories which we shall notice apply to both these classes, though more eminently to the primary. We begin with

I. *The primary laws of association.*

§ 153. We observe, (1.) that the theory is untenable which explains the phenomena of associations by the mechanical or physiological laws of a *bodily organ* which is assumed to be the instrument of the soul in representation.

Association not explained by bodily organization.

It has been held by not a few writers, among whom Bonnet was conspicuous, that the brain, or nervous system, is such an organ. As what we know in sense-perception was thought to be or depend upon certain vibrations, undulations, or oscillations of the brain and nerves, so it was held that the objects thus apprehended for the first time can be *re*-presented to the imagination or the memory, whenever these same oscillations or vibrations are resumed or repeated. Others maintained that every act of perception results in a permanent condition or disposition of certain of these fibres, which is active again in representation. Some held that, in addition to the oscillating fibres of the brain, there is also present a very delicate and sensitive fluid, intermediate between the brain and the soul. Those who held that the soul is immaterial, insisted that the brain and nervous system are its organs in representation, on the action of which the mind as completely depends for its images and remembrances in representation, as it does on the organs of sense for its objects in perception. Still greater plausibility was sought for this theory by the attempt to show that the soul itself has a special seat or organ in the brain, by the sympathy of which with the vibrations of the remaining portions all its phenomena can be explained. In view of the theory that the senses and the imagination were thus dependent upon the sensorium, *i. e.*, the brain and nervous system, etc., these powers were formerly ascribed to the lower or inferior energy, which was called the animal soul, or the soul in contrast with the spirit or higher and rational soul, to which the nobler and more spiritual functions were allotted. In modern times, since the various sensible qualities have been resolved into modes of motion, and many physiologists and some psychologists have resolved the capacities of the sensorium for different sensations into simple susceptibilities for more rapid vibrations, there has been a renewed disposition to make the representative power to depend on revived vibrations of the nervous energy. Such theories have, however, been usually carried out to the bald materialism with which they have a strong affinity.

We have already explained sufficiently how earnestly the cerebralists and associationalists of recent times reassert the same views, and seek to enforce them by the aid of the results of modern physiology. (§ 40.)

All these theories fail to be supported, by reason of a common defect. The structure of the brain and nervous system in no way indicates that they are capable of the vibrations or oscillations which are postulated of them. This structure is not entirely fibrous. What seem to be fibres, are not capable of the tension and relaxation which vibrations, whether rapid and forcible, or slow and feeble, would require. They are not sufficiently numerous to answer to the myriads of millions of states of thought and feeling which are represented in memory and the fancy. Not a single change of the kind alleged has ever been known to occur in connection with a represented object. We call the eye and the ear organs of sight and hearing, because, with the observed conditions and the varying states of these organs, sensations are present or absent, or vary in quality and force; but never has a nerve-movement been observed, or even conjectured, to which might be referred the remembered face of an absent friend, or the vivid picture of a once-visited scene. Nor has any vibration of fibres or nerves ever been known to be connected with any picture or remembrance whatever. No nerve-cell has been known to be formed in connection with a picture fixed in the memory, or a purpose decisively taken. Again, the theory, if satisfactory in every other particular, would fail entirely to account for the creative energy of the imagination. Representations of this sort are very abundant, and often very vivid and forcible; but how the most of these fantastic and gorgeous scenes could be provided for by any disposition of fibres or vibration of nerves, it is impossible to see.

What makes the theory plausible is the fact that certain conditions of the body are connected with a special activity of the representative power. In some of these states this activity is excessive, irregular, and even uncontrollable. When the body is in health and in a normal condition, *memory* both acquires and gives up its treasures with the ease and exactness of instinct; and *imagination* combines and creates, as if by the spell of an enchanter, so skilfully as to be herself surprised at her own work. Under the excitement of delirium, the elevation of enthusiasm, or the brief madness of passion, the power to recall and create seems almost to be used by another self; now mocking the vain efforts of the man to control the rush of his too affluent fancy, and now suggesting for his service or his delight unexpected stores of facts and fancies. It is vain, at times, that the soul essays to retard or to still the throng of unwelcome images that break in upon it like a succession of stormy waves. In sleeplessness induced by an elation of the nervous system, the rational soul seems to be separated from the imagination, and to become the passive spectator of its wayward caprices. We are wearied to exhaustion by the force and persistence with which these fancies at once bewilder and overmaster us. In delirium, the fancy seems to have completely overmastered the intelligence, paralyzed its functions, or frightened it from asserting its rightful supremacy.

These phenomena can be accounted for by two considerations: First, there is the general truth, that the soul is dependent for the measure of force which it has at command, on the force and normal activity of the powers which maintain the corporeal life. When the bodily force is weakened, the force of the mind is often weakened in every one of its functions—in sense, representation and thought.

Second, a disturbance of the functions and activities of the body is attended with an unequal action of the powers of the soul. This can in part be accounted for by the obtrusive influence of the sensations and other mental experiences which are the consequence of irregular bodily action. The soul seems to have at its command only a certain *quantum* of psychical energy, which may be evenly distributed among the various activities of which it is capable—as sense, consciousness, representation, and thought; or, if concentrated into one, it is in so far withdrawn from the rest. It has already been noticed, that we cannot exert the utmost energy in hearing and seeing at the same instant; still less can we employ sense-perception and the reasoning powers at the same moment and with the highest energy and effect. In extreme hunger or active pain, the sensations are so absorbing as to exclude all energetic spiritual activities, whether of thought or feeling. In still other conditions, the generally dormant vital and muscular sensations may be so positively obtrusive as to weaken the soul's capacity to fix the attention upon any other objects with steadiness and effect. And yet these muscular or vital sense-perceptions, though obtrusive and unpleasant as sensations, may be so vague and indefinite as *perceptions*, as to serve chiefly as the suggestors —under the laws of mental association—of other images. We ought never to forget that, in all conditions of our existence, so long as we exist as soul and body, these vague sensations of which the body in all its parts is the occasion, form the constant background on which are projected the more definite and distinctly remembered of our experiences. When these sensations become more than usually active, through an excited or a diseased condition of the body, they can suggest every image with which they have been connected in the past; and thus preoccupy the whole force of the soul's activity. The condition of the body may affect the whole activity of the soul, by simply introducing unusual *psychical* experiences, which operate according to purely psychical laws, both in withdrawing the attention from the rational functions, and in obtruding a throng of associated images. These considerations explain many cases of the singular and almost capricious dependence of the memory upon the varying conditions of the body.

The laws of association cannot be referred to any attractive power in ideas as such.

§ 154. (2.) The phenomena of association cannot be resolved into *any attractive force in the ideas themselves*, by which they suggest or revive one another. This theory differs from the one just discussed, in making the ideas, as psychical agents, to exert a force similar to that which was ascribed to brain cells or brain fibres.

Many of the explanations given of the phenomena of association, represent ideas as attracting one another somewhat as two drops of water, or two globules of quicksilver rush into one; or as if, when the larger drop or globule is divided, the one division draws the other after itself.

Thus Hobbes writes: "All fancies [phantasms] are notions within us, relics of those made in the sense; and those notions that immediately succeeded one another in the sense continue also together after sense; in so much as the former, coming again to take place, and be predominant, the latter followeth, by coherence of the matter moved, in such manner as water upon a plane table is drawn which way any one part of it is guided by the finger." (*Lev.*, p. i. ch. iii.; cf. *Hum. Nat.*, ch. iii., § 2; and *Elem. Phil.*, ch. xxv. Locke says: "Some of our ideas have a natural correspondence and connection one with another: . . . . . Ideas that in themselves are not at all of kin, come to be so united in some men's minds that 'tis very hard to separate them; they always keep in company, and the one no sooner at any time comes into the understanding, but its associate appears with it, and if they are more than two which are thus united, the whole gang always inseparable, show themselves together." (*Essay*, B. ii., c. xxxiii., § 5). Hume says: "These are, therefore, the principles of union or cohesion among our simple ideas, and in the imagination supply the place of that inseparable connection by which they are united in our memory. Here is a kind of *attraction*, which in the mental world will be found to have as extraordinary effects as in the natural, and to show itself in as many and as various forms. Its effects are everywhere conspicuous; but as to its causes, they are mostly unknown, and must be resolved into *original* qualities of human nature, which I pretend not to explain." (*Hum. Nat.*, B. i.,p. i.,Sec. iv.) James Mill (*Analysis of the Human Mind*, chap. iii.) says: "When two or more ideas have been often repeated together, and the association has become very strong, they sometimes spring up in such close combination as not to be distinguishable. Some cases of sensation are analogous. For example: when a wheel, on the seven parts of which the seven prismatic colors are respectively painted, is made to revolve rapidly, it appears not of seven colors, but of one uniform color—white. By the rapidity of the succession the several sensations cease to be distinguishable; they run, as it were, together, and a new sensation, compounded of all the seven, but apparently a single one, is the result. Ideas, also, which have been so often conjoined, that whenever one exists in the mind the others immediately exist along with it, seem to run into one another—to coalesce, as it were, and out of many to form one idea; which idea, however in reality complex, appears to be no less simple than any of those of which it is compounded," etc., etc. This view is accepted by J. Stuart Mill, and the doctrine of "*inseparable associations*," thus enounced, is with him *the* axiom, which is the "*open sesame*" of all metaphysical and psychological problems.

The most consistent and thorough-going advocate of this theory of the attractive force of ideas, as ideas, either in ancient or modern times, is Herbart. All the mental phenomena, and even the several powers of the mind, he accounts for by the actions and reactions of these ideas. Ideas are strengthened when they recur often enough to gather the force which blends them into one or arranges them in a permanent series. After being experienced, they remain in a condition of constant tension, ready on the slightest occasion to rush back into the possession or rather *the presence* of the soul; and again pressing hard to return as soon as a kindred object of perception or representation shall attract them back.

This theory is open to similar objections with the one which follows, with which it is intimately allied. We observe next, that

Nor into the force of relations as such.

§ 155. (3.) The conditions and laws of representation cannot be referred solely, or even primarily, to the force of certain *classes of relations* which exist between ideas. This theory is, in its principle, not superior to that which ascribes attractive force to the ideas themselves.

Aristotle enumerates three of the relations which are said to constitute the laws of representation, viz.: *Contiguity in time and space*, *resemblance*, and *contrariety* (*De Mem. et Rem.*, c. ii., § viii.). Hume asserts the three laws of association to be *resemblance*, *contiguity in time and place*, and *cause and effect*. Others increase this number to seven, viz.: *Coëxistence or consecution in time; contiguity in space; dependence as cause and effect, means and end, whole and part; resemblance or contrast; the being produced by the same power* or *conversant about the same object; signified and signifying; designated by the same sound.* Others, as Hamilton, contract them to two: *Simultaneity* and *affinity.* All these laws are founded in truth. They all describe facts of consciousness, although they fail as we shall see to recognize the ultimate principle or law of the mind's activity, in such cases.

Examples can easily be adduced of the representation of ideas under all of these relations. We begin with those of *place.* When I recall a single building upon a familiar street, I think at once of the building adjoining, and so on, of each that is next.

Contiguity of time is illustrated by the following: When a single event is thought of, which occurred upon some day of my life made memorable by joy or sorrow, that event suggests the others which occurred in connection with itself—either before or after—till the whole history of the day has passed in review before the eye of the mind.

Inasmuch as all objects adjacent in space must, if perceived with attention, be originally perceived by acts successive to one another in time, it may and generally will happen that when they are recalled as contiguous, they may also be recalled as successively perceived; and thus often the relations of *time* and *place* act *conjointly.* Thus, if I examine the interior of a large public hall or church, I may walk around it on my feet, drawing near to every part which I inspect; or, standing in one place, I may survey every object by successive applications of the eye, fixing the objects in memory by the relations of time. But these objects are also contiguous in place, and form together a whole of space.

The relations of *similarity* and of *contrast* serve to recall objects. If I see a house like the one in which I lived when a child—it is of no consequence when or where—it causes me to think of my early home. If I see a face that resembles the face of a dear but absent friend, it brings that friend to mind. The likeness may be of the whole to the whole, or of a part to a part; as of a door or roof (the part of a house) to a door or roof; or of a single feature in the face to another feature. So, objects that are unlike, especially such as are strikingly contrasted, recall one another. Cold makes us think of heat, light reminds us of darkness, joy of sorrow and sorrow of joy, sweet of bitter and bitter of sweet.

The relation of *cause* and *effect* is constantly recognized in our experience. The cause may recall the effect, or the effect the cause. Fire makes me think of heat, and ice of cold. The wound under which I suffer, recalls the blow which caused it.

Under cause and effect, and dependent upon it, is the relation of *means* and *ends*. Any instrument or contrivance suggests the use for which it was devised. Thus, a fire-engine makes us think of a conflagration; a locomotive, of the drawing of a railway train; a thumbscrew, or a case of surgical instruments, of torture or amputation. The thought of an end suggests the possible or necessary means. If a weight is to be raised, or a building is to be moved, we think of a lever, or a combination of screws and rollers.

To these relations three others have been added. *Operations or objects of the same power or faculty, suggest one another, and the faculty concerned. The sign suggests the thing signified, and the thing signified the sign. Objects accidentally denoted by the same sound are associated.* A little attention will convince any one that all these may find a place either under the law of cause and effect, or under the very comprehensive relation of contiguity of space and time.

Are not other relations supposable?

The attempt to increase the number of the relations that are conceived to operate as laws of association and conditions of representation, most naturally suggests the inquiry, whether there is any special charm in the three or four relations of resemblance, contrast, contiguity of space and time, and causation, which invests these alone with efficacy in the production of ideas. We ask at once, Why may not any other relations serve as well as these? Why, of the two objects that are connected by any relations whatever, may not each suggest its correlate? We find, in point of fact, that this is so—that objects connected by many special relations, as of *premise* and *conclusion, evidence* and *inference,* do recall each other.

The law of redintegration.

§ 156. (4.) Philosophers have with greater plausibility united all these relations under what they have called the *law of redintegration,* which is thus announced: *Objects that have been previously united as parts of a single mental state, tend to recall or suggest one another.* Redintegration, as here used, is equivalent to the complete restoration of the whole, on condition of the presence of one or more of its parts. This law was announced by St. Augustine, by Wolff, by Malebranche, by J. G. E. Maass, and is accepted with some qualification by Hamilton.

It is an interesting question, whether this law will meet and explain all the special cases of representation. If we concede that the three or four laws or relations enumerated by Hume and others comprehend every supposable instance, and attempt to resolve all these into the law of redintegration, we shall find the following results:

(*a.*) Objects contiguous in time present no difficulty. Indeed, the law of redintegration might be viewed as only another expression for the law that objects conjoined in time tend to restore or suggest one another.

(*b.*) Objects adjacent in space, as has already been observed, usually come under the relation and law of contiguity in time, and are therefore easily accommodated to the law of redintegration.

(*c.*) The most of the cases in which objects are recalled under the relation of cause and effect, will readily be solved by the law of redintegration. For in order to be connected as cause and effect so as to be recalled the one by the other, they must first have been united under this relation in a previous mental act; and if so, they come at once under the law of redintegration.

What is true of causes and effects, is still more obvious of means and ends. The same is true of premises and conclusions, data and inferences, or the so-called logical relations.

(*d.*) The relations of similarity and contrast present some difficulty. When I see a face never seen before, at once the thought flashes upon me, "The face is like the face of a friend long absent or dead;" or when I see a horse which strikingly resembles in color, form, or action, another horse which I formerly owned, and the image of that horse is called to mind, the objects that recall and those which are recalled, were never conjoined in fact. This seems to be insolvable by the law of redintegration.

Maass (*Versuch über die Einbildungskraft*), and others have sought to bring it under the same by the following solution: What we see in the resembling face, or the resembling horse, is some special and separable feature or peculiarity, one or more. Let this be called *a*, and let the remaining features or peculiarities be called *b*. Let all the observed features or characteristics of the same, both the resembling and the non-resembling, be called *A*. Let the face or the horse never seen before be designated by *B*. When *B* is seen, the part *a* is seen as a separable constituent, for by the supposition it attracts special attention. The first act is to perceive B; the next act, to notice *a*, the resembling feature; but *a* has before been conjoined with *b*, giving the total *A*. As soon as the past *a* is apprehended, it brings back its associate *b*, and *A* is therefore recalled. When, for example, I look at a portrait of Sir Philip Sidney, I am reminded of its likeness to the portrait of Queen Elizabeth, because of the ruff which is about the neck of each, which in this case is *the only common feature* and attracts at once the attention. The ruff—which is the same in both—brings back every thing besides in her Majesty's portrait—the head-dress, the features, the sceptre, the robes, etc., etc., till the whole is restored. If this solution is accepted, the law of redintegration is established as the one comprehensive and sufficient law of representation. And this would be, "*objects which have been previously united as parts of a single mental state, tend to recall or suggest one another.*"

The law of redintegration cannot be accepted for the reason that: The part of a mental state which is said to recall or tend to recall the whole, is not literally the same which has previously been an object to the mind. Every time the mind apprehends either a part or the whole, it has a new percept or image, whether partial or total. If, having seen two resembling horses together, I afterward see one, I am impelled at once to think of the other; or if the sight of a third resembling horse makes me think of one or both, there is to the mind in every instance a new object presented and pictured. The percept of the same horse taken in successive moments, or at long intervals, is mentally conceived not as the same, but as a similar mental entity or object. All its force to attract, or suggest, or recall another object, comes not from the sameness of the part or the whole *objectively* viewed, but from the similarity of the two or more mental percepts or mental images regarded *subjectively*, or as the products of the mind's similar activities. Whatever this tendency, or readiness, or force may be, it is derived entirely from the mind's own activity, and not at all from the sameness of *the objects* as parts or wholes. The mind thinks, or tends to think of *a* when it perceives or thinks of *b*, because it has previously acted in a *similar* activity, in whole or in part. When *a* occurs to it, whether in perception or thought, a certain form of partial subjective activity begins, which involves, by reason of the fact that the like activity has been previously experienced, a greater facility of completion.

The law of redintegration, as ordinarily phrased or enounced, is liable to the qualification which was noticed in § 154, viz., that no attractive force can be affirmed or conceived to pertain to ideas as such. Objects or ideas have of themselves no greater force or tendency to restore those which with themselves made up a mental state, than they have to attract one another. The force in the final analysis must come from and reside in the mind whose products they are.

§ 157. (5.) The real principle that explains all the phenomena and laws of association is to be found in the comprehensive general fact or law, *that the mind tends to act again more readily in a manner or form which is similar to any in which it has acted before, in any defined exertion of its energy.*

The real explanation. How enounced.

As the result of our analysis, we accept this as the principle which comprehends the so-called laws of association. We have seen that these laws are not physiological, but psychical; that the attractive force by which one idea is said to be able to recall another, does not lie in the ideas as such, viewed as separate from the mind's energy in producing or beholding them: nor does it lie in the relations as such under which the objects were connected in the mind's previous act of uniting them, nor in the power of a part of the mental state to reproduce its fellow-part or whole, but in the ultimate truth that, in whatever way the mind may act, it thereby is enabled to act in a similar manner a second time. Every original act is always complex, including objects separated and united, as parts and as a whole, by definable relations. If the mind cognizes a part of any of these wholes, it begins to act in a way similar to that in which it has acted before. The tendency to finish the whole of the act thus begun explains the principle that underlies the laws of association.

This comprehensive law enables us to explain not only the recurrence of two objects that have previously been connected in the same instant of time, but the return of those also which have followed one another in a consecutive order; as the words that form a sentence suggest each other, or the names that have been learned in a series, or the letters of the alphabet, etc., etc.

The reference of the laws of the representative power to the subjective force or energy of the mind, explains the influence of states of feeling, as well as acts of the intellect, upon the representative activities. The state of feeling in which I perceive or think of an object—*e g.*, a glorious sunset or an interesting story—is often as distinct to my apprehension as the object itself. It should follow that a similar feeling excited a second time ought as truly to tend to recall a similar object, as a similar object the feeling. That the feelings are potent instruments of memory, is confirmed by the experience of every one. It often happens that a feeling of disgust once occasioned by some object, can never be experienced again without recalling the object itself. This is often observed of the bodily sensations, as those of sea-sickness or headache. It is scarcely less conspicuous in the experience of purely psychical emotions when these are perfectly defined or are traceable to some determinate cause; like home-sickness or sudden fright. In such cases the experience of a feeling which is at all similar to the feeling in question, however dissimilar may be the occasion or exciting cause, will bring back the intellectual cognition with which it was originally connected. We have already explained that in such cases the feeling operates through the agency of the intellect.

This principle also serves to explain the predominance of certain associations over the intellect and character of different persons. If the tendency to reproduction and recall is an original force or law, then it is natural that the energy with which any individual act or state of the soul tends to be revived, should be proportioned to the relative force of the original act; in other words, to the attention which is bestowed upon its objects or parts, whether these are objective or subjective. An excited interest is the condition of concentrated attention; for, as has already been observed, aroused feeling awakens the intellect, detains its gaze, and excludes distracting objects. Hence, the intimate dependence of the memory and imagination of different persons upon the character and strength of the emotions, the buoyancy and depression of their spirits, etc. Hence, preëminently, the influence of those commanding purposes and prevailing habits which make and mark the individual man, upon the objects which he most frequently recalls and recombines, under his prevailing and dominant associations. That every man has his dominant associations is universally observed and confessed. The reason is, with the one person, that the favorite objects of the soul's activity are certain classes of objects with their relations; and with the other, objects that are very unlike them. But in every case, the associations by which each recalls objects, follow the energy with which he cognizes them. One man recalls objects and relations which never occur to another, chiefly because the one contemplates these objects and relations, and with intense energy, while they scarcely catch the notice or attention of the other. Open before two men the same landscape, the same picture, the same architectural design; tell them the same narrative, introduce them to the same companion, let them listen to the same poem, lecture, or sermon, and the active intellect of each will be busy in selecting objects from each, discerning them in special relations and fixing them for future recall.

It also explains why our associations with *objects perceived* are more energetic and permanent than those connected with objects remembered or imagined. That which is seen with the eye or heard with the ear, other things being equal, holds the attention more closely and longer than that which is merely remembered, or painted to the fancy. It is constantly present, firmly fixed, and held closely before the mind for it to return to as often as it will.

Associations with home.

The associations with home are a good illustration of this principle. When we merely *think* of the home of our childhood, it brings back a throng of recollections associated with its places and persons; but when we visit our home, we cannot repress them. They are connected with every apartment; they start up from every corner; they attend upon all our walks; there is not a tree, or rock, or stream, but thrusts into our very faces, and forces upon our attention, its throng of associate memories.

Objects of imagination have this advantage over objects of sense, that they are more free from unwelcome and unpleasant elements, and are subject more entirely to the creative power. But objects of sense stimulate the associative tendency to greater energy, and furnish it with the greatest variety of material.

Our principle also explains why certain conditions of the body affect the power to recall, both favorably and unfavorably. Disease may both hinder and quicken the energies of the soul to acquire, and, of course, to reproduce its acquisitions; for, in all cases, the tendency to reproduce is measured by the energy of the original activity; and this varies, as the body helps or hinders the mind to detain and concentrate its attention (cf. § 153).

The principle which refers the tendency to be reproduced to the original energy of apprehension enables us to understand why the mind represents only a portion, and often but a single element, of an object presented. We perceive a complex material object; we read a written page; we examine a fine drawing, engraving, or painting; we hear and understand an elaborate and convincing argument; we enjoy a succession of pleasurable sensations or emotions. But we bring away, or possess the power to recall, only a few parts or elements of each, but those are invariably the parts or features which we have energetically presented to our cognition. If we revive these speedily, we unite and preserve them by an act of greater energy.

It is essential to an act of knowledge that its objects be discerned in some relation. States of feeling even are moved and excited by the discerned relations of objects, as truly as by the apprehension of their unrelated existence. The relation is often quite as much an occasion of intellectual or emotional activity as the parts related. Sometimes it attracts the exclusive attention, and the entities concerned are set aside and overlooked. I may listen to several similar sounds from different musical instruments, or human voices; the sounds compared may scarcely be noticed, only the circumstance that they are similar. Twenty effects may be produced by a common agent or cause. The individual effects are scarcely observed, for the attention is occupied by the common relation by which they are connected. In hearing a person read, or in reading ourselves, we often do not notice the words; the mind takes up only the relations which constitute their meaning.

These facts explain why the relations of objects, and especially why three or four more prominent relations, figure so conspicuously as laws of association. The relations named are none other, as we shall see, than the comprehensive or general categories which connect and conditionate all our knowledge (§ 515). These relations are the laws of association, inasmuch as they are the instant and universal conditions of original cognition. Whatever we know energetically under these relations, we know a second time under and by means of one or more of these categories.

II. *The secondary laws of association.*

The secondary laws defined.

§ 158. The theories which we have considered thus far chiefly relate to what are called the primary laws of association. Other laws have also been proposed.

which are called *secondary*. The *primary* laws are conceived as explaining the tendency of certain classes of objects to recur to the mind. The *secondary* laws are conceived to regulate the recurrence of one individual object in any of these classes rather than another. They might with propriety be called *laws of the preference or precedence* of particular objects.

The secondary laws have been enumerated and propounded as follows: (1.) *Those objects are more likely to be recalled, other things being equal, which occupy the mind for the longest period of time;* (2.) *those also which are apprehended most vividly;* (3.) *those which are brought most frequently before the mind;* (4.) *those which were most recently present;* (5.) *those which are the most free from entangling relations;* (6.) *those which are contemplated with the greatest strength of emotion;* (7.) *those which are viewed with favoring circumstances of bodily health;* (8.) *those which are coincident with prevalent habits;* (9.) *those to which the original constitution of body or mind furnishes a special aptitude.* (cf. Dr. Thomas Brown, *Lecture* 37.)

How far reducible to the same principle with the primary.

A critical examination of these laws will enable us to reduce them to some general expression. Perhaps it will show that both the secondary and primary rest upon the same general principle. The first, concerning length of time, has already been shown to be necessarily involved in the operation of the general law for which we have contended, that an attentive or energetic apprehension of objects in their relations is a ground of their tendency to be recalled. The second is nearly coincident with the same fundamental principle.

The third presents ground for inquiry. Why does simple repetition give any advantage? We answer: A second look, especially if it follows that which preceded after a considerable interval of time, presents the object as divested of the distracting influences which novelty furnishes. Each new or repeated view, whether near or remote, also reveals some fresh relation either to a familiar or a novel object, and thus increases the chance of its being suggested to the mind a second time. For example, by one act the diamond is apprehended as the brightest, or the hardest, or the most costly of the gems; and so, when the gems are thought of, the diamond is suggested. At another view, its relation to carbon is discerned, and then the diamond will be recalled when charcoal, or marble, or carbonic acid are present to the thoughts.

The fourth law is, that an object contemplated recently, is, if other things are equal, more likely to be recalled than the same object if viewed longer ago. A countenance casually and hastily seen an hour since, may be recollected or recalled by another similar face within this short interval of time, but may be lost forever if the occasion which suggests it does not soon present itself. The fact is unquestioned, and it may perhaps be inexplicable. But obviously, it rather concerns loss or waste of power, than any positive force or tendency. If expressed in

11

the language or terms taken from the general principle which we have laid down as fundamental, it would be thus phrased: "the tendency of any act of the mind to be recalled or repeated is weakened by disuse, till, finally, it wholly ceases." Whether it is properly said to be weakened, or superseded, is an open question. This is true of the kindred question, whether any acquisition of the mind can be irrecoverably lost.

One palpable and prominent exception to this general tendency to weakness or loss may be urged, in the frequent cases of persons who in old age remember nothing so vividly as the scenes and events which occurred longest ago. Often the whole of the intervening life is entirely effaced from the soul, while the memories of youth and childhood are still vivid and distinct. Several reasons may be given for this plain exception to the operation of the laws already considered. Many of the remembrances of childhood have been recalled again and again through a long life. Though the events of childhood, as *realities*, were present to the mind longest ago, yet, as *thought-objects*, they may be the most fresh and recent. Nor should it be forgotten that the objects and events of childhood were contemplated by the mind at first with an almost exclusive and absorbing attention. The memorable occurrences of childhood were the absorbing subjects of thought for days before they occurred. They were reviewed with the fondest reflection after they were past. The learning to count ten or one hundred, the wearing of a certain dress; the beginning of school-life; the long anticipated, the often-reviewed and recited visit to some relative, the first considerable journey, the first party, the first composition—were most important occurrences in their time, and spread themselves along a large portion of the horizon of the infant life.

The fifth law (which relates to entangling relations) has already been provided for. If the points or features to which these relations,—and the thereby related objects,—are attached, are very numerous, the greater is the probability that the objects will be recalled, provided the relations, and the related objects, be discerned with equal energy of attention and ardor of interest. But if the multiplicity of relations divides and thus weakens the interest, the influence of their number is distracting and entangling. In illustration of the operation of this law, Dr. Brown observes: "The song which we have never heard but from one person, can scarcely be heard again by us without recalling that person to our memory; but there is obviously much less chance of this particular suggestion, if we have heard the same air and words frequently sung by others" (*Lecture* 31).

Upon this we remark: If the frequent repetition of the song has the effect to withdraw the attention from the first impression, and to exclude its being often repeated and revived, then it becomes less likely that the person who sung it for the first time will be suggested by the air; but if, every time it is sung by any one, that person is recalled, then the song will be more ineffaceably associated with him the more frequently it is sung.

The *sixth* and *seventh* have already been noticed and explained (§§ 152. 3). The *eighth* needs but a word. So far as facility of association depends on repetition, and so far as particular habits facilitate repetition, so far is this general fact resolved by the law concerning repetition. So far as habit, or easy repetition by habit, enables us to concentrate the attention with greater energy and interest, so far is its power explained by the strength of the single or repeated apprehensions for which habit provides.

The *ninth* law supposes that there are original differences and aptitudes in different individuals for certain classes of associations. This is doubtless true. But it should never be forgotten that these original aptitudes do not pertain to the faculty of representation or the so-called faculty of association *as such*, but that it extends equally to the power of presentation and intuition. Whatever we energetically observe or connect by relations, in original intuition, we revive by association. There is no special aptness for special associations, or for various and ready suggestion, separate from a readiness to discern special classes of objects and relations, and to discern them with interest and energy.

Apparent exceptions to the law of association.

§ 159. There are what seem, on the first aspect, exceptions to the universal application of the law of association. There are many cases when a thought seems all at once to dart into the mind, which has no apparent connection with any thought or thing that is present. We cite the familiar example recorded by Hobbes: "In a company in which the conversation turned upon the late civil war, what could be conceived more impertinent than for a person to ask abruptly, What was the value of a Roman denarius? On a little reflection, however, I was able to trace the train of thought which suggested the question; for the original subject of discourse introduced the history of the king, and of the treachery of those who surrendered his person to his enemies; this again introduced the treachery of Judas Iscariot, and the sum of money which he received for his reward" (*Leviathan*, p. i. c. 3).

This case is no more singular nor striking than the experience of any lively mind could furnish in every half-hour. If any person not absorbed with the objects of sense, or bent upon some present achievement, will break in upon his movements of reverie with the question, How did this or that thought occur to my mind? he will be surprised, and perhaps amused, at the series of strangely connected thoughts which introduced it to his notice. In many cases, the thought, though apparently abrupt and strange, will be found to have a real connection with the thought which it seemed to jostle and displace. There are thoughts, however, the connections of which we cannot follow. What ought we to believe in respect to these? Should we still hold that the laws of association govern their movement, though we cannot trace their presence or furnish the proof of their working?

In answer to this question, two opposite views have been maintained. The first is held by Dugald Stewart and others—that the mind is momentarily conscious of the presence of these intervening objects, though it cannot recall them in memory; that they are present long enough to act as media of association, but not long enough to leave any trace of their presence.

The second theory is urged by Hamilton, following a suggestion of Leibnitz, and agreeing with the school of Herbart. These all contend that, "though these intermediate objects may be present long enough to influence the train of consciously associated thoughts, yet the mind is in no sense aware of their presence; for it is unphilosophical to suppose an object present to consciousness without leaving some impression upon the memory. No analogous cases can be adduced, and the hypothesis must be rejected as groundless." Besides, it is urged, "another principle can be adduced to explain the phenomena—that of latent or unconscious modifications of the mind. In this we have a recognized and actually existing law, which is sufficient to account for all the facts." (*Met.,Lec.* xviii.)

Upon this argument we observe, that it is not true, as is represented, that there are no grounds on which to rest the first hypothesis. In the very case supposed, when one idea suddenly and strangely follows upon another, if we bethink ourselves at once, we can recall some intervening links. We say, if we bethink ourselves at once; for if the effort is made a few instants later, the clue will fall from our hands. At other times, when it seems to have totally escaped and eluded us, it can be recovered by persistent effort and determination. Now, the fact that in some apparently desperate cases we can succeed, demonstrates that the objects might have been—nay, that they actually were, present to the consciousness, though they seemed not to have been. We have a right to infer, then, on grounds of analogy, that they are so in all cases. The analogy of acknowledged and similar phenomena is wholly with the first theory. Moreover, analogy would seem to suggest and confirm the principle, that where there is a feeble activity of consciousness, there is a feeble hold upon the memory; and we conclude conversely, that where there is the slenderest hold upon the memory, there must have been the feeblest possible energy of consciousness. What is intended by the phrase *latent modification of consciousness*, is not altogether clear. If it be explained as only a very low degree of conscious activity, the two theories are in principle the same.

Representation unceasingly active. How it can be interrupted.

§ 160. The representative power tends to *unceasing activity*. The mind, if given up to the operation of the laws of association, would never cease to furnish itself with new objects. Each object last discerned would suggest another. This would call up its fellow, and the series of successive objects would suffer no interruption and would come to no end. It has been said with great effect—that, were the senses excited to action only long enough to furnish the soul with requisite material and fully to develop all its powers, and then to be sealed up forever, the spirit would have acquired material enough for its endless activity in simple representation. (Bishop Butler, *Analogy*, p. i., c. i.) We know from observation, that when the other activities are as nearly suspended as is possible, as in dreaming and reverie, the train of associated objects still rushes past the eye of fancy with a rapidity that cannot be measured. But strong as this activity is, and difficult of control as at times it may be, it does not often assume exclusive or supreme possession. There are two methods by which this activity is interrupted and turned aside. The one is *objective*, the other is *subjective*.

We consider, first, the *objective* interruption. Every new object of sense-perception introduces a foreign and diverting element. In such cases representation gives way to presentation or acquisition. We do not deny that both these activities may be exerted together, and that presentation and representation, may go forward side by side. It would seem from experience that this often happens. In waking gently from sleep, the images of the dream-world blend with the realities of the sense-world. Even in our waking hours, the hard world which the senses give us, is beset by the spirit-world in which we dream. The soberest world of the most prosaic and practical thinker sparkles with the images which the fancy interweaves into its homely fabric. Let this be admitted, and still it is true that the two species of activity cannot occupy the attention at the same moment with equal energy; and that the sense-world and sense-objects will break in upon the activity of the fancy. Let but a single object do this for a single instant, and a starting-point is furnished for a new train of thought in an entirely new direction.

The *subjective* interruption, diversion, and control of the representative activity of the soul, are still more important. The *ego* which at times may seem to be the helpless victim or the amused spectator of this moving diorama, is not always an idle or passive looker-on. It has but to detain any single object, and the object detained suggests new objects, to each of which it sustains many relations. By simply arresting the course of representation, its independent activity is as truly controlled and newly directed as if some object of sense had obtruded itself upon the attention.

But the mind can do that which is far more effective and important than to detain an object before its attention from impulse or passive excitement. It can exert upon every such object its higher activities of thought. If it cognizes the existence of the object, it discerns it as present, and as diverse from itself. It may remember it as having before been present. It may compare it with other objects, bring it into a new or a familiar class, name it, reason about it, make from it some induction, mould it into some imaginative creation, apply it in illustration and analogy, discern in it relations of beauty, learn from it some moral lesson, or find in it some manifestation of the divine. Each one of these activities will evolve a new product, which product may serve as a starting-point for a new series of representations. These activities are far more potent and effective than the merely passive services of the representing power, though they blend with them so intimately as not easily to be distinguished from them. As the mind mingles the thinking power with the activity of perception, when it seems only to see and hear with the organs of sense, so does it elevate and transform its acts of memory and fancy by the penetrating analysis and combining synthesis of rational judgment.

That is a most superficial and limited conception of the representative power and the laws of association, which resolves into them all the nobler and more important operations and products of the human soul. Such a view excludes individuality and self-respect—as well as the capacity for the higher achievements of science, duty and faith. (cf. § 40).

Besides this direct action upon the representative faculty, there is another which is exerted *indirectly, if possible with greater effect.* The action is direct when, in the ways described, the *ego* arrests and modifies the onward current of what would otherwise be passive tendencies. It is indirect so far as, by every such action, a greater facility or force is given to such tendencies for the future. Every present energy of attention, every special effort of creation or thought, gives additional strength to certain bonds of association, and imparts special facility to the mind in reviving their objects. This very circumstance enables us to apply the mind to similar objects with less effort and greater pleasure, till at last the mind has created for itself almost a new medium of life, a second atmosphere for its own easy and familiar action, which is purely the product of its own previous activities. The feelings provide for their own perpetuation and increased force as they direct to this or that intellectual activity. Hence, preëminently, every controlling or commanding purpose, whether morally good or bad, tends to perpetuate itself, and to secure the execution of its own behests. Under the constant presence and guiding control of such a purpose, all the trains of associated objects become its "ready servitors," which bring to mind, when needed, the facts and suggestions, the illustrations and arguments, the happy phrases and expressive words, which are required for thought, expression, and act. Various familiar phenomena illustrate the force of these indirect influences upon the representative faculty. The same material object suggests to different persons associations that are entirely unlike and even opposite to one another. The scene, the house, the apartment, which to one man is full of the deepest interest, is to another indifferent. To one person it recalls suggestions fraught with peace, affection, and joy; to another, memories of hatred, remorse, and terror. To the same man, on different occasions and in different moods, the same object will suggest different associations, according to the feelings of the hour or the purpose for which he is thinking. We may almost say without exaggeration, that in every present activity of the mind there is revived and indirectly made to reappear the whole of the man's previous history, as each of its acts and events have been taken up by the force of the soul's purely passive tendencies, and so incorporated into its very essence.

Law of association and law of habit.

§ 161. *The law of association,* according to the views of its nature and energy which have been enforced, *rests upon the same original principle usually known as the law of habit.* One object suggests another, because one mental state which is similar in part to another tends to be like it in every particular. This principle, when expressed in other language, is equivalent to, Any mental activity or experience, when it is repeated, is more readily performed.

Habit, Lat. *habitus,* Gr. ἕξις, is literally a way of being held, or of holding one's self. Thus defined, it must denote a permanent state of rest which has been reached as the result of action or growth, or a permanent form of activity, or of readiness or facility for any kind of activity. As such facility for action is universally observed to result from repetition of action, this last element is taken up into the conception or definition of habit. The acquisition of facility by repetition, supposes that some difficulty or hindrance has been overcome, whether the habits are purely psychical or corporeal, or whether they are both physical and mental conjoined; whether they are emotional or moral, or whether, as is often true, they are all three together.

Examples of bodily habits are furnished by a particular gait; the dexterous management of the hand in the use of a saw, a chisel, a hatchet, or a plane, in driving or in drawing; and the control of the limbs in dancing or gymnastic feats. The acquisition of such habits does indeed usually involve some psychical activity, and the gain of facility by repetition. But we may consider apart the formation of the body only to a new habitude, and for the moment have to do only with the changes in the states and functions of the body which *our senses observe* to be more and more readily made. We suppose, that at the outset the special use required is difficult, either because some habitual and undesirable adjustment or predisposition of the muscles has been attained, or because they are imperfectly or wrongly adjusted by nature. An effort is required involving physical tension or physical pain; as when we would bring the organs to utter the unused sounds of a strange language, or would bring the fingers or the limbs to painful or constrained positions. We may explain the obstacle or hindrance by a certain power or tendency of the reflex activities of the nervous system. The conquest may consist in the facility which it is possible to acquire, by a gradual assumption in the reflex motors of new forms of muscular adjustment.

We pass next to mental habits—*first*, those which are developed in connection with such bodily adjustments as we have supposed; and *second*, those which concern functions that are simply and purely mental. Side by side with the new adjustments to which the muscles are constrained with a more and more ready obedience, there must proceed a constantly increased facility in the mind's connection and control of the appropriate sensations, according to the ends which it intends to accomplish; *i. e.*, the mind in such cases furnishing the real beginnings of the new adjustments and growths of the body. The juggler and the gymnast, the mechanic and the artist, the dancer and the player on the violin or the organ, do not simply train the bodily organs to the requisite suppleness and aptitudes, but they acquire a surprising readiness of the mind to connect with every movement those sensations which indicate and regulate the activities to which the body is physically trained. If a mental facility supposes a mental difficulty, what is the nature of the difficulty? It is an original difficulty of mental application to certain mental objects, and, consequently in the ready mental combination of the objects which are concerned. This intellectual obstacle is usually increased, and in some cases wholly occasioned, by one that is emotional or moral.

In habits that are purely mental, as in the greater facility that is acquired by study in general; or the surprising progress which may be made in any special science, as the mathematics or the languages; or the still more unlooked-for dexterity which may be gained in certain intellectual feats, as of punning, rhyming, etc., etc., the difficulty lies in a reluctant or unwonted attention, and the dexterity pertains to the subjective tendency toward similar activities which is acquired by exercise. The difficulty and the capacity for facility are both assumed to be unquestioned and original facts.

When the habits are purely emotional or moral, so far as they can be conceived as such, the difficulty to be encountered is a natural or acquired tendency to excessive and abnormal activity in any emotion. This tendency can be overcome only by the frequent exercise of other emotions, till they act with normal readiness and strength. Leaving out of account the voluntary element, or rather supposing that this is rightly adjusted, it may be assumed that this original hindrance to the natural tendencies remains when the new habits are to be acquired. The completion of moral or emotional habits ordinarily involves also the training of the intellectual habits to the ready suggestion of new thoughts and very often of the body itself to readiness in appropriate actions.

Higher and lower laws of association.

§ 162. The laws of association are again divided into *higher* and *lower*. The lower are those which are presented to us in the acquisitions of sense and consciousness, and which are reproduced by the representative imagination and the uncultured memory. These are the relations of *time* and *space*. As they are more obvious and natural, they require little of higher culture or discipline. They are also developed earliest in the order of time, and are common to the whole race. The relations of *likeness and of contrast* form an intermediate class between the natural and the philosophical; being now present in the one, and then largely represented in the other. The higher are the relations of *cause* and *effect;* involving *means* and *end, premise* and *conclusion, datum* and *inference, genus* and *species, law* and *example*—all, in short, of the so-called *philosophical* or *logical* relations. All these are present in and control the higher imagination and the more developed processes of thought. The higher relations of thought and of the creative imagination are so diverse from the lower relations of sense, that they often supersede and displace, and sometimes even cross and contradict them. In sense-perception and consciousness, objects are conjoined, just as they happen to present themselves in space or in time. The mechanical memory or imagination servilely reproduces them under precisely the same relations in which they were originally presented and known. But thought and the higher imagination take the objects thus accidentally conjoined, and recombine and reproduce them under relations that are higher. Whenever objects are habitually conjoined under such relations, they will be persistently associated with and represented by them, so far even as to exclude the combinations presented to sense-perception. By such excess, those striking idiosyncracies of imagination and memory can be accounted for which are designated by the vaguely-used term, *absent-mindedness.* The *absent-minded* person is one who has become so habitually indifferent and inattentive to the objects which address his senses, through preoccupation from a roving imagination or abstracted thought, that his senses seem often to be unused, and his memory to be utterly untrustworthy. He becomes sublimely, or perhaps ridiculously, indifferent to the common relations of common objects and events.

As the higher may take the place of the lower relations, so the lower may exclude or displace the higher. The *constant* or even the frequent conjunction of objects and phenomena may in consequence be mistaken for their *necessary* relations or essential conditions or constituents. A savage, who should see gunpowder exploded by an electric spark, would associate the whole of the electric apparatus, and perhaps even the words and dress of the operator, with the occurrence of the explosion, and take the combination to be made by a necessary connection of things. The ignoramus who sees a conjurer perform certain manipulations, or hears him repeat the words of some incantation in connection with a surprising feat, unites the two by an association so inveterate as to believe the one is the cause of the other. The manifold and inveterate superstitions that have been so readily accepted and so tenaciously retained, are in this way to be explained. Startling or noticeable events have occurred together by a merely casual connection, which have been henceforward associated under the relations of cause and effect; as in the case of success in battle, the healing of disease, the removal of an epidemic, the termination of drought, the cessation of an eclipse, or the acceptable performance of some religious rite.

Nor are errors of this sort confined to uncultured and ignorant races or uneducated men. Men of quick association and ready suggestion, even if they attain the highest culture in many directions, often scorn that discipline to philosophical thinking of which they stand in special need, because, from the very quickness of their power to combine, they are most liable to mistake the suggestions of their various and ready wit for the sober and solid relations of thought.

The lower associations—those of constant or frequent conjunction—are most observable when they strongly affect our feelings. Objects which are in themselves indifferent, or which ought to be and would otherwise be positively offensive, *excite the intensest* liking or misliking, simply because they have been connected with objects which in their essential nature are fitted to please or displease us. The remembrance of a journey, or some other event of our personal history, is always unwelcome, because it was connected in our experience, and is therefore associated in our thoughts, with some serious disappointment or calamity. The sight of the surgeon who saved our life by performing a painful operation, is not always agreeable, however sincere may be our gratitude. Certain persons are very pleasing or very displeasing, because they bring to mind memories or thoughts which we cherish or reject.

A dress of the newest fashion may at first be singular and unattractive. But soon it is generally worn by those who are attractive in appearance, graceful and refined in manners, and high in social position. It is thereupon regarded as highly graceful and agreeable in itself, and no other is tolerable. It is not long before it becomes common, and this detracts somewhat from its factitious attractions. When it is worn obtrusively by the filthy and vulgar, and becomes conspicuous in connection with persons who are rightfully disagreeable, it is time that the fashion should change, or that some other novelty should appear, in order to relieve the associations of the fashionable world from the offensive taint of commonness and vulgarity.

The *moral influence* of accidental associations is still more worthy of attention, for their power for evil as well as their capacity for good. Pleasing manners, high intellectual culture, the attractions of wealth and position, may be and often are connected with libertine principles and easy morals, and thus become powerful aids and instruments of vice and corruption. The drunken revel may, by the force of associations of this kind, not only be endured as less disgusting, but it may be gloried in by the aspirant after high society, as the sign of gentlemanly breeding and fashionable life. The horrors of the first cigar and the first debauch are greatly alleviated by manifold suggestions that the experience of both are necessary to constitute the gentleman.

The force of casual associations is in no particular more conspicuous than in its influence upon *language.* A deed that is abhorrent to the conscience and offensive to the judgment and feelings of right-minded and plain-speaking men, is more than half reconciled to the moral feelings, and perhaps is installed among the virtues, by softening or dignifying the appellations by which it is named—that is, by designating it by words that suggest associations of respectability and honor. Men seek to keep down or to avoid associations of disgust or abhorrence by the device of euphemistic terminology. It is not always true that 'vice loses half its evil by losing all its grossness;' for the very grossness which is its natural manifestation and result, is sometimes the best defence of society against the corruption to which it tends.

The power of epithets and names to awaken pleasant or unpleasant associations is well illustrated in the history of parties and the practice of partisans. A party that is encumbered by an epithet or appellation of odious associations or disagreeable origination, hastens to disencumber itself of an appendage that is more fatal than the shirt of Nessus; while its opponents are as eager and determined that it shall retain the damaging reproach. The skillful application of epithets like *Whig* and *Tory*, *Malignant* and *Roundhead*, *Girondists* and the *Mountain*, *Conservative* and *Radical*, is often more efficient with the populace than the most convincing arguments or the most persuasive eloquence. Agreeable associations, through the subtle reaction of language, have not only palliated—they have even recommended the most contemptible follies, the most outrageous violence, and the most abominable crimes.

Even philosophy herself, though professing to be subject to thought-relations only, is by no means exempt from the influence of casual associations operating through this same medium of words. It is often more effective, even in the schools, to apply an epithet, as *sensuous* or *spiritual*, *empirical* or *rational*, *unselfish* or *utilitarian*, than it is to disprove an analysis or answer an argument: to give an opinion an odious name, or apply a contemptuous epithet, than to consider its evidence or refute its reasons. The soberest and best-governed men are more or less affected by individual associations in their tastes, their preferences, their manners, their reading, their companions, their politics, and their faith. We could not be wholly aloof or exempt from their influence if we would. We would not if we could; for, in so doing, we should forego much of our individuality, and of that which makes our individuality dear. But it is the interest and duty of every man so far to regulate the influence of such associations, that he do not become the easy victim or the abject slave of chance and arbitrary circumstances. Whatever is right and true cannot be disagreeable, when it is sustained, adorned, and hallowed by associations that are only attractive. Indeed, it is not till the reason and conscience rule so completely over the whole man as to transform and elevate even his individual and casual associations, that the education of the man is complete, and his character has attained that harmony and perfection of which it is capable.

CHAPTER IV.

REPRESENTATION.—(1.) THE MEMORY, OR RECOGNIZING FACULTY.

Having considered the conditions and laws of the representative power, we proceed to apply the results of our inquiries to the explanation of the principal modes in which its activity is exerted—to the so-called faculties of *memory*, *phantasy*, and *imagination*. The memory comes first in order.

The elements essential to an act of memory.

§ 163. An act or state of memory has already been defined to be that in which the essential elements of an act of previous cognition are more or less perfectly re-known, with the relations essential to each. These elements are not all recalled with the same distinctness, and hence there are varieties of memory; but it is essential to an act of memory that some portion of each of these elements and relations should be recalled and re-known.

The total complex of object and relations may be recalled more or less perfectly, or each of the constituent elements may be more or less vividly represented.

First: The *object* of the original act may be recalled with a greater or less completeness of its elements or parts, and this whether it be a thought-object, or a sense-object. Completeness or incompleteness in this particular usually attracts the attention, and marks the memory as strong or weak.

Second: The *original act* of the mind in the first apprehension may also be more or less perfectly recalled. I see a face in a crowd. I recall it perfectly, and know that I have seen it before; but I cannot revive a single vestige of myself as viewing it, only that I did thus view it I am certain by direct knowledge. And yet we must have this recollection of our previous psychical activity, or we cannot be said to remember it at all. This certain knowledge may vary from the vaguest possible impressions of our subjective state, to the most vivid and circumstantial review of each one of its constituent elements.

Third: The *time when* the object was previously known may be more or less perfectly recalled. If I remember an object viewed or experienced half an hour ago, I may recall the leading events which have happened to me from the present moment backward to the original act of acquiring this knowledge. If it was yesterday, or a month since, I can generally recall the events that were just before and after it, can connect it with the present by more or fewer intervening occurrences, and can fix the date so far as to know that it was in a certain month of a certain year; the attendants of which dates I can recover with more or less fulness.

In some cases, the event stands isolated in the dim and undetermined past. In others, it may not be wholly isolated from the events which preceded, accompanied, or followed, but yet it can scarcely be said to be united with the present by any connecting series of events that intervene.

Fourth: The *place where*, may be more or less perfectly recalled and recognized. *The place where*, is a phrase which denotes the adjacent and surrounding physical objects in their spatial relations, which form the background and the setting of every object perceived or every act of the person who remembers. Every object previously observed, every act of my own in observing it, when itself recalled, will bring back this accompanying setting more or less perfectly.

Fifth: The knowledge of the *real existence* or of the previous perception of remembered objects may also vary in the degree of accuracy or confidence with which it is held. For this simple knowledge no other explanation can be given, than that the mind is competent to its exercise. The question is sometimes asked, Why do we trust our memory? To this philosophers have sought to give an answer by enumerating certain *grounds* or *criteria*—as that the object must be *clear*, or that the image recalled must *represent* or *agree* with the reality. But all these criteria, or grounds, are merely other words or phrases, which express no more than the act of knowledge itself.

But does the mind always know, *i. e.*, remember, with equal certainty? Does it not sometimes distrust its own act in remembering? We answer: When we distrust our own act of memory, it is we ourselves who are not certain. We seek to be certain; sometimes we succeed, and pass from the condition of painful doubt into that of confident knowledge. The object which was vaguely recalled now stands vividly and distinctly before the eye of the mind. But the clearness and distinctness of the objects are not the real causes which effect, or the logical grounds on which we rest our positive knowledge. The term *distinct* and *distinctly*, objectively describe the subjective certainty, but do not account for or justify it.

"But do we not sometimes offer reasons to satisfy or prove to ourselves that what we remember must have been a fact?" We do often enumerate the circumstances which assure us that we cannot be mistaken, but not as logical *reasons* to justify the conclusion that we are in the right. We bring them up as particulars on which we dwell with attention, for the purpose of recreating a more complete and vivid picture of the past. In this sense we are said to refresh our memory—as a witness in court is asked or urged to do, when one or another circumstance is repeated in his hearing, or he is left to his own associations to revive the past. We may indeed urge this number of remembered particulars as reasons why *others* should trust our accuracy because our own remembrance is so full and detailed, but not as grounds or criteria for our own confidence. For this confidence we can give no other reason than that we find ourselves possessed of and using the power for this very function, which is *to remember*. And yet this act is exercised, as is every other act of the soul, with varying and unequal energy.

Memory technically defined. Relation of memory to representation.

§ 164. An exact and technical description of memory would be the following. Memory is a modification of representation. The representative power furnishes the materials for the memory, according to the laws of association. These objects being furnished, the mind in memory knows them by an act of recognition. More briefly, representation recalls, memory recognizes. The soul, in representation, is passive, blind, and mechanical, proceeding according to fixed and inevitable laws, by methods or processes which occur beyond or out of consciousness. The soul, in memory, on the other hand, is active, intelligent, and rational. The distinction between representation and memory, so far as our actual experience is concerned, is rather ideal than real, for representation passes into memory by an inevitable certainty, through the easiest, the most natural, and usually the most unnoticed transitions.

The psychologists of the associational school provide for only half the process—that of *representation*. The *recognition* they attempt to explain, but unsuccessfully, by the chemistry of association—*i. e.*, by the union or blending of a present with a past mental state. Representation and memory may, however, with propriety and advantage, be ideally considered apart. *Representation conceived apart from memory*, may begin with a *mental image*, and by the laws of its own activity call up another, and still another, till all at once the intelligence asserts, "The object now pictured I have known before as a reality." Or the object may be *material*, and perceived by the senses. In such a case, representation at once supplies a completing image or thought, concerning which memory pronounces, "This real object I have perceived before."

Memory, on the other hand, *as distinguished from representation, is an act of knowledge*. To know, requires objects, and the discernment of their relations. The objects of memory are peculiar, in that, as has just been explained, representation presents or suggests more or fewer of them. The relations under which they are known are, as we have shown at length, those of previous apprehension by myself in some determinate state of knowledge or feeling, at some previous time, and in some particular place.

But while we thus distinguish in an ideal way the passive and the active element in memory, both must be taken into consideration in order to explain its phenomena; for, in these phenomena, each of these elements modifies the other, and both must be present. The two are related in memory somewhat as sensation proper and perception proper are combined in the acts of sense-perception—the one is inversely as the other. In certain acts and powers of memory, the passive or representational element is prominent and conspicuous; in others, the active and rational is most apparent. In the two cases, we distinguish the memory as *spontaneous* and *intentional.* In *spontaneous memory, the object remembered, spontaneously occurs to the mind.* In *intentional memory it is distinctly sought after until it is found.* In spontaneous memory, the representative faculty is prominent, while the intelligence waits only to give its recognition to what is presented to its attention. In intentional memory, the intelligence is active, being aware that some object has been previously known, to recall which, it summons the energies of the representative power.

The distinction of these two kinds of memory is so obvious, and is so readily observed, that separate terms for the two have been employed in common life, and are found in many languages. The Greeks have *μνήμη* and *ἀνάμνησις*; the Latins, *memoria* and *recordatio* (cf. *Cic. de Prov.* 43); the English, *memory* and *recollection.*

The spontaneous memory.

§ 165. In the spontaneous memory, there are natural aptitudes and disabilities, which can only be referred to some original differences of the mental constitution. That such differences exist, is an unquestioned fact. For example: one person hears a series of unconnected names recited, and can repeat them all in the precise order in which they were uttered, while another can recall only now and then one. The eye of another runs down a column of figures, and he can copy the whole from memory, while his companion can scarcely recall a single one of the whole. One individual can learn a page of prose or poetry simply by reading or hearing it read but once, while another can with difficulty repeat correctly a single line or sentence. That these differences are natural, is manifest from this, that they cannot be remedied by any effort or art. No discipline of the attention, and no determination of the will, can enable one who is strikingly deficient, to acquire the power of this simple and effortless memory. That the defect lies in some original incapacity to fix the attention with interest upon the objects to be recalled, and not in the power of representation, is confirmed by observation as well as by the general law of the workings of the representative power. That the strength or weakness of this kind of memory is not owing to the physical strength or weakness of the organs of sense, but to the mental energy and the moral direction with which these physical instruments are applied, is abundantly manifest. Analogous to differences in the spontaneous memory—if, indeed, they are not examples of it—are the varying capacities to recall a musical air so as to repeat it, or to revive the image of the voice or manners of another so as to imitate them.

The relations which are employed in the natural memory are most conspicuously those of simple contiguity and succession. All memory begins with these relations, because our earliest energies and acquisitions commence with objects of this kind. In other words, there is a natural memory of space and of time, or, as we may say in a somewhat narrow sense, there is a natural memory of the eye and of the ear. In some persons the *memory of the eye*, while in others the *memory of the ear*, is conspicuous. Those who are remarkable for the memory of the eye, are such as can readily and vividly picture in the mind the details of the front or façade of a building, the outline and filling in of some remarkable tree, the features of the face of an acquaintance or friend, the page of a book as presented to the eye. Those distinguished for the memory of the ear, can recall successions of sounds—if they have a musical ear, of musical notes—strings of names, or words when connected in significant sentences. They can repeat dates of uninteresting events, and retail long stories whether tedious or amusing. Superiority in the one kind of memory is not necessarily accompanied by superiority in the other.

A good spontaneous memory, or, as it is often called, a good memory for facts and dates, is generally and correctly regarded as a great intellectual convenience, rather than as a decisive indication of intellectual power. It is doubtless true, that many persons are distinguished by natural memory, who are inferior in capacity for discrimination and reasoning. It has become a common observation, Great memory, little common sense. In such cases, the power of discerning the higher relations may be either originally deficient, or it may be neglected in consequence of the predominant use of the power of apprehending, and, of course, of recalling objects in the relations that are most obvious. A very energetic mind may be very limited in its apprehensions, and will, of course, be energetic though limited in its memory. It is noticeable, also, that persons who become eminent in those achievements which are proper to the higher intellectual powers and relations, are in early life usually distinguished for the strength and reach of their memory of both eye and ear.

There are not a few men who carry into the maturity of age, and into the most striking efforts of judgment and reasoning, a memory that is always clear, vivid, prompt, exact, and universal —a memory that never forgets a name, or loses a date, or is at fault in its recital of facts. Such are the men of universal knowledge, at least in some special department of study and research, like Scaliger in ancient learning and criticism; Pascal, "that prodigy of parts;" Niebuhr in history and statistics; A. von Humboldt in physics both celestial and terrestrial; Ritter in geography; Goethe in literature and art. The reason that in these exempt cases the higher or intellectual memory does not displace the lower, is that the employments or studies of the individual require him to be conversant with details as well as with their thought-relations, with facts as well as with principles. Hence, the higher memory aids rather than hinders the lower; the acquisitions of the quick eye and ear being fastened and fixed by the secondary processes of reflex thought.

The intentional memory.

§ 165 *a*. The phenomena of the so-called intentional or voluntary memory next require attention. They are characterized by the single feature, that the objects remembered, are sought for by a conscious effort or act. 'But how can this be possible? The very statement involves a contradiction in language and an impossibility in fact. If the mind seeks, intending to find or recover an object lost, then it already knows what it seeks for. In other words, the mind must already have remembered, in order to be put upon the act of endeavoring to recall.' In reply, we observe that, if every object remembered were in all cases remembered with equal fulness and vividness, then the objection would hold. If, in order to remember at all, the mind must recall with equal energy and success all which, in the nature of the case, is capable of being reproduced, then 'to intend to remember' would be plainly precluded by our 'having already remembered.' But this is by no means true. The object remembered may be considered as an object—whether *object-object* or *subject-object* is immaterial—out of all conscious relation to the mind viewing or caring for it, or as an object in such relation.

Taken in the first sense, the object is capable of being recalled vaguely in its general outlines, and confusedly in its details, or it can stand out before the eye of the mind with the sharpest outline, and inclose a perfect picture of distinct *minutiæ*. But the object of memory is more appropriately the object in some relation to the previous activity of the soul in some given place and at some given time. This more complex object admits also of every variety and degree, from the lowest up to the highest conceivable fulness and freshness. This, of course, provides for the possibility that the mind should, in its acts of recovery, go through all the intermediate steps of effort and intention, till the whole object, as objective and subjective, is fully represented and recognized.

In recovering the whole, we may begin with that which is eminently objective. We may set off with some object which we are sure, in our previous knowledge, had some relation to that which we seek—as the dates of some events that occurred before or after the one which we look for, the names which we have learned in connection with the one required; and we may dwell upon these till the date or name required occurs to the mind, and we recognize it with welcome. Or we may begin with the subjective element. We may recall ourselves in the act of being charged with certain duties or commissions—*where we were, what we were doing, of what we were thinking, how we were feeling*,—till by this means, the missing element reappears to make the recognition complete.

It has already been asserted, that in the *intentional* memory the active element is most prominent. This is true. But it happens, from this very circumstance, that the passive element is thereby brought into more conspicuous and striking contrast. It would seem to delight to tantalize us by the wantonness of its caprices, as now it flashes those very thoughts upon our mental vision which we are most desirous to hide out of sight, and then as provokingly hides those which we are most desirous to uncover. At one time we are disappointed by a strange and unaccountable forgetfulness of the most familiar objects; at another, we are surprised by the appositeness and the affluence of unexpected remembrances.

The sole and single function which the mind, as active, can exert, is to apply the force of its attention to the object or objects which it is certain have reference to that which is sought for. To these only have we access. These only we have at our command. Energetic and prolonged attention is all that the mind can do at the moment of remembering.

Memory as the power to retain, and to lose.

§ 166. Memory is sometimes defined as exclusively the power to retain, or the *conservative faculty*. So Hamilton treats it, and exalts this supposed power into a separate faculty co-ordinate with the power to reproduce and the power to represent. But when we inquire for the definition or statement of the function which this so-called retentive faculty performs, we find that no function of the sort is known to consciousness. Indeed, it is conceded by Hamilton, that whatever is done by this faculty is performed unconsciously.

No one holds that, during the interval, the mind acts upon the object, or with respect to it. It does not exert itself to hold it, or concern itself with it in the least. The expression *to retain* is purely metaphorical, and simply carries the thoughts over the period that intervenes between the moment when it was first apprehended, and the moment when it is known a second time. As Locke pertinently and truly observes, "This laying up of our ideas in the repository of the memory signifies no more than this, that the mind has a power, in many cases, to revive perceptions which it has once had, with this additional perception annexed to them, that it has had them before. And in this sense it is that our ideas are said to be in our memories, when, indeed, they are actually nowhere; but only there is an ability in the mind, when it will, to revive them again," etc. (Essay B. ii., c. x., § 2).

It is only by a metaphor that objects remembered can be spoken of as preserved in some repository or hiding-place, in drawers, pigeon-holes, or other compartments. Nor can the doctrine be maintained, that in the act of original acquisition the fibres of the brain are disposed in a certain position, which they retain, or at least retain the tendency to reassume. Nor can it be proved, as the followers of Herbart contend, that each object as apprehended, or the state of mind as excited to action by the object, is retained ever afterward in a condition of tension, which, on a fit occasion, springs forth into the presence of the conscious spirit. Now, if all these representations are figurative or metaphorical, the power to retain, or the doctrine of a retentive faculty, must be purely figurative also; the fact which it describes being merely that under certain conditions, and in obedience to certain laws, the mind can represent and recognize its previous knowledge. The mind that can do this in regard to the greatest number of objects, after the lapse of the longest time, is said to have the most retentive memory.

Cicero (*De Oratore*, i., 5), Plato and others, have compared the mind in preserving what it had known, to a tablet on which characters were impressed or engraved. Notwithstanding the cautious and accurate definition of Locke which we have cited, we find him, in the same chapter, indulging in such language as this: "The pictures drawn in our minds are laid in fading colors, and, if not sometimes refreshed, vanish and disappear." . . . . "In some, it [the mind] retains the characters drawn on it like marble; in others like freestone; and in others, little better than sand." . . . . "We oftentimes find a disease quite strip the mind of all its ideas, and the flames of a fever in a few days calcine all those images to dust and confusion, which seemed to be as lasting as if graved in marble." Again, the ideas are, "very often roused and tumbled out of their dark cells into open daylight by some turbulent and tempestuous passion." Hamilton justly observes, that, "of all these sensible resemblances, none is so ingenious as that of Gassendi to the folds of a piece of paper or cloth." But Hamilton does not notice wherein the truth and ingenuity of the resemblance mainly lies, viz., the circumstance that the mind, like the cloth, retains nothing but the capacity to assume the same folds and in the same combination and order which they had originally taken.

We observe here, that as the goodness of the memory may respect it as spontaneous or intentional; so we describe it in the one case as *ready,* and in the other as *tenacious.* The one does not exclude the other. If a person is able to recall every object that is required, at once, without effort or delay, his memory is called ready; but it is not necessarily implied thereby that he is deficient in the capacity to retain, but only that he is quick and apt to recall. On the other hand, when one is slow to recall, and yet sure to do so by the application of energetic attention if sufficient time is allowed, his memory is tenacious; by which is intended only that the object is certain to be recovered—not that there is a special capacity to retain, which may be possessed in eminent measure, to which may or may not be added another special capacity to recall.

The power to *retain,* in the sense explained, implies the power *to lose*, in the same sense; the capacity to *remember*, suggests that there is the liability to *forget*. The fact that we do forget, most men will not venture to question or deny. It is not, however, easy to explain why we forget, or to detail the process by which we lose an acquisition beyond recall. In one aspect of the case, it would seem that we ought never to remember—that the mind might be supposed to be limited to the contemplation of the new objects which the presentative power can bring before it. But when we have become acquainted with the possibility and the conditions of representation, it would seem that we ought to forget nothing, but that it must always be within the reach of every related thought to bring back all its correlates. A moment's reflection, however, must convince us that, were it possible for us to recall every object, the recall could never in fact take place simply for want of time. To recall the acquisitions of a few years, would require as long a time as to make the original acquirement, even if to represent were our sole occupation. But it is not solely for lack of time or opportunity that we do not recall. Often, when both are furnished, the related thoughts do not spontaneously present themselves.

The phrase *to forget* is variously employed—sometimes positively, at others comparatively; now absolutely, and then relatively; or, as *Stiedenroth* has it, "Forgetting admits of several degrees, or stadia. The first is a momentary displacement of an object apprehended which is yet certain to spring back as soon as the object displacing it is withdrawn. The second is a comparative withdrawal of the attention, as when we divert our mind from a painful sensation, or, as we say, forget it, in labor or play. The third is when an object will not present itself spontaneously, but we must bethink ourselves in order to recover it. The fourth is when we bethink ourselves in vain. The fifth is when it has vanished for so long a time that we question whether we can by any effort bring it back. The sixth, when we conclude that it is absolutely certain that we shall never recall it again." (*Psychologie*, Berlin, 1824, p. 82).

It is questioned by many whether this absolute forgetfulness is possible—whether, at least, we are authorized to affirm that the soul can lose beyond recovery any thing which it has known. It is certain that knowledge which has remained out of sight for a long period has often been suddenly recovered. Even acquisitions that were the least likely to be remembered, and which, previously, were never known or suspected to have been made, come up as though the soul were inspired to receive strange revelations of its capacities and acquirements.

Numerous examples have occurred within the observation of the curious, and not a few are recorded in history. The well-known and often quoted story, which was originally published by Coleridge in his *Biographia Literaria,* is in substance as follows: A servant-girl in Germany was very ill of nervous fever accompanied with violent delirium. In her excited ravings, she recited long passages from classical and rabbinical writers, which excited the wonder and even terror of all who heard them, the most of whom thought her inspired by a good or evil spirit. Some of the passages which were written down were found to correspond with literal extracts from learned books. When inquiries were made concerning the history of her life, it was found that, several years before, she had lived in the family of an old and learned pastor in the country, who was in the habit of reading aloud favorite passages from the very writers in whose works these extracts were discovered. These sounds, to her unintelligible, were so distinctly impressed upon her memory, that, under the excitement of delirious fever, they were reproduced to her mind and uttered by her tongue.

Rev. Timothy Flint, in his *Recollections,* records of himself, that, when prostrated by malarial fever, he repeated aloud long passages from Virgil and Homer which he had never formally committed to memory, and of which, both before and after his illness, he could repeat scarcely a line.

Dr. Rush, in his *Medical Inquiries,* says that he once attended an Italian, who died in New York of yellow fever, who at first spoke English, at a later period of his illness French, and, when near his end, Italian only. He records also that he was informed by a Lutheran clergyman, that old German immigrants whom he attended in their last illness, often prayed in their native tongue, though some of them, he was certain, had not spoken it for many years.

§ 167. Such facts illustrate the connection of the bodily condition with the phenomena of memory, of which a

Dependence on the body.

partial explanation has already been given (§ 153). They confirm two positions, to which daily experience and observation both testify. The first is, that the extent and reach of our memory is greatly affected by our bodily condition at the time when we acquire. Every object which we apprehend, when in a certain condition of health, we can afterward recall, and this we can do as readily and as easily as we breathe. On other occasions, if we are wearied by labor, exhausted by watching, or prostrated by pain, the book which we read, the conversation in which we take part, the incidents which happen, become almost a blank to us when we seek to recover them.

It is in place here to notice the circumstance, that certain parts of the day, and, with some persons, certain seasons of the year, are most favorable to the successful acquisition of possessions for the memory. In the evening, and especially late at night, the attention may seem to be as intently fixed upon the objects which are to be retained, as in the morning, and the intellectual force may appear to be more energetic. But it not infrequently happens that the acquisitions of the previous evening, which seemed to be so distinct and promised to be so permanent, have well-nigh vanished in the morning, and require to be reviewed to be made useful or sure. It is easy to see how, after the analogies furnished by these phenomena, can be explained the frequently evanescent character of the acquisitions which are made under the influence of wine or opium, as also the fact that the men of the strongest memories have often been either water-drinkers, or men of *strong heads*, not easily disturbed by stimulants.

The second position is, that, whether we can recall what we may be said to have acquired, depends also very largely—at times altogether—upon the bodily condition at the moment of our desire or effort to remember. Under the inspiration of joyous health or the stimulus of exciting disease, all that we have ever experienced, witnessed, or learned, comes back to us as if a good genius were pouring forth at our bidding all that we need or desire to recall. Again, in seasons of extreme weakness, we cannot recover the most familiar names, incidents, or dates, and our most common knowledge refuses to serve us.

It is pertinent here to refer to the many cases of the sudden and almost entire loss of memory, some of which are as striking as those of its development to unwonted energy. A lady of superior endowments and culture was for several days exposed to suffering and fear, in a storm at sea which terminated in the wreck of the vessel. A severe and protracted illness was the consequence, from which she slowly recovered. After her apparent restoration to complete health, it was found that the best part of her acquired knowledge was gone, and it was never afterward recovered. An attack of apoplexy has been said to efface all remembrance of the events of some definite period of the life.

Both classes of facts—those which illustrate the dependence on certain bodily conditions of both the power to acquire with effect the materials for the memory, and the power to recall them with ease—can be accounted for by the general views already expressed. The varying condition of the body through the several sensations of which it is the occasion, enters into the experiences of consciousness, and furnishes a most important element in them all. It is the constant background on which all the mental activities are projected, the never-failing setting with which every one of them must be accompanied. When these sensations are of a certain description, they are the normal and favoring accessories of the other actings of the soul. If they are abnormal, disturbed, or unpleasant, the mind is so absorbed or distracted by the presence of these obtrusive sensations, that it has little energy to spare for other objects, and no capacity to steady the attention upon them.

Again, the bodily condition may also present sensations which so far disturb and distract the attention, as to allow no time for the passive memory to respond to any call; may so hurry the mind from one object of present sense-experience to another, as to leave no opportunity for the representing power to thrust in a single mental image; or, again, these sensations may be so utterly dissimilar to any which have been before experienced, as to suggest no image of the past. Or, on the other hand, this complex of sensations may be most favorable to the easy and almost exclusive action of the passive or spontaneous memory, and may be so akin to the states which we would recall, as to be all luminous and living with objects that suggest those which we welcome or seek after.

To the question, whether the circumstances of the soul can ever so far be changed as to empower it to recover all the past, the analogies suggested by these facts would lead us to reply: (1.) Under no circumstances whatever can it be supposed that the soul shall recover what it has not in some sense made its own by the energetic action of its attentive consideration. That is not a proper object of memory to the soul, which has not been taken up into its life by its efficient acquisition. (2.) It is supposable that the conditions might be furnished of recalling all the past thus defined, under the actings of laws which are well known to us. We have only to suppose that a vehicle or subject of the required psychical experiences—call them sensations, if you will, and the occasion of them a new body—should be furnished, and these would of themselves give back every element of past acquisition or experience to which they might be analogous.

Varieties of memory; how explained.

§ 168. With the progress and development of the powers and activities of the soul, the memory itself advances through separate stages, each of which prepares the way for that which follows, and becomes its natural and logical condition. The memory of the infant differs from the memory of the child; the memory of the child differs from that of the youth; the memory of the man, in each of the several stages of active life, differs from that in the stage which succeeds it. In general, the memory of the person in active life differs from the memory of old age. The memory of the artist is very unlike the memory of the mathematician. The memory of the erudite and disciplined thinker differs greatly in its objects and its laws, from the memory of the person who has had little culture from reading or thought. Hence, there exist many clearly distinguishable *varieties* of memory; if we make nothing of the fact that every individual must have a type of memory which arises from those individual habits of thought and feeling which he can share with no other person.

Besides those varieties of memory which are common to all men in the successive periods of their life, there are the special peculiarities which result from one's pursuit or profession. The historian remembers facts and dates; the philosopher, principles and laws. The artist remembers landscapes and faces; the wit and the story-teller, never forget a successful jest or a capital anecdote. These habits of memory, as they are called, often grow stronger till they become fixed beyond the power of change. Persons distinguished for great intellectual power in certain directions, very often complain of a serious defect of memory which they cannot account for. Such one-sided habits and defects are not peculiar to the memory only, but pertain equally to all the activities of the soul: the condition of memory is energy in the original activities; these involve attention to the objects to be remembered; attention springs from an active interest in these objects; this prevailing interest follows the habits which constitute and express the character.

We return again to the fact that these varieties of memory are not only distinguished by the character of the objects remembered, but also by the method and relations under which they are recalled. The things which the child remembers not only differ from those which an older person recalls, but they are recalled in a child's order, and by the relations which are proper to a child. The same is true of the devotee to any study or pursuit so far as special intellectual habits are induced by such a study or employment. When the child recalls to itself or recites to others a series of incidents of which it has had experience, it depicts the whole, generally in the order of time, with little selection of materials according to their importance or their relation to any principle or purpose. The spontaneous memory of the eye or the ear, reproduces the past solely after the relations of time or place, with no reärrangement or selection of the same, such as would be suggested by the desire for the clearer apprehension of the hearer, or by the bearings of the story upon his intellect or his feelings.

This is very conspicuous in the memories, and especially in the narratives of uneducated persons. Thus, Dame Quickly recites the story of her wrongs in the following fashion: "Thou didst swear to me upon a parcel-gilt goblet, sitting in my dolphin chamber, at the round table, by a sea-coal fire, upon Wednesday in Whitsun-week, when the prince broke thy head for liking his father to a singing-man of Windsor; thou didst swear to me then, as I was washing thy wound, to marry me, and make me my lady thy wife." (Henry IV., 2d part, Act ii., scene i.; cf. S. T. Coleridge, *The Friend,* Sec. ii., *Essay* iv.) No finer opportunity is furnished for observing this variety in the order and method which characterize the memories of different persons, than in listening to the testimony of different witnesses in a court of justice, concerning the same transaction.

The memory of the young is usually more ready; that of the adult is more tenacious. This is, in part, owing to the greater physical vivacity of youth, which affects the actings of the soul. The vivacious old man is as quick to remember as he is to apprehend or judge; while the torpid and phlegmatic child is as slow in his memory as he is in his reasonings and inferences. The difference, however, is not merely a difference of temperament or animal spirits, but has its ground in the character of the relations which usually predominate at each of these periods of life. Objects that are recalled by the relations of space and time and of obvious resemblance, present themselves promptly, if they are remembered at all; but these relations are, from their very nature, limited to but few individual objects. Hence, the groups which are connected by such relations are sooner set aside and forgotten, and are displaced by others. The relations of thought, however, especially those which are founded on wide-reaching principles or laws, are in their very nature less obvious. But, on the other hand, the principles themselves are few, and are constantly before the mind. When these are once mastered, they are illustrated in every fact; they are exemplified in every instance. By means of them we can prophesy and construct the future as well as explain and interpret the past. These few bonds of association, when they control the memory, give to it perfect security in and command over its possessions.

The men of universal memory are those who combine most happily the ready memory of facts and events with the tenacious memory of truths and laws. They are those whose spontaneous memory is not displaced, but rather aided, by the development of the rational memory which sees in facts the illustrations of the higher relations of philosophic truth. They hold fast the acquisitions of youth and of old age by the permanence of principles which are as old as the universe and as new as the latest experiment by which they are verified.

The memory of the ancients, if we may believe all the stories which are told of the achievements of some of their more distinguished men, surpassed, in some respects, the average attainments of the moderns. It is not difficult to believe this to have been true, from what we know of the memory of those who most resemble them in the circumstances of their lives, and the discipline of their intellects. Their attention was far less distracted by a variety of objects than is the case with the moderns. The facts in science, literature, chronology, and history, which they were required to remember were far fewer than those which burden the memory of the modern scholar. More than all, they relied far less than we do upon writing, memoranda, and books, to preserve what they desired to retain. They committed their acquisitions to their own power to recall them. Conversation and repetition were practiced far more generally by them than by us. What was heard by the ear from the living teacher, was repeated and discoursed of by his interested scholars, till it became a part of their very being.

Development of memory. Its characteristics in the several periods of life.

The attention of the infant is at first occupied with the sensible world. It sees colors which delight the eye, it hears sounds which captivate the ear. It is long before it unites these separate percepts into individual objects, and still longer before it discriminates, by special attention, one object from another. Later still, it learns to notice with any effect its own inner experiences and activities. The relations of before and now are of still later evolution. But all these separate elements must be familiarized by attention before an act of memory can be at all definite and complete, inasmuch as, whatever suggestions of representation there may be, there can be no proper act of memory till all these elements are recognized.

The germinant memory of the infant must be exceedingly limited, because its materials are very scanty; the chief force of its intellectual life being expended in acquiring rather than in recalling. So far as it remembers at all, its memory is passive; intentional memory being as yet undeveloped, for the infant is the passive child of nature, and the stream of its memory runs side by side with the course of its objective life. The infant remembers, as animals remember, just that, and only that, which the objects of sense-perception recall to their thoughts.

The acquisition and the use of language opens the way for the higher memory, though obviously in its first beginnings. The right use of words, and of short sentences, requires that the child should connect names with distinctly discerned objects, and should express its wishes and thoughts by short sentences. But by-and-by the child finds that it forgets—that it has not the knowledge which it once possessed. It cannot recall the right name or phrase which it wishes to use, and which it knows it has previously spoken. It is impelled by its wishes to recall the forgotten object, and begins to practice the arts of the intellectual, or active memory. But these occasions and efforts are at best so infrequent, and of so little importance, that they train the intentional memory in a slight degree only.

It is by tasks imposed by others directly and indirectly, that the soul is disciplined to the exercise of this higher memory, and its various activities are developed. The child is taught written language. It learns the alphabet and spelling by the eye, or brief sentences or verses by the ear. Children are charged also with commissions to execute, with services of labor or courtesy which may not be forgotten, and with endless lessons from books to prepare and repeat.

By degrees, this pupil of others becomes his own taskmaster: he passes from the lower discipline of the memory, which others enforce, to the higher, which he imposes upon himself. The intentional memory, which has been trained by others, he cultivates for himself. He makes his own purposes; he proposes his own ideals; he knows what he must learn in order to accomplish these purposes and to realize these ideals; he appoints to himself his own lessons; tasks his own intellect to consider, and his own efforts to retain what he foresees he shall have occasion to know and to have at command. According as this training of the attention is more or less complete, so does his memory become more or less perfectly subject to his control, and from the passive spontaneity of early life, passes into the active energy of maturer years. This memory of manhood is also characterized by the predominance of thought-relations and of rational purposes. The spontaneous memory of early life is not thereby displaced; the original aptitudes of the memory of both eye and ear are not necessarily set aside. But just so far as one thinks and acts like a man, just so far will he remember as a man, and not merely as a child—that is, by the aid of those higher relations which thought requires, and which definite aims and rational activities necessarily involve. The memory of the man is not only intentional, but it is also *rational*.

When the man advances from the busy noon toward the quiet evening of life, his exclusive interest in the objects which have absorbed his manhood is relaxed, either through physical infirmity, or the success which satiates, and perhaps the disappointment which wearies a man with life. In place of an intent and absorbed devotedness to the present, there is a more frequent review of the past. Old scenes are described, old books are read, old companions are talked of, old stories are repeated. For this reason, recent objects are so readily forgotten, and the singular contrast is furnished in the memory peculiar to the aged—most tenacious of objects and events that occurred longest ago, and readily forgetful, if tenacious at all, of those that were most recent.

The education of the memory.

§ 169. The methods of education should recognise the wise arrangements of nature in developing and maturing the memory. In the earlier periods of life the spontaneous memory should be stimulated and enriched by appropriate studies. The child should learn stories, verses, poems, facts, and dates, as freely and as accurately as it can be made to respond to such tasks. During this early and objective period, it should learn as many languages as is possible in the circumstances, or as is desirable for its future pursuits. Especially should it learn those languages which can be taught in conversation, or acquired by contact with those who speak them freely and well. If the elements of the ancient languages are taught so early in life, they should be taught, as far as in the nature of the case is possible, by similar methods. But as the higher and rational powers awake to action, every acquisition that has been made by the lower and more obvious associations, should be secured against loss by recasting it and relearning it as it were, after the relations which are higher and more philosophical. English children who learn to speak French, German, or Italian fluently in early life, may lose their acquisitions almost entirely, unless these are fixed by a grammatical study of the same languages at a later period of life. The large accumulations of facts and dates, as in geography and history, which are made very early by many carefully-trained children, and with the greatest ease on their part, are liable to be effaced, and, as it were, swept clean out from the memory, unless they are secured against loss by reviewing and reärranging them under the new and higher relations which the development of the reason makes possible.

On the other hand, to anticipate the development of the reflecting powers, by forcing upon the intellect studies which imply and require these capacities, is to commit the double error of misusing the time which is especially appropriate to simple acquisition, and of constraining the intellect to efforts which are untimely and unnatural. The modern practice of occupying the minds of children with the reasons of things, *i. e.*, with laws, principles, etc., in the form of compends of astronomy, natural or mental philosophy, natural theology, etc.—is one that cannot be too earnestly deprecated, or too soon abandoned by those who would train the mind according to the methods of nature.

The cultivation of the memory; mnemonics.

§ 170. The cultivation of the memory is a subject which has been earnestly discussed by many writers, and is of practical interest to all those who are bent on self-improvement, or are devoted to the education of others. Many complain of a general defect of memory. Others are especially sensible of painful failures in respect to certain classes of objects, as names, dates, facts of history, sentences or passages from authors familiarly read. The question is often anxiously propounded, How can these general or special defects be overcome?

The conclusions which we have reached in respect to the nature and laws of memory, suggest the only practical rules which can be attained. These rules may be summed up in the directions: 'To remember any thing, you must attend to it; and in order to attend, you must either find or create an interest in the objects to be attended to. This interest must, if possible, be felt in the objects themselves, as directly related to your own wishes, feelings, and purposes, and not to some remote end on account of which you desire to make the acquisition.' It should never be forgotten, that in memory, the soul can recall no more than it makes its own—no more than, in acquiring, it constructs or creates as a spiritual product by its own activity.

The late Sir Thomas Fowell Buxton advised his sons in the following golden words: "What you do know, know thoroughly. There are few instances in modern times of a rise equal to that of Sir Edward Sugden. After one of the Weymouth elections, I was shut up with him in a carriage for twenty-four hours. I ventured to ask him, What was the secret of his success; his answer was: 'I resolved, when beginning to read law, to make every thing I acquired perfectly my own, and never to go to a second thing till I had entirely accomplished the first. Many of my competitors read as much in a day as I read in a week; but, at the end of twelve months, my knowledge was as fresh as on the day it was acquired, while theirs had glided away from their recollection.'" (*Memoirs* of Sir Thomas F. Buxton, chap. xxiv.)

Numerous devices have been contrived in order to aid the mind so to make its acquisitions as to secure them against loss, and to bring them readily to hand when required. They were not unknown to the ancients, as is evident from Cicero, *De Or*, ii.,86–88; *Ad Herenn.*, iii.,16–24; Quinct., *Instit.*, x., 1, 11–26. They all rest upon a common assumption or principle, viz., that it is possible, by means of arbitrary associations, so to connect what one desires to remember with a series or scheme of objects, artificially arranged or actually existing, that they can be readily and certainly suggested to the mind. Some teachers of mnemonics employ a scheme of geometrical figures, as squares or triangles. For example: if a person, in listening to a discourse or lecture, should, as the speaker proceeds, connect the leading thoughts or divisions with the panes of glass in a window-sash, or with the panels of a door, he would avail himself of the geometrical method, which addresses the eye, through the space-relations of visible objects. Often these systems have sought to aid the memory of dates by the letters of the alphabet; each presenting some number, and being employed in forming artificial syllables, such as could be readily attached to names of persons or places distinguished in history. Mnemonic verses and tables have been furnished for many of the important objects with which every student is expected to be familiar, as the names of the sovereigns of the great kingdoms and empires, grammatical paradigms and rules, logical formulæ, etc., etc.

A correct estimate of the value of all artificial memory may be summed up as follows: The natural, as opposed to the artificial memory, depends on the relations of sense and the relations of thought,—the spontaneous memory of the eye and the ear availing itself of the obvious conjunctions of objects which are furnished by space and time; and the rational memory, of those higher combinations which the rational faculties superinduce upon these lower. The artificial memory proposes to substitute for the natural and necessary relations under which all objects must present and arrange themselves, an entirely new set of relations that are purely arbitrary and mechanical, which excite little or no other interest than that they are to aid us in remembering.

It follows, that if the mind tasks itself to the special effort of considering objects under these artificial relations, it will give less attention to those which have a direct and legitimate interest for itself. Its energies, instead of following in easy obedience the leadings of nature, will be forced to efforts that are constrained and artificial. Whatever dexterity is acquired by these intellectual gymnastics, must be gained at the expense of that rhythmical power which always rewards those activities in which art follows nature. The wonderful feats of memory which are occasionally adduced as resulting from the latest new device in mnemonics, are the fruits of much time, labor, and enthusiasm. Had the same time, labor, and enthusiasm been expended in acquiring knowledge by means of the ordinary appliances, the acquisitions would have been many times more valuable for the culture of the powers and the uses of life. Perhaps even the number of facts recorded in the memory would have been as numerous.

There are occasions when the artificial memory is unquestionably useful. It may serve a good purpose in holding before the mind facts which it is important to remember, when neither the facts themselves, nor their relations, present attractions which are strong enough to fix or hold the attention. For the man whose intellectual force and interest are preoccupied, it is often difficult to apply the memory with success to such objects, unless they are arranged in some novel relations. The artificial memory comes to his aid, and offers the service and assistance of art to supplement the failing forces of nature ; to reënforce, and as it were, to renew the spontaneous memory by novel appliances.

But while we concede a certain advantage to the artificial memory under circumstances like these, we must still hold, with Coleridge (*Biog. Literaria,* chap. vii.), that, for the ordinary uses of the student, *sound logic,* a *healthy digestion,* and a *quiet conscience* are the proper conditions or arts of memory.

By *sound logic,* is, of course, intended a well-balanced and well-trained intellect, which by original structure and discipline, is capable of fixed attention, clear apprehension, and excited interest. Without these conditions, a strong and trustworthy memory is impossible.

A *healthy digestion* is also requisite; for if the digestion is disturbed, the action of the mind will be distracted by those vague sensations of depression and discomfort which are inconsistent with that harmonious interaction of the powers of the whole man, which is indispensable to a good memory. Even though it happens that persons in this condition are capable of extraordinary energy in their mental efforts, these occasions are yet certain to be followed by longer periods of listlessness and depression which exclude that repetition and review of the knowledge which are quite as essential as energy and interest at the time of the original acquisition.

A *clear* or *quiet conscience* is also a prime requisite, for a similar reason. Indigestion and intoxication of any kind disturb the memory by intrusive, uncomfortable, and exciting sensations. But the consciousness of guilt haunts the spirit with disquieting self-reproach, and fear of deserved punishment. Feelings of this sort do indeed often stamp upon the memory a few impressions that are ineffaceable. But for this very reason it is the more unfitted to attend with interest or enthusiasm to other objects, and its movements in all directions are enfeebled or depressed by distraction or constraint.

It is natural, in this connection, to notice the moral conditions of a good memory. The man who would have a strong and trustworthy memory, must always be true to it in his dealings with himself and with other men. He must paint to his own imagination, with scrupulous fidelity, whatever he has witnessed or experienced. He must never so yield to the bias of interest or passion, as to strive to persuade himself, even for a moment, that events were different from what he knows they actually were. He must seek to repeat to others the precise words of what he has heard or read, whenever he makes communications by language. Such a moral discipline to internal and external honesty, both implies and enforces a mental discipline to earnest and wide-reaching attention—an attention which does complete justice to every object that comes before it, and which neither slights nor omits any thing which ought to be brought to view. An intellect that is regulated and held to its duties by the tension of such a purpose, will act with the precision and certainty of clock-work. Its recollections will be trusted by others, because they are trusted by the person himself, and for the best of reasons—because he is true to what he remembers.

On the other hand, a person who is false to his fellow-men, will often weaken his confidence in his own intellect, and may end with an incapacity to distinguish falsehood from the truth. What he does not like to remember, he will persuade himself did not actually happen, or, at least, not in every particular as it spontaneously presents itself to his view. Then follows, by natural consequence, distrust of his own memory, because he is not sure that the materials are at hand with which he can correct his own omissions. The next step is, under the excitement of strong passion, to persuade himself that what he desires should be true, did really occur, or was really written or said. If he asserts this by his own word, he is the more strongly committed to believe it. At last, he becomes so false to the workings of his own memory, that he dares not trust it himself.

It is well to remember, that, while the liar has more pressing need of a good memory than any other man, he is of all men the least likely to possess it.

CHAPTER V.

REPRESENTATION.—(2.) THE PHANTASY, OR IMAGING POWER.

From perfect memory, we pass through the several forms and degrees of imperfect memory till we come to *the phantasy*.

Phantasy defined and illustrated.

§ 171. The phantasy, or imaging power, is that form of representation which brings before the mind's apprehension objects, or, more exactly, images, *as such*, severed from all relations of place, time, or previous cognition. The best example of the exercise of this power is furnished in dreaming. In what are called the abnormal or disordered states of the soul—as somnambulism, and the various types and degrees of insanity—the phantasy has a more or less complete control. Among the wakeful and normal states of the soul, reverie is the purest and the most perfect example of phantasy. The fewer the relations to the past or the present which the objects suggest, the more complete is the working of the phantasy. In earliest infancy this power may be supposed to be active, for the reason that the mind has not yet reached a condition in which memory proper is possible. In extreme old age also, when the incapacity to attend to single objects for a long continuance precludes intelligent and effective perception, memory, or thought, the phantasy may still survive, and actively call up the pictures of the past, simply as pictures, each recalling the next, according to the conditions and laws already explained. In the wakeful and earnest periods of the mind's activity, the exercise of simple phantasy is precluded, for the obvious reason, that at such times the mind is intent upon some rational object, which lifts it above the condition of the passive recipience or contemplation of pictures. And yet, with the higher activities, there are not infrequently mingled those approaching to pure phantasy. When one object suggests another in a train of associations, many may be recalled without a single distinct act of remembrance, and yet every one may be a transcript from some reality experienced in the past. Each is recalled, however, not as a remembered or recognized object, but simply as an image. When the higher functions of the soul are wholly, or in part, put in abeyance, as in fainting, fatigue, or sleep, or when there is bodily weakness, or any disturbance of the nervous equilibrium, as in fever, delirium or excitement from liquor or narcotics, or even in protracted sleeplessness, the phantasy asserts a more or less complete dominion. The mind is visited with throngs of pictures, which rush so rapidly by as to confuse it by their very swiftness, and to oppress it by a sense of its own impotence to arrest or direct their course. When this condition is permanent, the mind is said to be the victim of phantasy. Such a state is also called a state of *distraction*—which term describes the mind's incapacity to fix the attention or detain its flitting images long enough to allow the exercise of the functions of rational memory, invention or thought.

The interest of its problems.

§ 172. These conditions of the soul are grave problems to the psychologist. Three suppositions may be made in respect to them all:—(1.) These states may be said to be simply abnormal or irregular, recognizing and obeying no law. (2.) They may be set down as simply inexplicable, suggesting the existence of laws which cannot be discovered. (3.) They may be explained in great part by the usually recognized laws of the soul in its normal and wakeful condition. The probability is immensely in favor of the last. If the laws which govern the recurrence and representation of ideas have been fully and correctly set forth, they ought to explain the phenomena of the sleeping and disordered conditions of the soul. That they do so, is probable for the following reasons:—

The power of association is operative in them all.

I. The power of association operates very efficiently in all these states. In dreaming, somnambulism, insanity, etc., etc., its presence and powers are often most apparent. When we ask ourselves, Why did it happen that I had such or such a dream? it is often very easy to answer by a reference to the usually recognized laws of association. The strange and unexpected sallies of the insane, however wild and preposterous they may be, follow some law of association, though it often leads to the most fantastic result. There is always some method in their madness.

Deviations accounted for,— (1.) By changes in the relative proportion of the powers.

II. The deviations from the ordinary working of these laws can also, to some extent, be satisfactorily accounted for.

(1.) The powers of the soul ordinarily act in a certain conjunction with and proportion to one another. It is not surprising, that, when a single power acts alone, the phenomena should differ very greatly from those which result from the combined activity of them all. In the cases supposed, self-consciousness, rational activity, and the voluntary control of the bodily movements and the mental states, are all set aside; and the associative power asserts, to a very large extent, the possession of the soul. We ought not to be surprised, that a power ordinarily acting in connection with the wakeful reason and under its control, should manifest results unlike those which appear when these regulating elements are present.

(2.) Certain bodily states are known greatly to modify the actings of the soul, when the soul is wakeful and in health. It is according to the law of its being, that its action should be modified still more when the bodily affections become more efficient and obtrusive. It should not be surprising then, that under such physical conditions as sleep and cerebral excitement, even stranger psychical phenomena should be manifest.

(3.) The comprehensive law under which past mental states are reproduced, should be distinguished from the materials upon which it operates. While the laws of representation remain the same, the *conditions* under which they act, may vary enough to account for every variety of phenomena.

To the actual reproduction of an image, two conditions are necessary, viz., its actual previous presence to the mind, and the existence of an exciting occasion in something united with it as an element of the mind's previous knowledge or feeling.

In dreaming, insanity, etc., these conditions are peculiar. *First*, in the states of distinct and easily-remembered consciousness, are present many elements which are less distinctly noticed, because they are accessory and subordinate. In the states under consideration, those may be brought forward either as the materials of phantasy, or as the mediate suggestors of other materials. In every act of distinct perception, there is an extended background of such objects, standing out in the field of view with more or less prominence, but engrossing some share of the soul's energy. Any one of these objects, under possible exciting occasions, is capable of being recalled. In the normal states of the soul, the prominent or central object is usually recalled. In an abnormal state, one or more of the accessories may be represented. Under the feelings and purposes of wakefulness, a certain class of pictures and thoughts only may be certain to be thought of. In dreaming, another set may present themselves; in insanity, still another; and yet all of these may have been gathered from the mind's own experience. Again: there are many conditions of the soul marked by little energy of attention, as well as by the feeble influence of rational purpose, in which the phantasy greatly prevails. In walking, in driving for relaxation, in extreme fatigue, in the transitions from wakefulness to sleep and from sleep to wakefulness, in the many listless hours or seasons of reverie, there are multitudes of acts and objects which leave little impression, and are rarely, if ever, distinctly brought back to the rational and wakeful memory or imagination, but of which any one may be recalled under novel circumstances. Again: there are activities that have been experienced previously to the soul's conscious action. Some of these acts tend to be reproduced, and, under varying circumstances, may return either as a principal or accessory element. Again: the undefined *bodily sense-perceptions*, or sensations which are accessory in every mental experience, and are prominent in not a few—which form the background of many, and come into the foreground of many also, all tend to recur again.

The occasions which control the presentation and suggestion of images in these abnormal states of the soul are also peculiar. In sleep, all the organs of sense-perception are more or less quiescent, while the vital organs are active. In insanity, etc., the bodily condition and activities are irregular. In both, they are greatly unlike those which are present in wakefulness and health. These peculiar and morbid bodily states are manifest to the soul in the form of peculiar sensations, both vital and organic. Sleep, from the beginning to the end, is attended by a series of sense-perceptions unlike those experienced in wakefulness. Insanity, in all its forms and degrees, is attended by a nervous excitement or depression, which is revealed to consciousness by irritating and uncomfortable sensations. The sensations thus excited, become themselves, in turn, the excitants of images and thoughts kindred to themselves.

A third consideration should also be noticed. The *creative power* of the phantasy may have especial activity in dreaming and insanity. Whatever that power may be in its functions and products—if it be allowed that the phantasy is in any sense creative—if, in the waking and rational states, it is not tied to a simple reproduction of the past; if it has any liberty of origination, then it might be natural and credible that it should exercise this freedom more fully when unlimited by sense, reason, or will, than when constrained by these in the earnest activities of the wakeful and rational hours. That the creations of the phantasy of the dreamer and the madman have no correspondent realities, is obvious to all. The fantasies of "a madman's dream" are conceived by us as the most unnatural and the wildest of all unrealities. If the phantasy is, in its very nature, a creative as well as representative power, it is not surprising that it should create in madness and in sleep. If its creations are free in the one state, when reason is wakeful and the will is attent, and earnest purposes control, it is not surprising that, in those conditions of activity in which these influences are feeble, its products should be irrational and unnatural.

These considerations may serve as the foundations of a general theory of those various conditions of the soul's activity known as faintness, dreaming, somnambulism, and delirium. They are designed only to prepare for a more particular consideration of each. We consider, first of all, *sleep*, in the two following aspects:—

(1.) Sleep as a condition of the body, *i. e.*, sleep in its physiological phenomena; (2.) Sleep in its psychological experiences.

Sleep physiologically considered.

§ 173. We cannot understand sleep as a state of the soul, without considering the corporeal conditions which attend it. In order to interpret it psychologically, we must first examine it physiologically. In sleep, physiologically viewed, the organs of perception, and the nerves connected with them, are comparatively inactive, and seem incapable of performing their accustomed functions. Conversely, also, the soul can no longer control the organs of sense and of locomotion; or, more exactly, the soul loses, in a very great degree, its power to direct these organs.

On the other hand, the functions of the vegetative, circulatory, and respiratory organs, go on as usual, though in the case of some with a somewhat diminished energy. That in all these functions the whole tone of life is lowered, is manifest directly from observation, and is inferred from the greater sensitiveness of the body in sleep, to all those agencies which weaken or endanger the life. On the other hand, it is certain that the nutrition of the brain and the whole nervous organism, is greatly augmented in sleep, and that sleep is even essential to restore that waste of their material which wakefulness occasions. If wakefulness is protracted too long, by nervous restlessness, or excessive mental occupation or anxiety, it terminates in fever, delirium, or dementia, through a temporary disease or permanent lesion of the nervous organism itself. Hence, sleep is, if possible, more absolutely indispensable to the restoration of mental activity, than to that of any other human function. The incapacity of the organs of sense to be affected by impressions from without, as well as to yield to influences or directions from within, varies at different times. The want of control of the soul over its organs, also varies from the momentary loss of power which can suddenly be resumed, to that permanent impotence to speak or move, which is experienced in the most distressing nightmare.

In falling to sleep, the soul passes through many of these conditions, beginning with the slightest unconsciousness, and proceeding more or less gradually through more or fewer intervening stages. In awaking from sleep, it emerges from a condition of more or less complete insensibility to one in which the senses are fully refreshed and active; and more or less gradually, according as the occasion and manner of its waking is more or less gentle or violent. The same is true of the processes by which it loses and regains its command over the organs. The different senses, as has already been intimated, fall asleep at different times in various degrees, and awake also in unlike proportions. Thus, the sense of sight may be very obtuse when the sense of hearing is active, as is the case when a person watches by the bed of one who is ill, or in the instance of men who can find refreshment in sleep when reading or conversation is going on, and are able to recite when awake what has been read or spoken while they were sleeping. The miller sleeps while his mill is grinding, but wakes if it stops. Another person sleeps while it is still, but wakes when it moves. The watchman, when wearied, sleeps with all his senses, except the senses of touch and muscular direction. Soldiers sleep in every sense and organ of motion, except the legs with which they march continuously.

Sleep considered psychologically.

§ 174. The activity of the soul continues during sleep. It is not entirely suspended at any time, though its energy may now and then be exceedingly feeble. That it often acts during sleep, is confessed by all. Every dream involves some form of this activity. There is some diversity of opinion in respect to the question, whether this activity is constant, or whether it is not infrequently interrupted. Many have argued that this activity often ceases, from the circumstance that we are not conscious, and do not remember that we dream all the while that we are asleep; that we know that we dream more frequently when sleep is less complete, as soon after we fall asleep, or just before we wake; that in our deepest slumber it often happens that no signs of conscious activity are indicated to a looker-on; and that it is not necessary to the continued existence of the soul that it be constantly active. On the other hand it is urged that the soul is always active, because, on awaking, it is at once aware of its own identity, which involves the belief of continued existence during the interval of sleep; and when it wakes, it may recall or review a continued series of sensational experiences, if it cannot bring back an uninterrupted course of conscious activities. Moreover, it is urged that the fact that the soul does not recall all its dreams does not disprove that it dreams, for there are many waking states during the progress of a single hour, much more during a day, which cannot be recalled. There are also many dreams which we do not recall; as is obvious from the circumstance, that if, on awaking, we lay hold at once of the thread which is in our hands, we can trace our way backwards through the maze of even a succession of dreams.

That the soul acts with feebler energy when asleep than when awake, is obvious from the circumstance that in some of its powers it scarcely acts at all. This may be fairly inferred from that general dependence of the tone of its action upon the force of the body which is observed in wakefulness, which dependence, as may be fairly inferred from analogy, extends to its sleeping states. The only possible exception to this conclusion would be suggested by the fact that some of the powers —*e. g.*, the phantasy—may seem to act in sleep with greater energy than in wakefulness. With this exception, observation confirms what analogy suggests, that, in sleep, the general activity of the soul is greatly lowered.

The powers and capacities of the soul act with unequal and varying energy in different persons and in differing conditions of sleep. The representative power of the soul, as has already been said, is that which is especially prominent in sleep. The law or force under which it acts has already been explained as the tendency of the soul to act more readily a second time in forms and with objects which have previously occupied its energies. This tendency or force needs only to be supposed to be exerted without the regulating presence of the other faculties, in order to account for its greater apparent energy. All the so-called laws of association control the production and presence of the objects which make up the image-world of the dreamer. These objects are sometimes recalled under the relations of time and space, in succession or co-existence. Sometimes the relations of likeness or unlikeness control; at others, those of cause and effect. Very often, all these relations must be resorted to, to account for the presence of the various objects of which a single dream is composed.

This comparative irregularity and capriciousness pertains to the order in which these objects are presented to the mind. When the wakeful soul is intent on recalling some object to memory, all the operations of the representative power are controlled by this prevailing purpose. The multitude of varied objects which are presented by the associating power, are entertained or thrust aside by the judging and reasoning intellect, and so an order of their relative value is secured to the objects themselves by the mind's reaction upon them. Even if the mind gives itself up to reverie, it is constantly awake, or ready to be awake, to the suggestions of reason, of use, of beauty, or of rectitude.

There is also the rationalizing and sobering presence of the material world, with its obtrusive realities that cannot be mistaken; its permanent attributes, that cannot be changed; its eternal and superior laws, that can neither be resisted nor set aside. The perpetual presence of this fixed and orderly body of facts and truths, of itself gives reason and order to the fancies which it must in part control and regulate.

But in dreams there is an absence of judgment, or the judgments are partial, and the stream of images flows on, under the joint impulses given it by the energies of the mind's previous activity and the force of casual mental or bodily suggestions. The material world is withdrawn from the mind's cognizance as an apprehended fact; it is as though it were not, and never had existed.

The mind's *interpretations* of the images of fancy, and even of its bodily sensations, are often false and irrational. First of all, it judges the image-world to be a real world. How this is possible, it is not so easy to explain; that it is a fact, cannot be doubted. The mind is preoccupied by the action of the representative power. The first impulse, when a picture is presented of an absent reality, is to believe it to be real when there is no ground for the opposite belief. This is wisely provided in the constitution of man, to secure all those actions for which the knowledge or the thought of any reality is given. The mind, in dreaming, yields to this impulse. The mind, apprehending no real world with which to contrast and judge the imaginary, uses the little force which remains, to infer that the products of its shifting phantasy are themselves realities. They are believed to be real, for they excite all the emotions which such realities are fitted to produce. Delight is experienced at the image of a friend believed to be present, who is perhaps far distant, or long removed by death. Grief is felt at some distressing event which is simply pictured by the phantasy. The mind is not only incapable of discriminating the real from the fantastic, but it interprets the real to be itself a part of its fantastic world. It misinterprets the bodily sensations which it experiences, the sensations of cold or heat, of oppression in the stomach or the heart, and of pain or pleasure in any part of the body. Thus Dr. Gregory relates that, having occasion to apply a bottle of hot water to his feet, he dreamed that he was walking on Mount Etna, and found the heat insupportable. A person suffering from a blister applied to his head, imagined that he was scalped by a party of Indians. A person sleeping in damp sheets, dreamed that he was dragged through a stream. By leaving the knees uncovered, as an experiment, the dream was produced that the person was traveling by night in a diligence. Leaving the back part of the head uncovered, the same person dreamed he was present at a religious ceremony performed in the open air. The smell of a smoky chamber has occasioned frightful dreams of being involved in conflagration. The scent of flowers may transport the dreamer to some enchanted garden, or the tones of music may surround him with the excitements of a well-appointed concert.

The exercise of this judgment in respect to the higher relations of thought varies very greatly in the energy of its action, and the perfection of its results. There are many cases in dreams in which single steps, or parts of a series of steps in reasoning, are taken surely and correctly, while these processes are entirely disconnected with what went before or followed after, as if the rational powers had resumed for a single instant their full energy of function. In other cases, the reasoning may be correct and the data may be false, and yet the falseness of the data may not be perceived. In still other cases, the data may be correctly discerned, and the conclusions correctly derived, so that both premises and reasoning combine to a true and valid conclusion. Even the more difficult feats of the invention and arrangement of the materials of an argument, have been successfully performed in dreams. The creations of poetry, even to the selection of rhythmical words, and the composition of sermons and addresses, have been often effected. Difficult problems in mathematics have been solved and remembered; new and ingenious theories have been devised. Happy expedients of deliverance from practical difficulties have presented themselves, and brought relief from serious embarrassments.

Consciousness is ordinarily but feebly exercised by the soul in its dreams. It is often said to be absent altogether. By consciousness is understood the distinct apprehension of the psychical states, as the states of the individual *ego*, and not that transient knowledge of them which is essential to any intellectual activity. It is when consciousness acts as judgment, and recognizes the relations of psychical states, that its results remain in the memory. This form or degree of consciousness is usually entirely absent, or feebly exercised in dreams. The reason why it is thus feebly put forth, may be the same which accounts for the absence of correct interpretations of the semblances of the material world.

For the same reason the estimates of time are so extravagantly and even ludicrously erroneous. In our dreams, we occupy a year in making a voyage; we perform a journey, we witness a long procession, we climb a mountain, and yet the time actually expended is inconceivably short.

These erroneous judgments of time are the natural and necessary consequences of mistaking the phantasms of our dreams for real substances and events. We picture to ourselves the incidents of a voyage or a journey. We turn these pictures into realities, and they carry with themselves the estimates of time which would be required if they existed or occurred in fact. The weakening of the consciousness of the accompanying psychical states, withdraws any corrective influences which would be furnished by the more distinct apprehension of the time required for the experience of them.

The activity of the sensibilities in the dreaming state requires a moment's consideration. That we feel in our dreams, or seem to feel, will not be disputed. If we believe we are in danger, we experience terror; if we dream that we are safe or successful, we rejoice. In some cases, but not usually, the fear and happiness are as intense and as real as when we are awake. In other cases, we feel, but on the review are surprised that we felt no more. Our joy and sorrow are but the pale counterfeits of waking emotions. The intensity of the emotions depends on the strength of our belief and the time of its continuance.

Is *the will* properly active at all during our dreams? That we act, as well as know and feel, is obvious from experience. We seem to resist, to struggle, to speak, to sing, to walk, to run, etc. We strive to attend, to remember, to contrive, to compose, etc.; in other words, we seem to use our mental powers under some directive force for definite objects. It follows that the conative, or impulsive part of our nature—the capacities which fit for action, are employed in the dreaming state. If these capacities are properly called the will, then we use the will in dreaming. But if we mean by the will, the capacity to direct the impulses by a rational or a moral purpose, it is equally clear that the will is entirely dormant, or, at best, is only occasionally or feebly active. It is and must be inactive, because the appropriate conditions for its exercise are absent. The reason does not propose a distinct end which the mind retains in view. The reflective consciousness neither forms rules nor imposes them. The will cannot act as a rational or moral director when these essential conditions are withdrawn.

Somnambulism, or abnormal sleep.

§ 175. *Somnambulism* assumes three forms, which have certain features or phenomena in common, but which, in certain respects, are unlike. These forms are the *natural*, the *morbid*, and the *artificial*. The natural, is that which may occur in ordinary sleep. The morbid, is an incident or phase of active disease of body or mind. The artificial, is induced by the instrumentality of another person. Each of these forms or manifestations is subdivided into varieties, which pass into one another by scarcely distinguishable shades of difference.

Natural somnambulism is distinguished from normal sleep by the special sensibility of some—generally some *one*—of the organs of sense, and by special activity in the use of some of the organs of bodily motion. The appellation, *sleep-walking*, is derived from the act of walking in sleep, which for obvious reasons occurs more frequently than any other bodily activity.

A multitude of examples of natural somnambulism are recorded. One only will serve. "A young nobleman mentioned by Horstius, living in the citadel of Breslau, was observed by his brother, who occupied the same room, to rise in his sleep, wrap himself in his cloak, and escape by a window to the roof of a building. He there tore in pieces a magpie's nest, wrapped the young birds in his cloak, returned to his apartment, and went to bed. In the morning he mentioned the circumstances as having occurred in a dream, and could not be persuaded that there had been any thing more than a dream, till he was shown the magpies in his cloak."—*Dr. Abercrombie.*

The activities required in this case, were the sense-perceptions of sight to direct the movements and the active control of the legs and arms. Sometimes the sense of smell, or of hearing, or of taste, are observed to be unusually acute. The use of the voice is often observed. The mental powers are often excited with great energy, continuity, and success. Persons in the somnambulic state will recite passages from authors even in a foreign language, which they could not repeat when awake. Persons who are imperfectly proficient in a language, converse with far greater ease and correctness than they have ever been known to do in the normal condition. Some remarkable compositions have been written, and eloquent discourses have been spoken, which were quite beyond the ordinary capacities of the individuals from whom they came.

In the *magnetic*, or *morbid somnambulism*, such extraordinary mental power has often been observed as to be ascribed to inspiration from another mind, or to some miraculous deviation from the laws of nature.

The ordinary and the magnetic or ecstatic somnambulism, differ from each other, in that the ordinary is preceded and followed by ordinary slumber, while the ecstatic comes upon the patient and leaves him at once, usually in a condition of extreme disease. In their psychological features, the two forms of this affection may be considered as alike, differing only in the greater intensity of some of their manifestations. Both are also exaltations of phenomena which are occasionally exhibited in common dreaming and sleep.

In all forms of somnambulism, the representative power is the one most prominently and conspicuously active. The leading objects of cognition and feeling are the mind's own creations. The man lives and moves, he feels and acts, in and for a dream. Dream-objects are taken to be real existences, and these engross and absorb the chief energies, and direct to many of the actions. But the dream of the somnambulist is far more methodical and continuous than the dream of ordinary sleep. The mind apparently rests upon its objects for a longer time, and gives to them a more fixed attention than it does to the phantasmagoria of the common dream. Certainly it must do both of these, when it adapts speech and motion to its dream-world, as it does whenever it is prompted to speak, and walk, and lift, and write, at the rate required by its phantasms. Its sense-perceptions do indeed direct the motions and regulate the rate of many of its bodily acts; but it were a serious error to suppose that what it seems to see, or to hear by the ear, makes up the entire world, or the principal part of the world in which the mind has its being and performs its acts. Besides these sense-objects, there is a multitude besides, which make up the background, and the foreground even, of its field of view. In the case of the nobleman cited, in all his movements to and from the nest of magpies, his thoughts were occupied with many phantasms which he considered real, and with reference to which he performed the actions recited. These formed the connecting and the accompanying scenery of the sense-objects which he perceived. The fact that sense-objects were blended with them, served to steady and retard the progress of the dream, and thus to make it regular and methodical. The feats which the fancy performs, its powers of memory, its skill in invention, and its resources of creation, are only the natural results of concentrated attention upon a few, and these connected objects. But this exaltation of the fancy is purchased at the cost of its being limited to but few objects—to single and spontaneous trains of thought running in the courses started and traced by the muscular and vital sensations, or the few sense-objects to which the excited senses are awake.

The powers of sense-perception, so far as they are exerted at all, act with surprising energy and effect. It is not only a surprising thing that they should act at all in so profound a sleep; but that the organ should be more sensitive and the mind more acute than in the normal condition, is still more remarkable. But this is often observed in the somnambulist. The objects seen are often seen by the faintest light, and yet they are seen most clearly, because actions requiring acute vision of these objects are performed with precision and success. The touch must be acute, or the somnambulist could not walk so confidently in difficult and dangerous places, nor avoid obstacles so dexterously, nor perform so many nice operations, as in skilfully writing and playing on a musical instrument. The senses of smell and hearing are often unusually sensitive to odors and sounds.

The question has sometimes been raised, Whether the somnambulist really perceives with the senses? It has been argued that he does not, because he also dreams, and because his dreams furnish the greater number of the objects of his knowledge and feeling. It has been inferred that, when he seems to perceive, he only dreams, and that what seem to be the objects of his sense-perceptions, serve, through his interpretations, to form a part of the dreams in which alone he knows and feels. To this it is sufficient to reply that he certainly acts with reference to the real world, and that he really acts—*i. e.*, directs the motions of his legs and arms, and uses and modulates his voice. So far at least as he acts he must have real sensations.

But while his senses are often surprisingly acute, they are both limited and uncertain in their operation and in their results. He does not see everything in the apartment in which he is present, but only the table, or chairs, or the paper on which he writes, or the candle which he holds. It is only to those objects which have some relation to his thoughts and actions that he is sensitively alive.

The various observations that have been made, warrant the induction that the phantasy stimulates and awakens the organ of sense, and determines the mind to use it with wakeful attention. It is the soul itself that quickens the organ thus made ready by disease or weakness for this extraordinary activity, to that momentary excitement which is required to fasten the mind to its monitions.

This extraordinary exaltation of single senses is not without its *analogy* in the wakeful and normal conditions of the soul. The vision of the sailor, the lace-maker, the horologist, the hearing of the sentinel and the hunter, the touch of the blind, the machinist, and the musician, seem to the stranger to be something almost supernatural. The still higher exaltation of these sense-powers, in the case of the somnambulist, is on the same ascending line with these natural variations. It is only extraordinary in degree.

We come next to a subject still more interesting, and, at first sight, more puzzling, viz., the apparent increased excitement of intellectual power as manifested in achievements performed by the somnambulist, particularly when in the mesmeric or ecstatic condition. The first which we shall consider is the claim for him of the ability to perceive material qualities and objects without the medium of the organs of sense. For example: it is asserted that he can see near objects through the thickest bandage, and with the back of the head; that he can hear by the epigastrium, etc., etc.

In respect to the first claim, that near objects can be seen or heard independently of the ear and the eye, we need only observe that, provided many of the stories are neither false nor exaggerated, not one of them proves that the mind can have sense-perceptions independently of the nervous organism. If the story be received as true, that the person has seen (not remembered nor conjectured) through an interposed bandage or by the back of the head, it would still be true that the optic nerve and the retina might be so morbidly sensitive as to be affected by the light, even if the eyelids were closed or thickly covered. No fact is more clearly established than that, within certain limits, one part of the sensorium, or portion of a single system of nerves, can, under extraordinary excitement, perform the functions of another.

The second claim is of a power to see distant objects which no sense-power can reach, as objects immured in total darkness behind thick and solid walls. Such a power, or its exercise, can be explained by no known powers or laws of nature. There is nothing analogous to its possession or its exercise, in any thing which we know in the normal actings of the soul. Whatever the power may be which acts in this way, it is not vision. The person does not see the object, but if he discerns any thing, it is a phantasm, an image, or series of images which are purely mental. If there be any thing which he apprehends, it is a mental object, the production of his own soul. It exists while he beholds it, within and for his soul alone. If the object or scene has never been the object of his personal inspection, the pictures which he forms of it must be taken from materials within his own observation, or imparted by description. If it be the city of Pekin, or the Himalaya mountains, the picture is composed either of fragments of what he has seen of New York or Boston, of London or Paris, of the mountains of America, or Europe, or from some drawing or painting of the cities or mountains themselves.

The third claim for the soul, of a power to understand its own bodily disorders, as to their seat or cure, may be explained in part by the fact that the sufferer in the somnambulic state is far more keenly alive than when awake, to his own bodily sensations. If an organ is diseased, the disease will often be manifest by means of sensations which are prominent and unmistakable in the soul's experience. These are the data for its interpretations or inferences. The disease may have been an object of intense anxiety and earnest inquiry. The person affected may have more or less knowledge of the anatomical structure and of the functions of many of the organs. It will always be found to be true, in such cases, that the insight of the somnambulist in respect to the names of the organs and their functions, does not go beyond what he has learned by conversation or reading. Let him be ever so gifted, he will not learn the nature or the name of a single organ, or its office, or a single remedy, which has not been the subject of thought in wakefulness and health. If this is so, the case is reduced to extraordinary sagacity exercised upon data or knowledge communicated or impressed in an extraordinary manner.

Fourth, the exaltation of the higher intellect to the capacity to perform some very extraordinary achievements, remains to be considered. This is much more remarkable in the morbid than in the natural somnambulism. The somnambulist sometimes displays great acuteness of judgment. He sees resemblances and differences which had not occurred to him in his waking states, and which astonish lookers-on. He is quick in repartee; he solves difficult problems; he composes and speaks with method and effect; he reasons acutely; he interprets character with rare subtlety; he understands passing events with unusual insight; he predicts those which are to come by skilful forecast. How are all these phenomena to be explained?

We reply: By the excitement of the intellect from an intense interest in the subject-matter with which it is occupied, the concentration of the attention for a long time upon a few objects only and a few of their relations, and the previous familiarity of the mind with these objects and relations. That the mind occasionally acts with energy when in the dream-state, even in its highest functions, has already been noticed. That, when it thinks and reasons in somnambulism, it is animated by strong excitement arising from a strong interest in the subject-matter, is obvious to all, and will not be questioned.

Next, the attention is concentrated upon objects for a sufficient length of time to secure entire familiarity with them and their relations. The attention of the somnambulist is limited, as we have seen, to but few sense-objects. To all other objects except those which excite this or that sense, it is deaf and blind.

Last of all, his sense-objects and his dream-objects are ordinarily very familiar. They have previously been the frequent object of thought and speculation. The questions for which the person finds new answers, the problems for which he devises new solutions, the events or characters upon which he casts a new light, are not for the first time before his mind. The operations of his intellect are also all in the line of his previous efforts and training. The somnambulist does not for the first time appear as a mathematician, poet, orator, politician, or divine; nor does he display activities which have not been in their quality and kind, if not in degree, familiar to his use.

The gift of divination, or prophecy, which is claimed for the somnambulist, whenever it deserves consideration, is explained in part by the extraordinary sagacity which is developed in respect to subjects that are interesting and familiar to the mind. The somnambulist forecasts or prophesies, by reasoning upon the evidence before him. His attention being fixed and his interest being aroused, he applies his intellectual force to the subjects before him, and shows the same sagacity in foreseeing future results that he exhibits in interpreting events that are present, by the causes, the laws, and principles that are concerned in bringing them to pass.

One or two other features common to all the varieties of somnambulism remain to be noticed.

First, the somnambulist, when he wakes, usually, though not invariably, forgets his actions, perceptions, and thoughts during sleep. His dream, with all that it involves, is to him an empty blank. To many, this seems incredible; to others, it is an insoluble mystery. That it is not incredible, is established by the amount of decisive evidence which is adduced of its actual occurrence. That it is not inexplicable, appears from analogous phenomena in dream-life, as well as from the dissimilarity of the conditions of mental activity in the waking and the somnambulic state. The dreams of the profoundest sleep are rarely remembered, for the reason that the bodily condition, with all the sensations which it involves, is, in many respects, very unlike that which attends our lighter slumbers and our waking states. The sensations which accompany these varying conditions, as has been shown, are an essential element in our mental experiences. If the phantasy is active, they are the essential conditions of its activity in any determinate direction. For this reason, these bodily sensations direct the course and furnish the occasions for many of our dreams. But in somnambulism these sensations are more controlling and more unique than in any other dreaming or in any other sleep. Whatever else there may be which awakens and directs the phantasy is, if possible, still more unlike any other experiences of wakefulness or sleep. If the transition from ordinary sleep and ordinary dreams to wakefulness is often so abrupt and complete as to involve entire oblivion of all which we have thought, or felt, or done, it is less surprising that, when we awake from the sleep of somnambulism, whether the transition be sudden or gradual, it is so complete that the present presents few or no relations to the past.

These considerations both explain and confirm the second fact that has sometimes been observed, viz.: that the somnambulist, when he passes into a succeeding condition of abnormal activity, remembers the experiences, and, as it were, remembers the self of similar previous states. How this should be possible, most clearly appears from the principles already laid down: The objects of thought and memory, the motives and directors of action which were present in the previous condition, return to him a second time, and they bring with them their attendant experiences. When the soul passes a second time into the surroundings of his abnormal being, they are no longer strange, but he recognizes them as familiar, and, taking up new threads of memory, he recalls the preceding dream.

Some remarkable instances are recorded of alternating states, in each of which the acquisitions, the capacities, and the employments were unlike those in the other, and yet, as the similar states recurred at intervals, they were connected by continuity of memory.

The *artificial somnambulism* is peculiar, in that it is induced by the intervention of another person, who, by means of passes or other appliances, brings the subject into a sleep and dream, the processes and objects of which he directs, and from which he awakes him at his own will. Hence it is called artificial, as effected by another, in distinction from the natural, which is induced by ordinary sleep, and the morbid, which is the incident of active disease. It is also called the magnetic sleep. It originally received this appellation, because it was supposed to be produced by a magnetic influence, generated by or attendant upon all the animal functions.

There is still another condition called *hypnotism*, or the hypnotic state, which may be properly called the artificial sleep as distinguished from the artificial somnambulism—*i. e.*, the artificial dream. It is like somnambulism, as produced by the agency of another, and as being under the control of the producing agent. The connection of the mind of the operator with the mind and the actions of the subject, is not so manifest, or is not always carried so far as is claimed for artificial somnambulism. It is however so like it in every essential feature, as to deserve to be considered as at least a lower degree of its exercise.

For the purposes which we have in view, hypnotism and artificial somnambulism or mesmerism, may be considered as one. The states so designated have the following features: Artificial sleep; entire or total insensibility of some of the sense-organs; an unnatural excitement and acuteness of others; the capacity to maintain some relation with the operator, so that the sleep and the dreams of the subject are under his exclusive direction and control. All these phenomena, with one apparent exception, are analogous to those of the forms of somnambulism already considered. The production of the sleep is the result of an excitement of some of the sense-organs or parts of the nervous system, initiated by exciting and fixing the attention of a susceptible patient, by the aid of a strong will and the energetic activity of the operator. The physical and immediate cause of the sleep is common to all the cases. It is the congestion of the brain. The occasions or causes of the congestion are diverse. In natural somnambulism, it is an incident of ordinary sleep in a person of sensitive organism. In morbid somnambulism, it is an attendant of active nervous disease. In the artificial, the congestion is the result of the attention of the patient leading to excessive physical excitement of some part of the sensorium.

In artificial somnambulism, the feature which is at once the most distinctive and the most difficult to explain is the control of one mind by another. While the patient is inaccessible to communications from every other person, he is open both to communications and impressions from the operator. Not only is he open to communications from him, but he is also in a considerable degree subject to his control.

If, however, we consider the phenomena of natural somnambulism, or even those of the common dream, we shall find some striking points of resemblance. In both these conditions, great insensibility of certain powers is conjoined with extreme sensitiveness of others. The dreamer and the somnambulist are dead in some of their senses, and comparatively alert and active in others. The phantasy of both is active. To ordinary persons any approach into their inner life is entirely precluded. But to the observer who understands the habits, or can interpret the dream of another, it is not difficult to gain the attention, to institute and maintain conversation, to effect a communication with the thoughts, to give positive direction and control to the thoughts, and, through the thoughts, to the feelings. No feature of a person in this condition is so striking as the entire and helpless dependence of some of his powers on other persons for stimulus and guidance, and the passiveness with which both the senses and the fancy respond to their suggestions, and are controlled by their direction.

In the artificial somnambulism these conditions are intensified. The natural equilibrium is more effectually disturbed than in the state just described. The insensibility of some of the powers, and the sensitiveness of others, are heightened. This condition is induced by processes that bring the operator prominently before the attention of the subject, and connect him with the trains of thought which his phantasy pursues. The subject falls asleep with his eye fixed upon the operator, by obeying directions which fell from his lips, and following motions and signs which engrossed his own attention. When the sleep is effected, it is in its nature but partial. A portion only of his powers are awake, and, by concession, are morbidly and sensitively alive to their appropriate impressions. It is not unnatural, rather it is most natural and reasonable, to expect that these powers so sensitive would respond to the voice and even to the tones of the one person to whom the patient had passively surrendered in the beginning of the process; that indications which escape the notice of ordinary observers, should be intelligible and patent for him, and that, when these indications are conveyed they should control all his movements of thought and feeling. It is credible that the pictures before the fancy of the operator should be awakened in his own, and that his positive assertion should not only be taken as proof of their real existence, but should cause the subject to believe that his own senses perceive them, so that he should think he sees a mountain, a house, brilliant colors, smoke, flame, etc., etc., at the will of the operator who dominates over his fancy.

Hallucinations, apparitions, etc.

§ 176. Our discussion of the phantasy would not be complete, if we omitted to notice the phenomena of hallucinations, and spectral apparitions or illusions. A distinction should be made between the proper images of the phantasy, when mistaken for or believed to be realities, as by the dreamer and the somnambulist, and the actual vision of images in the formation of which the senses coöperate, such as occur to persons in a morbid condition when they are broadly awake, as also to those attacked by fever, or to such as suffer from the effects of certain narcotics or intoxicating drugs. One of the most remarkable cases of continued exposure to such visitations, is that recorded of himself by the celebrated Nicolai of Berlin in the *Transactions of the Royal Society of Berlin*, for 1799.

The case of Nicolai is by no means solitary. There are not a few persons of sensitive organization who occasionally see distinct images, visions, and phantasms of real objects, which have distinct form, distinguishable color, and a certain permanent endurance like objects actually existing. These phantasms, moreover, assume relations of place and motion to real objects. They are seated in chairs, they stand by the bedside, they look through the window, and have the dimensions which are suitable to their place and their distance from the observer. If the judgment of the subject of them is clear, and his self-command complete, he knows they are not real objects, even though he cannot remove them. (Cf. *Hallucinations, or the Rational History of Apparitions, Visions*, etc., etc., by A. Brierre de Boismont, Phil. 1853.)

These phantasms are much more frequent in transient delirium from fever, or permanent insanity. They are the almost invariable result of a variety of drugs, as opium, hasheesh (*Cannabis Indica*), and stramonium. They are the fearful attendants of that irregularity of nervous action which is the consequence of excess in the use of intoxicating liquors. It is noticeable that the excitement occasioned by each of these drugs, as also that *delirium tremens* is attended by phantasms of its own. These phantasms are not confined to vision alone. The other senses have their appropriate phantasms; the ear has sounds, the touch various feelings, and the nostrils distinguishable odors. None of these, however, are as definite, or as permanent, or as clearly distinguishable as the phantasms of vision.

It is important to distinguish these phantasms or apparitions from the images of the phantasy proper. Unless we do, we cannot clearly understand or interpret the phenomena of delirium, and certain other forms of mental aberration. Two agencies concur in their production—the action of the phantasy by means of the spiritual image, and that of the sense-organ which is appropriately concerned. It has already been observed, that when even a sense-object is imaged, especially if it be vividly and continuously pictured by the phantasy, as a sound or sight, the mind's attention to it tends to awaken a sympathetic activity of the sense-organ by which the object was originally perceived.

Again, in the sense-organism psychologically considered, there is a tendency to be excited or impressed a second time without a sense-object, in a manner similar to that which the presence of the object originally occasioned. Sometimes, in conditions of the system not known to be abnormal, this excitement goes so far as to produce in the mind all the effects of transient sense-perception. As a consequence, the mind has actual percepts without material objects, especially on waking from sleep. The mind sees colored spectra, and hears sounds when there are no material things or objects to be seen or heard. These occasional phenomena clearly establish the truth that the sense-organism, without the stimulus of an object, can be brought into a condition nearly allied to that to which it is excited by that object. Whether the excitement is mental or physical, is of little import, provided the excitement is furnished. Let, now, the sense-organism be in a condition of morbid sensibility, and let the phantasy be also morbidly aroused, and it is not unnatural that phantasms should take material forms or be invested with material qualities. But let the judgment itself be disturbed by more serious disarrangements of the nervous system; and the raving madness which sees nothing but phantasms where it ought to see realities, or which invests the real objects of sense with fantastic shapes and attributes, are fully explained (cf. §§ 78, 143, 150).

Insanity.

§ 177. It is no part of our duty to give a scientific theory of insanity. We have only attempted to explain the part which the phantasy has in the mental operations, under this condition of irregular psychical activity. We ought also to add, that it is by no means universally the case that the insane are haunted with phantasms. It often happens that insanity is the result of mere mental confusion or distraction, such as may result from the excessive rapidity or the excessive preponderance of certain organic or vital sense-perceptions. These may so distract or preoccupy the attention, as to preclude the possibility of a cool judgment or a controlled activity in respect to any matter whatever. In such cases, the phantasy, as well as the perceptions, are either so hurried and flighty, or so fixed and recurring, that the activities of memory, comparison, and judgment are all untrustworthy. Or, again, the mind, and not the body, under some overmastering passion, has given to the phantasy such complete control over the other powers, as to disturb the equilibrium of spiritual activity. In these cases the phenomena are purely mental.

The sense-perceptions are correctly made. The vision is disturbed by no spectrum. There are no special disturbances of the bodily sensations. But the mind is occupied with inferences incorrectly derived from its past experiences or its present condition. It is haunted with depressing images, or gloomy forebodings. Its distracted phantasy is so overpowered as to set at naught the testimony of the senses, the asseverations of trusted friends, the conclusions of its own better judgment, the principles, the faith, and the hopes which had been the soul's support and guide.

CHAPTER VI.

REPRESENTATION.—(3.) THE IMAGINATION OR CREATIVE POWER.

From the *phantasy*, the most passive form and exercise of representation, we proceed to the *imagination*, its most active and elevated energy.

§ 178. In treating of the creative imagination, we shall first consider the general characteristics, conditions, and laws, which are common to this power in all its phases and degrees of activity, and then the special forms in which it is manifested. Conditions and materials common to the imagination.

Our *first* duty is, to consider the conditions, laws, and characteristics which are common to the creative imagination. We ask, first of all, what are the *materials* which are furnished to this power from nature and experience, and which it is forced to make use of in all its creations? In answer to this general question, we would say:—

1. Space and time are always employed in these processes, and always appear in their products. The objects that are conceived, whether by the poet, the dramatist, or the inventor, as forming the scenes in which their personages, materials, or machinery are introduced, or within which they are conceived, are invariably subjected to the laws and relations of space. The acts and events which are described or imagined, all take place under the conditions of time. They precede and follow one another. They are either present, past, or future. The world of the imagi-

nation is always a world of imagined space and imagined time, as the world of reality is a world of real space and real time.

2. The necessary and universal thought-conceptions, and relations under which we cognize real beings, are always supposed and employed. Every being and thing which we imagine, we imagine more or less distinctly, as substance with attributes, as cause and effect under proper conditions, and as means and ends.

It is not intended that the imagination pictures those in their abstract form. They cannot be imaged, any more than they can be perceived by sense or consciousness. But as concrete objects can be perceived only under these relations, so when they are imaged, they can and must be imaged as connected by means of them.

3. The imagination is limited to the material qualities which nature furnishes. We cannot create or conceive of new colors by any exertion of creative energy. Hume and Tetens both suggest, that if the imagination were furnished with the colors *blue* and *yellow*, it could, by combining the two, image the color *green*, without ever having seen it. The mistake is twofold. The eye does not see the *blue* and *yellow* in the *green*, but the product which results from the combination of the two. The imagination cannot go beyond what the bodily eye furnishes.

In a similar way, the imagination is limited with respect to all the simple qualities of sense, to tastes, and sounds, and odors, and tactual feels.

4. In like manner, the imagination is limited to the spiritual phenomena and processes which consciousness reveals, as well as to the powers which these processes suppose. What it is to know, and feel, and will, we know by the varieties of our own experience; and what a being is who can exert these activities, we are taught by consciousness. In this way we learn what are the acts, and products, and capacities of spirit.

The power of the imagination to create new products.

§ 179. We inquire, *second*, What new products can be evolved and created out of these materials by the imagination proper? We follow the order of the topics already adopted.

(1.) In respect to space and time, though we cannot imagine objects to exist nor events to occur out of relation to each or to both, yet we can imagine them to bear relations to each, to which there is no type of reality.

The imagination can make changes in the *size* of objects. The types of animals actually existing, as of the horse, the man, the elephant, and the mouse, lie within certain extremes, the greatest and least of their kind ever known. The imagination scorns these limits, and it can give us horses of every size, from the ponies of Queen Mab up to steeds large enough for the uses of a giant. It can create men smaller than Lilliputians, and larger than the contrasted Brobdignags. It can make elephants smaller than mice, and mice larger than elephants.

Again, the *position* or *situation* of objects is determined by the character of their material and the laws of nature. Mountains hold a certain relation to vallies, streams to meadows, groves to lawns, houses to gardens, cities to harbors, roads, and rivers; so that, where we find the one, we expect to find the other. But the imagination acknowledges none of these relations or laws. While it must imagine all these objects as spatial, it can place them as it will in space. It can plant a garden in a desert a thousand leagues from a dwelling of man. It can build and people a city, without harbor, river, or road.

There are fixed *forms* of objects in nature, as the drooping elm, the aspiring pine, the umbrageous beech, the massive and gnarled oak. In rock and mountain, certain types are ever recurring. The same is true of the form of the horse, the deer, the dog, and of man himself. But the imagination can draw more graceful lines than nature has ever shaped, the material with which she works being more intractable, and the action of staining and decomposing elements being inevitable. Following her idealizing images, art has given us the Egyptian tomb and pyramid, the Chinese pagoda, the Grecian temple, and the Gothic cathedral, none of which are copied from nature, though all have been suggested by her forms.

In one aspect they surpass nature, for their lines are more consummately drawn, and their forms are moulded more perfectly. We even measure nature by what art has done, and commend her by epithets taken from art. We say of the stem of the pine or the elm, It shoots up like a pillar. We call the forest a "pillared shade." We say of a man, He stands like a statue;

or, He is an Apollo, for graceful strength; She is a Venus, for beauty.

In *time*, also, the imagination has boundless range. It must represent all actions and events, as either *now, before*, or *after*, yet it can do as it pleases as to which shall be *now, before* or *after*. Nature, in respect to time relations, acts after its own laws and within its own limits. The imagination can override them all, and accordingly she can make Puck "put a girdle roundabout the earth in forty minutes," and Uriel "glide on a sunbeam," "swift as a shooting star."

There are also special creations which the imagination forms and constructs, of which space and time are assumed as the only required conditions. Let all material existences be conceived to cease to be, leaving only an empty void within any limits which may be supposed, and in that void which is feigned, the imagination can construct the surface with its ever-varied outlines, and the solid of every conceivable form. These are purely mental constructions, and exist only for the mind and by the mind which forms them. Their form may be suggested by certain material things with which we are conversant. But the *line*, the *surface*, and the *solid* constructed by the mind, are far more perfectly drawn and moulded than any that nature has ever furnished in material objects, or than art has imitated with material instruments.

The imagination can also sweep all actual events and phenomena from the line of time, and then plant along its course the shadows of events that shall only symbolize or represent its successive intervals or instants. It can also group and combine these as it will. Real events, as they precede and follow one another, may incite to these acts of pure construction; but the acts and the products which they excite and suggest are to be referred to the creative energy of the imagination. What relations these hold to the distinctions of number, will be discussed in the proper place (§ 280).

(2.) In the world of matter, the imagination can create no new material, but it can divide and combine the parts of the material things with which it is familiar, so as to form new existences.

The head and trunk of a man it can fit to the shoulders and body of a horse. It can form a mermaid—part woman, part fish. It can provide men, women, and children with wings, and turn them into angels and cherubs. It can represent any animal with a human head. It can add to the head of a man the ears of an ass, and give to another the mouth and nose of a puppy. It can connect the part or the whole of any plant with the part or the whole of any animal, making a cabbage to sprout from the hump of a camel, or a rose-branch to nod from the head of a horse, as we see delineated in some quaint pictures and engravings. It can recombine and reärrange the parts of inorganic things as it will, making a rock to be balanced upon a roof-ridge, and a bridge to stand dry in a desert. There is no limit to the grotesque and fantastic combinations which can be made with the parts and the wholes of material objects. Though the imagination cannot invent a single new sensible or material quality, it can connect such qualities as nature has never combined, making flaming red dogs, bright yellow oxen, woolly horses, talking mules, musical jackasses, golden mountains, rivers of wine, ponds of beer, and fountains of hot coffee.

(3.) In respect to spiritual beings, the imagination is limited by similar constraints and invested with a similar freedom. A spirit has no visible or extended parts; therefore, as a spirit, it cannot be divided and recombined; but a spirit may be connected with any kind or form of matter, may be imprisoned in trees, may animate a cloud, may dwell in an animal form, or "leap like Minerva from the head of Jupiter!"

Not a single new spiritual capacity can be invented or imagined. The loftiest and the purest of spirit-creations, simply feel, desire, and will. The humblest and the most degraded can do no less. We cannot invest the highest archangel with any endowment other than these. We cannot refuse to the lowliest animal some poor analoga to some of these functions.

In respect to the limitations and the conditions of the exercise of the intellect, the imagination has the widest range of creative power. It can conceive the intellect of a God that creates all that it discerns, and discerns whatever it creates, without condition or process, by an all-penetrating and all-comprehending intuition. It can also imagine the intellect of an idiot, struggling to free itself from the gross obstructions of a diseased body, and fixing its painful attention in the first beginnings of knowledge.

In respect of feeling, it can, on the one hand, imagine pure love glowing with the energy of seraphic fervor, or simple hatred raging with fiendish malignity; and, on the other, the most imperfect and feeblest actings of either.

There is no limit to the variety of spiritual beings with which the imaginary world can be peopled, nor to the variety of the conditions of being and acting to which they can be subjected. The graceful Titania, with her frolicsome and mischief-making fairies; the hideous Caliban, in body and spirit the very contrast of the wonderful Miranda; Satan and Abdiel; are examples of the variety of spiritual creations which the imagination can construct out of its limited materials.

(4.) We have seen that the imagination cannot step without the charmed circle of thought-conceptions and relations. Some of the examples of what it can do within that circle by newly conjoining attributes of material and spiritual beings, have already been given. It cannot conceive of beings, except as substances and attributes, but it can join any attribute, of any intensity and compass, to any substance. It cannot break them from that connection which binds all real beings and events as causes and effects; but it can make any existence to serve as the cause of any other as its effect, and thus can reverse the whole order of actual being by its capricious and fantastic combinations; or it can enlarge the bounds of science by its happy suggestions of undiscovered powers and laws, and the appliances of art by applications, before unimagined, of familiar agencies to new results. All things in the world of fancy must be conceived as fitted for some end, but the adaptations may be imagined as wildly as the caprices of a madman's dream, or as wisely as the perfect fitness which we believe has been arranged by the all-wise God.

With this view before us of the materials to which the imagination is limited, and of the products into which it transforms them, we are prepared to inquire, *third*, How does the imagination effect these changes; or what is the precise work which the imagination performs in its creative function? We observe, in answer to these inquiries, There are three different methods in which its creative power is shown. (1.) The imagination can recombine and arrange the constituents of nature in new forms and products. (2.) It can idealize and apply the relations of objects to extension and time. (3.) It can form and employ an ideal standard for the intensity and the direction of the activity of natural or spiritual agents, and for the material objects and acts which symbolize them. We will consider these acts in their order.

The combining and arranging office of the imagination.

§ 180. The examples already cited both prove and illustrate the fact, that the imagination very largely acts in the way of reuniting and reärranging the materials furnished to experience, and they also suggest the limitations under which this function can be employed. It is obvious, also, that the so-called parts of objects, and objects treated as parts, are as minute and numerous as any species of analysis can separate.

There are sense-parts and sense-wholes, representative-parts and representative-wholes, and thought-parts and thought-wholes. A whole, as a building or tree, may be a part of the landscape with which it is connected; while it is still a whole with respect to its doors, windows, roof, etc., and whatever else makes it quantitatively complete. This is an example of sense-wholes and sense-parts. Again, the several properties or relations of the dwelling or the tree, its form, dimensions, color, smell, etc., are thought-parts, which can be combined into new wholes, by taking away and adding, as we have already seen. If these new wholes are individual, they are formed from representation; if they are generalized, they are the work of thought proper, or logical wholes in the larger sense of the word. The synthesis of the creative imagination reaches as far and is applied as widely as the analysis of sense and thought can go. The imagination may reunite into varying products all that perception and consciousness separate or distinguish, and under every one of the relations in which they apprehend their objects. These relations are its only limits and laws.

The idealization of the relations of space and time in art, and mathematical science.

§ 181. We have already referred to the fact, that the imagination, in every work of art, goes beyond, and outdoes the perfection and refinement of nature. The forms which sculpture moulds, and which drawing outlines, are, as we have seen, more perfect than any which nature produces; certainly, they are more perfect than any which the senses can discern, or which nature can furnish as models. These constructions cannot be explained by any process of analysis, or selection of the parts of real objects, whether this analysis is called mental, or is performed by sensible instruments. The lines and shapes of grace which have been copied in marble or drawn upon canvas, in respect to delicacy of transition and ease of movement, far surpass those of any living being or actually existing thing. They are suggested by, but not copied from, any such beings or things. The story that the Grecian painter assembled from every quarter the most celebrated beauties, that he might borrow some charm from each, and combine all together in a perfect work, could never have been true. While it is true that nature, in some respects, far outstrips and surpasses what art can do, it is true, on the other, that the imagination, in her province, can go far beyond the attainments of nature. As we have already said, we even measure nature by some of the achievements of art. We apply the ideals of the imagination still more frequently to try and to test what spiritual achievement furnishes.

Those peculiar products which are employed in mathematical science, and which are known as geometrical and numerical quantities, cannot be made by any process of separation or combination of the parts of material objects. In matter there are no points, lines, surfaces, solids, and spheres, such as geometry conceives and reasons of. The unequal faces of a material cube, the rough edges formed by two adjacent faces of a solid, the obtuse corners in which three adjacent faces terminate, are none of them these objects of thought, nor are they wholes from which these can be evolved or separated as elements or constituting parts. The line is not part of an edge, nor is the surface a part of the material face. If they were parts which could be separated by actual sense-perception from a whole, they must exist in that whole, or be distinguished as one of its material constituents.

If it be said that these are distinguished and separated in the mind, that the process of analysis or abstraction is mental, it is still true that the mind can only separate what it first discerns. These objects cannot be discerned by bodily sense, nor can they be represented by simple imagination. They must be created by the mind, for the mind to behold, when the mind beholds them. It may be properly said to construct or to create them—first, in individual examples and applications, and then by rapid and easy generalizations. An individual point, line, surface, triangle, solid, sphere, is first constructed by the mind in relation to and by suggestion of a rude material occasion, and the product is then generalized by the ordinary processes and conceived as resembling every similar creation, so that whatever is true of the one, is readily affirmed of all.

What is true of geometrical, is true also of numerical quantity. Numbers symbolize the relations of objects contemplated in a series, as constituting a whole, divisible into unit parts. In order to conceive of number, the mind must first view objects in all these relations. But in nature, so far as the senses can know, there are no equal parts constituting divisible wholes. Whether the ultimate molecules or atoms of matter are or are not equal, none such are discerned by the senses. The successive mental states which consciousness observes and by which it first apprehends and measures the successive portions of time, are none of them observed in actual experience to be equally long or short. All these must be idealized in the imagination before they are separated by its analysis and combined in its creations. We proceed to

3. The formation of an ideal standard for psychical acts and states.

§ 182. The spiritual acts and states of which we are conscious, differ from one another in respect to the direction which they take—*i. e.*, in respect to the objects on which they terminate, and hence to the quality of the affections—as well as in respect to the energy or intensity with which they are performed. But none ever reach a perfection in either respect which is so complete as can be conceived. Whatever or however we know, feel, or choose; we can conceive it possible to surpass what we actually do or experience. A perfect standard is created by the imagination. It cannot be derived from the parts which we observe in ourselves or others, because the parts are no more perfect than are the wholes. Consequently, whenever we perceive dimly and believe that we might perceive more clearly, or whenever we would feel warmly or purely, or choose rightly, and our feelings or choices do not satisfy our tastes or our conscience, we then create for ourselves an ideal standard of spiritual achievement.

In respect, also, to the expression of these ideals in material forms, the imagination creates and applies the ideals which it always aims but always fails to reach. Whether the medium of expression be language—the language of gestures, of looks, of tones, or of articulate speech—or whether it be lines, or color, or solid form, as employed by the draughtsman, the painter, or the sculptor, it is all the same. The use which we can make of the medium is never so perfect as our ideal of what is possible. As we have noticed already, every such medium, physically regarded, falls short of the psychical perfection which we can conceive—*i. e.*, create—in the mind. When this medium or material is required, not only to set forth an idea of simple outline, form, or color, but to represent another ideal of thought, feeling, and passion, then it is found to be doubly true that the ideals which the mind can frame, rise far above the reality which the voice or hand can execute. Hence it is that the ideal excellence of the poet, the orator, the actor, the musician, and the artist, is ever higher than his achievements—that the one flees before the other like its shadow, and can never be overtaken.

The ideals of science and of art, of achievement and of duty, are the products of that form of psychical activity which is properly called the creative imagination. It is *imaginative*, because the representative or imaging power is conspicuously prominent in its functions. It is *creative*, because there is no counterpart in nature from which its objects and products are literally transcribed or copied. But this is not all. The *reason* and the *feelings* are conspicuous, and both rational and emotional relations are recognized and controlling. The *creative function* is rendered possible *by the union of the thinking power with the imaging power;* the joint action of both resulting in those ideal products which address the intellectual and emotional nature.

The ideals of the mathematical imagination are only possible when the imagination has been disciplined by thought. One chalk or pencil line is narrower than another, one of the laminæ of mica is thinner than another. As we divide these lines and cleave off these laminæ, we seem to approximate to the ideal line and the ideal surface, simply because the senses and the imagination are less distracted and occupied with sense or imaged properties. The imagination selects, therefore, the line or surface whose thickness is least obvious to the senses, to suggest or represent the sole relation to space with which the intellect is for the moment concerned; or, which is even more satisfactory, it takes for a point an object whose dimensions are the smallest discernible to the senses or picturable to the imagination, and considers it simply as moved or movable directly to another point like itself, and thus *constructs* in the imagination the mathematical line. That is, it begins with an object or an image as far removed from sense as possible, and uses it so as to suggest the various relations which extended matter holds to space; or, to speak more exactly, to other matter extended in space. By the imagined motion of this line, it proceeds in a similar way to construct the surface, etc., etc. The so-called approximation of the actual to the ideal line and surface, consists in the more facile suggestion of the relations in question by means of one reality rather than by another.

The *ideal* of the *artist* depends on the relations of outline, form, color, etc., etc., to æsthetic pleasure; whatever may be its sources and kinds. He brings the line, the model, or the picture, as nearly as his materials and skill will allow, to a condition in which there shall be no drawbacks to the pleasurable effect which is sought for. As long as a single distracting or inconsistent feature or property is prominent, so long is his ideal unreached. As this will always be the case from defect of materials or defect of skill, so long will it be true that he can never make his work absolutely perfect, and that his ideal of what he imagines might be possible, will never be reached.

The *ideal* of the *inventor* is some agent, or combination of agencies, that are freed from the limitations which pertain to the ordinary machines or instruments. These he illustrates to himself by fondly and sometimes obstinately conceiving of his model only in those relations of adaptation and capacity which he knows it to possess, and overlooking or denying other limitations to which it is liable.

The ideals of *psychical* and *moral attainment* suffer under limitations of another sort. We select the most satisfying example of the actual which we can find, and fixing our attention upon those of its relations which we desire to contemplate, and withdrawing it from all defects and limitations, we make the example an ideal of the psychical power or the moral excellence which we wish exclusively to contemplate.

If the ideal excellence is contemplated as an attainable end of our being, or is enforced by the authority of conscience or the will of the Supreme, then that which was a conceivable ideal is viewed in still other relations. It is accepted as possible: though an ideal of the imagination, it is enforced as reasonable and obligatory.

The result of this analysis is but another illustration of the interdependence of all the powers upon one another, and especially of the higher functions of the imagination upon thought and reason. It enforces and explains the near affinity of the imaging with the thought-power. It also indicates the advantage which language and music may have over painting and sculpture in expressing and suggesting what color and form cannot convey.

These truths also enable us to understand and explain how it happens that all ideas, however refined and elevated, are in some sense founded upon and related to the actual experience of each individual. A person born and nurtured upon a plain, who had never seen a hill or a mountain, can scarcely imagine the charm to the eye and the excitement to the mind which such scenery imparts. One who has never been upon the sea, can neither picture to himself nor to others the wild sublimity of an ocean tempest. The oriental, basking in the heat of a tropical sun, and always surrounded by the fruits, the foliage, and the flowers that such a sun alone can nourish, cannot form an ideal picture of an arctic winter. Nor can the Scandinavian, out of the pale sunlight of his brightest days, or the most luxuriant vegetation of his starveling summer, construct an adequate representation of the exuberant life, and the glowing intensity of a tropical landscape.

The actual life of every painter and every poet, in the materials which it furnishes, must largely determine the direction and characteristics of his imaginative power. From the writings of Dante, of Milton, of Scott, and of Bunyan, as well as from the pictures of Raphael and Murillo, of Gainsborough and Wilkie, one can easily conclude as to the place of their birth, the kind of education which they received from the books and men and scenery with which they were conversant.

§ 183. It follows that the imagination is capable of steady growth, and requires constant cultivation. This training and growth are not, however, occasional, but constant; they are not the results of separate efforts, which are consciously directed to some definite ends of creation, but are the consequents of an activity which is spontaneous, irrepressible, and often excessive. Indeed, in all minds the creative imagination mingles more or less prominently with the other mental operations, always modifying and sometimes greatly disturbing the acting of these powers and their results. In sense-perception, the imagination too often selects for itself what it will see or hear, and brings a report accordingly of what it thinks it has seen and heard. After the desires are grown strong and the character is fixed, the shaping spirit of the imagination enters largely as a modifying influence into the perceptions. In the observations of consciousness, and the reports which it records of what it has seemed only to observe, the same influence and the same effects may be traced of its creative energy. The observation and the record are both disturbed by the power to notice what we are anxious to find, and to leave unobserved, or to imagine that we cannot see, what we do not wish to find to be true. In the act of recalling for ourselves or communicating to others what we may have actually observed or experienced, the creative imagination often intrudes, consciously or unconsciously biassed by the desire to please ourselves or our fellow-men. The frequent and strange untrustworthiness of the memory, can be accounted for only by the selecting or idealizing activity of the imagination, when it seems to be simply recalling the actual past. Inasmuch as the thought-power, in its various acts of reaching general conceptions and conclusions, chiefly depends on the fidelity of the representative power in reproducing the actual; whenever it creates instead of recalling, all the results of thinking must be disturbed. In this way the imagination may and does enter very largely into the acts of generalization, inference, and deduction; disturbing and misleading all.

The imagination is capable of growth and culture.

Is developed from the earliest till the latest periods of life.

§ 184. More generally we may say, this creative power is developed at the earliest period of our existence, and is busy in all ages and conditions of our human life. Childhood, in some of its aspects, is the most literal, and the most observant of reality; yet even then the shaping activity of the imagination is always busy, filling the real world with another of fancies and dreams. The most trivial and unsuitable objects are sufficient to excite its action. The rude and unfinished toy is more acceptable to the child than the more costly and elaborate, because it leaves more room for the constructive power. It is all the better if the greater part of the work is left for this to complete and supply. The sports and plays of childhood are little romances, prompted and acted over for the simple exercise and delight of the imagination. In later years the imagination is always busy. The interest which each man takes in the position in life which he holds or aspires after; in his employments, his friends, and associates; or the dislike and disgust which he conceives for each and for all; arises from the ideal lights with which the imagination invests them. The eye of the painter looks every landscape into a picture, and idealizes every face that it beholds.

> The lunatic, the lover, and the poet
> *Are of imagination all compact.*
>
> *Midsummer-Night's Dream.* Act v.

This constant activity of the creative power explains its rapid growth, and its development into the capacity for sudden and surprising achievements.

Whenever an occasion calls for the manifestation of the power thus trained and matured, it acts as by the force and with the promptness and precision of apparent inspiration. Whether the exigency be that of the artist, the poet, or the inventor, the creative power formed by the ceaseless activity of years meets its requirements from the resources that it has been gradually providing. These resources may consist in part of the countless creations which it has shaped in connection with its perceptions and reveries, and which are again summoned back by the memory when first these images are needed; or, the resources brought to the exigency may be the dexterity which has been acquired by use, and which dexterity consists in the power of so controlling the associating power that it shall yield the very materials which are wanted for the imagination to work upon.

In no other way can we explain the rapidity, the precision, and the success with which the constructing and inventive power seems to act when it is tasked to its utmost energy and produces its finest results.

Special applications of the imagination. The poetic imagination.

§ 185. The fact has been noticed, that the creative imagination is present by its actings with all the other powers of the soul, and determines the character of their products. We have also seen, in our analysis of ideals, that the converse is true as well. All these powers are present in varied proportions and energies in those activities which are recognized as the acts of the imagination, and give a varied character to what are called its products, whether they appear in the form of poetry, fiction, the fine arts, philosophy, ethics, or religion.

Of these, the *poetic imagination* is the most interesting, and invites to a special analysis. *Poetry* may be defined, that use of the creative power which is employed for the gratification of the emotional nature in the production of pictures more or less elevating in their associations, which are fixed and expressed by means of rhythmical language.

The sources from which the poetic power derives its materials are as numerous and extensive as the universe of matter and of spirit, and yet but few of these materials subserve the proper aims of the poet. While the poet may lawfully appropriate truth of every kind, provided it serves his purpose, yet it is pre-eminently that truth which holds or may be made to assume some relation to man which is of use in poetry.

This human truth, which these pictures suggest, illustrate, or enforce, must be that which is within the comprehension and reach of all men. It is not the truth of the schools, nor of any special and limited society, not that which is capable of being conveyed in abstract or technical words or understood by a select few after a special training, but it is the truth which is open and intelligible to all men upon certain implied and easily recognized conditions. This is the first of the three characteristics which are recognized by Milton in his brief description of poetry as " simple, sensuous, and passionate."

Poetry should indeed be *simple,* because its products are designed for the use of all men ; and its images, thoughts, and words should be easily comprehended by all who have attained certain advantages of culture, and have been trained to a certain degree of thought and feeling. It should also be *sensuous*—that is, it deals with images, not with generalized and scholastic language. It presents pictures to the mind's eye, not refined and subtle reasonings to the thought-powers. It introduces action into every scene. It is eminently concrete and picturesque. It should also be *passionate*—*i. e.,* its simple and pictured truth should come from a soul that is animated by warm and elevated emotions. The presence of feeling as a requisite of all that composition which is called imaginative, is not always recognized so distinctly as it deserves to be. Without feeling, and, in general, without feeling of a higher kind, the mere power to create is of little worth, and its results are of little interest. Indeed, without it the power will not be so matured into a predominant energy, or be so regulated, as to become a ready instrument at the service of its possessor. But with it, the creation of the kind of pictures in which the emotions delight, becomes a pastime and an occupation, and poetry is to the poet its own " exceeding great reward." Inasmuch as only the higher emotions act with a steady and intellectual pressure in the refined occupation of poetic culture and composition, the images which association presents and the imagination detains and reconstructs, are of an elevated character; they assume the lofty and ennobling character of ideals in the better sense of the word. Hence it becomes so generally true that poetry is almost necessarily elevating in its nature and influence. Hence it has been held to have something in it that is divine.

The ends of poetry are not always elevated. Poetry may serve simply or chiefly to amuse. When this happens ; when its pictures are employed for this end, and the associations under which they are present, and the emotions which they excite, are not especially ennobling, the poetic imagination is, in the language of later critics, called the *fancy.* When the aims are higher than simple gratification, and therefore involve more elevated associations and feelings, it is dignified as the *imagination* by eminence, and so designated. The adjective *imaginative* follows very closely this higher sense of the word. In this activity the image-making power simply plays or sports with images for their picturesque effects and the amusement which they give—or arranges them for ends of illustration or pleasure. Though it abounds in images, it lacks the loftier attributes of the higher imagination.

§ 186. It is peculiar to the poetic imagination in all higher and lower forms that language is its medium. It is not essential that this language should be metrical; though a rhythmic movement, and the regular return of similar syllables in measured accent greatly heightens its effects. The poetic power is also shared by the novelist, the dramatist, and the orator. But poetry must always employ language, and for this reason it essentially differs from painting, sculpture, and even music. Painting and sculpture create images indeed, but they fix them permanently upon the canvas or embody them in marble. But poetry can only suggest them by words; it portrays its images only, as by words it wakens in the imagination of another, images similar to those which the poet himself conceives. If the imagination that receives is feeble, slow, and perverse, it is in vain that the poet tries to excite it to follow his lead. But if it is strong, quick, and sympathizing, it may be aroused by the words of the poet to finer creations than even the poet himself has known. The suggestive power of words gives to the poet a marvellous advantage in the greater breadth of his field and the variety of his effects. The painter and sculptor apparently present all their work to the eye. It is true that this work is better appreciated by one eye than another. In one sense it takes an artist to interpret an artist; but even with this allowance, the range of their indications is narrow, and the possibility of manifold suggestions is limited. But words have a capacity to suggest more than they directly convey, and hence to take up into their import a multitude of pictures according to the variety of uses to which they are applied. The word whose literal import is prosaic, trivial, or mean, when used by genius in a new application, becomes poetic, picturesque, and elevating. The material which in common use is cold, conventional, and dry, has capacity, by dexterous combinations, to awaken delightful

Its medium is language.

imagery, and to kindle exalted associations. In this way language itself becomes permanently enriched and elevated by the fact that it has been employed by men of poetic genius.

The philosophic imagination.

§ 187. The relation of the imagination to thought has been the subject of much discussion, and has given rise to no little diversity of opinion. Many have contended that its influence is unfavorable to the operations of the intellect in the discovery of truth; that it distracts the attention, biasses and misleads the judgment, and disqualifies for any of the reasoning processes. On the other hand, the fact is undisputed that the men who have been most distinguished in philosophy, especially as discoverers or inventors, have been remarkable for reach and glow of imagination. Striking examples of the combination of the poetic imagination with eminent philosophical genius are numerous. We name *Plato*, *Kepler*, *Galileo*, *Bacon*, *Newton*, *Leibnitz*, *Davy*, *Owen*, *Faraday*, and *Agassiz*. A moment's reflection will show how this must necessarily happen. The objects of present observation must always be limited in number. They must reappear in the form of representations. The facts with which the philosopher has to do must come to him in the form of images, when he would discern their various relations and subject them to the processes of thought. It is important that these should be readily represented. This can only happen when the associative power is wide in its range of relations, and quick in its activity. These qualities almost invariably accompany, if they do not necessarily involve, great energy of the creative power.

But whatever may be thought of the importance of a vivid imagination, as furnishing the materials for the philosopher, to invention it is entirely essential; indeed, without an active imagination, philosophic invention and discovery are impossible. To invent or discover, is always to recombine. The discoverer of a new solution for a problem, or a new demonstration for a theorem in mathematics, the inventor of a new application of a power of nature already known, or the discoverer of a power not previously dreamed of, the discoverer of a new argument to prove or deduce a truth or of a new induction from facts already accepted, the man who evolves a new principle or a new definition in moral or political science—must all analyze and recombine in the mind things, acts, or events, with their relations, in positions in which they have never been previously observed or thought of. This recombination is purely mental. Every discovery is, in fact, a work of the creative imagination.

It is true the power of thought must attend the operation. Unless the representations and combinations are made and regulated with reference to the ends of thought, they will be made in vain. But the range of these pictured objects must be wide; every one of them must be vividly conceived, that all the attributes, and analogies, and relations may come before the eye of the mind. The more vividly this presentation is made, provided the processes of analysis and comparison go on with equal energy, the wider is the field of discovery and the greater is the chance of success. The world of images is also far more plastic than the world of reality. Its materials come and go more quickly than real objects. More can be crowded at once into the field of view. The mental analysis and synthesis required, can be more rapidly performed upon the shadows which the mind summons to its service, than upon the things which it can slowly call up and slowly survey.

But there are special reasons why the peculiar type of imagination which the poet requires is closely allied to that which is essential in philosophic genius. To the higher imagination, as required by poets and orators, there is always requisite the power to interpret the indications or analogies of the beings and phenomena which they observe. The intensity of interest that fixes and holds the mind in the patient attention of the philosopher is closely allied to that strongly absorbed and controlling enthusiasm which holds the poet to the images which his fancy summons or creates. Both dwell in such a world with an enthusiasm which is not easily understood by others. That which maintains the interest of each, is the passion of each for the image-world which he recreates. That which gives to each his mastery over this world, is the familiarity which results from long-continued practice in calling up its objects and in moulding them at his will. Such a mastery, arising from such a continuity of effort, can only be attained by that passionate interest which is the secret of genius, whether genius labors for the ends of scientific or poetic truth; whether the end for which it labors is the truth of science that addresses the intellect, or the truth of feeling which controls the heart.

In the communication of scientific truth there can be no question that a large measure of imagination is of essential service. He who would amply illustrate, powerfully defend, or effectively enforce the principles and truths of science, is greatly aided by a brilliant imagination. This, of all other gifts, is the best security against that tendency to the dry and abstract, the general and the remote, to which the expounder of science is exposed by reason of his familiarity with principles which are strange to his pupils and readers, and which need to be continually explained and illustrated by fresh and various examples. The philosophic writer or teacher who is gifted with imagination is more likely to be clear in statement, ample in illustration, pertinent in the application, and exciting in the enforcement of the truths with which his science is conversant, whatever may be its subject-matter.

The practical and ethical imagination.

§ 188. The *practical* and *ethical uses* of the imagination are numerous and elevated. These are sufficiently obvious from the single consideration, that the standards by which we regulate our aims and estimate our achievements must always be ideal creations. They are continually formed and reformed by the imagination. These ideals, so far as the particulars of the character and the life are concerned, may vary both in their import and in the vividness with which this import is conceived. If they are consistent with the conditions of human nature and human life; if they are conformed to the physical and moral laws of our nature, and to the government and will of God, they are healthful and ennobling. Such ideals can scarcely be too high, or too ardently and steadfastly adhered to. But if they are false in their theory of life and happiness, if they are untrue to the conditions of our actual existence, if they involve the disappointment of our hopes, and discontent with real life, they are the bane of all enjoyment, and fatal to true happiness.

It is not what we actually attain or possess that makes us happy or wretched, but what we think is essential, or possible, or just for ourselves to attain. The ideal standard by which we measure and judge our attainments in all these respects, is a most important element of satisfaction or discontent. It is of little consequence what a man has, if he imagines that he must have something more in order to be truly happy. If his ideal contemplates self-sacrifice, suffering, and evil, as possible conditions of good, he will be still more secure of a happy life. If it reaches forward to another scene of existence, and brings before him the blessedness of a character perfected by suffering and made fit for the purest and noblest society conceivable, his happiness on earth may even be augmented by disappointment, sorrow, and pain.

If, on the other hand, these ideals are factitious or unreasonable, they become the source of constant wretchedness. If a man, to be happy, must be as rich or as fashionable, as successful or as accomplished as he dreams of, all his actual enjoyments pass for little or nothing till his ideal desires are gratified. These are the standards by which he measures his good. If he fails to realize these, he cannot be satisfied.

The ideals we frame of life and happiness must involve a more or less positively ethical character. We cannot imagine what we are *to be* and *to become* in fortune and success, without proposing more or less distinctly what we *ought to be in character* and to perform in action. Hence, in a certain sense, what a man aspires to become, has already *ethically* decided what he is. His aims and standard are the reflex of his wishes and his will, as well as the assurance of what he can achieve in the future.

The ideal standard of duty may be constantly corrected and improved. From his own experience of the effects of acts or habits, or his observation of these effects in others, a man may supply what he has omitted to observe, or correct that in which he has erred, and so advance to a higher and more perfect rule of feeling, of manners, and of life. In this way a community may rise or sink, may advance or go backward. Every man may also advance the ideal of others by his good life, by the realization in himself of what is worthy, and his more perfect manifestation of it in appropriate and beautiful acts. The contemplation of fictitious characters, elevated and ennobled by ideal beauty, has served to quicken and enforce the ethical ideal of thousands of susceptible minds. The poet, the novelist, and the dramatist, may quicken the fervor, and instruct the minds, may elevate the tastes, and reform the lives of all their readers.

Relation of the imagination to religious faith.

§ 189. The relation of the imagination to religious faith is interesting and important. The objects of our faith, by their very definition, have never been subjected to direct or intuitive knowledge. And yet the imagination pictures these objects as real and most important. What are the materials out of which it creates them? Whence the suggestions which it idealizes into more refined and spiritual essences? By what authority does it invest these creations with verisimilitude and impose them upon the assent of the intellect, as representing the most real and important of all truths? What analogies are there between the finite and the infinite which authorize the imagination to use the one to symbolize the other, and which justify its faith in its own symbolic creations?

Of the Divine Being as Infinite, we have no direct experience. All our direct apprehensions of spiritual attributes and relations are of the limited only. It is by the limited that we reach the unlimited even in thought.

Conceding that we can think the infinite, can we also image it? We cannot. The sphere of the imagination is only the finite. All the pictures which it can construct are of limited objects. It is by means of such pictures only that it can image its concepts of the infinite, if it attempts to image these at all. That it can adequately picture them, no man believes. What is pictured by the image, is some limited example of some real being which suggests or exemplifies the thought-relations required.

These thought-relations are: existence, power, knowledge, origination, foresight;—all which, we say and believe, are both finite and infinite. But when we seek to image these as infinite, we select some finite examples that illustrate these attributes; we choose an image to give life and reality to the analogon of that which we believe to be unlimited in respect to its sphere and energy.

But these utmost efforts of the imaginative power to reach the infinite and the absolute, are always attended by the belief that they fall short of the reality; that no enumeration of finite objects, however interesting in themselves, or significant they may be, are at all adequate to illustrate the divine; that no continuation of space or of time can express the divine eternity; that no quanta of dependent beings can fitly represent the Being who is self-existent. To have the materials that shall enable a man fitly to image the infinite, one must himself be infinite. There are, indeed, analogies between the created and the creating spirit; else the one could not know the other in any sense or to any degree. But these analogies are too few and too inadequate to enable or authorize man to penetrate into the secret things which belong to God, or to make conceivable the divine by any images which man applies so freely and so rationally to limited things. The imagination is not easily content to use the analogies which are placed at its command, and to refrain from using those which it may not lawfully employ. It would fain go further than it can or ought. To do this, has been its constant temptation. To refuse to go as far as it may and ought, is weak and unphilosophical; but to attempt to go further, is always irrational, and, it may be, impious.

In respect, also, to the capacities and experiences of the spirit-state,—when separate from a human body or any material organization—the imagination is limited in the materials of its working and the products which it creates. We know the soul only in its connection with the body. To image any of its acts or states without a constantly present background of bodily sensations, is to imagine a mode of existence that seems to us imperfect and unnatural. We cannot imagine the soul without the body by which to know and act, and without material objects to act upon. If we attempt this, we bring to our aid some attenuated matter for the soul's habitation and instrument, and we surround it with a world of objects that wear the forms of material things. But here the question continually presents itself, How far can we image that world by this, and the soul's experiences in that world, by its experiences in this? Can we properly imagine either? May we apply the pictures drawn from this life to illustrate or make conceivable the scenes and events of another state? We not only can, but we must; yet ever with the caution, that the images which we use be not allowed to suggest more than the data authorize.

It should not surprise us to find that the imagination, when it rises into faith in the objects of the unseen world, invariably uses pictures that are borrowed from the world of matter, and phrases all its language from materials furnished by this imagery. It cannot do otherwise. However lofty its conceptions may be, however soaring its aspirations, undoubted its beliefs, or ardent its hopes, all these must be pictured and expressed in the images taken from that world of matter which is adapted to a soul that knows and acts through a material organism. If there be a revelation that is conveyed by human language or addressed to the human soul, it must in this respect be accommodated to the capacities of the soul that is to understand and accept it. The fact that a revelation must be conveyed by such a medium, does not disprove that it is possible, or at all detract from its importance or authority. It cannot be argued against its divine origination or supernatural confirmation, that it conforms itself in this respect to the nature of the being to whom it is given.

If, however, we regard the necessary limits of imagination and faith, we shall not expect that either will do more for us than lies in the capacities of either. We shall not confound the images of analogy with the intuitions of direct knowledge. We shall not mistake the accessories of illustrative imagery for the realities of the concepts or truths which this imagery sets forth. We shall not revel in sense-pictures of the fancy, as though the sensuous in them were literal truth. We shall not be imposed upon by pretended seers, because, forsooth, their pictures of the unseen are so minute, so copious, and so beautiful, or so confidently set forth; overlooking the circumstance that these visions may be merely the residua of a too luxuriant fancy, or the creations of an excited and perhaps an insane imagination. The recognition of the human limitations in the divine, will teach us to interpret the divine aright, while it may save us from accepting as divine that which is only limited and human.

PART THIRD.

THINKING AND THOUGHT-KNOWLEDGE.

CHAPTER I.

THOUGHT-KNOWLEDGE DEFINED AND EXPLAINED.

Thinking and the thought power defined.

§ 190. The third kind of knowledge of which the intellect is capable, is *thinking, or thought.* The term *thought,* when used in this special or technical sense, is applied to a great variety of processes, which are familiarly known as *abstraction, generalization, naming, judging, reasoning, arranging, explaining,* and *accounting for.* These processes are often grouped together, and called the *logical,* or *rational* processes.

The importance and intimate relationship of these processes is seen by their place with respect to the higher knowledge and attainments of man. It is by thought only that we can form those conceptions of number and magnitude which are the postulates and the materials of mathematical science. By thinking, we both enlarge and rise above the limited and transient information which is gained by single acts of consciousness and sense-perception, as we lay hold of that in both which is universal and permanent. By thought, we know effects by their causes, and causes through their effects: we believe in powers, whose actings only we can directly discern, and infer powers in objects which we have never tested nor observed: we explain what has happened by referring it to laws of necessity or reason, and we predict what will happen by rightly interpreting what has occurred. By thinking, we rise to the unseen from that which is seen, to the laws of nature from the facts of nature, to the laws of spirit from the phenomena of spirit, and to God from the universe of matter and of spirit, whose powers reveal His energy, and whose ends and adaptations manifest His thoughts and character.

These processes give us the most important part of our knowledge, and qualify us for our noblest functions. Thought makes us capable of language, by which we communicate what we know and feel for the good of others, or record it for another generation; of science, as distinguished from and elevated above the observation and remembrance of single and isolated facts; of forecast, as we learn wisdom by experience; of duty, as we exalt ourselves into judges and lawgivers over our inward desires and intentions; of law, as we discern its importance and bow to its authority; and of religion, as we believe in and worship the Unseen, whose existence and character we interpret by His works and learn from His Word.

But what it is to think, and how thinking should be defined, may be more easily understood by a concrete example. We take a familiar object, as an *apple*, and proceed to *think* it, in the various processes already named.

First of all, we know it as a being or a something, as distinguished from nothing; next, we think or know this being as possessed of and distinguished by attributes or properties which we can separate in thought from the being to which they belong. We go further: we observe in other objects—apples—attributes like those which we discern in this; we see the objects to be similar in color, form, taste, etc. In this way we form the mental product called a general *notion* or *concept* of the apple, or of apples in general as we say, which we can analyze and define. To *abstract* and to *analyze*, is to think. Next, we restore, or think back, these general concepts to the individual apples, and in so doing, we divide them into *higher* or *lower*, *wider*, or *narrower classes*. *Classification* is involved in thinking. As we proceed, we mark and fix what we have done by language. We give names to each of these attributes, to the concepts and things formed and denoted by several attributes united, and to the classes and sub-classes into which they are separated. Thinking is necessary to language. Next, the apple holds relations to space and time. It is both extended and enduring. The perception of the apple conditionates or involves the knowledge of both *space* and *time*. By thought and imagination we are enabled to separate the object perceived from both time and space, and to construct in space the various geometrical figures, as well as to conceive and define them by their necessary attributes or properties. Moreover, all sorts of entities, whether things existing, or thought-things, whether attributes or beings, can, by the common relations to time in the mind that thinks them, be thought in the relations of number. Again, the object—the apple—is believed to be produced from a tree, beginning as the germ in the blossom, and gradually expanding into the ripened fruit. It is known also to be dependent upon the agencies of heat and moisture acting together with the living tree. Thought, connects these as *cause* and *effect*, and finds in the phenomena thus connected, the relations of the powers and laws of their causative agents. We proceed to a higher act of thought knowledge. By observing the powers and conditions in any class of apples, their habit of growth, the soil, situation and temperature favorable to their successful cultivation, we infer that the same are required in all cases for this kind of fruit, and confirm the suggestion by experiment. But we do not rest with the *induction of powers and laws.* We observe that the apple is useful and pleasant as food. We notice that it is the product of cool climates, and can, with proper care, be preserved through the winter. We do not merely observe and record these as facts, but we connect them by the relation of *adaptation*, or fitness to the wants of man.

The nature and processes of thought might be illustrated by an example selected from the world of spirit. By consciousness, we know only individual states of perception or feeling. But we detain or repeat one and another; we observe their likeness or unlikeness; we form concepts; we group them in classes which divide the individuals to which they belong; we fix and record the products of our acts by a name; we find common causes, powers, and laws for similar phenomena; we discern the adaptations of spiritual objects to one another and to the world of matter, and thus bind together the world of matter and spirit, in the unity and harmony of one comprehensive plan; the thinking of man interpreting in these ways the thoughts of God.

From this particular example of thought we derive the following definitions: To know by thinking, is to unite individual objects by means of generalization, classification, rational explanation, and orderly arrangement,— *Thought-knowledge is that knowledge which is gained by the formation and application of general conceptions.*

Thinking is a species of knowledge; but knowledge has been

defined as the apprehension of objects in their relations. Thinking, is the apprehension of objects as *generalized* and their implied relations.

Some persons may question the propriety of designating these several processes by the terms *thinking* and *thought*, for the reason that these words sometimes signify to imagine, or believe on insufficient evidence.

On the other hand, it should be remembered that thinking and thought, in the best English usage, denote, in a general sense, the higher as distinguished from the lower operations of the intellect. There are no single words so appropriate as these, which can be set apart to the technical service and designation of the operations of the rational faculty; no other terms for these operations are in actual use whose common signification is at once so comprehensive and so definite as are these.

Appellations for the power of thinking, etc.

§ 191. If it be difficult to find an appropriate term to stand for all these higher processes, it is almost as difficult to find or select an appellation for the *power* which qualifies us to perform them. The *intelligence* and the *intellect* have been thus appropriated, but they are also used for the capacity of the soul for every species of knowledge, the lower as well as the higher; for the power to know by sense and imagination, as well as the power to know by general conceptions. The *understanding* is sometimes employed in this very general sense, and sometimes limited to a single and special function, as by Coleridge and others, after Kant. The *judgment* is used, likewise, in a wider and a narrower sense. The *reason* seems better fitted than almost any other term, and yet the reason is used for the very highest of the rational functions, or else in a very indefinite sense for all that distinguishes man from the brutes. It remains for us to choose between the *rational faculty* and the power of thought, or briefly, *thought*. For brevity and precision we prefer *thought*. It is scarcely necessary to observe that, like perception and representation, and many subordinate terms, thought is used at one time for the *power*, at another for the *act* of thinking, and at another for its *products*. Thus we say indifferently, "Man is endowed with *thought* as well as with sense:" "Sits fixed in *thought* the mighty Stagyrite:" "A penny for your *thoughts!*"

The power of thought may be considered *in two* aspects: *as a capacity for certain processes or functions; and for eliminating and generalizing certain fundamental conceptions or relations.* In the one of these aspects it performs the several acts which we have enumerated, of generalizing, judging, reasoning, etc., the most of which are usually called logical processes, because they are more or less intimately related to deduction or reasoning. In the other, it is viewed as the discoverer of certain native conceptions or intuitions, and the propounder of certain first truths, or first principles; which are also called necessary and universal propositions, axioms of reason, or, metaphysical conceptions and metaphysical truths.

Hamilton refers these two processes to two faculties, *the elaborative and the regulative*, the one of which *elaborates* or works over the materials furnished by the lower powers, according to the conceptions or rules which the other furnishes or *prescribes.* In this he follows Kant very closely, who calls the logical faculty, the *understanding*, and the power which controls its beliefs by *ideas*, the *reason.*

It is more satisfactory to consider the two in conformity with the analogy which we discern in the other powers of the soul; the one as the capacity for certain definite acts or processes of knowing, which we consciously exercise and employ; and the other as the unconscious source of these conceptions, according to which the material of knowledge must arrange itself by the very constitution of the thinking power.

The thinking power, viewed as the capacity for certain processes, thinks in various methods, and matures certain products, the two being often denoted by the same word. These several products are called *the forms of thought*, or thought-formations. These forms are the *concept*, the *judgment*, the *argument* or *syllogism*, the *induction*, and the *system.*

As the discerner or the discoverer by intuition of certain necessary conceptions or relations, the thinking power is said to know or assume certain *forms of being*, according to which it performs its operations, and constructs its products or *forms of thought.* These are called indifferently, *forms of being* and *forms of knowledge*, for the reason that the mind can only know what is or exists, and according to the relations in which it exists. Some of these forms of being or forms of knowledge are *time* and *space*, *substance* and *attribute*, *cause* and *effect*, *means* and *end*.

Relation of thought to the lower powers.

§ 192. The power of thought, as a capacity for certain psychological processes, is dependent for its exercise and development on the lower powers of the intellect. These powers furnish the materials for it to work with and upon. We must first apprehend individual objects by means of sense and consciousness, before we can *think* these objects. We can classify, explain, and methodize only individual things, and these must first be known by sense and consciousness.

These lower powers are not only necessary to furnish the objects for thought to work upon, but they are developed earlier than the higher powers. The infant must go through a training of the eye and the ear for months, before it begins to name and classify with effect. It is the conscious subject of a multitude of mental states, before it gathers the most obvious under a general conception. The discipline of attention must be for a long time enforced, before the developed mind can learn to apply the commonest concepts or to affix the simplest names. The conceptions of cause and effect, and of means and end, are not developed till the intellect has become still more mature.

To the development of thought, the representative faculty is also largely subservient. The individual object must not only be apprehended in order to be thought of, but it must be recalled again and again. To thought, the discernment of similarity is required; and in order to this, the past must be frequently confronted with the present, and the present must be compared with the past. Objects striking for their likeness or their difference, must be recalled by the memory and revived to the imagination, in order that like objects and like phenomena may be grouped and arranged in the rudest classification. If the classification is to be perfected to anything like scientific exactness, the memory and imagination are to be tasked still further in order that one's *thought-products* may be just to the reality of things.

But while the thought-power, in its various operations, is thus shown to be developed later than the several forms of direct cognition, it should not be supposed that it springs into perfect and mature energy by a single bound, or that the acts of infant perception are not affected by its rudimental activity. The human intellect is a unit, and the action of one power is tinged or modified by the feeble energy of all the others. The sense-perceptions of the infant may seem to be more feeble and less mature than are those of the young of the brute. The higher powers may meanwhile seem to lie torpid long before they are called into distinct activity. But before they are revealed to the conscious subject of them, or are expressed in the simplest forms of language, they give direction and character to the perceptions of sense. They impart to the human eye a cast of dawning intelligence which distinguishes it from the keener eye of the dog or the eagle.

Concrete and abstract thinking.

§ 193. Thinking, again, may be distinguished as *concrete* and *abstract*. In concrete thinking, we know of thought-conceptions and relations only in their application to individual or concrete objects. We should say more exactly, we know individual objects under or by means of the relations which thought furnishes. In abstract thinking we separate these conceptions and relations from any and all individual objects. We consider them apart by abstraction, and sometimes treat them as though these conceptions and relations could have an independent existence. In concrete thinking, we proceed as we have described in § 189.

In *abstract thinking*, we separate or abstract from every individual object the generalized conceptions which we produce by thinking, as also those by means of which we think: as the concept, the judgment, the argument, the inference and the system, on the one hand; and substance and attribute, cause and effect, means and end, on the other. We even abstract and generalize our very acts or processes of thinking, and view them apart from the individual examples or cases in which they actually occur. We ask, What is it to conceive, to generalize, to judge, to reason, to infer—nay, what is it itself to think? We discuss the nature and origin of these conceptions, and their relations to one another, to the objects to which they are applied, and indeed to all our knowledge.

Concrete thinking is performed by every human being whose powers are fully developed. All men freely apply its conceptions and relations. By means of them they know sensible and spiritual objects, so far as they know these at all. A stone or an apple, a horse or a dog, a house or a church, a spirit or a person, each and all are known as beings, and are distinguished and defined by certain attributes or properties. One of these acts upon another, as a cause producing an effect, etc. In myriad examples, objects are familiarly known by us as substances and attributes, as causes and effects, as means and ends. In the concrete form, all these conceptions are present in the language, and familiar to the minds of the most uninstructed men.

But when these conceptions are abstracted, and viewed apart from individual beings, they are not made familiar to the mind without a special discipline. It is only a few men who possess the tastes or the training which qualify them readily to deal with or rightly to understand thought-conceptions when abstracted from individual things. Skill in using, and discrimination in understanding them, can only be acquired by concentrated and patient efforts.

Relations of thought to language.

§ 194. Thinking is aided by language, and, to a great extent, is dependent upon it as its most efficient instrument and auxiliary. But thinking is not constituted by, but, on the contrary, itself originates and gives form and law to language.

The connection between thought and language is so intimate, that we shall have occasion to refer to it again and again. One or two general remarks in respect to it, seem here to be in place. The reason why thought requires such an instrument and assistant as language, is, that the objects of thinking are generalized objects, and to such objects there are and there can be no realities actually existing. The results or products of our thinking are not manifested by any changes which are actually affected in material or spiritual objects. It is only by language—the sound to the ear, and its symbol for the eye—that the *products* of thought activity can be fixed so as to be the *objects* of recall and future use. Hence words spring into being as fast as definite conceptions are formed. Hence it is as natural for man to speak as it is to think, and man "speaks because he thinks." The name fixes, preserves, and exhibits the transient concept as in a crystal shrine, both hard and clear. The proposition embodies the judgment for the use of the man who first thinks it, and who utters it to stimulate the thinking of others. In applying names, we must enter somewhat into the nature and properties of the objects for which they stand. In defining terms, we must be guided to their meaning by observing the things to which they are applied. In accepting or rejecting propositions, we must think of the relations of the objects which they concern.

It follows, also, that the study of words must be a study and discipline of thought. To master a language that is rich in its vocabulary, requires that we contemplate the nicer shades of thought which are expressed by the endless variety of the conceptions that are embodied in its words. If it is complicated in its structure, we must discriminate the delicate relations which this syntax expresses or suggests. No language can be *dead* to the intelligent student. Its delicate tissue reflects the varying shades of thought, feeling, and opinion that run through every part of the fabric, like threads of silk and gold.

But, on the other hand, words in no sense constitute thought, as some hastily infer. Language is simply thought expressed, though the thought is made permanent by being expressed. It is formed by the thinking power, because this requires for development and perfection a sensible expression of its inner processes, and seeks a permanent embodiment and record of their results.

CHAPTER II.

THE FORMATION OF THE CONCEPT OR NOTION.

The processes involved in forming the concept.

§ 195. Thinking has been already defined as that series of processes by which we form and apply general *notions* or *concepts*. It is obvious that the first act in this series of processes is to form or develop these products. We begin with the *concepts* of *material objects*, such as a stone, an apple, a horse; and observe that such objects must be perceived, in part, at least, before we form general notions of them. We do not insist that the process of perception should be complete before the act of generalizing begins. It is necessary, however, that a percept should go before the concept in the order of time, as it is the foundation for it in the relation of logical subordination. A general notion requires individual objects to which it can be applied; and individual objects in the material world can only be known by perception. The mind begins to form concepts as soon as it notices that several perceived objects are different as individuals, and yet are in any one respect alike. Before generalization, they may be known confusedly or known vaguely. As soon, however, as they are distinguished as not the same, and yet as united by a common likeness, the process of generalization has begun. This process is possible even with single percepts. If ten patches of red color, of the same form, dimensions, and intensity, were presented to the eye, the mind might gather, or conceive, or grasp them together, by their common redness, and form a general notion of them; thus uniting them as one by the single similarity of color. If these ten red discs of color, by the use of the remaining senses, are afterwards known to be ten red apples, *i. e.* if other points of likeness are perceived, the generalization is more complex in its materials, but the process is the same.

The process involves acts of *analysis,* of *comparison,* and of *generalization.* The mind must notice that which is common, and distinguish it from that which is diverse. This act is an act of comparison. Its appropriate object is likeness. It discerns a quality as similar. It takes this similar to be the same, and, so regarding it, finds it in every one of the individual objects. This similar something, conceived as common to many objects distinguished as individuals, is a general conception, notion or *concept.*

The mental acts which we have described, are familiarly known as follows: The act of analytic attention by which the similar element in each one of any number of objects or phenomena is separately observed or noticed, is usually called *abstraction,* because the mind draws it away from the other parts or relations. Kant and Hamilton say that *abstraction* refers to that *from which* the mind withdraws itself, while it *prescinds* the element to which it attends. Thus, in the example cited, the mind *prescinds* the redness, and *abstracts* its attention from all the remaining attributes.

The next step is, to perceive by *comparison* that the several objects to which we thus separately attend, are alike. The next step is, to consider these several *similars* as the same, the one something which is common to all the individuals perceived. This is to *generalize*—to make general—more properly, mentally to think or affirm a common something of all these individuals. The similar *red*, or *round*, or *sweet*, or *bitter*, is made one, and, as one, is regarded as common to each of the different individuals. Which of these acts is first performed, is immaterial—whether the mind seems to generalize before it abstracts, or the reverse; or whether it analyzes, compares, and generalizes all in one. It is all the same as to both process and product, whether we separate the redness from the first apple which we perceive, before we apply it to the many, or are stimulated by observing many red apples to notice and abstract that which is alike and common, or whether the points of difference excite us to generalize the one or more elements in which the objects are alike.

Again, when this common something has thus been generalized from like objects, it can be applied to—*i. e.*, affirmed or *predicated* of—any and every other object to which it is appropriate. Thus, *spherical*, after being thought of a single class, as of apples or balls, may be thought of all objects that are *round*—as of the vast spheres which are hung in the heavens, or of globules so minute as to be indiscernible by the naked eye.

It has been already observed, that these processes develop and presuppose the distinction *of substance and attribute*—*i. e.*, of being and distinguishing relations. The individual apples of which we think the redness are beings, the redness is their common attribute. What is the nature of, and what the authority by which we make this distinction, we do not propose here to inquire. For our present purposes, it is sufficient that we call attention to the fact that it is fundamental to the process of forming the notion, and that it must be assumed as real, and be firmly believed by the mind. (Cf. § 323.)

The product, its nature and appellation.

§ 196. The product of the processes considered, is called a *concept* or *notion*. We employ these terms because they may be made precise in their import and technical in their use. *Conception* is sometimes used; but conception is, in our English philosophy, used indiscriminately for any and every object of the mind's cognition, or else is arbitrarily limited, as by Dugald Stewart, to the individual object of representation, and thus made equivalent to *image*. *Abstract general conception* (or even *general conception*) is sufficiently precise in its import, but is too cumbrous for common use. *Concept* and *notion* have each, in their etymology, a special signification appropriate to one aspect or feature of the product to which both are applied. *Concept* signifies something grasped or held together, and refers us to the act by which different similar attributes are treated as one, or the same act in which separate individual beings are united as one by their common attribute or attributes. *Notion*, on the other hand, indicates that which is or may be known by certain signs or marks, *notæ*—*i. e.*, constituting, defining, and distinguishing attributes. Both terms may be properly employed as technical and scientific designations.

The reality of any such mental product or thought-object has been questioned, chiefly by those who have misunderstood or misconceived its nature. Its import or nature has been imperfectly or vaguely estimated even by many who have believed in its reality. It is only by explaining its nature, both *negatively* and *positively*, that its reality can be vindicated and established.

The *concept* is *not* a *percept*, nor is its object an object as perceived. This last is strictly individual; the concept is uniformly general. In order to prove this beyond question, we have only to ask what the mind knows when it sees a man, and what it thinks of when it utters the word man, and applies it in thought to the human species. No one can doubt that the two objects of cognition are diverse.

The *concept* is *not* a mental *image*, or the object of the mind's cognition in representation. We recall an individual percept, one or many; or we form, by creation, some image unlike any which we have in fact perceived. Both are clearly distinguishable from that which the mind thinks or knows, when it uses a general term.

We state *positively*:—The concept is a purely *relative* object of knowledge. This is its distinctive feature, that it holds definite relations to objects of sense and consciousness. As a mental product and mental object, it is relative, being formed by the mind and understood by the mind as indifferently common to single objects; which objects only enable the mind to understand its import. The individual things to which it relates, give to it all its significance and utility. Without these, it is a *no-thing*, an unintelligible and unreal fancy. This peculiarity of the concept is implied in its various appellations. It is called a *general*, that is, capable of being thought of many individuals, which are thereby grouped into or conceived as a class. It is called also a *predicable*, by its very nature capable of being affirmed or thought of single objects. It is a *universal—i. e.*, as pertaining alike to all the individuals to which it belongs.

Again: as being this common and relative thing, the concept *respects only the similar attributes* of individuals, or such as might be supposed to be alike. It respects those elements which analysis can separate as individually distinct, and comparison can unite as alike. Attributes, properties, and relations, are the only objects which it respects. These are first discerned, then compared, then united into a single thought-object. Herein lies the difference between the act of a brute and the act of a man in perceiving objects that are alike. In one sense, the brute may perceive what is similar as readily as a man; in some cases, even more quickly, for his senses may be more keen. If he has been ill-treated by any other animal, or frightened by any object, every thing like either will be avoided at once. But the brute does not attend and analyze as does a man. Hence he cannot discriminate so as to abstract; or, at best, the degree and range of such efforts must be very limited. His power to compare and discern the like and the unlike would for this reason be lame and feeble,if no other could be suggested. Should it be granted that the brute can discern similar attributes, it has no power at all to conceive or think the *similar* as the *same*. It cannot form and use a concept as founded on attributes and as common to individual beings. Hence, the brute is incapable of language. He may utter sounds and cries which instinct extorts and to which the instinct of the hearer responds, and thus the voice and ear of the animal tribes may serve some of the useful and social ends which language accomplishes in man; but the brute is incapable of using words as the signs of concepts, because he is incapable of *thought*. He cannot form and use a concept, and therefore he can neither speak nor understand a single word. Even the parrot, that miracle of talkers, is incapable of language, and never utters what deserves to be called a word.

We observe still further, that all which the concept contemplates or signifies, is the common attributes which are discerned in the individuals to which it is applied. These attributes are its proper and sole import or signification. The concept, as such, is not at all concerned with the number of individuals in which these attributes are found, or with anything else which may be true of them. It is all the same to our thinking and to the concept which we form by thinking, whether the tree of which we make and use the notion, is here or there; is high or low; is the tree which we have often seen and admired, or the tree which is ten thousand miles distant; is the tallest of the cedars of Lebanon, or of the firs of California, or the most dwarfed that exists on the coldest mountain summit. It is even indiffer[illegible] whether it actually exists or not; it is only essential that it be formed by the mind from the actual constituents of every object that is properly called a tree.

Concepts as concrete and abstract, as simple and complex; their content and extent.

§ 197. Concepts are distinguished in their application, as *concrete* and *abstract*. The concrete notion contemplates attributes, and is applied to beings existing. The abstract notion treats attributes as though they were themselves such beings. *Man* and *human* are concrete; humanity is an *abstract* notion. The concrete notions are applied directly to an actually existing being, for purposes of classification and language, which need not here be explained. The abstract *humanity* is applied to designate a being that is purely fictitious, but which, in language and in thought, is treated as though it possessed actual existence. The attribute is conceived as a being, in having attributes affirmed of it; as when we say, humanity implies liability to error. It has adjectives prefixed to it, as in the phrase, *our original humanity.* It is divided into classes: humanity is either *refined* or *degraded*, etc. In short, it is capable of being treated in every way, as though there were living beings called humanities. But when we analyze the real meaning of language, and the thoughts of those who use it, we find that the only beings distinguished by the mind are the living men who are endowed with human attributes.

Concepts, again, are still further distinguished as *simple* and *complex*. Those notions which are made from a single attribute, are *simple*. Those which are made of more than one, are *complex*. Simple notions are called, by Locke, *simple ideas*. They cannot be analyzed or decomposed into any constituent elements. The mind directly discerns them by its various powers of knowledge. Such words as *white, whiteness, green, greenness*, etc., etc., are usually given as the names of simple notions. It would be more exact to say that we treat these notions as simple, because we do not ordinarily distinguish in thought, or by language, the discernible shades of white and green. Those which are properly simple, would be such shades of color as can be distinguished from every other. On the other hand, *chalk, chalky*, are complex notions, because they signify more than one attribute. So, *man* and *human* are complex spiritual notions, for they contain many attributes.

No thing or being actually existing is represented by a simple notion. A grain of sand or a mote in the sunbeam, is complex, for it has form, dimensions, color, weight, etc., etc. Nature gives us no simple ideas. She touches us through too many avenues of knowledge. She leads us to observe varied attributes in every existing thing. We, in our thinking, analyze and separate her complex objects, and reconstruct and recombine the elements which, at her prompting, we have abstracted and generalized. In this way we separate and reconstruct the elements or attributes of material objects as nature exhibits them to us, as of plants, and animals. Thus, all the concepts which are expressed by the general terms that form the staple of every language, are constructed by the mind. They are passed from one mind to another. They are fixed in words and recorded in books and literature. The names of the objects that human art and skill has constructed for use or beauty, likewise stand for the complex of simple notions which we observe in these objects. The artificial creations, such as are conceived by human invention and spring from human society, the crimes which are defined by human law, the offices and relations of government, the signs and proofs of property, the rights and duties of men, all these are complex notions, which are made and sustained by civilized men, and interest most profoundly their hopes and fears. These are still further removed from the notions and terms more usually conceived as abstracta, but, like these, they are susceptible of being so analyzed as to be carried back to living beings. But these all are complex notions, and some of them are exceedingly complex in their constituent elements. If we consult a dictionary, and run the eye down its lists of words, we shall be surprised to find how large a portion of them stand for these artificial creations, these complexes of abstracted properties.

Still further, notions are technically distinguished by their relations of *content* and *extent*, or, as they are often termed, their *comprehension* and *extension*, their *depth* and *breadth*.

These relations grow out of the very nature of the notion, as has been shown by our definitions. A notion cannot be a notion, unless it has these two relations. It can neither be formed nor used unless both these relations are considered.

The *content* of the notion *is the attribute, or attributes*, of which it consists. It is its *contained attributes* considered *as a unit or whole*. Those notions whose content we have the most frequent occasion to consider, are complex notions. Every simple notion has a proper content in the single attribute which, when conceived as common, is made a concept. Such complex notions as *chalk, snow, milk, felony, burglary, theft, man, spirit, body, soul, legislation, monarchy, republic, a state*, etc., have so manifestly a sum of contained attributes, that it is with especial propriety that we speak of their *content*. These constitute their meaning or import. When these are fully stated, the notion is *defined*. They are also called the *essence*, or *essential* constituents, of the notion, because they make up or form its being as a thought-product or thought-creation.

The *extent* of a notion originally and properly signifies the number of individuals to which it is applicable. If we could know, by actual enumeration, how many horses or men there are at any time existing, their sum would be the extent of the notion horse. We rarely, however, have occasion to consider individuals; for these are divided again and again into larger and smaller groups, to each of which there is a fixed notion and name. These divisions are effected by adding to the content of the notion which includes a greater number of individuals, an additional attribute—in the case of the horse, an attribute of color, perhaps; and we have a new content, *white horse, black horse,* etc., giving an extent of fewer individuals. In many cases, we designate the concept thus newly-formed by a separate name, as *pony,* for a small horse, *charger, hunter, roadster,* etc. So trees are divided by means of notions, whose content is given as *deciduous* and *non-deciduous.* The latter are divided into *pines, firs,* etc.; the firs are again divided into *hemlocks, spruces,* etc., each having some attribute not belonging to the content indicated by the word *fir* or *fir-tree.* In consequence of these divisions or groupings of individuals into broader and narrower classes, the extent of the notion in actual use always stops short with subordinate groups, and does not carry us down or back to the included individuals. These individuals are always intended, however, and the subordinate classes are said to constitute the *extent,* because they, in their turn, are applicable to and comprehend individuals.

As the *content* of a notion is exhibited by *definition,* so the *extent* is shown by *division.* This division is effected as the indirect consequence of adding to the content of the notion a new attribute, which immediately narrows its extent. The adding a new attribute, or new attributes, for this end, is called *determination,* or the act of bounding off, or limiting.

It follows that, as the *content* of a notion is increased, its *extent* is diminished. Hence the maxim: the content is inversely as the extent. Both propositions are true, the greater the extent, the smaller the content; the greater the content, the smaller the extent.

Classification, its origin and different species.

§ 198. In forming the notion from, and applying the notion to, individual objects, the intellect *classifies* these objects; that is, it groups them into divisions which are *broader* and *narrower* in their extent; and of course *higher* and *lower* when ranked according to their place in a system. This consequence follows from the fact that nature has so constructed individual beings that they are capable of being grouped into larger and smaller divisions, by means of their resembling attributes; and from the desire in the human soul which meets this fact of nature by connecting objects in an orderly arrangement.

The first efforts at classification are necessarily rude and imperfect. *Children* when left to themselves group together objects in singular combinations and discern resemblances between things which older people never would think of connecting. In the poverty of their language they apply the words which they possess, to the strangest uses, on the very slightest and the most whimsical analogies. They soon learn better, as we say. That is, they take from older persons the conceptions and classifications which have been made before them. In other words, they think over again the concepts that are made ready and presented for their use, in the words of which they learn both the import and the application. In learning to talk they are constrained to fall in with those classifications which previous generations have made before them, and have recorded in the language which they have left behind.

Savages do not classify under the same restraints. When novel objects are presented to them, they usually seek out some concept or word already known and familiar, and extend it to the novel object by some resemblance, however forced or violent this may be. The goats which Captain Cook carried to the Pacific Islands were called by the natives *horned hogs:* the horse on a like occasion was called a *large dog*. The *dog* and the *hog* being the only quadrupeds with which these savages were familiar, these novel animals were taken into the only concepts and names that were ready for their reception. When the Romans first saw elephants, they called the animal *Bos Lucas* or *Lucanus*, a Lucanian ox, from the province in Italy where they were first seen.

The *classifications of science* differ from those of common life in being founded on a more exact observation, and directed by the special rules which are furnished by scientific principles. These may be certain assumed ends or known powers or laws of nature which were discovered long after the classifications had been perfected which are recorded in the words of common life. The classification of animals into vertebrates, articulates, mollusks, radiates, and protozoans, and the subdivision of the vertebrates into mammals, birds, reptiles, and fishes, are very different from those represented in the words horse, ox, whale, snake, hawk, quail, robin. Neither the so-called natural nor the artificial systems of botany give us what we know under the household names of the lily, the rose, the pink, and the violet. And yet these common names do as really classify their objects as do scientific names. To classify is no secret of science, no process reserved for the select few who are initiated into a magic art, but it is as universal and necessary as the act of thinking. The classifications of common life may be as rational and as useful for the ends of common life as are those of science for its special uses. They are founded on the obvious appearances of objects to the senses and the mind. They are adapted to the uses of men of ordinary culture. Indeed what wealth of thinking does every cultivated language embody and represent! Each one of its words has gathered into its subtle essence the results of the repeated and refined observations of the men who perhaps by successive efforts at last reached the concept which each single term enshrines. In like manner the technical nomenclature of a single science when finished and arranged, is a transcript of all the discriminating thoughts, the careful observations, and the manifold experiments by which the science has been formed. It represents in brief, all the most careful definitions and the most complete and best classified divisions which the devotees to its special objects have perfected by their labors.

Classification is nearly allied to *systemization.* The division of objects into classes which are *broader* and *narrower*, has a close affinity with their orderly arrangement in classes which are *higher* and *lower*, through a succession of divisions and subdivisions. Both result from the application of notions in their extent to existing objects or to objects which are conceived to exist.

Classification and systemization, are the characteristics and consequences of all thought-knowledge and preëminently of scientific knowledge. They are indispensable to enable us to grasp individual facts and to retain our observations. They are an intellectual convenience and an intellectual necessity. But they do not constitute the whole of thought or the whole of science. Though scientific knowledge is of necessity classified and arranged knowledge, yet much more than this is true of it.

We have entered within the threshold of our analysis and comprehension of thought-knowledge, but the light which shines from the inner sanctuary casts its radiance only upon those objects which are the nearest to our view. It remains for us to consider other acts, involving profounder relations in the constitution of the universe, in the methods and forms of our thinking, and in the products which this thinking evolves.

How much do we gain by knowing by concepts?

§ 199. It will not be amiss, however, to ask at this stage of our inquiries, what addition do we make to the knowledge which we gain by perception and consciousness, by superinducing upon it the acts or processes of thought which we have thus far considered? What do we know more about an object seen or experienced, by generalizing its attributes, determining its class, or assigning to it a name? We may answer this question by asking two or three others. What more does a man know about a single apple by calling it an *apple*, a *fruit*, a *plant-product*, an *organized being*, than he does by seeing, feeling, tasting, and smelling it? We answer, its *common relations*, *i. e.*, properties, attributes, and uses. When we think or intelligently say of a sense-object, it is an apple, we both think, and impliedly say of it, it is like a multitude of other sense-objects, in many most important respects, as of color, taste, size, etc. When we think or know it to be a fruit, we enlarge still more widely the sphere or extent of the objects to which it holds relations. So when we think it to be a plant-product.

That was no inconsiderable act which was signified by the record which describes the various living animals as brought to Adam that he might name them. The capacity to name them implied an insight into their nature. For this reason it must of necessity be true, if we suppose the original man to have been endowed with the requisite discernment, that "whatsoever Adam called every living creature, that was the name thereof." It seems to be a trifling thing for the child to be able to affix suitable names to the objects and beings which first attract its attention. At first thought the act is trivial, mechanical, parrot-like, as it were, to attach an articulate sound to one or more similar objects; but when we view it as implying the power of intelligently applying this name to a still larger number of objects which are in many respects unlike and yet alike, it becomes an act of the gravest import. It indicates an important development of the soul's action, and the evolution of a new product. When the child asks, What is it? meaning thereby, What is it called? it really asks, What is the nature, what are the relations, of the object to which the name belongs, as it learns one by one what these relations are, and notices in what they are alike, and in what they are unlike.

That was no slight achievement of Aristotle, to seize upon, bring out and establish the truth that the concept of an object either declares *what it is*, or at least indicates the direction which must be taken in order to find this. The concept is the permanent *what*-ness or *what-sort*-of-ness, which may be thought of the things to which it is applied. It is the τὸ τί ἦν εἶναι, *i. e.*, its real and permanent nature. To ask what a thing is, according to Aristotle, is to take the first step and perform the first of the processes which are essential to its complete mastery. It is to propose the first of those questions, the answers to all of which carry the mind through the entire circle of scientific knowledge. Aristotle also recognises the intimate connection of the concept with the word, calling the two by the same term, ὁ λόγος.

§ 200. The *what* which the concept and the word both propose to communicate, is not the direct observation which presentation gives, but the higher and more comprehensive knowledge which thought aims to achieve. It is not the knowledge that *a being is*, but the analytic and comparative knowledge of *its relations*.

Relation of knowledge by concepts and by intuitions.

CHAPTER III.

THE NATURE OF THE CONCEPT.—SKETCH OF THEORIES.

In the preceding chapter we have considered the nature of the concept in a general way, so far as was required in the analysis and explanation of the psychological process by which it is formed. We return to it a second time for more careful consideration.

§ 201. The nature of the concept and its relation to real or existing objects has been the occasion of endless speculation, of fantastic theories, and of sharp and persistent controversies in every period distinguished by philosophical inquiry. *Socrates* was the first to insist upon the importance of forming concepts of the objects of our knowledge in order that the permanent and essential might be eliminated from that which is accidental and transitory in individual objects. But he taught little or nothing in respect to the nature of the *concept*, or of that in the object to which the concept is the *counterpart* or *correlate*. *Plato* took up the inquiry where Socrates left it; insisting more abundantly than he upon the necessity of this higher knowledge, and showing that in attaining it we must define and divide—must go from the individual to the general, by successive inductions, and so on from one step to another, till we reach that which exists of and by

The doctrines of Socrates and Plato, and Aristotle.

itself—that which is alone the permanent object of [true] knowledge. This is the idea, ἡ ἰδέα or τό εἶδος. But what this idea is, and what are its relations to the concept, he does not accurately teach; where it exists he does not assert; whether in the object itself, or in the mind of the Creator, or in the mind of each thinking man, he does not define. He seems to teach that ideas, or *the idea*, have an existence and essence separate from all these, that they are eternal and incorruptible, existing before all temporary and perishable beings, and imparting to the perishable and phenomenal in these beings all their dignity and interest. Ideas are realities, things and events are their shadows. But whether by these representations, he intends only personification and poetic fiction, or exact scientific definition, is not always easy to decide.

As against Plato, *Aristotle* insists that the only real beings or substances are existing beings or things, the πρῶται οὐσίαι, or primary entities, as he calls them. He is distinctly aware that there are other sorts of beings besides these. The δεύτεραι οὐσίαι or second entities are distinctly discriminated from the πρῶται οὐσίαι, or individual beings. He aims to show in what sense the former are so called, and how they are related to real beings, or, in modern phraseology, to show the relation of concepts to real existences. This he does by distinguishing between matter and form. Matter cannot exist without form. Every existing being has some determinate form. There can be no form without matter. The one requires the other. The two are correlates, seeking each other, as Aristotle figuratively speaks, by a natural appetency. The form only is conceived by the mind. What the mind conceives of a being is its essence, τὸ τί ἦν εἶναι. In modern language the concept is made up of the essential qualities that are common to several individuals, omitting those which are undiscriminated; these last being matter.

Aristotle set out with the determination to avoid those personifications which so abound in Plato. But he did not entirely succeed. Should we concede that he was not himself betrayed into hypostasizing these metaphors, he did not secure his disciples from this error. So it happened that the ideas of Plato and the forms of Aristotle were both regarded as actual realities, and as such, furnished fruitful material for the subtleties and controversies of their earlier disciples and commentators, in the decadence of the Greek philosophy.

Porphyry's questions and the scholastics.

§ 202. It was, however, among the scholastics of the middle ages that such discussions became conspicuous, in the schools of *the Realists, the Conceptualists, and the Nominalists.* The immediate occasion of these discussions and controversies was furnished by a passage from Porphyry, in the preface to his Introduction to the *Categories* of Aristotle. This Introduction was translated from the Greek by Boethius, and a brief passage proposed the problem for the different sects which we have named—who received their appellations from the different solutions which they gave to it. "Mox de generibus et speciebus, illud quidem, sive subsistent, sive in solis nudis intellectibus posita sint, sive subsistentia corporalia sint an incorporalia, et utrum separata a sensibilibus posita, circa hæc consistentia dicere recusabo. Altissimum enim negotium est hujus modi et majoris egens inquisitionis." In other words, the questions which naturally suggest themselves concerning Universals are the following:

'*Have Universals a separate existence, or do they exist in the mind only? If they have a separate existence, are they corporeal or incorporeal? Are they separable from sensible objects or do they subsist in these only?*'

The *extreme Realists* answered these questions in the spirit of Plato, or rather of the doctrine which Aristotle ascribed to Plato, viz.: that Universals have an existence that is separate from and independent of individual objects. They even contended that they exist before these, in rank and creative power, certainly in point of time. These views were formulated in the motto *Universalia ante rem.*

The *moderate Realists* adopted the creed of Aristotle that Universals have a real existence, but only in individuals. Their motto consequently became *Universalia in re.*

The *Conceptualists* and *Nominalists* agreed in this that individuals alone have real existence; and that Universals, both genera and species, are formed by the mind, by bringing together many similar objects and designating them by common terms.

They differed in that the *extreme Nominalists* held that the name only is general and is employed to avoid an indefinite number of proper names which would be otherwise required; while the *Conceptualists* interposed a concept between the name and the objects collected into a class. The motto of both Conceptualists and Nominalists was *Universalia post rem.*

The differences of opinion that ripened into these separate philosophical sects began to be manifest in the ninth and tenth centuries. It was not, however, till the second half of the eleventh that different philosophers and theologians were commonly known by these appellations, and that the doctrines themselves became the occasion of earnest and bitter strife. These divisions reappeared at intervals and were not finally terminated before early in the fourteenth.

Modern Philosophers.

§ 203. In modern times the diversities of opinion in respect to the nature of the concept have been as great, and the controversies well nigh as active as they were among the schoolmen. The same questions have in fact been agitated, and the same difficulties encountered, with this difference—that the form which these questions have taken has been more generally psychological, rather than metaphysical. This was no more than was to be expected from the general course of modern philosophy. But in the recent German speculations, the logical and metaphysical direction of thought has preponderated over the psychological and inductive.

Hobbes, a nominalist of the extremest school, says, *Human Nature* (c. 5, § 6) "The universality of one name to many things hath been the cause that men think the things themselves are universal; and so seriously contend that besides Peter and John and all the rest of the men that are, have been, or shall be in the world, there is something else that we call man, viz.: *man in general*, deceiving themselves, by taking the universal or general appellation for the thing it signifieth." * * * "It is plain, therefore, there is nothing *Universal* but *Names*." In *The Leviathan* (p. i., c. iv.) he says: "There being nothing universal but names, for the things named are every one of them Individual and Singular, one Universal name is imposed on many things for their similitude in some quality or accident."

Locke, on the other hand, who was a *Conceptualist*, says in his *Essay* (B. IV., c. vii., § 9), "Does it not require some pains and skill to form the *general idea of a triangle*, [which is yet none of the most abstract, comprehensive, and difficult,] for it must be neither oblique nor rectangle, neither equilateral, equicrural, nor scalenon; but all and none of these at once. In effect, it is something imperfect that cannot exist [*i. e.*, in fact, or actually]; an idea wherein some parts of several different and inconsistent ideas are put together. 'Tis true the mind in this imperfect state has need of such *Ideas*, and makes all the haste to them it can, for the conveniency of communication, and enlargement of knowledge." That he was not a Realist appears from the following (B. III., c. iii., § 11 sqq.): * * "It is plain by what has been said, that *General* and *Universal*, belong not to the real existence of things; but *are the inventions and creatures of the understanding*, made by it for its own use, and *concern only* signs, whether words or ideas." "When therefore we quit particulars the generals that rest [remain] are creatures of our own making, their general nature being nothing but the capacity they are put to by the understanding, of signifying or representing many particulars." He argues at length against the Realistic doctrine of permanent essences or species. "Whereby it is plain that the essences of the sorts, or (if the Latin term please better) species of things, are nothing else but these abstract ideas." "To be a man or of the species man, and to have a right to the name man, is the same thing. Again, to be a man, or of the same species man, and have the essence of a man, is the same thing."

G. W. Leibnitz. To these doctrines of Locke, Leibnitz, in his *Nouveaux Essais*, takes the following exceptions: He denies that the essence of the species is only an abstract idea, and asserts that the generality of such ideas consists in the mutual resemblance of individual things, and *this resemblance is a reality*. (*Nouv. Ess.*, B. III., c. iii., § 11.)

Berkeley, *Introduction to the Principles of Human Knowledge*, thus attacks the doctrine of Locke. After describing the doctrine as commonly received, he proceeds: "Whether others have this wonderful faculty of abstracting their ideas, they best can tell; for myself I find, indeed, I have a faculty of imagining, or representing to myself the ideas of those particular things I have perceived, and of variously compounding and dividing them. I can imagine a man with two heads, or the upper parts of a man joined to the body of a horse. I can consider the hand, the eye, the nose, each by itself abstracted or separated from the rest of the body, but then whatever hand or eye I imagine must have some particular shape and color. Likewise the idea of man that I frame to myself, must be either of a white, or a black, or a tawny, a straight or a crooked, a tall, or a low, or a middle-sized man. . . . . But I deny that I can abstract one from another or conceive separately those qualities which it is impossible should exist so separated; or that I can frame a general notion by abstracting from particulars in the manner aforesaid." And yet Berkeley, in another passage concedes the power of abstraction so far as this: "A man may consider a figure merely as triangular, without attending to the particular qualities of the angles or relations of the sides. So far he may abstract. But this will never prove that he can frame an abstract, general, inconsistent idea of a triangle." In respect to generalization also, he concedes the following: "An idea, which considered in itself, is particular, becomes general by being made to represent or stand for all other particular ideas of the same sort. To make this plain by an example: suppose a geometrician is demonstrating the method of cutting a line into two equal parts. He draws for instance, a black line, of an inch in length. This, which is itself a particular line, is nevertheless, with regard to its signification, general; since as it is there used, it represents all particular lines whatsoever; . . . . and so the name *line*, which taken absolutely is particular, by being a sign is made general."

Hume agrees with Berkeley, adopting nearly his language. The only difference between Hume and Berkeley is, that Berkeley makes the particular idea to *represent* the general, while Hume adds that it becomes general by being annexed to a term which is customarily conjoined with many particular ideas, and readily recalls them. In other words, Hume introduces his doctrine of the association of ideas to explain how one idea and term can represent several objects, and become general. This last doctrine has been expanded and re-applied by later writers.

Reid, in criticising both Hume and Berkeley, does not give his own views in the form of a statement precisely defined. He seems scarcely to know what his own opinion is. In respect, however, to the question under consideration as to the nature of the concept, he lays down some important distinctions which are quite in advance of the doctrines previously admitted. He observes (1) that a general idea must be the product of an individual act of the mind, and in that sense and so far, is an individual, and not a general, entity. (2.) "Universals cannot be the objects of imagination when we take that word in its strict and proper sense." "Every man will find in himself * * * that he cannot imagine a man without color, or stature, or shape." "I can distinctly conceive universals, but I cannot imagine them." (3.) "Ideas are said to have a real existence in the mind, at least while we think of them, but universals have no real existence. When we ascribe existence to them, it is not an existence in time or place, but existence in some individual subject; and this existence means no more, but that they are truly attributes of such a subject. *Their existence is nothing but predicability, or the capacity of being attributed to a subject.*" *Essays on the Intellectual Powers.* Essay V., c. vi.

Dr. Thomas Brown, (*Lectures:* 46, 47) avows himself to be a conceptualist, and contends that all the nominalists have either in fact admitted or unconsciously implied the truth of this doctrine. He distinguishes three steps or elements in the generalizing process (1) "the perception or conception of two or more objects, (2) the relative *feeling of their resemblance* in certain respects, (3) the designation of these circumstances of resemblance by an appropriate name." He criticises some expressions of the conceptualists as incautious, particularly the use of the word *idea* to express "the feeling of resemblance," because this word "seems almost in itself to imply something which can be individualized and offered to the senses." "The same remark may, in a great measure, be applied to the use of the word *conception*, which also seems to individualize its object." "The phrase *general notion* would have been far more appropriate." "Still more unfortunate is a verbal impropriety in the use of the indefinite article." "It was not the mere general notion of the nature and properties of triangles, but the general idea of *a* triangle of which writers * * have been accustomed to speak." This has exposed the doctrine of general notions to ridicule, such as Martinus Scriblerus is made to use against Locke.

Sir William Hamilton, (*Lectures on Metaphysics*, Lec. 35) criticises Brown severely for misrepresenting the nominalists, in asserting that they overlook the fact that resemblance in individual objects is the ground of applying to them universal names. Hamilton then labors earnestly to show that discerned or predicated resemblance is individual, and not general; inasmuch as if likeness exists between a pair of objects, it must be an individual relation of likeness.

In his logic, however, and in all the treatment which he gives to the concept, Hamilton proceeds upon the hypothesis of Conceptualism, in the manner in which Reid qualifies and explains it. Indeed, it would seem that his peculiar doctrine of the syllogism and deductive reasoning can have no meaning on the theory of Nominalism. And yet he would almost have us believe that he is a Nominalist, and "that the opposing parties are really at one."

John Stuart Mill, in his *Logic*, B. i., c. 2, and his *Examination of Sir William Hamilton's Philosophy*, chap. 17, earnestly advocates Nominalism. Names are names of things, but while they *denote* things, they also *connote* the attributes of things. Thus horse (or chalk) denotes every individual horse (or piece of chalk), but at the same time it notes or marks, *i. e.*, connotes all that is peculiar to every horse, or to the class horse. Instead of the term *concept*, or *general abstract notion*, Mill would use *class name*. The mind, whenever it uses the class name intelligently, must have some individual object before it, either *perceived* or *remembered*. It need not, however, direct its attention to every part of this individual object. It need think of, *i. e.*, attend to, only those parts which the name connotes. It need not think of all of these even, but only of those which it has occasion to use for its immediate purposes.

Of the modern German philosophers, *Kant* should be named first, not only in the relation of time, but on account of the influence which he has exerted upon all subsequent philosophy. Kant distinguished very sharply between individual and general objects of knowledge, and in the spirit of this distinction introduced many technical terms which are not only still retained in the German systems, but have been adopted by English thinkers. Kant's terminology is not only a permanent monument of his own activity, but it has served to fix some very important distinctions in the minds of speculative men. Kant says very little psychologically concerning the nature of the concept as the product and object of the mind's activity, or concerning its relation to the objects of sense. Speculatively, however, he treats this topic very fully. First of all, the concept, *der Begriff*, is the product and object of the understanding, as the percept *die Vorstellung—der Sinnliche Gegenstand*,—is the product and object of the action of sense. The image *das Bild, das Schema*, is the work of the fantasy, both reproductive and productive. The percept is individual and so is the image proper. The concept is general and definite. The *Schema* is intermediate between the two, being indefinite and movable, and in a certain sense general (cf. § 149). The percept, the image, and the *Schema* are all directly apprehended by the mind. The concept is mediately apprehended and mediately applied, requiring, to be used, that it should be imaged in an individual object, or applied to some individual. Knowledge by concepts is preëminently mediate knowledge.

In the *concept, the matter* is distinguished from *the form*. The matter is furnished by the senses, *the form* by the understanding; before, however, the two are brought together, the sense-matter must become a *percept* in the forms of *space* and *time*. *E. g.* The matter of the *orange* is furnished by *all* the senses. This matter becomes the percept *orange* by taking certain relations of space. It becomes a *concept* by being viewed by the understanding as a being with attributes which are distinguished from each other, and yet are common to many individuals, involving the recognition of diversity, similarity, and production or causation. These and other such *forms* are given by the understanding itself; which, in acts of thought, as it were, covers over or invests the matter of the senses with any or all of them. It would seem from these doctrines, that Kant was eminently a conceptualist, inasmuch as he insists so much upon the concept as the medium of thought, and so often repeats the assertion that thought is knowledge by the medium of concepts. But he does not declare himself such. His discussions are all logical and metaphysical rather than psychological. Though a theory of the powers and processes of the soul is constantly implied by him, it is rarely presented in the psychological form.

§ 204. Kant emphatically gave that ideal direction to philosophy which reached its terminus in the extreme doctrine of Hegel, who makes the concept everything and the individual nothing, who evolves the real world from the concept, to which he ascribes an infinitude of elements and a power of self-development, adequate to produce the countless varieties of individual things. Should it be said that this is a misconstruction of his doctrine; that he treats only of the relation of concepts to one another, and of individuals only so far as they are *conceived* or turned into concepts, the result is the same, so far as our position is concerned; which is that he does not concern himself with the relation of the concept to the individual, nor with the nature of the concept as a product of the mind, or as a representative of concrete being, but regards it as an all-sufficing and independent entity. Hegel may therefore be called a *logical realist*.

G. W. F. Hegel.

CHAPTER IV.

THE NATURE OF THE CONCEPT.—GENERAL NAMES.—LANGUAGE.

THE brief review which we have taken of the various theories of the concept will enable us to see more clearly and to define more exactly its real nature as a mental product, and its relations to the objects from which it is formed, and to which it is applied. Every false or defective theory is founded upon some important truth. The consideration of defective or exaggerated theories is most useful in enabling us to ascertain the truth in all its relations, and thus to develop it completely, as well as to distinguish it from errors of excess or defect. In the light of our historical sketch, we observe:

§ 205. 1. The concept, as a mental object or product, is to be distinguished from the *mental act* by which it is originally produced or recalled. Such an act is necessarily *individual*. The concept produced or recalled is *general*.

Essential characteristics of the concept.

2. The concept, as a mental product and a mental object, implies that *the distinction of individual beings and their attributes is accepted as real*, and must therefore be admitted as possible.

3. The attribute must first be known or apprehended as *related* to a thing or being. It is always held by the mind as attributable to or predicable of some being or thing. Its import, or what is thought of by the mind, is not the being as such, but the being as related, or the being together with a something related to it.

4. The attribute thus related, is next viewed in the relation of *similarity* to other individual attributes, constituted and known like itself. When the individual *red* is compared with other individual *reds,* there is added to its import its likeness to each and all of these.

5. The use of the concept thus formed to *classify objects* enlarges its meaning still further. The capacity of the concept to be a classifier, arises from two circumstances: the fact that the attribute which is its germ, is common to more or fewer individual beings, and the fact that these attributes are distributed in gradation. Whenever it happens that one attribute, as *red,* belongs to more beings than another attribute, as *sour;* then the *red* may denote the larger class—*i. e., the genus;* and the sour, the smaller or subordinate class—*i. e., the species. Sour,* in such a case, may be the *differentia* of the species—the *sour-reds.* If *oval* were universally present with the species *sour-reds,* it might be a *property;* if *hirsute* were sometimes present and sometimes absent, it would be an *accident* of the same species. The application of any attribute in all or any of these class-relations, obviously adds to its import. When a concept is used to classify, an additional relation is thereby taken up into its meaning, and this meaning is thereby so much enlarged.

We distinguish what may be called *generalization*—the use of the concept as general or as common to more or fewer individuals, from *generification*—the arrangement of these individuals into higher and lower classes. *Generification* simply recognizes the fact that these concepts are distributed in gradation, some belonging to more and others to fewer individuals, and that consequently these are classed according to their extent into *genera* and *species.* The process and the product in the second case, both imply and are built upon the process and product in the first. In the first, we bring the individual under the general, by the direct act of forming the general from the individual in the way described. We know the individual under this concept or general name. In the second, we perform the reflex act of employing the general to divide all the individuals to which it belongs into classes as wider and narrower, or higher or lower.

§ 206. 6. The mind, whenever it uses a general term intelligently, must understand or conceive the import which belongs to it in some or all of the particulars which we have enumerated. We do not intend that the mind consciously distinguishes and dwells upon each of these relations, but that, in forming and applying such terms, it must in some sense have recognized them all. The question in dispute between the different parties is, what object the mind thinks of or has before itself when it uses general terms. Our previous analysis has, we think, established that it thinks of all these thought-relations, and that they all enter into the distinctive import or meaning of *the concept as such.* The conceptualist is right, if what he contends for is that the mind must impliedly have formed a concept of one or more generalized attributes, as often as it employs a general term. If the nominalist contends that the concept is *only* a general name—*i. e., a name* which the mind applies to many objects—he is manifestly in the wrong. What the mind considers, is not the name, but the *meaning* or *import* of the name.

How far the conceptualist and nominalist are both right.

7. The nominalist is right when he urges that the mind cannot conceive or acquire knowledge of the import of any concept, except by means of some individual example of the qualities or relations which it includes. We cannot know what single sensible attributes signify, as *red, sweet, smooth,* etc., without the actual experience of the sensation which each occasions, or of one that is analogous. So is it with the concepts of simple acts and states of the soul, as *to perceive, to imagine, to love, to choose.* The same is true of the concepts that are clearly complex, as *house, tent, knife, tree, horse, meadow, mountain, valley, township, legislature, authority, wealth, value, rent, wages, feudalism, civilization.* Of all these concepts, the elements must first have been made intelligible to the mind in some concrete example—*i. e.*, by being observed, experienced, or thought, in some individual being or agent.

We cannot know a quality or qualities, a relation or relations, except as exemplified in some individual being or thing, for the reason that these can neither exist nor be known except as belonging to beings or things. We cannot know what *red* is, except by the inspection of something red; what *imagining* or *remembering* are, except as an individual spirit imagines or remembers; what *equality, identity, height,* or *depth* are, except as some object is known as equal to another or identical with itself, or as high or low as compared with another.

The theory of the nominalist also finds ready acceptance, because names are so prominent and efficient in aiding thought. Experience teaches that, without the help of names the mind makes little progress in forming or applying its concepts. The use of language, and of spoken language even, is found to be almost essential to successful thought. Without language, the discriminations of attributes are few, the generalizations are narrow and limited, the power to enter into and receive the thoughts of others is almost dormant.

Many have gone so far as to conclude that, without words—*i. e.,* names—we cannot think at all. Experience with deaf-mutes, who have acquired little even of the language of signs, disproves this extreme conclusion. These show, by their actions, that they generalize—*i. e.,* form concepts—to a limited extent. They classify and arrange observations, they analyze and compare attributes, they apply principles in deduction and infer them from data. But while these facts show that it is not impossible to think without names, they also prove conclusively that without such aid, it is impossible to think with much effect. As soon as they learn to form and use names by the mastery of signs and written language, their power of thought is greatly quickened, and their stock of concepts is rapidly increased. But the language of the eye alone, which is the only language at their command, is immeasurably below the language of the ear in the fineness and variety of its material, as well as in its capacity for ready assimilation and recall. Still, the surprising acquisitions made by deaf-mutes, in spite of all the disadvantages under which they suffer, decisively prove that the mind is not restricted to any one kind of material out of which to form for itself a language; that words, in whatever form, are only the signs of thought, and are not essential to thought itself.

§ 207. 8. The truth that every *concept* is capable of being referred to an individual thing or *image*, and every individual or image can be *thought* into a *concept*, reconciles the strife between the conceptualist and the nominalist.

The imaging of concepts.

The conceptualist, in insisting that the concept must ignore and neglect the individual and its characteristics, often seems to overlook the dependence of the concept upon the individual thing or image as the originator of its materials, and the explainer of its import. Locke says, positively, "the general idea of a triangle" "must be neither oblique, nor rectangle, neither equilateral, equicrural, nor scalenon, but all and none of these at once." "In effect it is . . . an idea in which some parts of several different and inconsistent ideas are put together." The nominalist asserts that the only ideas which we can frame or mental objects which we can think of, are individual. Bishop Berkeley insists: "The idea of man that I frame to myself must be either of a white, or a black or a tawny, a straight or a crooked, a tall, or a low or a middle-sized man;" plainly implying that we can form no other thought of man than of one possessing these and other individual characteristics. And yet he concedes that, "An idea, which, considered in itself, is particular, becomes general by being made to represent or stand for all other particular ideas of the same sort." But how the individual can *represent* particular ideas, he does not explain, and seems never to have considered.

This remark brings the point in dispute to a distinct issue, in the questions, "How can one individual represent other individuals? Or, How can the individual explain and illustrate the general? A concept is general, an image is individual, how can you think the one into the other?" The sides of every individual triangle must have a definite length, and the angles a definite measurement and relations. Every individual man has in like manner a definite height, form, color, etc. We think these into concepts, not by overlooking the individual relations of each, but by considering their likeness to similar attributes in other objects; the sides and angles, not in their individual relations, but simply as sides and angles—*i. e.*, as bounding a figure and as being contained within two lines. We do not so much *leave any thing out of view*, as we *add* the *new relations* of likeness which the formation of the concept involves. An object viewed without thought-relations, is an *image*. An image with these relations added, becomes a *concept*. It is true that, when we *think* the image into a concept, we give special attention to fewer elements; but we need not overlook or omit any in regarding these few. Least of all do we introduce into the concept elements that are *inconsistent* or *incompatible*, and conceive—*i. e.*, *image*—a triangle which is neither rectangular, acute, or obtuse, as Locke asserts is necessary and as Berkeley objects is impossible.

Different images illustrate the same concept.

It is curious and instructive to notice here, that every man images the concepts which he employs or hears of, by examples that are peculiar to himself, and which are derived from his individual experience or observation. If his experience or education is marked by very striking peculiarities, the concrete examples suggested to him by every concept and name will be as peculiar. An Esquimaux, a Chinese, and a European, would picture very different objects to the imagination, on hearing or reading the words *state*, *legislation*, *wealth*, *money*, *wages*, *civilization*, *fashion*; and even the more concrete terms, *house*, *city*, *ship*, *oar*, *sail*, *knife*, *feast*, *procession*, *township*, and *meadow*. And yet their concepts denoted by these words are substantially the same, inasmuch as the more important and essential relations of objects are common, however greatly their individual characteristics may differ.

This circumstance explains how there may be a community of thoughts, with a very diverse experience. The nature of things and the nature of man remains unchanged. The same powers, laws, and ends are perpetually reappearing, the same principles are continually illustrated, under forms the most unlike.

The truth represented by realism.

§ 208. 9. *The realist* emphasises the truth that every real concept should suggest or express some one or more of the *essential properties* and *unchanging laws* of individual beings. He is not content with the view of the nominalist, who makes the general term a mere class-name for the simple convenience of language, nor with the view of the conceptualist, who regards the concept as a chance-assemblage of attributes. He insists that concepts proper ought to signify and represent those objects and those attributes only which are *permanent and constantly occurring.* This is the truth that has given currency and influence to the realistic theory, in spite of the extravagant and metaphorical language, and the insufficient arguments by which it has been stated and enforced.

All individual objects of nature exist under constant conditions, and are produced by permanent forces, according to fixed laws and ends. These constituents, conditions, causes, laws, and ends of individual objects are often called *their inner truth, their essential nature, their true meaning, their real and permanent being.* The individual mass of earth or ore, the single crystal, leaf, herb, tree, fish, bird, reptile, quadruped, and man, have accidental relations of position, form, size, color, or taste; they exist here or there for a longer or shorter period of time, but these relations are of little importance for the higher ends of knowledge and of practice. It is to reach and to impart the knowledge of permanent *elements, causes, laws,* and *designs,* that concepts are formed, classes are arranged, and names are given. As we have seen already, many of the earliest classifications and concepts are rude and unsatisfactory for scientific purposes, because they are founded upon attributes that are superficial and narrow in their significance and indicate few or none of the permanent elements and laws of being. These are gradually outgrown and displaced by others which as soon as discovered suggest more comprehensive agencies and laws.

The classifications of botany.

No better illustration can be adduced of the differing import of different kinds of concepts and classes, than is furnished by the history of botany. Linnæus hit upon the convenient expedient of classing the different individual plants by the number of the stamina that appear in their flowers and of subdividing the classes into orders by the number of pistils. The device was convenient, because all plants have flowers, and the number of the stamens and pistils is in most cases constant, and presents a ready means for their division and subdivision into classes and sub-classes. To a certain extent this division signified something—so far at least as the number of stamens and pistils was found to indicate other common characteristics of importance, and seemed to point to deeper qualities and laws. But this was by no means universally the case. The classes and orders that were founded upon the number of these organs, were concepts of little interest, because they signified nothing in respect to the structure or the germination, the growth or the habits, the flower or the fruit. Hence the Linnean system was abandoned for a system of classes and of nomenclature, which was founded on indications of greater practical and scientific significance.

The mistakes of the realists have been twofold. They have, both in language and thought, confounded the subjective concept, which is a purely psychological product, with its objective correlate—the related elements which it represents or indicates; and have often called both by the same name, and invested them with the same properties. They have used a highly metaphoric terminology to express the nature of universals, and their relations to individual beings. The *ideas* of Plato and the Platonists, present from eternity in the Divine mind; the *forms* of the Aristotelians, incapable of existing apart from matter, yet essential to every material thing and species; the *substantial* and *essential forms* of the schoolmen, as well as their *universals ante rem* and *a parte rei;* the *forms* and *ideas* of Kant; the *notion* of Hegel,—self-moving from the empty yet posited nothing, and self-developed by constant growth into all the fulness of *the idea,* with a capacity claimed for this notion to pass into the objective, giving the world of material being, and then to return to itself so as by self-conscious affirmation and distinction to blossom into spirit and thus complete the circle of absolute knowledge;—all these are examples of the exaggerations and personifications of realism in its endeavors to express a most important truth.

This subject has, of late, assumed a very great interest and importance among naturalists, in connection with the question of the permanence of species in the natural and vegetable kingdoms. Certain naturalists contend that none of the so-called species are permanent, either in the plan of nature, or its actual divisions; that every one of them has been developed by evolution from previously existing types, which owed their form and apparent permanence to certain conditions or laws that were but temporary in their action and transitory in their results. In this way Darwin, (*Origin of Species, etc.,*) Huxley, and others, reason from certain varieties produced within species, that all species existing at present, have been themselves developed. Herbert Spencer, by a broader application of the same general assumption, makes every type of existence, both material and spiritual, to have been developed from lower forms, which are held in being till forms still higher and more exalted shall displace them. On the other hand, *Owen*, *Agassiz*, and *Dana* find that the classifications of science must assume a more permanent and firmer foundation for the species which they accept, in the action of permanent forces after the fixed types that are contemplated in the unchanging plan and the manifested thoughts of God. In this assumption they express the scientific truth of the bold metaphors of Plato, who taught that by definition and division, we find in the temporary and phenomenal those eternal and real ideas which exist in unsoiled and unalloyed purity in the mind of the Deity alone. (Cf. Agassiz, *Essay on Classification.*)

Value of naming and of language.

§ 209. 10. The reasons why language aids our thinking are the following.

(*a.*) The name is both a *sensuous* and an *individual* object. It presents to our sense-perceptions a definite object, which we can readily evoke, distinctly apprehend, and easily and unmistakably repeat. What it represents, is indeed abstract and general, but the name itself is an individual object of sense-perception.

The word addresses a single sense, the ear or the eye singly, or the two combined. In either case it is ready to appear when called for. The winged word flies to our aid, and the ghostly product of thought is at once embodied before the senses.

(*b.*) The word is the sign, not of the whole of the individual thing or being which might image or exemplify the concept, but of a *portion* of its attributes or relations. In consequence, words present a greater variety and refinement of objects than exist in the world of nature. The words *red fruit*, *acid-fruit*, *currant*, *cherry-currant*, may all be imaged or exemplified by the same sense-object, viz., the fruit before us. *Red* stands for a single one of its properties; *fruit*, and hence *red fruit*, for several; *currant*, for more; and *cherry-currant*, for even more. So the terms *company*, *organized company*, and *legislature*, may all be exemplified by the same body of individuals discerned by the senses, while each of the words represents more or fewer of its attributes or relations.

To fix and represent a single attribute by a word, is also necessary for the service of communication which language performs. Another mind could not be brought to direct its attention to the attribute and property which we with difficulty discern, unless the attribute were represented by a name. This, however, does not weaken, but rather confirms the service of the word to thought, in rendering its acquisitions permanent and ready for use.

(*c.*) Names enable us to add to our stock of logically dependent concepts. One concept is dependent upon and grows out of another. One concept, when formed, enables us to form another, and is often the essential condition of the existence of the second.

(*d.*) Names aid most efficiently in rapid thinking, by sparing us the necessity of dwelling on the entire import of the word before us. In conversation or rapid discourse, as well as in reading by the eye, only enough of this import is attended to to satisfy the present occasion—all else is omitted. Even whole sentences, when they are familiar, are received as the signs of single concepts or relations, viz.: those which the present occasion requires. This can only happen when the language is familiar to the eye and the ear, so that, as the eye and the ear each catch enough to identify the word or phrase, the mind also catches enough of the import to satisfy the present occasion. Were not the words addressed to the senses, and capable of rapid formation and reception, they could not serve in this rapid application.

The relation of symbolic to intuitive knowledge.

§ 210. 11. The analysis which has been given of the nature of the concept and its relations to the individual object or image, explains more exactly the relation of what is called *symbolic, mediate,* or *logical* knowledge, to that which is *intuitive, immediate,* and *experimental.*

We have already spoken of this distinction in a general way. We return to it again, for the sake of greater exactness. Knowledge by concepts is symbolic, mediate and logical. Knowledge by direct apprehension, whether in connection with consciousness or perception, is called intuitive.

When I perceive a sense-object, as a man, a house, or tree, or am conscious of an individual state of spiritual activity, or discern with the mind's eye a mathematical figure, I know intuitively each of these objects. When I recognize either as belonging to a class, or give to either a name, I am said to know it by means of the concept or name; and these concepts or names are said to be the *media* or *symbols*, which I employ in knowing. This distinction, as thus stated, originated with Leibnitz, and much has been made of it by later thinkers, as Kant and other German philosophers, as also by Hamilton, Mansel, and Morell among the English.

The grounds for this distinction have been explained already in the positions, that every concept supposes an individual concrete, either real or imaginary, in which it is exemplified, and no person can conceive the import of the concept except as he resorts to this concrete for interpretation and explanation. When I pronounce such words as *white*, *red*, *sweet*, *sour*, etc., I presuppose that the person to whom I address them has known by experience, *i. e.*, by intuition, what they signify; that he has either seen these colors and tasted these tastes, or those which are in some respects like them. If he has had no intuitive or analogous experience of them, my words convey to him no meaning. The same is true of all the so-called simple ideas of Locke, which are the constituent elements of all those which are complex.

It should be remembered, however, that language may be used either for philosophical thought on the one hand, or pictorial and emotional effect on the other. In the one case, the mind is occupied with the more abstract and general relations of objects. In the other, those which are broader and more obvious are employed, often solely for the excitement and gratification of the emotions. In both cases, use must be made of the objects and images of individual experience. But in the first, the relations concerned are less dependent upon the individual images which happen to be suggested, because to convey or awaken general relations is the chief end. The individual examples by which each individual hearer or reader verifies or illustrates these concepts and their relations, is of less importance, provided he understands their import.

But even here intuition is far better than symbolic knowledge; rather should it be said, intuition *with thought* is far better than symbolic knowledge without intuition. The most careful definition of a *mountain*, *the ocean-surf*, a *cataract*, a *giraffe*, a *palm-tree*, may convey impressions far less satisfactory, and far less accurate, than the inspection of a moment might furnish, provided the inspection leads to thought—*i. e.*, to the formation or verification of concepts. With the concrete before us, our concepts are more exact, because we see for ourselves. The concrete also furnishes the material for any new concepts which we ourselves may form directly from their objects.

The defects of mere words and of the images which they awaken in comparison with actual intuition are still more striking when the objects are *described* rather than *defined*, and for the purposes of vivid impression and excited feeling. One is forcibly impressed with these defects, when he reads a description of a scene in nature with which he is personally familiar; especially if he reads it with the scene actually before him. However graphic or complete the description may be, it is but a lifeless outline when compared with the fulness and vividness of the reality, or with the throng of images which are awakened in the memory. The impressions received from words by one who has never witnessed the reality, are but as thin and pallid shadows, when contrasted with full and glowing intuitions. The most exact description of the falls of Niagara is a very different thing to one who has recently seen the cataract, or who reads with his eye open upon the scene, from what it can be to one who has never seen its wonders. If a person has never seen any waterfall, it is still more impotent to instruct the mind.

These facts bring to light very distinctly the truth that language operates to a very great extent by suggesting the images and remembrances which have been gained by the experience and observation of each individual person. Besides the direct office of instructing the mind, it serves to awaken a multitude of kindred images and facts which are suggested by them. Words which to one are dead and meaningless are to another full of life and import. Words meant only in kindness may awaken images of sorrow and pain. The reader of poetry must have somewhat of a poet's power to receive and re-create. The student of philosophy must have something of a philosopher's reach and insight, to understand and judge what he reads.

There is a large class of facts and truths, as well of scenes and events, to which language can do but scant justice. These are those to which the facts and events which we know and have experienced are only remotely analogous. Language is feeble to convey to the inhabitant of a plain or a prairie, the impressions of mountain scenery; to the stranger to woods, the grandeur of an aboriginal forest; to one who has always lived inland, the glory and the beauty of the ocean.

When the means of finding analogies are still more scanty, the communication by language is still less successful. How curiously do we endeavor to anticipate what may be the scenes and objects to which another life may introduce us! But how feeble is our power to imagine these, because our stock of analoga is so scanty! We desire most earnestly that description in language may convey to us the desired information. But language may be to a large extent inadequate, because all the images of which language can avail itself must of necessity be taken from the scenes of the present state of being.

It is sometimes asserted that the Infinite Spirit can have no common relations with the finite,—that all our conceptions of the infinite, being finite, must therefore be inadequate and unworthy; and that, consequently, all attempts of language to convey knowledge from the higher to the lower must be forever impossible, because the media—*i. e.*, the images and concepts—must both be finite. This is urged against the possibility of any communication from God through the forms of finite nature, or by the media of human speech. It may be granted that, to the mind, in its studies of nature, the *images* that are suggested or presented, and the language founded on such images, are wholly inadequate to express the divine, because they are finite; it may be granted even that the concepts of spiritual relations must necessarily be interpreted and illustrated by images taken from finite objects, and that so far there are essential defects in all our imaginations concerning God: yet it may remain true that there are relations of similarity and analogy between the finite and the infinite spirit, which render it possible that the one should be understood by the other, and that the language which describes the one to the other should convey actual truth.

The infinitude of God may not exclude personality, which itself establishes a likeness between man and God. Personality may involve similarity of knowledge in respect to all the higher relations of truth. A common sympathy may rest upon a similarity of emotional capacities, while similarity in the still higher endowment of a personal will, may render possible a similar moral goodness. These likenesses or analogies, may coexist with the greatest disparities in every other respect. The one being may be infinite and the Creator; the other may be finite and created; and yet the One, by indications through his works and communications by his word, may make himself truly, if not perfectly, known. The imagination of the finite may be inadequate to *picture* the infinite, and yet the thinking of the finite may apprehend the relations by which the infinite first thinks and therefore creates, and in creating manifests himself to the created.

CHAPTER V.

JUDGMENT, AND THE PROPOSITION.

Judgment implied in the formation and use of the concept.

§ 211. The processes already considered, and which are involved in forming and applying notions, are alike in this; they are all acts of *judgment*. The mind cannot *know*,—much less can it *think*, without judging. To think, is to judge. Even in forming or evolving its notions—that is, in providing itself with the materials for what are usually called acts of judgment—the mind must *judge*.

The truth of this assertion is evident from the following considerations.

(1.) It is evident from an analysis of the act itself. If we retrace the steps which we have taken in forming concepts, we find that we cannot know attributes, except as we affirm them of individual beings. An attribute without a being is inconceivable in thought and impossible in fact. Suppose we meet with a series of unknown and unnamed objects, each of which has some attribute or property, that is unfamiliar and even without a name: or suppose the attribute to be familiar and nameable, while the objects are unnamed. We think and say of each of these objects, it is yellow, red, or green. In thinking or saying thus we in fact perform a process which can only be represented by some proposition, one element of which is affirmed of another: *e. g.*, *x.* is *yellow, red, or green;* or if each is as yet unnamed, x [individual] is y [general]. The nearest and best expression of this act which we find in any form of language is the impersonal verb, as, *it shines, it lightens, it rains,* in the use of which the unnamed being is present to the senses, and the attribute is judged or affirmed of it.

(2.) It is still further implied in the truth already developed, that every notion is by its very nature and essence *relative,* i. e., *related to* individual objects or actually existing things.

(3.) The same is evident from the consideration of the meaning of names, or notions in language. A name is the verbal symbol of a concept or notion. But to be a name, it must be a name of some object or objects; some object must be called by it; it must be applied to some thing or being. But these acts imply judgment.

(4.) It is implied by the nature and definition of knowledge. An act of knowledge has already been shown to be necessarily and universally an act of judgment, whether it takes the form of presentation, representation, or thought. Every such act implies the apprehension of an object as existing; and more, its existence in some relation. If it is true that knowledge by perception and memory implies judgment, much more does knowledge by thought; forasmuch as *the general* with which thought has to do, can by its very essence and nature, be only a relative and a predicable entity.

We conclude then that wherever there is a notion, there is an implied act of judgment. Every such notion has been formed by judgment, and is capable of being expanded into a judgment. It is an organic thing, representing in its very essence the act which gave it being, and capable of being developed into similar though more complex products. It is like a seed, which is a miniature plant, having come from a plant and being ready to spring into a plant; or it is like the cell which is the organized and organizing element of development in vegetable or animal life. We do not judge by a mechanical and superinduced act of the intellect, which, finding two names or notions, proceeds to fasten them together; but it is of the very nature of the notion, that it can be applied or united to some object. This natural and necessary act of union or synthesis is an act of judgment. The true doctrine may be stated thus: *every concept is a contracted judgment; every judgment is an expanded concept.*

Judgments are psychological and logical.

§ 212. The judgments by which concepts are formed, are called *primary*, and *psychological* judgments. They are distinguished by the circumstance that their subject is *an existing and individual thing.* Judgments in which *concepts* are affirmed or denied of one another are *secondary* and *logical.* The secondary, comparative, and logical judgments are all founded on those which are primary, natural, and psychological. To be convinced of this truth, we need only to consider the expression of judgments in language, and to trace the order of progress by which logical judgments, *i. e.*, judgments consisting of concepts, come to be reached and understood.

The secondary judgment, when its subject is an individual being, differs from the primary in this, that the subject is denoted by means of a common term. Instead of saying *it*, we say *this orange.* If the subject is a universal, as *all oranges,* the mind gives the result of its observations or inductions, by using the concept in its largest extent.

When purely mental entities are treated of, whether a fiction of the imagination, as *the centaur*, or a mathematical construction, as *the triangle,* or an abstractum, as *virtue,* they are treated as actually existing beings.

How the subject of a judgment is expressed in language.

The fact has already been established, that the concept, by its very nature, contemplates attributes only; and that concepts, like *man, human, humanity,* so far as their constituent attributes are concerned, stand for precisely the same content of attributes. When they are expressed in language, however, *man* and *human* differ in this, that the one word, *man,* denotes a being to which these attributes belong, and the other, *human,* denotes the attributes only. By what process the mind comes to be possessed of these two sorts of words, we need not here inquire. But when it does possess them, it cannot but use them. Instead of thinking or saying, *it* is yellow, or, *it* rains, the man says, *orange* is yellow, *cloud* rains. Soon he learns to say this in three ways; *this* orange is yellow, *some* oranges are yellow, *all* oranges are yellow, according as he uses the general name for one, a part, or all of the beings for which the orange stands. In order to do this, he applies special terms to denote these three relations, viz., the words *the, this,* or *one; some* [*a few* or *many*]; and *all.*

The fact that a concept has the two relations of *extent* and *content,* fits it to be used both as the name of one or more individuals, and as an attribute only. When a concept is used to denote beings, it is used in the relation of extent. When it is used to denote attributes, it is used in the relation of content. In the secondary judgment, a concept used in its extent only is employed as the subject, taking the place of the individual intuition; the notion as content is retained as the predicate: and the natural judgment in which only one notion is used, becomes a secondary judgment in which two notions appear. The two species of judgment are, however, essentially one and the same, inasmuch as both express what is essentially involved in the act of thinking, viz. · the act of affirming a concept of an existing being or thing.

The signification of the copula.

§ 213. The *copula* expresses the act of judging or affirming, whatever is the kind of judgment or the relation affirmed. It makes no difference whether it is or is not expressed, it is still present as an element in every judgment. The act of judgment is the same whatever be its verbal expression, whether subject, predicate and copula are condensed in a single word, as, *pluit*—or expanded into two, as, *it rains*—or into three, as, the *clouds are raining.*

The copula does not require or imply that there should be an actually existing material or spiritual thing or agent, of which the attribute is affirmed or thought. The being may be an imaginary being, as *a centaur*, or a mathematical entity, as *a triangle*, or an abstractum, as *whiteness*, or *virtue*, or *legislation*; and yet one or more attributes may be asserted or thought of each. All that the copula properly signifies is, that the concept has this or that attribute, one or many. Whether the concept is of a real being or a thought-being is left to be determined by other sources of knowledge. If a centaur is spoken of, we know it has only imaginary existence; if a triangle, that it is a mathematical conception or construction; if virtue or legislation, we know we must go back to concrete beings, to find the realities of which these are *abstracts.*

Classes of judgments. Judgments of content.

§ 214. It has been established that every notion is a contracted judgment and every judgment is an expanded notion, and also that every notion has two relations—the relation of *content* and the relation of *extent.* It follows that notions can be expanded into two kinds of judgments: *judgments* of *content* and *judgments* of *extent.* Each of these forms of judgment require special illustration.

We begin with the Judgment of Content. This is the form of all original and natural judgments. It is by a judgment of content, *i. e.*, of a common attribute or relation—that every notion is originally formed. In this form judgments most frequently occur in language. Objects are observed and their common attribute or attributes are *thought*, i. e., *judged*, of them, and the judgments when expressed in words are those propositions which abound in every language. It is only by a reflex act that the mind develops and employs judgments of *extent.*

These natural judgments of content, serve the purposes of common life and of common intercourse. For the ends and uses of science we need to go further and to employ propositions of *definition.* In such propositions we assert not merely one or more attributes for information, but we indicate for distinction, *the attributes* which make up or constitute the entire *content.* To satisfy the ends of science we must express what is called the whole content, since if we state only those elements which are common to this concept and many others, and omit one or more that is peculiar, we do not distinguish it from others. If we define a circle as a curvilinear figure, the circle is not distinguished from an ellipse. If we define man to be a two-legged and featherless being, this is true also of a plucked chicken. Hence the rule by which we try and determine a good definition: The proposition which expresses it must be convertible.

The content was called by Aristotle and the Scholastics *the essence,* i. e., the attributes or elements which make the notion to *be* what it is as a notion.

Aristotle, however, also recognized in the essence that which existed really and permanently in the objects to which the concept belonged, rather than the attributes themselves which constitute the concept. He applied essence metaphysically rather than logically, to the objective correlate of the concept, rather than to the concept itself as an intellectual or purely subjective product.

A proposition of content properly expresses only *logical truth.* It very often implies, however, *real existence.* Propositions may concern existing beings or notions of beings to which there is no corresponding reality. The proposition as a definition only, expands the content or essence of the concept, without deciding whether any corresponding reality exists in fact. When for example we define the centaur we give the attributes that make up the conception without asserting or even knowing whether such a being exists. When we define a triangle we state the essential constituents of the concept produced by the constructive imagination, knowing that it has no other existence. When we define man we both define the concept and believe the concept is realized in actual fact. The definition of centaur implies only thought-essence or logical truth. The definition of man implies both logical and real truth. The copula *is,* in the one case signifies *is defined as* or *consists of;* in the other—*is defined as* and *really exists.*

In very many cases we readily interpret the meaning of the copula and the character of the judgment and definition, by our knowledge of the subject-matter. In other cases we have no such knowledge as qualifies us to determine whether the definition is really true, as well as logically consistent. Suppose any one of the following concepts is to be defined: *virtue, duty, inalienable right, natural liberty, tyranny, a sovereign state.* It is of essential importance to know whether the definition concerns only the concept as a mental product, existing in and for the mind only, or whether there are actual relations and activities of human nature, to which the concept corresponds. In the first instance we should need to consider only, whether the concept is correctly defined as the term is ordinarily used, or as this or that school of philosophers or politicians have conceived it. In the second, we should inquire, whether it answers to a truth of fact, *i. e.*, whether the concept has a corresponding reality.

Scientific truth implies both *logical* and *real* truth. *Logical* truth is but another name for logical consistency. A dexterous logician, if suffered to frame his own concepts and construct his own propositions, may easily frame a system which shall have sufficient truth to give plausibility to all that is defective by omission, or false by positive error. Every definition should therefore be scrutinized both as to its consistency and its truth. It should always be remembered that a proposition may be logically true and yet really false, while science requires that the definition should not only be logically consistent and logically complete, but also really exhaustive and actually true.

§ 215. The proposition of extent is the natural consequent of the proposition of content. The proposition of content is first in time, because the knowledge of the individual goes before the knowledge of the general. As soon, however, as a single attribute is affirmed as common to many individuals, then this common attribute can be employed as itself dividing or separating these individuals into a class by themselves. As soon as we think, This house is white, it is possible for us to refer the house to the class of white objects. But because every generalized attribute may classify the objects to which it belongs, it does not follow that the mind recognizes it in this relation, or expresses the relation in language. It is not till the adjective, white, becomes a noun, that we use it as a classifier, and think or say, *whites, i. e.*, white men, are *English, French*, etc., etc., or *white things* are *so* and *so*. It is not till we

Judgments of extent.

turn back upon our thinking, and recognize the fact that these attributes divide into classes the beings to which they belong, and even go further and notice that some classes of objects are wider and some narrower than others, that we have occasion to think of these notions in their extent, or to expand them into propositions expressing this relation.

Propositions of extent like those of content, strictly considered, only assert logical truth—*i. e.*, the subordinate classes into which the concept is divided. But they often imply the real existence of the objects to which both the comprehending genus and the included species belong.

Propositions of extent, whether used in common life or for the purposes of science, are clearly distinguishable from propositions of content. It is, however, easy to confound the one with the other; and easy to interchange the one with the other. Indeed we are often tempted to translate the propositions which express the one into those which express the other. We cannot say that man is an animal without implying that he possesses those attributes which are involved in the concept and the term animal. Whenever we assert that man is a species of which animal is a genus, we must ascribe to man certain attributes. Conversely, we cannot assert certain attributes of man without placing him in a distinct class. These facts are not at all inconsistent with the truth that at some times we use propositions with sole reference to their *content*, and at other times with exclusive respect to their *extent*. Indeed, the use of propositions of extent is a necessary condition and consequence of logical *division*. But if *division* is distinguishable from *definition*, then are propositions of extent clearly distinguishable from propositions of content.

As *definition* gives complete propositions of *content*, so *division* gives exact and complete propositions of extent. Both processes are involved in the beginnings of thinking. They are only carried forward to their completed perfection when we reach the precise and comprehensive knowledge which science attains. Both are the necessary conditions of the formation and use of general terms, and are the constant accompaniments of language. Both are perfect in their ideal aims whenever the definitions in any branch of knowledge become precise and true, and the divisions orderly and exhaustive.

§ 216. It is a superficial view to regard scientific knowledge as different in kind from common knowledge: to reason as though the man of science has developed intellectual powers which are peculiar to himself, or has discovered special processes or rules having no relation to those which are natural to all men. The powers employed by the true philosopher and the uncultured are the same. The common man thinks as really, and in his way as effectively and as sagaciously, as does the philosopher. Scientific and common knowledge.

Often it is not easy to find the dividing line which separates common from scientific knowledge. We cannot say, in the history of any branch of knowledge, Here common knowledge ceases and science begins: At this point he who knows as a man, begins to know as a philosopher. Of some sciences it is true, that at a certain period of their development, common terms are exchanged for those which are technical, and a scholastic, sometimes a repulsive nomenclature takes the place of words which are familiar from use and warm with grateful associations. Even objects that in the earliest classifications had been grouped together by affinities so close that they seem to have a necessary and unbroken relationship, are strangely separated, and find themselves suddenly in a new and unpleasant society. Plants and trees apparently the most alike are thrown into the most distant groups, and those which are apparently the most diverse and dissimilar are inexplicably brought together. In those sciences which are less technical in their definitions and classifications, the lines of transition and division are not even suspected. We cannot find the place where science in its technical form begins, and formally takes its leave of common knowledge. In *Psychology, Ethics, Politics, Law* and *Theology*, common terms are in a great measure still retained; they are only employed with a more careful definition and a more exact application.

Science when viewed in the light of our analysis, is simply *knowledge by concepts carefully defined in order to a complete division and a methodized arrangement of the things or objects to which these concepts are applicable.* In forming scientific notions, the mind discovers relations and attributes which it had never observed before. In looking more patiently, it observes more closely. As it proceeds to use and apply the notions already attained, in the processes of deduction and induction which are yet to be explained, it discerns still other relations of likeness and unlikeness. As it proceeds in its triumphant course it still continues to define and divide. Science began when it formed the first proposition of content. This involved a proposition of extent. It will have finished its course and completed the circle of its possible triumphs, when it shall have exhausted all that is knowable by these two processes, each involving the other—when it shall have arranged in systematic order, everything which can be known, by *complete* and *subordinated divisions as the result* of *true and exhaustive definitions.*

CHAPTER VI.

REASONING.—DEDUCTION OR MEDIATE JUDGMENT.

Nature and importance of reasoning.

§ 217. The process of thought or mode of thinking which we are naturally led to consider next in order is *reasoning.* That *to reason* is a function of the thinking power, will be questioned by none. By many it is esteemed its special and almost its sole function, a function which absorbs all the rest into itself. Many make the capacity *to reason* to be the exclusive and distinctive endowment of man, striving to account for all the other thought-processes by resolving them into this.

Reasoning, also, like every other act or mode of knowing, is itself an act of judgment. It is distinguished from judgment proper by being *mediate and indirect;* whereas *judgments proper are immediate and direct.*

The acts of judgment proper have already been explained as acts in which a general notion is thought or affirmed of an individual being, or a concept, by direct inspection and comparison. When, for example, we judge of ten apples, that they are red, or oval, or round, or of equal or unequal weight, or of similar taste or odor, we perform acts of *direct* or immediate judgment. But when we reason concerning them, that because they are red, or similar in odor, therefore they taste alike, we judge *indirectly* or mediately; we consider, not only the apples themselves, but the relation of one of their properties to another. This truth is implied in the remark that in judgment we compare *two notions*, and discern or pronounce that the notions agree or disagree; whereas in *reasoning* we compare *two judgments*, and declare or discern that the judgments agree or disagree. If we distinguish the *process* of reasoning from the *product* or *result*—as in the other acts of the intellect—we should call the first *reasoning* and the second an *argument*. The latter is exclusively limited to *deduction*.

Reasoning, inductive and deductive.

§ 218. The process called reasoning is two-fold, *inductive and deductive*. It is known by the two names, *induction* and *deduction*. These two are sufficiently distinguished by the following definitions: In *deduction* the mind begins with general propositions, and reasons to those which are particular or individual; in *induction*, it reasons from individual or particular to general judgments.

In *deduction* we assume or imply that the mind is already furnished with judgments or beliefs that are more or less general, and proceed to derive from them, those which are particular or singular. In other words, we apply the predicate of a general proposition to a particular or individual, to which we had never applied it before. For example: 'we ought in every act to consult the wishes of our parents; therefore we ought to do this in choosing our business in life.' In *induction*, on the contrary, we proceed from singular or particular to general propositions or truths. We observe that one or several pieces of iron-ore, with certain characteristics, are magnetic. We infer that every similar piece of iron-ore is magnetic.

Both these processes are called processes of reasoning. The means employed, *i. e.*, the grounds or foundations of each, whether they are general or particular propositions or individual facts, are called *reasons*, sometimes *data*. But to reason *par éminence*, is to perform the process of *deduction;* and *reasons* or *grounds* of belief are preëminently those *general principles* or *truths* from which we derive or deduce particular conclusions. Hence, when we use the words *to reason* and *a reason*, we are usually understood to have in mind the deductive process. On the other hand, we say freely that we reason by induction or inductively; and no phrases are more common than *inductive reasoning* and *reasoning* by *induction.*

These two processes are usually combined together in every case in which our knowledge is enlarged by what we call reasoning. When we use examples of reasoning for the purpose of illustrating the nature of the process, we seem to be able to separate deduction from induction. But whenever we reason with the express design of adding to our knowledge, or of increasing our confidence in that which we already possess, both processes are called into requisition. If, for example, we should reason deductively, to prove to a person who did not already believe it, that a particular act of obedience, or perhaps of resistance, to the government, was obligatory; we should use the process of induction to prove that such an act was distinguished by the characteristics or criteria which showed it to come under the duties of a loyal citizen.

In many cases of induction, also, the process of deduction is brought into requisition. We can scarcely suppose that Franklin established the identity of lightning with machine electricity, without asking himself many times over what would be the consequents in fact, if his hypothesis should prove true. We know that Sir Isaac Newton drew certain inferences from the supposition that the law of gravitation was real, when combined with a false datum in respect to the earth's diameter; and because observed facts did not coincide with the theory, he did not accept the theory which his so-called induction had already reached.

Induction and *Deduction,* like the analysis and synthesis of which they are special forms, accompany each other in all the higher processes of thought. The two blend together so intimately that it is often difficult to sever them, or to find or trace the line where the one begins and the other terminates.

Thus far we have considered Deduction and Induction together. We proceed to study them apart, chiefly from a psychological point of view—beginning with Deduction.

The forms of deduction.

§ 219. Our chief inquiry is, what is the proper conception of the *deductive* as an intellectual *process;* and incidental to this, what is the nature and what are the results of the *product* which it evolves. To answer this question satisfactorily we must consider, first of all, the forms of language in which the process is expressed and its results are preserved.

These forms are two, the *Enthymeme* and the *Syllogism*, or the abbreviated and the expanded syllogism. The enthymeme consists of *two* expressed propositions, which are connected by *because* or *therefore.* The syllogism consists of *three,* of which the first two are simple assertions, and the third is introduced by *therefore.* For example, *M is a* {*usurper,* / *lawful ruler,*} *therefore he* {*cannot exact obedience;* / *ought to be obeyed;*} *or, M* {*cannot exact allegiance,* / *ought to be obeyed,*} *because he is* {*a usurper;* / *a lawful ruler;*} are examples of the two forms of the enthymeme. {No usurper can / Every lawful ruler} {require allegiance; / ought to be obeyed;} M is {a usurper, / a lawful ruler,} therefore M {cannot require / ought to be} {allegiance, / obeyed,} are examples of the expanded syllogism.

In the *enthymeme,* the first proposition may be either the *conclusion,* or it may be the *reason.* In the syllogism, the first proposition is called the *major premise;* the second, the *minor premise;* and the third, the *conclusion.*

The two premises of every syllogism must have one term common to both, which is called the *middle term.* In the examples given—*lawful ruler* and *usurper* are the middle terms respectively of the two syllogisms. Unless there is this middle term, there is no force or convincing power in the argument: if we introduce two middle terms, there is no conclusion. The middle term must also have the relation affirmed to the other two. If we substitute *worthy* or *unworthy person* for *lawful ruler* or *usurper,* the conclusion will be false.

Every enthymeme can be expanded into a syllogism. The syllogism when expanded expresses in separate propositions the truths which the enthymeme implies. There is in every enthymeme the suppressed premise of a syllogism. When we reason in the examples given, M is a lawful ruler, therefore he ought to be obeyed, or M ought to be obeyed because he is the lawful ruler, we believe and imply in the argument—though we do not assert—that *every lawful ruler ought to be obeyed.* This is the major premise of the syllogism into which the enthymeme is by this addition expanded. The difference between the enthymeme and the syllogism is only a difference between a contracted and an expanded form of expression; or between an elliptical and a fully explicated sentence. It is a difference of language only, and not in the least a difference of thought or of the relations of thought or knowledge; what is expressed in the one, being implied in the other.

The syllogism not *a* but *the* form of deduction.

§ 220. The syllogism is the only form which fully expresses in language all the processes in the act of deduction. Some have contended that it is one of the forms of deduction, but not the sole form appropriate to it. Thus, Principal Campbell in his *Philosophy of Rhetoric* contends that the syllogistic is only one of the possible methods of reasoning, while there are others which are in many cases greatly to be preferred to this; and J. S. Mill, in his *Logic*, urges that it is not a form of reasoning at all, but a convenient expedient for recording and referring to the results of our experience in particular or individual cases. It is obvious, for the reasons already given, that it is a form into which all deductive reasoning may be phrased, and it is the one and the only form in which all the materials considered and the relations involved are fully stated in language. When for example we supply the premise that had been suppressed in the enthymeme, we do not add that which is superfluous to the process through which we have gone, or to the argument which the process implied. We simply express in language what we had thought or were ready to think in fact—that which if we had not believed when we drew our conclusion, we should not have reached it at all. Thus, if we did not believe that all lawful rulers ought to be obeyed, we could not reach the inference that M ought to be obeyed because he is the lawful ruler.

Again; In the syllogism the process of reasoning is *fully* expanded and complete. Any additional propositions, whether connected with either of the premises or with the conclusion, are seen at once to be a premise or a conclusion of another syllogism. If for example we enlarge the premise, "all lawful rulers ought to be obeyed," by the reason "because it is the will of God, or an obvious duty," we introduce an additional process of reasoning, the object of which is to prove that the first premise is correct. If we add a reason for holding that M is a lawful ruler, as "because he has been properly commissioned or fairly elected," we do the same for the second premise. If we annex to the conclusion an additional remark, as "therefore M ought to be obeyed, and to disobey him is a serious crime," we simply introduce a second conclusion, which requires another argument to support it.

Every argument, whether positive or negative, whether the propositions are universal or particular, can be expressed in the form which has already been stated, by changes in the phraseology or the position of the terms, without affecting the sense or the force of the argument.

This is demonstrated at length in every treatise on formal logic. A few examples will suffice for our purpose. If we make the first premise negative by substituting "no lawful ruler should be disobeyed," the real nature of the argument is not changed. The same is true if in the second premise we substitute "M rules lawfully" for "M is a lawful ruler"—a proposition of content for one of extent.

If we change the form of the first premise by inverting the order of the terms, that is, by *conversion*, which we can do with the negative premise and retain its full meaning, we bring the middle term into the predicate of each of the premises; but the argument and its power to prove a conclusion are the same.

If we convert the second, or minor premise, we bring the middle term into the subject of each premise, but this does not alter the strength of the argument.

If we transpose the order of the premises, the relation of each part to the conclusion is the same, whatever may be the order in which the two are uttered. All these changes can be made in the arrangement of the parts of the syllogism, without affecting the nature or force of the argument.

The dicta or formulæ of the syllogism.

§ 221. The rules for testing the validity of the syllogism may all be founded on the maxim, usually called the *dictum de omni et nullo*. It is as follows: *whatever is predicated of a class either affirmatively or negatively, may be affirmed of whatever is contained in or under the class.*

For this dictum, later logicians have substituted the maxim, *Nota notæ est etiam nota rei, repugnans notæ repugnat etiam rei.* This is adopted by J. S. Mill in his *Logic*, I., c. ii, § 3. It is the same in principle with the dictum of Aristotle. The same is true of the special construction of the syllogism proposed by Hamilton, by which the propositions are stated in relations of quantity, and the *dictum de omni et nullo* is displaced by *whatever is a part of a part is a part of the containing whole.*

In another form this dictum would be founded on the fact that the middle term, as it is a concept, stands to other notions in the two relations of *extent* and *content*, and would read thus, "A notion that is, or is not, in any *extent*, may, or may not, take to itself the notion which is of its *content*." The last formula has the advantage of stating concisely both the likeness and the difference between an act of judgment and an act of reasoning. For in an act of judgment, as we have seen, a concept may be expanded either in the direction of its extent or of its content. So far as the single act of judgment is concerned, the notion is viewed in only one relation, that of its extent or of its content, as the case may be. But in an act of reasoning, a notion, *i. e.*, the middle term, is viewed in both these relations at once, and the result is that a relation is developed and observed between notions, which had not been discerned before.

But neither the relations of a *genus* to a *species* nor those of *a part to a whole*, nor those of *extent* and *content* combined, give to the premises of the syllogism the power of demonstration. They *suggest* and they test the validity of a syllogism, but they do not explain that in the deductive process which gives it convincing power over the mind. No syllogism is valid to which the *dictum de omni et nullo* cannot be applied, but it does not follow that the maxim expresses the real ground of our faith in the psychological process which we call deduction. The relations of both major and minor terms to the extent and the content of the middle, may be the only relations that need to be expressed in language, and yet may not develop or exhibit the real relation which leads to our assent to the conclusion.

In point of fact, every attempt to explain the deductive process, as such, by these relations, has failed, and the failure of these attempts has perpetually exposed the doctrine of the syllogism to suspicion and contempt. Cf. Locke, *Essay*, B. IV., Chap. 17, § § 4–8; G. Campbell, *Phil. of Rhetoric*, B. I., Chap. 6; D. Stewart, *Elements*, P. II., Chaps. 2, 3 & 4; J. S. Mill, *System of Logic*, B. II., Chap. 3; S. Bailey, *Theory of Reasoning*.

The real error or defect consists in making the essence or import of both induction and deduction to consist in classification and the apprehension of class relations. If induction consists only or chiefly in establishing general facts by extended observation, then deduction must by consequence signify the recognition of what must already have been known in the formation of the class. If induction is a synthesis of individuals into a comprehensive whole, then deduction must be an analysis of this whole into its parts. If the synthesis has been carefully made, then the analysis is unnecessary because it is superfluous. According to this view of the two processes, deduction is only subsidiary to induction, and when we seem to perform the process of demonstration or proof, it is *the inductive* and not *the deductive* element which gives it any value or force.

Deduction rests on the relation of reason to consequent.

§ 222. The relation which is characteristic of the deductive process is that of a *reason* to its *consequent*, or of a *ground* to its *inference*. It is by means of this relation that we know objects by means of this process of knowledge. This relation is suggested to the mind in the syllogism by the relation of *a whole* to *a part*, but it is not therefore resolvable into this relation, nor should it be confounded with it. When we say, *all magnets attract iron ; this is a magnet ; therefore it attracts iron ;* the word *all* suggests or indicates that there is some reason founded on *the nature or properties* of the magnet, which forces us to believe that this particular magnet will do the same. This relation finds expression in language by *because* in the enthymeme, and by *therefore* in the syllogism. *Because* signifies *by cause of. Therefore* means *for*, i. e., *on account of that*, viz., that which had been previously stated in the premises; *there* being equivalent to *the foregoing*. Both words signify *by reason of*.

In other words, in order to explain the process of deductive reasoning, we must assume that every thing that exists and occurs, whether in the material or spirit world, exists and occurs under the real relation of causation, or constituent elements and laws. Every phenomenon and every thought-creation in the universe exists by the workings of powers with which finite agents are endowed, in obedience to fixed conditions and laws, in order to accomplish rational ends or results. Every such existence is an effect; material things, spiritual agents, nay, even mathematical and logical concepts. The nature and the constitution of these effects are all explained by the *causes, conditions,* and *ends,* by, under, and for which, they are conceived to exist and to act. *Any one of these elements,* when applied to explain their existence, or to confirm our knowledge when we seek explanation or proof, is called a *reason.* When such a reason is discovered to *explain* or account for a fact or phenomenon, the process is called induction. When it is applied to *impart* or *confirm* knowledge concerning a fact or truth in respect to which the mind seeks to be informed or convinced, the process is called deduction. To know by either or both of these processes is to know by a reason, —it is to reason, *ratiocinari;* it is reasoning, *ratiocinatio.*

For proof of this we appeal to the process of reasoning itself. In doing so, we should not employ any of those trivial examples which occur in most books of logic, but rather select some example of the process of deduction when it is of actual service, *i. e.,* when it is employed to relieve the mind from doubt, or to answer its questionings as to what is true. In every such case we shall find that the mind has no direct access to the object before it, and can gain no immediate or intuitive knowledge. It is the cause, the effect or the law, the end or the means,—one side or term,—to which the mind has any means of access. But it knows or may know that under the law of causation this is necessarily connected with the other term. The use of this relation for the relief of doubt or the acquisition of knowledge, is reasoning. When the relation of causation is applied to this use it passes into the relation of reason and its consequent. The necessary connection involved in causation when thus applied gives to deduction convincing force. This discerned necessary connection between *a cause* and *its effect, means* and *end,* etc., etc., is what we call the force of demonstration or deduction.

That the deductive process and the syllogism are founded on the relation of causality was distinctly taught by *Aristotle.* He remarks, *Anal. Post.,* II., 2: τὸ μὲν γὰρ ἄἴτιον τὸ μέσον, which means in this connection, the middle term, is causal in its significance. To the like effect is the passage, *Anal. Post.,* II., 12, τὸ γὰρ μέσον αἴτιον. Aristotle distinguishes between *the cause* of *being* and *the cause* of *knowing,—ratio essendi* and *ratio cognoscendi,* i. e., between *the cause* and *the reason,* but he does not show how the one is related to the other.

The later Greek logicians being more occupied with the forms of the syllogism and its application to the detection of fallacies than with its psychological import, left very much out of view this important hint of their great master. The scholastics committed the double error of believing that the syllogism was the sole instrument of acquiring new knowledge, or of discovery properly so-called, to the neglect of induction; and of supposing that the formal relations of the syllogism constituted and measured all the relations of things. Hence the axioms were so generally received in the Continental schools, that the principles of identity, of contradiction, and excluded middle—the so-called laws of thought—are the only criteria of real truth and actual knowledge, and that the process of deduction itself can be explained by these axioms.

Leibnitz is a distinguished and notable exception to this nearly uniform course of speculation. He asserts that, for the purpose of philosophy, besides the principle of contradiction another is required, viz., the principle of the sufficient reason. But the principle of the sufficient reason of Leibnitz is explained and applied by himself indifferently, alike to the causes of actually existing phenomena and the reasons of demonstrated truth. That is, the *ratio essendi* is not distinguished from the *ratio cognoscendi*, and of course there is no attempt to show the relation of the one to the other. It is not surprising that a principle so imperfectly enounced did not take a permanent place in the schools of philosophy. Even Wolf himself, Leibnitz's professed disciple and expounder (*Ontol.*, § 70 sqq.; Met., § 30 sqq.), attempts to resolve the law of causation and the sufficient reason into the law of contradiction. The tendency of modern philosophy has been to consider the law of the sufficient reason as extra-logical (*Hamilton, Dis.*, p. 603), or to derive it in both forms, of real and logical cause, from the relations of concepts to concepts, instead of founding the *ratio cognoscendi* on the *ratio essendi*, i. e., on the relations of things; thereby inverting the processes of nature and destroying confidence in the grounds of knowledge and of faith.

The relation of logical reasons to causes and laws.

§ 223. The conception of the *logical reason* is wider in its range and application than that of the real cause on which it is founded. The real cause is usually prior to the effect which it produces. The mind in apprehending or observing its actual workings, assumes or supposes the cause, in order to anticipate or explain the actual effect. But in applying this relation for the purposes of reasoning, the mind may begin with the effect and conclude to a cause, as properly as when it begins with the cause and reasons to an effect. Either involves the other in a connection of thought; either can be made to imply the other in the order of deduction or reasoning.

The reason and the cause *coincide*, when from an actual cause, (the conditions and laws being included or supposed,) we reason to the certainty or reality of the effect. Thus the fire did or will fall into a vessel of gunpowder, therefore an explosion did or will occur. They diverge, when we reason from the effect to the cause, *i. e.*, when the effect is made the reason for our belief in or knowledge of the cause: as the vessel of gunpowder exploded, therefore heat in some form was present. The known effect is in this case *the reason* for the believed or proved *conclusion.*

In a similar way we reason both forwards and backwards from the means to the end and from the end to the means, making either the end or the means the reason, and the means or the end the conclusion. So in moral action we reason from the motives forward to the act or purpose, and backward from the act or purpose to the impelling motives, making either the reason for believing the other, with such reservations as the nature of their mutual activity requires.

The distinction should also be noticed between causes, *i. e.* powers, and laws. Laws designate those permanent circumstances or relations which, though not separate agents themselves, modify the production of the effect, so that with or without these, the effect does or does not actually occur, or the energy of the effect varies as these circumstances vary. The best example of a law as distinguished from a cause or agent, is the law of gravitation —according to which the force varies inversely as the square of the distance. For the purposes of reasoning, however, the law may be viewed as giving efficiency to the cause; *i. e.*, the power in question, *e. g.*, gravitation, is known or manifested as a cause which we can apply in deduction, so far as or when it obeys certain laws.

Geometrical reasons.

§ 224. When we employ reasons to prove geometrical truth, the grounds of the process and the conviction which it imparts are found in the nature of the materials conceived as necessitating certain products or effects in a way similar to that in which an existing agent, whether matter or spirit, brings to pass its results. The triangle, square, cube and sphere are regarded as possessed of certain properties, which, when subjected to certain changes, or brought into certain combinations, make the existence of certain other properties necessary. The *ratio essendi*, or the conceived properties of the geometrical figures in space as constructed by the mind becomes the *ratio cognoscendi.* The geometrical figure is regarded as having *causal* efficiency, the effects or consequences of which cannot be set aside.

Thus: two triangles are similar, *i. e.*, their sides and corresponding angles are equal, because they are the halves made by the diagonal of a parallelogram. The reason is found in the properties of the parallelogram. But these properties are determined by the constructive acts of the mind, *space* being assumed as allowing the mind to conceive or construct certain figures. These figures when constructed are divided, *i. e.*, new figures are constructed—they are compared with each other—they are superimposed upon one another—in short, there is a series of consecutive acts passing into effects, the acts determining the effects, and the effects being determined or defined by the mind's acts and the material, viz., space, with which it works. We reason from these acts, *i. e.*, from that conceived as the cause to the effect, or from the effect back to the cause, precisely as when the cause and effect are material.

The same is true, when we reason from the essential constituents of a logical concept; or construct what some logicians call immediate syllogisms, *e. g.*, conclusions of logical *conversion*, etc. These last scarcely deserve to be called proper, as the process is merely formal. But if they are so regarded, then the parts and the whole, from one to the other of which in such cases we reason, have been previously fixed by the thinking power, or the power to generalize at all. These logical products, as *wholes* and *parts*, *positives* and *negatives*, etc., are regarded as causal of certain results to any objects brought into certain relations with them. They are reasoned of as though they were actually existing beings with causal properties obeying unchanging laws. By the same rule: We say, some islands are surrounded by water, because all islands are surrounded by water. Any special act of duty can only be performed by a moral being, because duty in every case is the act of such a being.

CHAPTER VII.

REASONING.—VARIETIES OF DEDUCTION.

The varieties are three; these subdivided.

§ 225. The sameness of the process of deduction enables us to understand the diversity in the *several varieties of deductive reasoning.* These are determined by the differences in the subject-matter upon or about which the process of deduction is employed, so far as this subject-matter occasions a difference in the character of the reasons upon which the reasoning depends. Material forces and reasons differ from the psychological and moral. Both these are unlike the mathematical. Those which are purely logical differ from all the others. These differences in the subject matter also require a special preparation in each case, in order to make it ready for the application of the deductive process proper.

The varieties of deductive reasoning usually recognized are *the Probable,* the *Mathematical,* and the *Formal.*

Probable reasoning is again subdivided into three, the *physical,* the *psychological,* and the *historical,* according as the subject-matter is physical beings and phenomena, spiritual agents and their manifestations, or those combinations of the two which make up human history. It is often called *applied reasoning,* because its materials are facts known by observation and induction, and its processes are *applied* to the materials thus acquired or furnished.

Mathematical reasoning is threefold, according as it is concerned with continued or discrete quantity, or as it combines the methods appropriate to each. It is *geometrical, arithmetical* and *analytical.*

Formal reasoning concerns itself with *pure concepts abstracted from all beings and phenomena,* and with the relations which such concepts involve. It is sometimes technically styled simply logical deduction, and its arguments are called *immediate,* or purely logical, syllogisms.

Probable reasoning.

§ 226. In *probable* or *applied deduction,* we may for the present assume that the premises are furnished by *induction* and observation. In applied reasoning as defined, *induction* is always necessary to furnish *major premises*, because there can be no reasons, if there are no general or universal powers or laws. For minor premises in these cases, *observation* often suffices, because it often furnishes individual facts or events. When these minor premises affirm any thing of a class of generalized objects, *induction* may be required as well as observation. This description of reasoning is called *probable*, sometimes *problematical* and *moral*, simply because the subject-matter depends on causes which are contingent and is not necessarily true. Its reality cannot be proved by demonstrative evidence. As such it is contrasted with the mathematical and formal, the subject-matter of which is in no sense a real being or event, and is dependent on *no contingency* for its existence or occurrence, but on the properties or relations of mathematical and logical concepts. The terms probable, etc., do not, however, imply that the processes involved are less valid or convincing, or that the premises or conclusions are less trustworthy.

But whether the reasoning process, as such, relates to facts of matter, to facts of spirit, or to facts of history, it rests upon reasons in the way already explained. The facts are reasoned out whenever the power or law with its conditions is employed to *prove* that they must have occurred inasmuch as the causes exist which require them; or whenever facts or events known to exist are *explained* by being referred to such agencies or laws.

Thus, the suspended weight let loose, it is reasoned, must fall, because the force of gravitation is always in action; or *the reason* why it fell, or why it ought to be believed that it fell, is that this power was acting at the time, under certain of its laws.

In the sphere of spirit, *I reason* that at the thought of Hannibal I shall always think of Fabius, because the two, by association, have become permanently fixed in my thoughts. By a reference to the operation of this power under its laws, I *explain* the fact, that I thought of Fabius a moment previous.

The student and interpreter of history reasons concerning the events of the past when he seeks to *explain* them by their appropriate causes and laws, or to *forecast* the future by means of the great forces or agencies,—the so-called principles—through which the course of events and the results of important movements in society can be interpreted.

Deduction is more satisfactory and convincing when applied to material than when applied to spiritual phenomena, because the agencies known in the one sphere are more numerous than in the other, and because the laws according to which these agencies produce their results are capable of being expressed in mathematical formulæ. Hence, in many of the physical sciences we apply the rigor, the certainty and the variety of geometrical deduction, as in *Mechanics, Optics, Navigation, Theoretical Astronomy* and *Chemical Analysis.*

Mathematical reasoning, materials of.

§ 227. This introduces into the sphere of pure deduction a second element, viz., *the mathematical,* which in many of the physical sciences, is combined with that which is *contingent* or *problematical,* but which in the pure or abstract mathematics, gives character to what is called by eminence *mathematical reasoning.*

The objects or entities with which *mathematical reasoning* is concerned, are constructed by the mind itself on the suggestion of, and of course with reference to, certain material things and occurring acts, which are related to one another in space and time. Hence these entities themselves have certain definite relations to space and time, which are called their properties.

We find ourselves, at a certain stage of intellectual development, possessed of the concepts which are employed in geometry, arithmetic, and algebra—as *the Point, the Line, the Superficies, the Triangle, the Square, the Circle, the Cube, the Sphere, the Cone,* etc., as also *the Unit, the Sum, the Difference, the Multiple, the Divisor* and *the Ratio.*

These are properly called concepts or general notions. The individual objects of which these concepts are affirmable are, as it would seem at first, individual objects of sense or spirit; as when we affirm a line, or point, or superficies to belong to a block of ivory. On second thought, we are sure that the mathematical point, line, or surface, cannot belong to any material object as such, for the reason that there are no perfectly *even* or *sharp* edges or *even* planes in any material object. Nor are there in nature any perfect units, exactly the counterparts of one another.

These individual entities are then generalized, and become concepts; having a content and extent, and being capable of definition, division, and classification. The individual and the general are however scarcely distinguished by the mind itself. Indeed, in the mathematical processes the mind passes so quickly from the individual to the general and returns so readily to the individual as not always to notice for the moment with which it has to do, whether with the lines and triangles as individuals, or with them as the representatives of all conceivable lines and triangles.

It is another marked and distinctive peculiarity of these relations, that they are clearly and entirely distinguishable from all other generalized properties. The length, breadth, etc., of any material object cannot be confounded with its sensible qualities, nor can the relations of number be mistaken for those properties of matter or spirit of which sense or consciousness takes cognizance. Not only are they clearly separated as classes, but each member of the class is sharply separable from every other. The line can not possibly be confounded with the surface, nor the sum with the difference.

Definitions and axioms.

§ 228. These concepts, like all others, can, as has been explained, be expanded into *propositions* of *content* and *extent.* Mathematical propositions of *content* are the definitions which state the attributes that constitute the essence of each of the complex concepts which we form by mathematical construction, as the square, the triangle, the cube, etc., etc. The best and most satisfactory definitions are those which bring directly before the mind *the act* or *process* by which the concepts are supposed *to be constructed.*

Such definitions we sometimes phrase in the language of command, as, *draw* me *a line, move a plane,* etc. For this reason they are called *postulates, postulata,* i. e., concepts which may be constructed and assumed without dissent. The definitions of the concepts of number scarcely need to be given. We assume at once that all men know what they signify. When an explanation of them is required, we refer directly to the process of numbering, *i. e.,* we *count* by a series produced by the constant addition of units. Mathematical definitions also state the entire import or essence of their concepts. We are certain that the definitions of a triangle and square are exhaustive. Such concepts are in their very nature transparent: we can see through them as through crystal water to the bottom of a mountain lake. We know that the properties enumerated perfectly distinguish each concept from every other. The definition does not indeed express all that is true of its concept as related to every other in every conceivable combination, (else reasoning or analysis could not add to our knowledge,) but it gives all that is essential to enable the mind to distinguish it from every other, and adequately to define *what* its content is.

Mathematical propositions of *extent* are such as these: Triangles are plane or spherical; and each of these is acute, obtuse, or right-angled. For the same reason that mathematical definitions are exhaustive, mathematical divisions are known to be complete. As the first are exhaustive, on account of the limited number of the elements involved, it follows, that all the subdivisions which depend upon such elements, can be easily compassed, and confidently enumerated by the mind.

Hamilton pertinently observes: "Mathematical, like all other reasoning, is syllogistic; but here, *the perspicuous necessity of the matter necessitates the correctness of the form;* we cannot reason wrong."

Axioms are prominently employed in mathematical reasoning. Axioms differ from definitions in this, that they state the necessary relations that are involved in the nature or application of all the concepts of quantity as such, whereas the definiton expands the content or extent of some special concept.

Axioms are of two species, the *analytic* and the *synthetic.* Examples of analytic axioms are such propositions as the following, '*the whole is greater than its part,*' and '*things that are equal to the same thing are equal to one another.*'

They are called *analytic* propositions as contrasted with synthetic, because, as it is contended, they evolve or explicate in the predicate what is impliedly known or assumed in the subject.

There is another class of axioms, such as these: *Two straight lines cannot inclose a space: Two or more parallel lines, if produced ever so far in either direction, can never meet.* These examples apply to geometrical quantity only. These are clearly synthetical propositions. Whatever may be true of those of the other class, in axioms of this sort the predicate contains matter which the subject does not imply. And yet these propositions are self-evident and intuitively true. They cannot and need not be demonstrated.

The question has been earnestly agitated whether the axioms or the definitions are the foundations of geometrical reasoning. It has been very generally held that the axioms are the real *principia* upon which such reasoning depends: that is, that they are the unproved but assumed major premises of which, with certain minor premises furnished by the definitions, all those syllogisms are constructed that make up the demonstrations of geometry.

It is obvious that the only kind of axioms to which this question can apply, is the first of the two classes above cited, the so-called analytic axioms. Those of the second class, all would concede, are as truly *principles* as are the definitions; *i. e.*, they are as well fitted to serve as major premises for syllogisms.

The method after which the demonstrations are conducted by Euclid, has lent a decided support to this view. In all these demonstrations, these axioms are constantly cited as major premises for the truth of the conclusions which are derived from them. The arguments are in substance as follows: All things that are equal to the same thing, are equal to one another. The case of the equality of the two lines or angles A, and B, to a third C, is a case of the kind. Therefore, this is a case of their being equal to one another.—A is equal to B.

Against this doctrine cf. *Locke, Essay*, B. iv. c. vii. § 10. *Reid, Essays on the Intel. Powers, Essay* vi. chaps. v. and vii. *Principal Campbell, Philosophy of Rhetoric*, B. i. c. v. § 1. *Dugald Stewart, Elements*, Part ii. subd. i. c. i. sec. i. (1) and (2).

For our present purpose, it is of little consequence to determine whether the axioms or the definitions are or are not the foundations of geometrical deduction. In the one case we begin our series of deductions with certain general truths that are more extensive than, and are prior to the subject-matter of geometry. In the other we find our first propositions in the definitions, or the additional truths which the definitions introduce and make possible.

The construction of geometrical figures. Auxiliary lines, etc.

§ 229. It is more important to observe that what is called geometrical demonstration is *very far from being a process of pure deduction*. As preliminary to this and coincident with almost every one of its steps, a process is carried forward of *preparing the materials concerning which we reason,* so that they can be brought into comparison. This is ordinarily termed the construction of the diagram or the drawing of *auxiliary lines.* In some cases these constructions are very easy and simple, in others they are difficult and complex. In all cases they task the power of ready invention, and fertile suggestion. The preparation of the diagram for the demonstration of the 47th prop. 1st book, of Euclid's Geometry, is no inconsiderable achievement of inventive skill and sagacity.

It ought to be observed, that in order to be certain of the possibility of drawing some of these lines, and of the character of the figures which will result from them, we can neither depend upon the axioms or definitions, nor upon the results of previous reasoning processes, but must rely solely upon our direct intuition of the properties and relations of the figures which our postulates enable us to draw, and which our definitions describe. We know, for example, by intuition only, that we can connect the opposite extremities of a square or rectangle, and that the diagonal thus drawn will divide the rectangle into two triangles with a common base. In constructing a rectangle, we must presuppose the space which we circumscribe, and some of the consequent relations to it and to one other of its bounding lines. So soon as we divide this space, we add to this knowledge also, by direct inspection or intuition. The same is true whenever we add to or divide any construction, whether one that is original or superinduced.

It should be noticed, that in all cases of complicated geometrical construction, the completion of the diagram is the result, to a large degree, of a *tentative* process. We draw a line, and then observe whether the new relations brought into existence by this construction may serve as connecting links between the proposed conclusion and its proof. The new constructions which we form for each new theorem, furnish fresh material for yet other processes of deduction, and thus enlarge the material by successive syntheses, to which our deductions can be applied. The new truths which these new constructions enable us to discover are intuitively assented to, both in their conditions and their evidence. They are axiomatic, and similar to the axioms of the second class which we have already considered. The number of such axiomatic, *i. e.,* obvious truths made possible by the endless variety of geometrical constructions, is well nigh unlimited. With every new construction, some new relation is evoked, and its truth is intuitively assented to.

Moreover, in geometrical reasoning the several quantities must be *measured* by or with one another. The diagrams are constructed, and the needful auxiliary lines are drawn solely in order that the parts may be so prepared that one may be compared with another. As the triangle is the simplest figure that can be constructed, the original measurement to which, in the last analysis, all others are reduced, and by which they are tested, is that of two triangles. In Playfair's Geometry the first act of demonstration and that to which all the remaining attach themselves and are referred, is that of the fourth Prop. by which two triangles are superimposed on one another. The possibility of comparing two triangles being established, we have the means of comparing all those plane figures which can be resolved into equal triangles. This may be considered another *auxiliary* step in geometrical demonstration. It is obvious that this or any act of measurement is not deduction proper.

§ 230. After the material has been prepared we proceed to apply to it the processes of geometrical demonstration. How we do this can be understood most satisfactorily by an example. Geometrical reasoning explained by an example.

In the fifth proposition of Playfair's Geometry, B. I., it is proposed to prove that the angles at the base of an isosceles triangle are equal. The first step is to prepare the diagram by producing the two sides, A B, and A C, indefinitely towards D and E.

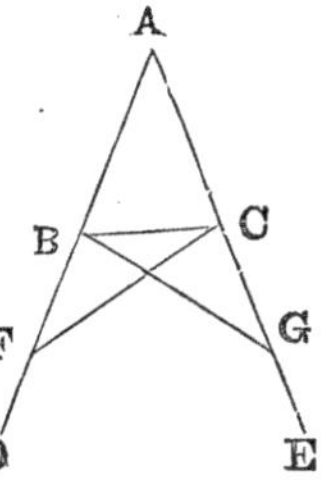

In the lines thus drawn, the two points F and G are taken at equal distances from A, and B G and C F are joined. It is manifest, '*to the eye*,' as we say, that we have two pairs of triangles, A B G, A C F, B C G and C B F. The first two have the two corresponding sides equal—the one by construction, the other by the addition of equals to equals—as also the included angle common. By deduction from the conclusion of the fourth proposition, the bases C F and B G and the several angles are proved to be equal. These two conclusions give, in the two smaller triangles, one side of each equal; by subtraction of the equals A B and A C from the equals A F and A G, the sides B F and C G are equal; that the included angles included between the equal sides of each are equal was proved from the fourth proposition. It follows by the same syllogism upon the same premises, that the angles B C F and G B C are equal. These equals are, then taken from the equals A C F and A B G, and the remainders are equal. These are the angles at the base of the isosceles triangle.

It will be seen that the syllogisms employed are either five or two, according as we consider the axioms to be or not to be the foundations of geometrical deduction. There are three cases in which the axioms, if equals be added to or taken from equals, are employed in what, in form, appear to be syllogisms. In the other *two* the conclusion of the *fourth* proposition is made the major premise, and the conclusion is regularly deduced. In all, we have a general proposition for a major premise, a particular case for the minor, and the conclusion made up of the major and minor term. That is, there are in all these cases, formal syllogisms; but there is this difference; in the one case the axiom adds no force to the belief of the conclusion, because this would be equally clear to the mind without it; in the other, we are referred to the nature of the concept or construction—as of two triangles equal in two sides and the included angle—as necessarily involving equality in the remaining side of each. The reason for the conclusion is the properties of such triangles as constructed by the mind, by means of the known properties of space. It would be a trivial fiction to say that the relation of equality requires that two things equal to the same thing should be equal to one another; but this must be said, if the axiom is a reason for the special applications of itself.

But again: we demonstrate or deduce in this way by these two concatenated syllogisms, that the angles at the base of this individual isosceles triangle are equal to one another. But we see at once that it must follow that whatever is true of this or any isosceles triangle must be true of every one. Hence we generalize this conclusion directly, and make it ready to be used as the major premise of another syllogism. This is the last step in the process of a geometrical demonstration. It is not by *induction* proper, however, that we pass from the individual to the general, for the reason that the properties and relations of space which are used in an individual construction in space, do not like those of matter indicate one another with more or less *probability*, but each requires the other by an unavoidable necessity discerned by intuition.

The processes of arithmetic and algebra are scarcely considered processes of deduction at all, not because deduction is not present and actually performed, but because it plays so inconsiderable a part in reaching the results. The chief concern of the mind in performing problems of this sort, is to invent such combinations and to apply such methods of dealing with them, as will bring to pass the result—which is usually to establish a new equation between elements that can be evolved from the data. The mind seeks to change the expression of the quantities given, so that they can be advantageously compared. The mind deduces only when it applies some rule or principle, or uses a formula previously determined to be true of all members or all objects similarly situated with the individual case. Both these processes are similar in principle to the expedient of devising *auxiliary lines* in geometry. The result is readily generalized.

Immediate syllogisms.

§ 231. The third species of deduction is the formal or purely logical, such as is employed in *immediate syllogisms*. Here the reason for the conclusion is found in some one of the necessary relations of the concept, whenever such a relation can be applied or viewed as a cause necessitating a new relation expressed in the conclusion. Inasmuch as there are several such essential relations, a variety of such deauctions is possible. Syllogisms of this sort are called by *Kant syllogisms of the understanding*, because the understanding is defined by Kant to be the logical faculty. These conclusions are sometimes styled *immediate*, in contrast with those which are *mediate*, because they are built upon a single proposition, or more exactly because no *middle term* is present or provided in the ordinary acceptation of the word. The major premise is derived from an expansion in language of those relations which necessarily belong to the concept, and may be expressed in purely formal propositions. These arguments are usually treated in books of logic under the title of the *Conversion* and *Opposition* of Propositions, and often are not treated as syllogisms at all.

The following is an example, usually cited as of *subaltern opposition: All islands were originally attached to a continent;* therefore, *some islands, or this island,* e. g., *Ireland, was originally attached to a continent.* The argument in this form is an *enthymeme.* In order that it may be expanded into a *syllogism* the major premise is required: it becomes—*whatever is true of all islands is true of some islands; it is true of all islands that they were attached to a continent; therefore it is true of some islands that they belonged to a continent.*

We assert, *No man is perfect;* therefore, *some men, or this man is not perfect:* the major premise being *whatever is denied of all men is denied of some men*, or *this man.*

In *conversion* we conclude from *All men are mortal,* that *some mortals are men.* From *No man is perfect,* that no *perfect being is a man,* and so on throughout the cases that are possible, the major premise in each instance being a periphrastic proposition, as *the predicate affirmed of all men may be the subject when limited by some,* etc.

It might seem at first that the proper major premise in such cases, should be the more general axiom, as in the first example; *whatever* is *true of any whole* is *true of its parts.* But on a second thought we correct ourselves by observing, that in such a case no middle term can possibly be devised to connect the major with the minor. The same is true, only more eminently, of what are called the laws of thought—as the laws of *identity*, of *contradiction*, and of the *excluded middle;* no matter is furnished in such propositions, by which we can proceed to a conclusion. They are not laws of thought in the sense of being major premises for deduction. They are rather generalizations of the particular processes which the mind performs, and of the relations which they involve. They are simply rules for logical consistency (cf. § 270).

The force of the argument in all these cases is found in the essential nature of the concept, as involving certain relations, *e. g.*, of *the whole to its part, of the subject to its predicate, and of the positive to its negative.* But the nature of the concept is but another name for the *properties* or *relations* which the mind necessarily conceives every concept as possessing, which the mind must necessarily think it as being, or as able, in other relations, to effect or evolve. The purely *logical properties* or *relations* are viewed as causes of what is expressed in the conclusion, like *physical causes* and *mathematical relations,* and so far forth are used by the mind as the reasons of the conclusions which it accepts.

Two elements in most acts of deduction.

§ 232. The foregoing analysis of the varieties of deduction requires us to distinguish that part of the process which is *preparative* or *auxiliary,* from that which is *simply* and *strictly deductive.* That which is characteristic of each one of these varieties is derived from the elements and materials which these subsidiary processes furnish for deduction. But in actual reasoning, the two operations are so intimately blended together, that it is not easy to distinguish the one from the other. For example, in probable reasoning, the force and conclusiveness of the argument may seem to turn chiefly upon the *facts of observation and testimony* which establish the minor premise, or the *inductions* which support the major, and very little upon the act of bringing the two together in the relations of an argument. As soon as the auxiliary and preliminary steps are taken, the conjunction of the parts as major and minor, naturally occurs to the mind, and, with it, the inevitable conclusion. In geometrical reasoning, as we have seen, the establishment of the conclusion sought for, depends almost entirely on the skilful suggestion of the appropriate *auxiliary lines,* and the *orderly concatenation* of the several arguments, so that the result may spring forth of itself. In common life, the issue of the reasoning depends upon the establishment of certain facts, in connection with certain principles. Upon the proof of the facts and the enforcement and illustration of the principles, the reasoner expends the resources of memory and invention, of wit and eloquence. The facts being established and the principles received, the argument enforces itself.

Skill in the *invention* of *middle terms*, or media of proof, is an art in respect to which men differ more widely than in respect to merely logical consistency, or the capacity to derive conclusions from their premises. Upon skill and aptness in this, is founded very largely the estimate in which the ability of a reasoner is held. But this affluence of invention and skill in selection must also be attended with a ready tact in forecasting all the results of a multitude of deductive processes, when applied to all the cases which invention suggests. There must also be present the capacity to hold the attention evenly and steadily in long and closely-connected series of deductions, all which capacities come only from the special development, and usually from the patient and practiced training of the philosophical powers. When these habits are matured by such training, the soul learns to act with the precision and rapidity of intuition. It must so act in order to reason with success when pressed by a powerful antagonist, in the haste and excitement of debate, or under the unexpected and ingenious assaults or defences which are elicited in an active controversy.

The establishment of the *principles* or the *reasons* which are involved and required in an argument, is often the point of chief importance. Inasmuch as the deductive power is prominently employed here, the *logical* faculty, or power of analytic and consistent thinking is especially tasked, and superiority in this is necessarily manifest. The power readily and surely to fall back upon principles, and to apply them to special cases with aptness and force, is the power which distinguishes the *reasoner* from the man of extensive knowledge, the man of fertile invention, the man of ready wit, or the man eloquent in description and appeal. To this power must be superadded, *as it is always supposed*, the capacity to proceed with logical clearness and rigor from the reason to the conclusion. The last marks the *logician proper*, as he is contrasted with and distinguished from the *reasoner*.

Deduction adds to our knowledge. In what sense?

§ 233. This analysis also enables us to answer the question which has been frequently agitated, whether *deduction adds to our knowledge*. Many have contended that it does not and cannot. They urge, that if we know the major premise, we already know the conclusion; that when we assent to the major, *All men are mortal,* we have already decided the question, whether *Peter is mortal,* and that whatever advantage there may be in employing an argument, the argument does not add to our stock of knowledge. We do not, it is urged, gain by it any new truth.

To this argument, in the form in which it is urged, we might reply, in the *first place,* that if we substitute for "*we know already,*" the phrase "*we might know if we would think or reflect,*" there would be less reason to object to it. The design of reasoning is often to lead a person to *reflect* or *think* concerning the application of the facts or principles to which he assents. When a man institutes a process of deduction, or follows one presented by another, one of three things may be true. *First,* he may never have accepted, through ignorance or want of thought, the *major premise,* or, at least, not so distinctly as to be ready to apply it in every particular case. But he may be induced to accept it for the first time by the excitement of the occasion—*i. e.,* by the use or application which is now to be made of it. *Second,* he may never before have accepted the *minor* so as to be able to connect it with the general truth, even though it had already been familiar to his knowledge and assent. *Third,* he may have accepted both *major* and *minor,* but may never have thought of the two together so as to perceive that relation between the two which involves the conclusion.

In the *second* place, an argument is usually addressed to a person who has not accepted a conclusion, by a person who has accepted it. The one who uses the argument, knows this conclusion to be true. The person to whom it is addressed has not assented to it. The argument is used to make him give this assent. In some sense of the phrase, it adds to the knowledge of the person whom it convinces. It ordinarily does this by leading him so to reflect, that he enlarges his knowledge or his belief. *First,* it may be, he is led to accept the major; *next,* he assents to the minor; and *last* of all, he is induced so to connect the two, that he himself is convinced, and of himself accepts the conclusion.

Reasoning is, in fact, constantly employed to enlarge the knowledge of men. It would be idle, as it might seem, to contend that the student of a system of geometry does not increase his stock of knowledge, or that all the knowledge which he gains is acquired by *induction* or *intuition*. Deduction *is* constantly employed as a means of instruction in all departments of science, and it would seem with the greatest advantage to those whose knowledge it augments.

But knowledge is as truly concerned with the apprehension of relations, as with the cognition of facts. New or additional knowledge is as properly the knowledge under new relations of facts already known or very familiar, as the acquisition of new facts by observation, testimony, or induction. Deduction applies reasons to facts or events, in order to establish their truth, or explain their existence or occurrence. It is often required, as we know, to convince ourselves or others that a fact or event must have been true or must have occurred. The man that is convinced by such a process of the reality of any fact, must thereby have gained new knowledge of its relations.

Or, again, the process is applied to explain why it occurred; the fact or event being admitted, the reason for its occurrence is asked for. When such reason is given by the application of the deductive process, the fact is known in a new relation. The knowledge of the fact as explained by its reason is certainly new knowledge. Deduction applies general causes, elements or properties, as *reasons* to confirm or explain events and facts. It not only adds to our knowledge, but it adds knowledge of a kind which is eminent for its worth and dignity—thought-knowledge of the most exalted character—knowledge in the light of the principles and laws which govern and explain all individual facts and events.

CHAPTER VIII.

INDUCTIVE REASONING OR INDUCTION.

Inductions properly and improperly so-called.

§ 234. We have seen that, in order to perform those processes of deduction which relate to facts and events—the processes called probable reasoning—the mind must be furnished with major premises or general propositions.

The process by which we gain the truths thus applied, is called *induction* or *inductive reasoning.* We proceed to inquire: What is the nature of this process? What are the conditions and grounds of its exercise? What the assumptions on which it rests? What are its applications to human knowledge, and what the rules for its successful use?

Induction is usually defined as the *deriving generals from particulars;* and in this is contrasted with deduction, in which we are said to proceed from generals to particulars. This definition is correct so far as it goes, but it is by no means precise or exhaustive. There are many processes conceivable, in which we derive generals from particulars which are not processes of induction. For example: We observe ten oranges, and, noticing them one by one, perceive a common likeness of qualities. We gather the results of our observations into the general judgment or proposition: all these oranges are slightly oval, or light yellow, or yellow mottled with green. It is obvious that such a judgment, though general and derived from particulars, has not been gained by induction. This is further obvious from the fact, that such propositions cannot be applied in deduction. To seek thus to apply them, would be an idle form, attended by no advantage, and leading to no conviction. If all that we know or had learned was simply: all swans hitherto observed were white, or all men observed or reported have died, we should already have included in the major premise the truth of the conclusion, and it would be idle to expand the knowledge already gained into a form of deduction. With such general propositions as premises, deductive reasoning would be either superfluous or impertinent.

"If induction," says *Galileo,* "must go through every individual instance, it would be either useless or impossible; impossible if the number of cases were infinite; useless, because then the universal proposition would add nothing new to our knowledge."

And yet inductions like these—so-called—have been named by some the only perfect or truly logical inductions. (Cf. Sir Wm. Hamilton, *Logic, Lec.* xvii. §62; *Lec.* xxxiii. § 108; *Appendix* vii.) It is sufficient to observe that, if such inductions are exposed to no error, they contribute no truth. They are safe but useless, for they admit of no application, except as a convenience for the memory.

That which is properly called induction is a process of another character. Examples of it are such as these. I observe a certain number of oranges, and, noticing their characteristics, infer or believe that all oranges have certain peculiarities of form, internal constitution, habits of growth, etc., etc. In like manner, I infer all swans are and must be white; not merely all the swans that have existed, or those which have been observed or described, but the whole species in the past, the present, and the future. In such cases we take the examples which we have observed, to stand for or represent the entire class.

It follows that *judgments of induction* differ from *simple judgments*, in certain important particulars. To return to our first example; we see *ten* oranges with certain well-defined characteristics. We bring them under their appropriate concepts, and judge or affirm these concepts of the individual objects. In *induction* we proceed further: we add to these simple judgments yet another, viz., that what we have found to be true of these, may be received as true of all others like them. The ground of the first judgment is facts observed and compared. The ground of the second is what is called *the analogy of nature. A judgment of induction is* then *a judgment of comparing observation, enlarged by a judgment of analogy.* The judgment of observation is founded on *observed similarity.* The judgment of analogy is founded on an *interpreted indication.*

What is usually called *experience,* includes *acts of induction.* Simple observation and judgment do not constitute what we usually call experience; for this imports not only that we have made and preserved observations, but also that we are capable of applying their results in parallel cases. This implies the power to discriminate between cases that are, and those that are not similar. Without this power or discipline, observation or bare experience would be possible but useless. For it would enable us simply to attain and retain our knowledge of the past, but never to apply it to the future.

In view of these considerations, the questions return upon us with augmented interest and importance: What is the ground, what the nature, and what are the rules for a sound induction? They are questions which have often been asked, and not always very satisfactorily or thoroughly answered. As preliminary to the development of the correct answers, and to a satisfactory theory of induction, we may profitably consider a few examples in which the process has been successfully applied.

Inductions of common life and inductions of science.

§ 235. The inductions of common life have already been noticed. They differ from the inductions of science, in that their results are incapable of being reduced to universal statements to which there are no exceptions. Nor do they result in the discovery of ultimate properties, agencies, and laws. Their results are seen in the common sense and common prudence which are essential to the performance of the common acts and duties of common life. Uncommon skill and readiness in interpreting such indications is termed acuteness, discernment, sagacity, and tact. Less than the usual capacity to make such inductions quickly and correctly, is denominated slowness and stupidity. The average capacity is called *common sense,* in one of the significations of this term.

The *second* class of examples of the process of induction is furnished by the discoveries of science. The inductions of common life are in one sense discoveries, but the indications are so readily interpreted and the inferences are derived with so great unanimity and universality, that the intellectual process (or processes) by which they are made, attracts little attention, and is, therefore, not readily analyzed. But when some new and wonderful agent in nature is brought to light, or some new law of its acting is established, and especially when the power or law is applied to some brilliant or useful result, we inquire with the greatest interest, How came the discoverer to think of that? How did he satisfy himself that what he thought was true? In such cases we are more likely to find answers to our questions, inasmuch as the steps of the process have often been slowly made, and the considerations which have led to them can be distinctly reproduced.

We select, first of all, the brilliant discovery by *Franklin* of the *identity of lightning with electricity.* With the electrical agent, or, as it was called in his time, the electric fluid, Franklin was entirely familiar. He was so far master of the methods of developing it in sufficient quantity or intensity, as to be able to produce its ordinary and obvious phenomena, as well as to exhibit phenomena that had previously been unknown. He had the electrical machine and the Leyden jar, and could produce at pleasure the electrical light, and the report following the connection of bodies in opposite electrical conditions. With these, then somewhat novel phenomena, he had become entirely familiar in observation and thought; as familiar as men in common life are with the aspect or form of a fruit, or with the expression of a gentle or vicious animal. He had also closely observed the phenomena of lightning, and had noticed similarities which had never been thought of before. The wave-like sheet and the zig-zag line and the loud report were seen to resemble the less impressive phenomena of the machine and the Leyden jar; and it occurred to his thoughts that the similarity of the phenomena indicated a common agent or power as their cause. This suggestion was strengthened by the thought, that clouds might be to clouds, or clouds to the earth, as the opposite surfaces of the Leyden jar. The mere observation of similarities like these might have satisfied the mind of Franklin, that the power or fluid in the heavens must be the same with that which could be accumulated by the machine from the earth. When at last he succeeded in bringing the power in question to affect a small quantity of matter, when he made it to run along an insulated kite-string, to emit a spark, to charge a Leyden jar—in short, to exhibit not only similar but the same indications with machine electricity, the induction could no longer be doubted. The decisive experiment proved the correctness of his thought.

Dr. Black was led to the discovery of *carbonic acid gas*, by observing that caustic lime increased in weight when changed into common lime, and by inferring that this weight must be derived from some agent in the atmosphere. This suggested the thought that the other alkalies, being like caustic lime in other properties, were like it also in this. The experiment was tried, and the suggestion was found to be correct. This put him upon the inquiry what the agent was which entered into combination with all these substances. The inquiry resulted in the separation of *carbonic acid gas* as a newly-discovered agent, and the determination of its properties and laws.

Dalton is *said to have discovered the law* that chemical combinations are effected by the union of their constituent elements in fixed proportions; and that, when a larger portion of an agent, as oxygen, enters into such a combination, it is invariably a multiple of a smaller. He was led to this by the knowledge that, in some cases, a combination in such proportions had in fact been observed. Being a teacher of mathematics and accustomed to mathematical relations, he generalized the result of a few chance observations into a universal law; it "being irresistibly recommended by the clearness and simplicity which the notion possessed."

One of the most instructive instances of modern discovery, is that achieved by *Sir Humphrey Davy*, of *the metallic bases of the alkaline earths.* The similarity in appearance and in many chemical properties between such alkalies as potash, soda, and lime, and the clearly identified oxyds of metals, had led to the suggestion, that they were similar in chemical constitution—*i. e.*, that they all were oxyds of metals. But the metals believed in do not exist in nature in a separate state, nor had they ever been exhibited in separate form by any agent of decomposition hitherto employed. The suggestion that there were such metals, and that they might be evolved, was confirmed by all the indications required as evidence, except their actual production. The application of the galvanic battery to chemical decomposition, and the triumphant success which had attended its use, led Davy to try it upon the hitherto intractable and irreducible potash. Under the solvent power of this wondrous agent, the knot which had never before been unloosed was in an instant untied. At the magic touch of this new instrument, the little globe of the newly-discovered metal leaped into view, and the *happy suggestion* was confirmed and accepted as an *undoubted fact.* It scarcely needed an experiment to convince the sagacious interpreter, that similar metals were encrusted within *common lime* and *soda.* The discoverer was almost as certain before as after the battery was applied, that *calcium* and *sodium* would in fact be evolved.

In the last series of discoveries we notice the following order and progress of thought and experiment. *First,* the oxyds of metals were observed to be like the alkalies in certain important properties. But the metallic oxyds were known to be produced by chemical changes; copper, iron, etc., constantly undergoing this process before our eyes. The two substances being alike in certain particulars, it was conjectured that they were alike in others. If pure potassium could have been found in a separate state, the readiest way to determine the point would have been to oxydize the metal and see whether the result would be potash. The next thing was to *de*-oxydize it. This was accomplished by the agency of galvanism. The fact that galvanic agency could decompose chemical compounds so intractable, suggested that possibly there were none which it could not overcome. If this were so, it would follow, that the force which held them in union, must be electric. This was established by its appropriate evidence, and is called by Whewell, "the highest generalization at which chemical philosophers have yet arrived." *Hist. Inductive Sciences*, B. xiv. c. 10. The mental process is precisely that which is common to every case of Induction. Certain objects are seen to be alike in certain properties or laws. It is believed or judged that similarity in these particulars indicates likeness in others. Potash is like iron-rust in certain respects; therefore it is like iron-rust in being the oxyd of a metal. All chemical compounds are strikingly alike in certain particulars. Certain of these are separable by the electric force; therefore all are separable by this agency. But if separable by it, they are held in union by the same force.

From discoveries of this kind we pass to those in *astronomical physics*—to the discoveries of *Copernicus*, *Galileo*, *Kepler*, and *Newton*.

Copernicus began by *discovering*, as it is said, the heliocentric theory of the solar system. The way in which he was led to adopt and defend it, is described by himself. He had found in ancient authors, accounts of Philolaus and others who had asserted the motion of the earth. "Then I began to meditate concerning the motion of the earth; and though it appeared an absurd opinion, yet, since I knew that in previous times others had been allowed the privilege of feigning what circles they chose in order to explain the phenomena, I conceived that I also might take the liberty of trying whether, on the supposition of the earth's motion, it was possible to find better explanations than the ancient ones of the revolution of the celestial orbs."

"Having then assumed the motions of the celestial orbs which are hereafter explained, by laborious and long observation I at length found that, if the motions of the other planets be compared with the revolution of the earth, not only their phenomena follow from the supposition, but also that the several orbs and the whole system are so connected in order and magnitude, that no one part can be transposed without disturbing the rest, and introducing confusion into the universe."

In 1609 *Galileo* constructed his telescope, and very soon discovered the satellites of Jupiter. This at once confirmed the Copernican theory, by opening before the eyes of men another system subordinate to the solar, of heavenly bodies revolving about their primaries, thus giving an analogon of the greater. The subsequent discovery by the same instrument of the phases of Venus, at once confirmed the new theory of the revolution of the planets about the sun, and answered an objection against it by explaining why Venus did not appear larger when nearer the beholder.

Copernicus furnished *the suggestion*, by reflecting on the known fact, that the apparent places of objects may be accounted for by the motion of one or both, and that the simplest solution or theory was to be preferred. *Galileo*, by his telescope, prepared the way for *the experiment*, by enabling observers, in a certain sense, to observe for themselves whether it was the sun or the earth which moved.

Kepler prepared the way for the discoveries of Newton, by his determination of the orbits of some of the planets, and the law of their motions. *Newton* had been himself familiar with the law by which, in obedience to terrestrial gravity, bodies fall to the earth's surface. The first thought which led to the extension of this agent to the celestial bodies occurred to him in 1666. "As he sat alone in a garden, he fell into a speculation on the power of gravity: that, as this power is not found sensibly diminished at the remotest distance from the centre of the earth to which we can rise, neither at the tops of the loftiest buildings, nor even on the summits of the highest mountains, it appeared to him reasonable to conclude that this power must extend much further than was usually thought. 'Why not as high as the moon?' said he to himself; 'and, if so, her motion must be influenced by it; perhaps she is retained in her orbit thereby.'" Upon this suggestion, he proceeded to the calculation of the deflection of the moon from a tangent to its orbit in a single second; it being assumed that the moon was at the distance from the earth which was then received. The result disappointed him; for he found that this deflection would be thirteen feet, which did not correspond with that required by the supposition that gravity deflected it. He laid his calculation aside. The subsequent discovery that the course described by a falling body is an ellipse, and that the distance of the moon from the earth could be correctly ascertained, enabled him to accept his theory on the ground that it coincided with actual fact. The distance of the moon had previously been computed on an assumed but mistaken diameter of the earth. A more accurate measurement of a degree upon the earth's surface led to a correction of the distance of the moon, and Newton's theory was henceforward accepted as a demonstrated truth. He first conjectured that the extension of a known force from the earth to the heavens, is possible and rational. He asks, "*if so*" "*what then?*" following out his *induction* by a mathematical *deduction*. He then, by other mathematical calculations decisively tested this deduction, and the conjectured agent was established as a *vera causa*, and its laws were carefully computed; the true theory of the heavenly bodies was forever settled.

The examples cited are sufficient to illustrate the nature of the inductive process. They have been taken from the physical sciences, not because these differ essentially from those which concern moral and political subjects, but because they illustrate more strikingly the steps of induction. The objects with which they are concerned are more interesting to the majority of men. The effects of discoveries in them are more obvious. The experiments and observations which have led to them are more brilliant and startling. Many of their results are permanently fixed in the arts of life, both useful and ornamental. Some of them are continually brought to our thoughts by engines and instruments which materially contribute to the convenience and comfort of man. The telescope, the prism, the quadrant, the hydraulic press, the steam engine, the galvanic battery, are all permanent memorials of what these processes have wrought, and they prompt to eager inquiries after the operations by which they were first constructed in thought.

The attentive consideration of these examples proves that induction in science is substantially the same process with induction in common life—that in both cases it is a process of interpreting indications.

Why are the indications of science more difficult?

§ 236. This assertion prompts the inquiries, Why then are the processes of common induction so easy and those of science so difficult? Why is the progress to common sense so easily and rapidly made in the infancy and childhood of the individual, and why have the advances of science been so difficult? Why so long delayed?—why, even now, is it true that in respect to so many branches of knowledge the race is yet in its infancy? To these questions the following answers can be given.

We notice *First:* that in science, the properties observed,—which are the *indicia* or indicators of others,—are less obtrusive than those used in common life, and are often far removed from common observation. To be apprehended even, they require closer attention than men in common life are able to give.

Many of these properties can only be apprehended by some nicely constructed aid to the powers of sense, or some costly and ingeniously devised apparatus; to the production of which special inventive sagacity is required, which sagacity itself must be the fruit of many men or generations which have gone before.

Second: The inductions of common life are founded on superficial and partly inaccurate observations. Those of science rest upon the sharpest analysis. The common observer observes facts and detects principles in regard to things or powers in the gross, both as they are combined and operated in nature. He does not go far beyond the things and phenomena which the common necessities of life require men to distinguish. The scientific observer continually aims to detect and separate, by a refined and acute analysis, powers and agents which are never divided except by artificial appliances,—and some of which are never parted even by these. Hence *the experiments of common sense* and *the experiments of science*, are very different.

Third: Many of the inductions in science are far more general and comprehensive than the inductions of common life. Many of the subtle agents or laws which science detects, are far more general and extensive than those which observation discerns.

Consequently they furnish the grounds for more varied inductions. They can be applied to explain a greater number of individual phenomena. They suggest very many possible theories. They incite to a manifold greater number of experiments. When any such comprehensive power or attribute is established, it can be used in a large number of deductions.

Fourth: One of the distinguishing peculiarities of scientific inductions is found in the circumstance that they are so widely and severely *mathematical.*

The relations of space and number are capable of being affirmed of every material agent, and hence when any one is found to exist and act according to such relations, we have at once the occasion and means of a very comprehensive generalization. The language of mathematics is the most precise and intelligible, the most easily communicated, and the most readily understood of all language. The tests of measure, weight, and quantity are the most easily applied of all tests. The sciences of space and number are also capable of the clearest, the most convincing, and the most fruitful of deductions, and hence so far as they can be legitimately applied, they can most readily test experiments and record their results.

Fifth: Science is necessarily more a *growth than any other species of knowledge.* One discovery not only in fact prepares the way for another in the actual history and order of man's attainments, but by the necessary dependence of one discovered law or agent upon another. The discovery of the law of universal gravitation was in the nature of the case impossible without the aid of *pure Geometry, Algebra, the Calculus, and the laws of Mechanics.* Optics, with the use and the invention of the telescope, had been in part developed before and in part perfected by Newton, before they could be applied by him to this particular discovery. In almost every great induction, many of the sciences and arts are laid under contribution. All the previous steps are presupposed when a single forward step is to be taken.

This is true only to a very limited degree of the inductions of common life. The well-qualified and well-trained man can with no great difficulty develop of himself much that the race has gained by common sense and observation, or can appropriate and master it with ease. The common sense of to-day in a refined and educated community in England or America readily appropriates the products which the common sense and experience of another generation had matured and preserved in language, traditions, manners and institutions. For all these are taken up by the mind with marvellous ease, and require but little of that discipline, which the mastery of the circle of those sciences which are necessary for success, imposes upon the *discoverer*. The difference is slight between the common sense of Socrates and the common sense of the honest and independent observer of the nineteenth century, compared with the immense disparity in the amount of positive knowledge possessed by the student of Physics in Socrates' time and in our own.

These considerations we think sufficiently explain the differences which exist between the inductions of science and those of common life, and establish the truth that the process is substantially the same in each. These differences are fully accounted for by the difference in the subject-matter, without requiring any difference in the process of interpreting them.

Induction in both combines an accurate *observation* of properties and a sagacious *interpretation* of what they indicate. But precisely here arises the most interesting and vital of questions, "On what ground or by what evidence do we proceed from the known to the unknown?" We can safely reply, it is not upon the ground of simple experience. For a long time it was believed that all swans are white, for the reason that no swan of any other color had been observed or heard of. "Mankind were wrong," says J. S. Mill, "in concluding that all swans are white: are we also wrong when we conclude that all men's heads grow above their shoulders and never below, in spite of the conflicting testimony of the naturalist Pliny? We have no doubt what is the correct answer to this question. But why are not men wrong in rejecting such a story, and in believing with assured confidence, that wherever men exist, their heads are not beneath their shoulders? Why is a single instance, in some cases, sufficient for a complete induction, while in others, *myriads of concurring instances*, without a single exception known or presumed, go such a very little way towards establishing an universal proposition? Whoever can answer this question knows more of the philosophy of logic than the wisest of the ancients, and has solved the great problem of induction." *Logic*, B. iii. c. 3.

If we seek to answer this question, we say it is more *credible, or reasonable to believe* that swans should vary in color than that men should vary so greatly in form. But why is it more credible? Some would deem it sufficient to reply that in most species of animals, individuals which are alike in every other respect differ in color,—in other words, that it is a generally observed law that color is very variable, while some constant outline or type of form is uniformly observed in every species, or at least has never admitted a deviation so monstrous as would be implied in having the head beneath the shoulders. This would be Mill's answer to his own question. But this does not fully explain our confident assurance that it is altogether incredible that a species of men should be so constructed. We cannot admit the supposition for a moment, for the decisive reason that men so formed could not perform the functions of men with any convenience or success; that such a form would offend both the eye and the mind, and would be entirely incompatible with the ideal of beauty and convenience to which we assume that nature would certainly conform.

Considerations of *convenience* and of *adaptation*, and even of *beauty* and *grace*, go far in such a case toward deciding the question. They give that weight and force to those " single instances which in some cases are sufficient for a complete induction," and take away all force from the " myriads of concurring instances " in other directions. It must be on the ground of such relations assumed *a priori* to be true of the whole universe of being and to hold good of its properties, powers, and laws, that we proceed in all our judgments of induction. These direct the mind in interpreting the indications furnished by observation. These prompt to the questions which we ask of nature in our experiments. These suggest the hypotheses by which we account for the phenomena. These confirm all the theories which we finally accept as true.

The a priori relations assumed in induction.

§ 237. We inquire next, what are some of the *truths or affirmations* which the mind *assumes* in all its inductions, and by which it regulates its inquiries into the properties and laws of the physical universe? We call these in the present stage of our discussion *assumptions*. We do not by this imply that they are not valid and true: they are logically necessary to the inductive process when it is analyzed. We need not here inquire whether they are all ultimate and original to the mind. It is enough for our purpose to ascertain that they are *a priori* in relation to the ordinary processes of inductive inquiry. Some of them are as follows:

(1.) All the objects with which the mind concerns itself in its inductions, are known as *substances* and *attributes*. It is with the properties or attributes of matter and mind, as exhibited through phenomena, that these inquiries are exclusively occupied, whether they are known as qualities, powers, or relations. Beings are known to the philosopher by their attributes or relations; it is by these, that they are distinguished, classified, and named.

(2.) Induction assumes and implies the reality of the *causative energy*, as necessary to explain the origination of every begun existence, and of all occurring phenomena. Whether it investigates the powers of nature or the laws of nature, it proceeds upon this as a necessary assumption. A *power* in any being or agent is its *capacity* to *produce an effect* under appropriate conditions and according to definite laws. The power of heat to expand metals, of a burning body to explode gunpowder, of oxygen to corrode metals, of the soul to know objects knowable, and to care for objects desirable; all express and suppose a single common relation, viz., the relation of *an energy that is causative of effects.*

That this relation is real, is assumed and implied in all our investigations into the unknown. This is true, if our inquiries respect the ascertainment of the unknown originator of a known effect, and result in the discovery of such elements as *oxygen* or *hydrogen*, or of such metals as *potassium* and *aluminium*, or of such agents as *gravitation* and *electricity;* or if we are still on the quest, and the cause or power sought for is not yet evolved. The same is true if our inquiries are directed to the determination of the *laws* or the *precise conditions* under which an ascertained cause produces a given effect, or to the more definite statement of the relations—*mathematical* or otherwise—under which these conditions vary with a varying effect, as in the determination of the laws of gravitation, of chemical affinity, or of mental perception, association, desire, and volition.

(3.) *Time and Space, with the relations* which they hold to extended objects and succeeding events, are also assumed in induction. So also is the possibility of the mathematical constructions which are conditioned by Time and Space; in other words, the reality and nature of *geometrical* and *arithmetical* quantities, their relations to one another and their varied applications to concrete objects and phenomena. These are not only assumed, they are put in the fore-front of the whole scheme of modern inductive philosophy. The processes of mathematical investigation are made the models for all scientific investigation. Their results are the instruments of measuring all physical forces and of formulating all physical laws.

Gravitation was scarcely determined to be a force, till its mathematical relations were expressed in the law that it is a force varying inversely as the square of the distance. The laws of falling or projected bodies are expressed by means of the geometric curves in which they move, and by the numbers which describe their velocity. The pressure and flow of fluids are reduced to mathematical expressions. Chemical affinity is comprehended under the wide-reaching principle that different elements unite in definite numerical proportions, which has furnished the foundation for modern chemical symbolization. The entire theory of astronomy is a combination of mechanics and applied geometry. Modern researches respecting light, electricity, and heat, have dared to propound the theory that all these are different modes of motion, the rate of whose vibrations determines these subtle and marvellously potent phenomena. They have at least demonstrated that the varying phenomena of these so-called forces or agents are attended by motions that can be made the test of their presence and the measure of their intensity.

So extensively have mathematical relations been applied in modern induction, that it has been gravely urged on the one hand that spiritual phenomena and forces can in no way come under the inquiries of science, because, forsooth, they cannot be subjected to mathematical relations, and, on the other, that they can and must be subjected to these relations in order that any science of spirit may exist.

(4.) Induction assumes that properties and laws which are known, *indicate* and *signify other powers and laws;* that in these indications nature is honest and open in her dealings with man; in other words, that she is consistent with herself, or uniform in her methods of revealing or suggesting what man is prompted to interpret or explain. For example, we judge that a certain form or appearance in a fruit indicates a certain flavor; that a particular aspect of stem and branches signifies a habit of leaf and fruit; that a given expression of countenance betokens a peculiar disposition or temper in man or beast; that striking similarities of attributes in metals indicate a similar capacity to be oxydized; that obvious and pervading similarities in phenomena prove that electricity in the earth is the same agent as the cause of lightning in the heavens; that the same power which is pervasive enough to affect bodies near the earth, is probably or at least possibly—in part or solely—the power which holds the moon in its changing path around the earth.

It is implied in the honesty or, which is equivalent, in the significance or interpretability of nature that she is also uniform, or *self-consistent* with herself from time to time; or in other words, that her laws and methods are *permanent.*

In other words, induction requires that we assume that nature is constant and uniform in her agencies, operations, and laws; also in her methods of making these known to the mind of the inquirer into her secrets.

It might here be asked, Why do we believe this to be true? Is the assumption groundless and ultimate, or is it founded upon some reason? It might be said that otherwise we could not know or interpret nature at all: If nature were not thus honest and uniform, the human mind could have no knowledge except of individual things, or the knowledge acquired to-day could not be relied on for to-morrow. But it might still be inquired, What necessity is there that we know and generalize? or more broadly, By what right do we presume that the objective universe is so constructed that the human mind may know it? We say, "If it were not so, it would not be adapted to the mind: The mind would feel impulses and use activities which would find no corresponding objects: It would be impelled to modes of action in generalizing, interpreting, in explaining and forecasting, to which tnere would be no corresponding realities. If this answer is appropriate or valid, it suggests another assumption, viz.:

(5.) *Nature adapts objects and powers to certain ends.* In other words, physical forces are regulated and controlled by *design.* The application already made shows that this principle is assumed. This will be still more clearly manifest from the examples previously cited. When Copernicus proposed to himself to try whether, on the supposition of the earth's motion, it was possible to find a *better explanation* of the revolutions of the celestial orbs than those currently received from the ancients, we ask what he would conceive to be a better explanation, and find an answer to our own question, in the reasons which led him to prefer his own. These reasons were, that this theory supposed greater simplicity and symmetry in the mechanism of the heavens, than the older theory furnished. But why is a neater and more symmetrical theory to be preferred? Because it is *better adapted* to satisfy the mind of man,—because this mind thus reflects: Were I to provide for the motions and appearances of the heavenly bodies, with given materials, viz., force, motion, etc., I should hold and move these bodies by the simplest possible arrangement of motions, and the most economical disposition of forces.

Newton, reflecting on the force of gravity, inquires within himself, "Why may not the force which extends beyond the tops of the highest mountains also extend as far as the moon, and why may she not be retained in her orbit thereby?" His own question implied the answer: "If this single force, known to exist, would explain the movements of the solar system, it is more rational to believe that this force actually exists than to adopt any other explanation." This involves the assumption of a wise adaptation to the designed effects of the force or forces conceived to be at command. It is by a reference to the same assumption that we explain the general laws of philosophizing which Newton has laid down. The rule that real and sufficient causes of phenomena are to be taken to explain phenomena, whether it is or is not interpreted as coming under the more *general law of parsimony,* is only an enunciation of the truth that if an element, or force, already known to exist, can be employed to evolve, produce, or accomplish an effect, no new force will be provided or is to be supposed. If we ask upon what this assumption rests, we reply, that any other arrangement would be *bad economy—an unwise adaptation of means to ends.*

Underlying all inductive inquiry, we find the assumption of a *twofold adaptation in nature; first,* of the several parts or forces to one another, and *second,* of the indications of nature to the mind that interprets them. But in assuming that nature thus adapts her forces to ends and also that the human mind is competent to discern these ends and to interpret the skill and success of nature in accomplishing them, we imply—

(6.) That the human intellect in induction, judges the constitution and operations of nature by referring to what it would itself consider to be rational and wise. In other words, induction assumes that the *rational methods of the divine and human intellect* are similar, and that the human intellect is therefore capable of judging of the principles and aims by which the universe was constructed and its laws can be known. More briefly expressed, *Induction is only possible on the assumption that the intellect of man is a reflex of the Divine Intellect; or that man is made in the image of God.*

The three rules of induction.

§ 238. The so-called rules or methods of induction are three: The *method* of *agreement, the method of difference,* and *the method of concomitant variations.* They are briefly stated as follows: (1.) If in all cases of an effect or phenomenon, one condition is uniformly present, that is the cause or includes the cause of such a phenomenon or effect. (2.) If, in every instance in which an effect does occur, one single condition is present, which is uniformly absent whenever such effect does not occur, this constantly present or absent condition is presumed to be its cause. (3.) If, whenever an effect or phenomenon is marked with peculiar energy, any condition varies with proportional intensity, this varying condition is the cause of such an effect.

Properly conceived, these are rules for testing or proving inductions, or rules for experiment: they cast no light upon that which is most essential in the inductive process. An experiment is a nice analysis or observation, made for an express design. Analysis, *i. e.*, discriminating attention, is the condition of all observation of qualities and causes. It begins with sensible perception, and without it, generalization and classification are impossible. The analysis used in induction is peculiar only in being directed to those properties and laws which are less obvious, and often guides to a special search for those which the senses cannot directly detect, but which the mind divines.

The rules for this search are not different in fact from those which the simpler inductions of common sense and of common life require and employ. It is only because the relations upon which they are employed are less obvious, and the discriminations are more difficult, that these rules need to be distinctly considered and formally applied, and that the formal recognition of them by Bacon and Newton contributed so largely to the advance of modern science.

The conditions of a successful hypothesis and discovery.

§ 239. Their design is to *test a theory, hypothesis, or suggestion* which the mind has already formed. The experimenter upon nature must come to her with his question formed and the answer anticipated, before he applies the methods of agreement and difference. Lord Bacon says abundantly that it is the *prudens quæstio*, or the wisely-suggested question, which directs the experiment to an anticipated result, and which very often confidently predicts the result before it is actually established or proved.

If now, *the question* suggests and guides the experiment, and if the anticipation predicts the fulfillment, we ask, *What suggests the question?* What are the grounds on which, or the methods by which the mind forms its *hypothesis?* When for example, Newton anticipated in thought the solution of the motions of the solar system by gravity, or Davy believed that he could bring out from the brown and earthy potash the brilliant potassium, what were the grounds upon which, and the rules after which, their minds proceeded? The question may be more generally stated: What are the conditions of successful *invention* and *discovery?*

To this question many would reply, 'No answer can be given. The power to read the secrets of nature is a gift of nature. It can be improved by exercise; it can be formed and developed into *tact and skill;* but what are the methods by which exercise can form or mature it, is quite beyond the reach or power of analysis to trace out or describe.' There is some truth in this

18

view, though not to the full extent of this representation. Analysis can at least separate and describe the essential elements of the process, and can so far describe the conditions of successful achievement.

(1.) The first condition is, that *the attention be directed* to that class of objects and powers already known, which are to indicate and suggest the unknown. The discoveries of science are founded upon powers and relations which are overlooked by the great majority even of cultivated men. The sagacity which we seek to explain, is always exercised in respect to that subject-matter to which the discoverer has given special attention, and with the peculiarities of which he has become specially familiar. The chemical discoverer is a chemist. The discoverer in physics is a student of physics. As we have already observed, Franklin had become familiarly acquainted with electricity and lightning, by long-continued attention to the phenomena of both, before he thought of their identity. It was not till Newton had meditated long and frequently on the forces of the universe, that he was in a condition in which it was possible for him to anticipate the theory of universal gravitation. Davy must, of necessity, have been familiar with all the chemical facts already ascertained, in order to conjecture the unknown base of potash. It is plain, that if the philosopher is to interpret indications, he must first observe and attend to them.

(2.) It is implied in attention to objects that their relations should be carefully regarded. For the purposes of knowledge, and especially of science, relations are all-important. The relations most important to science are those of likeness or unlikeness leading to classification, the relations of number and magnitude which are the conditions of mensuration, the relations of causation and design which are essential to reasoning.

In respect to the power of apprehending relations with facility and success, men differ greatly. In simple judgments of comparison, one man discerns similar and dissimilar qualities when another can discern neither likeness nor difference. Likenesses and unlikenesses of form are likewise detected by the quick eye of one man, which can scarcely be made apparent to the slower and less acute observation of another. To whatever causes these differences of power may be ascribed, whether to a finer sensuous organization, or a more refined and discerning spiritual nature, the fact cannot be doubted that they exist. They are, in part to be ascribed to training and opportunities, in part to the interest or necessity which enforces the application and the energetic action of the powers, and, in part, to original aptitudes and capacities.

(3.) The next condition of success is an *acquired familiarity* with the special modes of indicating the unknown which are followed in *any special sphere* of observation or scientific inquiry. The florist marks indications in flowers which are unmeaning to other persons, and learns to connect them with what they indicate. The cultivator of fruits gains the same sagacity with fruits. The sportsman alone learns by experience to understand the significance of certain actions of his game. The keen and discerning eye in every department is trained by what it is accustomed to, and gains some definite impressions in respect to the methods of nature in accomplishing her objects, and in indicating her powers and laws. The devotee of any special science soon gains a familiarity with the movements of nature within his own sphere. He enters, so to speak, into her spirit.

The literal import of this language is as follows: The physicist and chemist, the botanist and geologist, learn by degrees that in their several spheres certain properties are far more prevalent than others; that they are very often present and manifest; that certain combinations of elements and agencies are, so to speak, favorites with nature. Certain powers are very limited in their application, and of course are manifest in a smaller number of phenomena. Others show themselves in a great variety of existences, and explain a great number of phenomena. Just as far as discovery or experience proceed, just so far do they mark off certain powers and laws as more, and others as less extensive. This is the simple result of experience often repeated in respect to a sufficient variety of cases; this experience matures into familiarity with what may be called the preferences, or favorite methods, according to which nature conducts her processes and manifests her powers.

(4.) The next step towards discovery is *the use of the constructive imagination.* All the steps previously considered are acts of experience. The act now considered is an act of mental construction or combination. It relates to facts as supposed, or conceived to be possible or probable by the mind. The objects, relations, and methods of nature being all mastered by quick and attentive observation, must be marshalled by the memory and placed at the service of the imagination to be re-arranged and re-combined.

Let a complex substance be presented for that analysis in thought which precedes the test of experiment: or let some unexplained phenomenon be proposed to be accounted for. The first effort is to present to the imagination every known element or agent, and to ask which is more likely to be the one which we require. Or if none that are known will meet the exigency, what unknown element or agent—and acting by what laws—may be supposed to solve the problem?

To be able to answer these questions the memory must be quick to suggest all the powers and agents that are known, in all their known relations. The presence or absence of a single essential fact may determine the question whether a discovery shall or shall not be made.

It is not enough, however, that the memory suggests all that she has gathered, unless the imagination reconstructs and recombines in relations as yet untried and unknown. The imagination takes all the materials at its command, all the powers and agents which are known to exist, with their laws and relations, and connects them in new constructions. It makes these combinations not to amuse or illustrate, not to convince, instruct, or to persuade, but simply to conjecture what is best adapted to meet the exigency.

What is called *accident*, too, very often combines with memory and the imagination, and, at times, determines a great discovery in science, or a grand invention in the arts. The *Marquis of Worcester* happens to see the rising and falling of the cover of a tea-kettle, and forthwith he commences a course of speculation in respect to the laws of the agent which furnished this force; and thus sets in motion the course of discovery which has given to science and art steam power with all its applications.

But thousands and tens of thousands of men had observed the same phenomenon which attracted the attention and excited the inquiries of the Marquis of Worcester. His previous knowledge of science and his familiar acquaintance with scientific relations alone enabled him to turn his knowledge to the use of discovery. The promptness and vigor with which the associative faculty avails itself of such an incident decide the question whether it shall be received as a productive seed or whether it shall fall upon the barren rock.

The *curiosity* of the investigator is a most important condition of failure or success, for it determines whether or not the intellect shall be effectively applied to the objects and relations which alone prepare the way for new knowledge. Perseverance and tenacity hold the attention and the memory to the question which may have been started; they task the memory to give up all its past acquisitions, and stimulate the imagination to perseverance in its efforts to reconstruct them.

(5.) To success in induction, the power of *sure and ready deduction* is also essential. The real nature and reach of any theory which is suggested by the memory or constructed by the imagination, cannot be understood until the most important consequences and applications are derived from it in the form of conclusions. The law of gravitation was no sooner suggested to the imagination of Newton, in the question, "*why not*," and sanctioned by the approving answer, "*it is very probably true*;" than the additional thought, "*if so, what follows*," led him to an act of deduction.

The power of wide-reaching, sure and rapid deduction, is an important element in the qualifications of the successful discoverer. A severe training in the discipline of the Syllogistic Logic, and the linked demonstrations of Geometry, as also in the subtle calculations of Numbers, is an admirable if not an essential preparation for success in discovery.

(6.) The conditions previously described being all fulfilled, the reason then judges which of all the various possible suppositions which the imagination suggests, gives the most satisfactory solution and is most probably true.

The choice between hypotheses.

§ 240. But by what standard? What are the grounds and tests of probability? The history of Induction shows that these differ in different cases. Sometimes the known existence of some agent or law, or its very extensive prevalence in the economy of nature, is the deciding circumstance in its favor. We always assume that nature works the most diverse effects by the fewest possible elements or forces. Sometimes it is what is loosely termed analogy.

But analogy and the want of it pertain to very different qualities and relations; sometimes to those which affect the senses immediately, as the eye and the touch, sometimes to those which are more remote from direct apprehension, as to mechanical or chemical effects or mathematical relations. Which analogies shall be decisive in such cases is determined by the importance attached to each in the general or the special economy of nature, or by what is called the congruity with her methods in similar departments.

In the application of these and of similar criteria the intellect *appeals*, so to speak, *to itself*. The interpreter of nature continually asks himself thus: Given, certain elements, powers, and laws, how should I indicate them? or how should I apply them? Or, in the reverse order: Given, certain ends, effects, and phenomena, which of the known forces at command would a rational being employ for this or that object, if he aimed at an orderly, an intelligible, or a beautiful universe? Or, if no one of the forces known is adequate to explain the effects of phenomena, what unknown force or element is required to account for them, so as best to fulfil their objects, and what must be the properties and what the laws of such an agent?

The language so often used, that man is *the interpreter of nature*, that nature has her *methods*, her *economies*, and her favorite ways, implies that in all these judgments, there is a belief in the constructive or arranging processes of another mind.

When Kepler exclaims, "*O God! I think thy thoughts after thee!*"—when Agassiz catches and repeats the same sentiment, in asserting that all just and *thorough classification is but an interpretation of the thoughts of the Creator*, they simply express in other language the assumption on which every sagacious anticipation or felicitous theory is founded, viz., that the *rational methods of the Divine and human intellect must be the same*. This of course, includes the assumption, without which the principles, maxims, and methods of the inductive philosophy have no meaning and no foundation, viz., that the universe of matter and mind has its ground and explanation in an intelligent originator. In other words, *Induction rests upon the assumption,—as it demands for its ground,—that a personal or a thinking Deity exists.*

It follows that the most successful theorist and the most sagacious questioner of nature is the man who takes the wisest views of her indications, by the appropriate signs of her economy in the use of given forces, and of her adaptation to the ends of harmony, beauty, and perhaps of beneficence; and who has been most accustomed to reflect upon the actual methods by which these various workings of nature are accomplished in varying cases, as in mechanical effects, chemical combinations, vital forces, and spiritual endowments. He is the wisest interpreter of nature, who through nature has entered most intimately into the thoughts of God.

The place of experiment.

§ 241. (7.) Last of all comes the *experiment* to test the *theory*, however sagaciously conjectured—to answer the *question*, however ingeniously proposed. Though we must assume that the methods of the divine and the human intellect are the same, yet we must concede that the elements and powers, the laws and methods of the universe, *i. e.*, the thoughts of the Creator, can, as yet, be conjectured by the created intellect only to a limited extent.

Even of the facts which have been observed and known we are not always sure that we have considered all in all their relations, when our theory was constructed. We therefore bring the judgments founded upon these limited data to the revisal of the Infinite Mind. We question nature whether our thoughts correspond with her own. We correct the answers which we had devised by the decided responses which our experiments elicit.

While, then, on the one hand, man, in constructing his wise questionings and in framing his sagacious theories, may claim a likeness to God; he concedes his human limitations in submitting his theories to the test of experiment. Rightly conceived, every scientific experiment is an act of reverent worship.

CHAPTER IX.

SCIENTIFIC ARRANGEMENT.—THE SYSTEM.

Scientific arrangement. System in its lower import.

§ 242. We have already considered the several processes of objective or concrete thinking, and the products which they evolve. The processes are *analysis; generalization; classification; judgment,* in the two forms of *definition* and *division; and reasoning, by deduction* and *induction*—giving us as their products, the *concept;* the *class;* the *proposition;* the *argument;* and the *principle or law.* The combination of these several processes and their results in a complex result or *product,* is *scientific arrangement,* and the product is *the system.*

Scientific arrangement or method may be defined in general, as the gathering of individual objects into a *synthetic* whole, by any one of the analyses and generalizations of thought. When any number of such objects are united into such a whole, that whole may, in a certain sense, be called a *system.* This is not, however, the usual signification of the term. We employ it in this sense simply to call attention to the truth, that the process of classification is the beginning of systemization. This is the first condition or step of the synthetic process which terminates in the system proper.

Inasmuch as every concept has the two relations of *extent* or *content* either dormant or developed, that arrangement of individual objects in *these two directions* which follows from the application to them of both the content and the extent of a notion is *more properly a system.* When several notions of a more or less comprehensive *content,* or a more or less widely applicable *extent,* are used to define and divide the individual objects to which they apply, these objects are brought into a system; or the mind is said to take a systematic view of their several properties, and to class them as mutually related to one another. Their properties are seen to be more or less extensively the same; the classes in which they are grouped or gathered are said to be higher or lower, and the several classes are arranged into a hierarchy or a subordinated whole.

Inasmuch, also, as every concept results from, represents, and may be expanded into, its propositions; these twofold propositions of content and extent express, when properly arranged, the systematic arrangement or method of the objects to which such propositions can be applied.

Every *concept*, as well as every proposition that respectively defines and divides, and thus arranges and subordinates, the objects to which each belongs, indicates or suggests some property or power or law of the beings to which it is applied. Most *names* of things indicate that they belong to some permanent class, and are possessed of properties that are fixed in the designs, and are perpetuated by the laws of nature. The most important propositions of definition and division simply expand and apply these permanent properties and laws.

System in its higher significance.

§ 243. The more important of these properties and laws are those which are discovered by *induction*, applied in *deduction*, and verified by experiment, after the methods which have been explained. When so discovered, and applied, and established, they are used to explain or account for the less obvious events and phenomena in the universe of matter and of spirit. The properties, principles, and laws which are thus inferred in induction, applied by deduction, and verified by tests of fact,—as they are respectively established,—serve also to define and divide the beings and events which they concern; but by notions that are constituted of the more refined elements, and that divide beings into the more comprehensive and significant classes. Hence result *scientific systems*, *i. e.*, systems founded on principles more profound and wide-reaching than those which direct the classifications of common life.

It follows that scientific arrangement and systemization,—the concepts and terms,—are applied with pre-eminent propriety to the methodical arrangement which is founded and effected by these *more recondite properties and more extensive laws*. Such properties and laws are said pre-eminently to *explain* the operations of nature, and to enable man to *predict* phenomena, as well as to control events and results by art or skill. Such arrangement gives the system, in the pre-eminent sense, when many of these more subtle and significant laws and properties are arranged in order as higher and lower, *i. e.*, as more and less comprehensive in import and extensive in application.

It is important to observe that the terms *scientific method* and *system* may be applied to a narrower or wider range of beings or events, and may be founded on generalizations which are narrower and wider, or on inductions which are more or less profound. They may include a single kingdom of organic or inorganic existences, or may embrace all material things. They may define and arrange these according to the more obvious properties and laws which are open to common observation, or may employ those properties which appear to hasty observation to be very remote, and which are reached only by the most sagacious conjectures, and the most skilful experiments. They may include the domain of spirit only, or extend to the kingdoms of both matter and spirit, and arrange the two domains by the properties and laws which can be established as common to the two.

Systematic arrangement and scientific method are also freely applied to *abstracta*, or those artificial products which are the creations of the human intellect; to those concepts which *law*, *ethics*, *theology*, *politics*, and *political economy* familiarly employ, as well as to those abstract forms and rules which *grammar*, *logic*, and the *mathematics* prescribe. But all concepts are derived from propositions, as their originators and vouchers. A system of *definitions*, properly subordinated and derived, is therefore essential to every scientific system of concepts, terms, rules, and principles, and should always be justified by the concrete examples and existing beings from which the concepts are derived, and by which the principles are tested.

PART FOURTH.

INTUITION.—THE CATEGORIES.—FIRST PRINCIPLES.

CHAPTER I.

THE INTUITIONS DEFINED AND ENUMERATED.

§ 244. Thus far we have inquired what are the processes and products of knowledge, when the knowing power is employed in the form of direct activity. The critical and speculative stage of our studies.

We are now to turn the power in upon itself; to inquire what are *the relations* which it necessarily assumes in all those operations. In doing this we enter upon the last and highest stage of our inquiries—which is properly called the *critical* or the *speculative*. It is *critical* because it analyzes these operations for the purpose of testing their trustworthiness. It is *speculative* because it aims to find the ultimate elements and foundations of all science and all knowledge.

This critical analysis of the power of knowledge is the last and highest form of the mind's activity, because it supposes the complete development and discipline of all the other powers. The mind must be trained to analyze everything besides, before it can successfully analyze the processes and products of its own power to know. The mind must reach a high degree of psychological development, before it is prepared to *comprehend* its processes and products under their most comprehensive logical relations. The power of thought must be disciplined by exercise upon many objects and in manifold methods before it can be competent to analyze the most general relations that are assumed in the several operations of knowledge and are the rational foundations of its confidence in whatever it knows. It must have studied these operations of the intellect familiarly, before it can ask itself what relations each of them imply. As the thought-power is at once the analyzing and generalizing power, so the study of these relations is regarded as intimately related to it.

This critical examination of the power to know, involves a philosophical scrutiny of the grounds and trustworthiness of all knowledge and belief. It convinces us that the relations or principles which we receive and trust as axioms in one kind of knowledge, are to be trusted in another. It shows us, moreover, that we are bound to believe and follow them wherever they lead us, because we cannot know any truth without them. It sets aside objections that are derived from the denial of these relations by showing that they are not only fundamental, but are always applicable It disarms *skepticism* of every kind, whether it be *philosophical, ethical,* or *theological,* by showing that the relations which the human mind must apply in its lower knowledge, it cannot refuse to trust in their higher applications.

These inquiries conduct us from the field of psychology towards and into the fields of both logic and metaphysics. It is not practically easy to draw the lines which determine the boundaries of each. The critical analysis comes first in time, and is appropriate to psychology: logic and metaphysics avail themselves of the results which this psychological analysis gives.

Strictly speaking, in *psychology* we show by analysis that we constantly require and employ these cognitions, while in *logic* and *metaphysics* we inquire *what they are,* and what are their relations to the other objects of knowledge. Inasmuch, however, as it is impossible to separate the analysis of a process from an analysis of its product, the psychological will often seem to encroach upon the logical and metaphysical sphere.

These ultimate facts and relations are not gained by any of the processes of the intellect which we have thus far considered. They are not perceived by sense-perception, nor felt by consciousness; they are neither reproduced in memory, nor represented or created by the phantasy; they are not generalized from simple experience of material or spiritual objects; they are neither proved by deduction, nor inferred by induction. Their truth and validity are not *apprehended by,* but they are *involved in* these processes. They are developed and brought to view in connection with these processes, and are assumed in them all.

They have been referred to a separate faculty.

They have sometimes been referred to a special and separate faculty. This so-called faculty has been designated by various appellations, as the *reason, com-*

mon sense, judgment, intuition, faith, the intelligence, the regulative faculty, the noetic faculty, ὁ Νοῦς as contrasted with *ἡ Διάνοια, die Vernunft* as contrasted with *der Verstand.* But it has been generally conceded that the word *faculty* is not employed in its usual signification. Thus Hamilton observes (*Met. Lec.*, 38), the term "faculty is employed, not to denote the proximate cause of any definite energy, but the power the mind has of being the native source of certain necessary or *a priori* cognitions."

The appellations by which they are known.

The cognitions or beliefs themselves "have obtained various appellations." They have been denominated: *Intuitions, categories of thought, first principles, self-evident or intuitive truths, primitive notions, innate cognitions, metaphysical or transcendental truths, ultimate or elemental laws of thought, primary or fundamental laws of human belief, pure or transcendental or a priori cognitions.* They are called *intuitions* because they are discerned by reflex analysis to be present in all our *knowledge*, and *categories of thought* because as generalized conceptions they are of universal application as the foundations of thought and science.

It will be observed that some of these appellations designate propositions, which affirm the reality and authority of these relations, and others the relations themselves in the form of concepts. The distinction is purely formal. It is a matter of terms and not of thoughts, of language only, but not of things. It is true in this as in all other cases, that it is from or through a proposition, that each of these concepts is derived. The concepts of cause and effect and of causation, those of means and adaptation as well as those appropriate to extension and duration, are first gained through propositions expressing beliefs.

Not *first* in the order of time, but in logical importance.

§ 245. It is often convenient to generalize these as propositions. In such cases we call them primitive judgments or first truths. In naming them first truths or primitive judgments, it is not intended that these truths or judgments are acquired *first in the order of time*, or that the mind's assent to them is prior to its other acts of knowledge. That they cannot be acquired or assented to first of all, is evident from the unquestionable fact that, by very many they are never acquired at all. The majority of men never think of them, much less do they assent to them. Even the majority who attain to not a little culture, do not reach a clear and intelligent conviction that these propositions are true.

It was forcibly urged by Locke that such propositions as "*whatever is, is*" and "*the same thing cannot be and not be at the same time,*" cannot be innate, for the plain reason that men at their birth, and in all the early period of their existence are entirely incapable of understanding the meaning of the conceptions and terms of which these propositions are composed; and if they cannot understand the constituent elements, much less are they capable of asserting that one of them is true of the other. It might be further enforced by the consideration, that the mass of men are incapable of that analytic abstraction which is necessary to detach the universal from the individual example in which it is realized. Or, if we concede or suppose that the causal attribute or relation could, by analysis, be distinguished from the individual example of *cause* or *effect*, an additional act of generalization would be necessary to qualify the mind to assent to the general truth, "EVERY *event must have a cause.*"

They are, in fact, attained last in the order of time.

These truths, instead of being the *first* to be consciously possessed and assented to, are the *last* which are reached, and by only a few of the race are ever reached at all. Experience proves that long courses of training are required, to bring the intellect into a capacity for analysis and generalization, which may enable it to understand and assent to them. The mind must be exercised to some extent in philosophical studies before it can comprehend their import and application.

Various significations of a principle.

§ 246. These truths or judgments stand first in the order of *rational* or *logical importance*. Hence they are called *first principles:* principles or truths *a priori*, as opposed to knowledges *a posteriori*. As *concepts* they are called *categories*, *pure cognitions*, etc.

The term *principle*, which is so often used in this connection, is variously employed, and admits of many senses. It may be generally defined as any thing with which the mind begins in an act of rational or logical combination, or more generally still, as the constituent of any synthetic product. The word *principium*, ἀρχή, is, literally, a beginning or starting-point. Inasmuch as there are as many beginnings as there are processes or progresses to different ends or results, so the word principle is used in the following special meanings.

1. Any *constituent element* of an existing thing, whether it is material or spiritual—whether it is a being, act, or product, is a principle. The materials which we bring together, or think belong together so as to constitute any existing object, are sometimes called principles. In a similar way, the simple concepts that make up any complex concept or general notion whatever, are called principles.

2. Any *causal agent* in matter or spirit, is called a principle, because the cause is looked upon as originating and beginning the effect. Thus we say of a machine, it has the principle of motion within itself. It is not uncommon to apply it to the capacities of the soul, viewed as causes of its functions or activities Thus, we say, there is a principle in man's nature by which he is able to distinguish truth from falsehood, or right from wrong.

3. All *general propositions* which are admitted or used as *premises* in deduction, are also principles. They are so called, because the mind *begins* with one of them in the process of its reasoning.

4. All *generalizations* from *induction*, as well as all collected observations from experience, are called principles, for the reason that they are used to explain and account for the occurrence of particular events or phenomena. The mind begins with these in all its rational solutions. Hence the powers of nature and the laws of nature, as well as observed facts when generalized and supposed to indicate some concealed law, are freely called principles.

5. Those *general truths* which are the starting-points of the reasonings or communications of *any special science* or *art*, are called, with eminent propriety, principles ; because, in imparting or demonstrating the science, the teacher begins with these as facts, or reasons from them as premises. Hence the fundamental maxims or assumptions of mathematics, of logic, of law, of ethics, of politics and political economy, are called the principles of each of these sciences.

6. But the appellation of principles is applied with prëeminent propriety to any one of those *universal concepts and relations which are implied in any of the different kinds of knowledge*,

because it must be assumed or supposed as a beginning or element to make that knowledge conceivable.

7. If there are other objects of knowledge usually called *infinite* and *absolute,* which are necessarily implied in the special and limited relations, and are their necessary correlates, these prëeminently deserve to be called principles, as they are in rational order and dependence *before,* and are the grounds and explanation of, *all other objects of thought and knowledge.* Whether there are such, must be decided by our subsequent inquiries, and will be discussed in the appropriate place.

The relation of intuition to experience.

§ 247. Our knowledge of these truths is *occasioned* by, but is not *derived* from *experience.* This is most happily expressed in a sentence quoted by *Hamilton* from *Patricius; cognitio omnis a mente primam originem, a sensibus exordium habet primum.*

Indeed, the most sagacious thinkers coincide in this opinion, that our higher and *a priori* knowledge, while independent of experience as the source of its evidence and authority, is dependent upon experience as the occasion of its development. Thus *Leibnitz,* in criticising *Locke* for asserting that all our knowledge is derived from sensation and reflection, says: "The senses, although necessary for all our actual cognitions, are not, however, competent to afford us all that our cognitions involve." *Reid* also observes, in defence and explanation of Locke's real meaning: "I think Mr. Locke, when he comes to speak of the ideas of relations does not say that they are ideas of sensation or reflection, but only that they *terminate in* and *are concerned about,* ideas of sensation and reflection." *Essay* vi. c. i. The doctrine of *Kant* upon this subject is uniformly as follows: "We must then first of all observe, that although all judgments of experience are empirical, *i. e.,* have their ground in the immediate perceptions of the senses, yet conversely it is not true, that all empirical judgments are for this reason judgments of experience, but in addition to the empirical element, and in general in addition to that which is given to sense-intuition, particular concepts must be furnished, whose origin is *a priori* in the pure understanding, under which every percept must be subsumed and so changed into true experiential as distinguished from empirical knowledge." *Proleg.* § 18.

Victor Cousin also repeats himself to the same effect abundantly in the following strain: "The idea of body is given to us by the touch and the sight, that is, by the experience of the senses. On the contrary, the idea of space is given to us, on occasion of the idea of body by the understanding, the mind, the reason; in fine, by a faculty other than sensation. Hence the formula of Kant: 'the pure rational idea of space comes so little from experience, that it is the condition of all experience.'" "Now the idea of space, we have just seen, is clearly the logical condition of all sensible experience. Is it also the chronological condition of experience and of the idea of body? I believe no such thing." "Take away all sensation; take away the sight and the touch, and you have no longer any idea of body, and consequently none of space." "Rationally, *logically,* if you had not the idea of space you could not have the idea of body; but the converse is true *chronologically,* and in fact, the idea of space comes up along with the idea of body." *Elements of Psychology, translated by C. S. Henry,* chap. 2. *Cours de l'Histoire de la Phil. du* 17*e siècle. Leçon* 17.

The several stages by which these categories are developed in experience are the following:

(1.) The *first* act or stage is the cognition of any *concrete object,* of which any attribute involving an intuition might be affirmed, or exemplified. The object may be material or spiritual, it may be a being or an act, as these are commonly distinguished. For example, it may be a fruit, a piece of marble; the combustion of wood, the explosion of gunpowder, the shooting of a star, the running of a horse; a remembered occurrence, a sally of imagination, a fixed purpose, or the ego of our conscious acts.

It is conceivable that these and the like objects may be cognized for an instant, without the perception of any relation.

(2.) The *next* step or stage is the apprehension *of these objects as related* in one or more given ways. The fruit is known as oval in form, as large or small in size. The color, taste, and feeling of the fruit are thought of it as qualities or properties. The combustion and explosion, the remembering, the imagining, are known as acts of the material or spiritual agent or as effects of which these agents are the causes, or as the ends to which other acts are adapted, and for which they are designed.

This second stage is reached by the whole race, not to the same extent or perfection in all, but so far that all may be said to achieve this kind of knowledge.

Material objects are known by all men as long and short, round and square. Events are known by all as before and after. One object or act is known as the cause or the end of another object or act. The words which express and indicate the more familiar of these relations are accepted in the language of all men. They are spoken by all, and understood by all as signifying these relations.

(3.) The *next* stage or act is when the relation is *abstracted from the beings* to which it belongs and is generalized into a concept higher and more extensive, which is treated as a separate entity. Thus long, short, etc., are contemplated as length or shortness; round, spherical, etc., are known as roundness and sphericity; past, present, and future are known as time relations; the power to produce this or that effect is abstracted and generalized as the causative relation; the individual fitness to accomplish this or that end is generalized and abstracted as the relation of adaptation.

This third stage is more rarely reached. For the common purposes of life men have little occasion to view these attributes and relations as separate entities, and still less to carry them to the higher degrees of generalization. Practical men have little need to consider or to speak of the relations of time and space or substance or cause, when separate from concrete objects and events, and when generalized in abstract language. Even thinking men, who may be well disciplined and practised in intellectual activities of other kinds, have few motives and little inclination to deal with such entities in their more abstract forms.

(4.) The *fourth* stage of experiment and assent is the critical consideration of the processes of knowledge, and the discernment of these *relations as essential elements in all these processes* and as the fundamental principles which are implied in them all. It is manifest that this stage is reached only by a few, and by those only whose attention is directed to the critical examination of their intellectual processes, and to a speculative consideration of the principles which they involve.

(5.) The *last stage* or act of distinct knowledge is the recognition of *the correlates*, usually called *infinite* or *absolute*, *which are required by these relations when* they are generalized and reflected on. Thus the relations of extension when apprehended as belonging to every material object, *i. e.*, to the universe in its parts and as a whole, imply Space as their correlate; those of duration imply the correlate of Time; the universe conceived as a single effect implies a single causing agent—the universe conceived as a designed effect requires that this agent should be intelligent.

These correlates *Space*, *Time*, and *God*, are conceived as the *conditions* of the possibility of the universe, and the ground of its reality, and are therefore the *first principles* of every thing that is and can be known.

It is manifest, for the reasons already given, that if it be assumed that there are such correlates to these finite beings, the consideration of them as the real and the necessary principles of all beings is not within the reach of the majority of men. It requires a capacity for the highest analysis and abstraction of which the human mind is capable. It supposes an interest in and a capacity for wider generalizations than most men exhibit. Few men attain to these ideas through processes that are purely speculative. Fewer can give the philosophical reasons by which they reach and on which they receive them.

All men may have the capacity to assent to truths concerning them when propounded in terms that are not philosophical, and enforced by reasons that are not abstract and speculative; but the number is exceedingly small of those who can analyze the processes by which they are seen to be necessary, or assent to them as the grounds of all being and of all knowledge.

This review of the several stages by which these truths are developed to the mind's assent, serves to explain and confirm what has already been asserted, viz., that though first in authority and in logical dependence, they are the last in the order of time; and that though all men manifest a practical belief in these principles, when exemplified in the concrete, yet but few understand or assent to them when stated in a speculative form.

It also enables us to understand how it is possible that they should be discovered and tested in a variety of methods suited to the condition of each of these classes, as also why the criteria which satisfy one class of minds should neither reach nor convince minds of another class.

What is most important, it explains why the evidence for their truth and universal acceptance which is furnished by the *language* and the *actions* of men is more decisive and satisfactory than that which comes by speculative analysis or philosophical argumentation.

We have seen that all men reach the *second* stage of knowledge, so far as to apprehend many objects in one or all of these necessary relations to some other object, *i. e.*, as substance or attribute, as cause or effect, as means or end, etc. This recognition of these concrete relations, they express by their language in appropriate concrete terms, as by the *noun*, the *adjective*, the *verb*, etc., in the various forms of flexion and construction. Few men reach the *third*, and the number is therefore small who reflect upon the relation of causation when it is generalized from individual instances, or who ask themselves whether it is universal and necessary to the mind.

And yet the very language which all men use is a constant profession of their faith in their reality and importance. Almost every sentence which they frame and word which they employ is a voluntary acknowledgment, that these intuitions are necessarily accepted by all men. When they act, every one of their expectations and deeds is a more decisive avowal that these principles are absolutely certain, and never admit an exception.

This review also explains how it can be that men may reject truths in theory which they admit in fact. In other words, it explains the apparent paradox that there may be truths which men always recognize in their actions, but deny or question when they are phrased as speculative or philosophical propositions.

Such propositions must always be expressed in the language of the Schools, that is in language which is *abstract* and therefore to a certain extent *technical* in its signification. They must be defended by philosophical evidence, the evidence that is appropriate in the Schools; which often rests upon principles with which the mind is by no means familiar, and is enforced by methods of reasoning to which it has not been trained or wonted.

We are justified in appealing from the philosophy of men to their words and actions. What all men inadvertently confess in their casual assertions, what they imply in the very forms of their language, what their actions unbiased by their theories show that they recognize, what their expectations from others show that they believe that their fellow-men also accept, what is assumed in all investigations and reasonings without the attempt to give any reasons for its truth,—these are all taken to be or to involve universal and necessary truths of Intuition, however difficult it may be to define them correctly, to reconcile them with the dicta of a received philosophy, or to show their place in any order of systematic arrangement.

The Three Criteria of First Truths.

§ 248. The philosophical criteria of the categories and first truths are usually stated as three: '*their universality, their necessity, and their logical independence and originality.*'

(1.) First truths are *universally* received. If they are not universal they can be neither necessary nor logically independent and original. But in what sense are they understood and by what evidence can they be shown to be universal? Surely not in this, that all men actually assent to them when propounded in a scientific form and phraseology.

This as we have seen is from the nature of the case impossible, inasmuch as all men are by no means capable of understanding the terms and grasping the conceptions which enter into them. But all men can believe them in the concrete, *i. e.*, in every individual case in which they are exemplified, without knowing that thereby they presuppose knowledge, which, when stated in its abstract form, would involve the principles in question.

(2.) First truths are also *necessary*. Truths to be universal and primitive must be necessary, *i. e.*, the intellect must be constrained by the constitution of its being and the spontaneous workings of its nature to receive them as true. It cannot know objects of any kind except under these relations and according to the connections which they involve. Should it attempt to do so, or to prove that it does not employ and recognize them, it would make the effort of knowing without them, and of proving that it did not, by using these very relations in its efforts and its arguments.

(3.) First truths must be *logically prior* to, and *independent* of, all other truths. Each one of them is the most generic concept of many similar individual relations. It can be itself resolved into no other, and can be proved by no other.

This is what Buffier must intend, when he says, "they are propositions so clear that they can neither be proved nor attacked by any propositions more clear than themselves." Hamilton means the same when he calls them *incomprehensible,* defining the term to signify, that of which we know the fact, but cannot give a reason. Hence they are called self-*evident* truths and *intuitions,* because they need only to be seen or apprehended to be believed. *The act* of *critical or speculative intuition* is not an act of sense-perception nor an act at all analogous to it; but an act of knowledge which is direct and original and is the necessary condition of all other acts of knowing.

It follows that these truths are neither discovered by induction nor generalized from experience. That they are not the results of induction has been shown by the nature of induction as revealed in the analysis already given of the process. It has been shown that the process itself involves certain assumptions as true; or the belief of certain relations as original and self-evident. Unless we begin by assuming that these relations are valid and original, we cannot confide in the process of induction itself. Indeed, without these assumptions, the process can have no meaning.

That they cannot in any way be generalized from experience has been shown by the analysis already given of their relations to experience. J. S. Mill, in his *Logic,* contends most earnestly that all the so-called original necessary truths, including the postulates of mathematics, are derived by Induction through experience. The considerations already adduced are decisive against his theory. President M'Cosh entitled the earlier editions of his able work, *Intuitions of the mind Inductively considered,* but he used Induction in a general and popular sense.

Nor can they be regarded as the highest premises for comprehensive syllogisms, obtained by successive processes of regressively evolving the premises or assumptions on which narrower syllogisms are founded. This view has been countenanced, if it has not been taught directly, by philosophers of very high authority. Cf. Dr. Thomas Reid, Essays, VI. c. iv. Aristotle, *Anal. Post.* i. 3; cf. i. 22. Cf. McCosh, *Intuitions of the Human Mind,*

Part i. B. i. c. ii. § 1 (6). Buffier, *Traité d. prem. ver.* Dessein, etc., § 6.

It is, however, one thing to show that without first truths no deduction is possible, and quite another to show that such truths must be employed as the ultimate premises in the most comprehensive deductions. The analysis already given of the deductive process has shown that it rests primarily upon the relation of reason to conclusion, which in its turn rests upon the relation of cause to effect. It has also shown that the materials for deduction are all derived from induction, or mental construction—as in mathematical or purely logical reasoning. First truths, or intuitive relations are implied as in one sense the support or foundation of the processes of deduction, but not in the way of serving as ultimate premises.

Were we to consider the process of deduction solely in its logical relations, we should clearly see that these truths could serve no use as premises. Nothing could be proved by such universal and wide-reaching propositions as *every event must be caused*, etc., etc. For as soon as you interpose the minor, 'this explosion is an event,' you make no progress towards additional knowledge in the conclusion : you know already that this explosion was an event : you could not have known it at all without having already decided that it was one of the things that are caused.

For the purposes of deduction, all such principles are barren and useless. Nothing can be derived from them. From their very nature, they are simply statements concerning those relations or elements, that are present in every act of our higher knowledge. It is only because they are present as an essential and necessary element in all these processes that they must of necessity be conditions of deduction.

They are independent of one another.

§ 249. These intuitions or categories, are in the strict sense of the term *logically independent of one another*. Their apparent dependence upon one another arises from the limits of the human intellect, which prescribe a certain order in the familiar acquisition of these concepts and in the frequency and extent of their application.

The observation is very common that by a logical necessity we must think of being before we think of its relations or attributes ;

of time before we think of space; of all these before we think of cause, and of these together with causation before we think of design; or, as expressed in other language: Being is fundamental to all other categories, and must be presupposed before and as the condition of them all: and in a similar manner the less must precede the more dependent till the entire circle is complete.

But no one of these categories can be developed from another. If it could be it would not be primitive and original. Nor can one be explained into or resolved by another. None of them is properly complex, for if this were so, each of the constituent elements would be original and primitive, but not their constituted whole. They cannot be dependent in the relation of *content;* for the import of one cannot be resolved into that of another. Nor is one more *extensive* than the other, so far as the real objects are concerned to which each may possibly be applied. Every object that exists must be conceived as existing, as diverse from others, as related to others, as whole or part, as in time and space, as capable of number, etc., etc. Were the mind capable of attending to all these conceivable relations of every existing object by a single intuitive act; were it not dependent upon the slow processes of observation and induction to learn which is related to which as *cause* and *effect, power* and *law, means* and *end,* these relations would be equally extensive in their application, and would all be co-ordinate with one another in the view of the human as they are before the divine mind. But inasmuch as the human mind proceeds in its knowledge step by step, some of these relations are familiarly and far more *extensively* applied than others. Some of them are applied to objects of imagination and thought, while others are more rarely affirmed even of things. The relations of dependence between them are *chronological* and *psychological* but not *logical.*

Hegel's development of the categories.

This attempt to develop the categories from one another was carried to its extreme by Hegel, who began with *being,* and making being to be equal to *nothing, i. e.,* to have no content, sought by what he called its *becoming, i. e.,* the independent and necessary movement of the concept, to evolve all the categories from one another, not only of thought but of material and spiritual existence, in a self-completing and perpetually repeated circle. This self-evolved and self-completing circle of necessary concepts was conceived by him as the *Idea,* and all these together constituted the *Absolute, i. e.,* the sum total of mutually-related possible, and conceivable thoughts and things.

Hegel's mistake was twofold. He attempted to derive things from thoughts, or *real* from *logical* relations, instead of finding all *logical*, *i. e.*, all *generalized* relations in those which are real. He attempted to derive one category from another, instead of explaining the apparent dependence of one upon another by the order in which they are developed to, and the extent in which they are applied by, the mind through its psychological limitations.

Divided into three classes.

§ 250. The categories or intuitions may be divided into *the formal, the mathematical*, and *the real*. The *formal* are those which are involved in any act of logical knowledge, whatever be its object-matter—whether it be real, imagined, or generalized—whether it be an actually existing or a purely mental creation. They are essential to the most abstract form of knowledge, and appear in all its objects or products. The *mathematical* are those which grow out of the existence of space and time and suppose these to be realities. The relations included under this definition are not exclusively used in the sciences of number and quantity, but inasmuch as they are fundamental to these sciences, we distinguish them by the epithet mathematical; using it to designate all the time and space relations and those directly dependent upon them. The *real* are those which are ordinarily recognized as generic to and fundamental of the so-called qualities and properties of existing things, both material and spiritual. We do not, however, by using the term *real*, imply or concede that the *formal* and the *mathematical* are any the less real—but that they are not limited so exclusively to objects really existing.

CHAPTER II.

THEORIES OF INTUITIVE KNOWLEDGE.

A complete sketch of the various theories which have been held in respect to the nature, origin, and authority of primitive notions and intuitive judgments, would include the most important portion of a complete history of *Metaphysics* or *Speculative Philosophy*. Such a sketch would be entirely out of place in the present work, and will not be attempted. We shall only endeavor to group and critically examine, under a few comprehensive titles, those theories which have any present interest for modern thought, or which are still maintained in modern schools of philosophy.

The theory of a direct mental vision of first truths.

§ 251. 1. It has been extensively taught that these original ideas and first truths are discerned by direct *insight* or *intuition* independently of any relations to phenomena. The power to behold them is conceived as a special *sense* for the true, the original, and the infinite; as a divine Reason which is permitted to gaze directly upon that which is eternally true. Such are the representations of *Plato, Plotinus*, etc., among the ancients. Thus the Platonizing and Cartesian divines of the seventeenth century, as *Henry More, John Smith* of Cambridge, *Ralph Cudworth*, and multitudes of others, freely express themselves. *Malebranche, Schelling, Coleridge, Cousin*, and others, have given sanction to such views more or less clearly conceived and expressed. Those who combine with philosophic acuteness, the power of vivid imagination and eloquent exposition, not infrequently meet the difficulties which attend the analysis and explanation of the foundations of knowledge by these half-poetic and half-philosophic representations.

It is manifest that the representations which they give are not true when literally interpreted. No direct inspection of primitive ideas and principles is conceivable. It is not by withdrawing the attention from, but by fixing it upon, the facts and phenomena of the actual world, that the truths and relations of the world which is ideal and rational can be discerned at all.

The theory that they are discerned by the light of nature.

§ 252. 2. Many of the earlier philosophers and theologians of modern times, following the Scholastics of the middle ages, were accustomed to say that these ideas and truths are discerned by *the light of reason* and *the light of nature*, that they shine forth or are evidenced by their own light. The use of this language is in part to be traced to the often-repeated maxim of Aristotle that some truths cannot be demonstrated, but must be accepted without proof; in part by a Platonic interpretation of the passage in the gospel of John (i. 9), in which *the Word* is said to enlighten every man who cometh into the world.

It is obvious that the phrase is figurative and expresses only the fact which remains to be explained and accounted for, that these truths are neither generalized from experience nor deduced by logical ratiocination; that they are no sooner thought of than they are assented to, and that upon them as original assumptions rests the validity of all generalization and deduction.

That they are innate or connate.

§ 253. 3. The doctrine has been earnestly held and taught that these ideas and beliefs are *innate in* or *connate with* the soul. This is well known as the doctrine which *Descartes* is supposed to have taught, and to the refutation of which Locke devoted the first book of his *Essay*. It is that the intellect finds itself at birth or as soon as it wakes to conscious activity, to be possessed of ideas to which it has only to attach the appropriate names, or of judgments which it needs only to express in fit propositions. Whether this doctrine as thus stated and defined, was ever held by any one may perhaps be questioned. Even Descartes himself seems, when pressed, wholly to abandon the doctrine in the form in which he had propounded it and made it the foundation of the most important conclusions.

On the other hand, it would be conceded by many, and can be defended as true, that the capacity to evolve these ideas and these truths is born with man and forms an essential feature of his constitution as man. Not only is man endowed with these capacities, but he is furnished with tendencies which impel to their exercise, and after which these conceptions and judgments are surely and necessarily developed so soon as the mind applies the necessary attention or awakes to the requisite conditions. Even before these conceptions are generalized they are assented to in the individual and concrete, in the most important kinds of knowledge.

§ 254. 4. From the doctrine of innate ideas and the school of Descartes, the transition is natural and direct to the *views held by Locke* and *the several divisions of his school.* These are naturally grouped together, though the interpretations of the meaning of Locke are very diverse, and the several schools that are named after Locke, hold opposite and incompatible opinions. It will be found, however, that they can all be traced to Locke, either as they are sanctioned by his direct authority or were derived from some of his principles by logical deduction or natural growth, or as they were devised to supplement some of his supposed oversights or defects.

The views of Locke and his school.

Locke, as is well known, rejected the doctrine of innate ideas and protested most vigorously against it, in the first book of his *Essay.* This protest was of the greatest service to philosophy in delivering it from the vague and fantastical assertions upon this subject which had been allowed before his time. It has been questioned and may be doubted, whether any sober and considerate thinker ever received the doctrine in the form and sense in which Locke rejected it. But it is certain that many philosophical writers have expressed themselves in language which warranted the interpretations which Locke thought it necessary to refute.

But Locke did not guard himself against serious oversights in this polemic. He did not distinguish between those positive ideas of objects and acts in both matter and spirit which make up the materials or facts of knowledge—and the *relations* between these materials, which, if possible, are more important than the facts which they connect. Nor did he conceive at all the difference between an idea as *acquired* by experience and as *occasioned* by experience. He did not discern that a relation which is developed by experience to conscious apprehension, must be implied or assumed to make experience possible. He did not distinguish between innate ideas and innate dispositions or capacities to develop and assent to truths which involve original ideas. To correct these oversights, *Leibnitz* subjoined his well-known reply to the adage, *"nihil in intellectû quod non prius in sensû"—"nisi ipse intellectus."*

Locke asserts positively that all our ideas are obtained through two sources, *Sensation* and *Reflection:* Sensation gives the knowledge of sensible objects and their qualities; Reflection gives the knowledge of spirit and its operations. He was careful to add that except through these two sources we have no ideas whatever. What Locke intended by *ideas* admits here of a question similar to that which was noticed in connection with *innate ideas.* Did he mean positively to exclude from ideas those necessary relations by which the mind connects all the objects of matter and spirit which it observes or experiences? It is probable that this distinction was not in his mind, and that for this reason he did not provide against uncertainty or ambiguity of interpretation. It was not unnatural that different constructions should be put upon doctrines thus announced, and that according to these diverse interpretations, there should spring up among his followers different schools of philosophy.

One class of those who called themselves his disciples, by greatly limiting or almost setting aside his definition of reflection, interpreted him as teaching that all our positive ideas are of material objects, and perverted his principles so as to make him teach a materialistic philosophy. *Condillac* thus applied his doctrine, and derived from it the conclusion that all our ideas, whether those of sense or spirit, are simply *transformed sensations*. "Locke distingue deux sources de nôs idées : les sens et la réflexion. Il serait plus exact de n'en reconnaitre qu'une source, parce que la réflexion n'est dans son principe que la sensation elle même, soit parce qu'elle est moins la source des idées que le canal par lequel elles découlent des sens."—*Traité des Sensations*. This doctrine in the form in which it was taught by Condillac and by others of the French school, was long since abandoned, but tendencies to the same doctrine, if not to the same opinions in respect to the nature and origin of mental activities and their products, retain their hold most tenaciously among many modern psychologists, such as J. S. Mill and Alexander Bain with others.

Hume (*Treatise on Human Nature*, Part III., § § 2, 3, 4, 14, 15; *Inquiry concerning the Human Understanding*, § 7,) applied Locke's dictum in respect to the sources of knowledge, to the analysis of the relation of causation, or as he called it, of the ideas of *Cause and Effect*, and of *Necessary Connection*. He first demonstrates, as it is easy to do, that these *ideas* are not to be gained from Sensation. He then inquires whether they can be gained by Reflection, or the conscious experience which we have of the exercise of power in the production of effects by volition. To this he answers in the negative, experience giving us only the invariable succession or the constant conjunction of these internal ideas.

How then, he asks, does it happen that we connect objects as causes and effects, and what is the meaning of the combination? We certainly do thus connect them, and we give to them as thus connected the names respectively of causes and effects. To his own question, he replies: Objects which are observed to be always conjoined, we invariably *associate* in our minds: When we observe the one we cannot avoid thinking of the other: The principle of association is that which explains, and it is the only mental law that explains, the combination of objects and events as causes and effects.

The solution applied by Hume to the single relation of cause and effect, has since his time been applied to the explanation of other of the so-called necessary truths or primitive cognitions. *Dugald Stewart* used it to account for the belief that every visible or colored object involves a belief in, and an apprehension of extension. Dr. *Thomas Brown* carried it still farther, applying it to a great number of relations. *James Mill*, in his *Analysis of the Human Mind*, was the first to find in the doctrine of inseparable or indissoluble associations a solvent for all necessary beliefs and original conceptions. *John Stuart Mill*, his son, in his *Logic* and *Examination of the Philosophy of Sir William Hamilton*, has applied this principle in detail to all the so-called original and necessary truths with the conceptions which they involve; persisting in attempting to show by this single formula that mathematical conceptions and axioms are generalized from experience, that the universal and necessary belief in causation is itself the product of induction, which again results from associations that cannot be overcome or separated. *Herbert Spencer*, while on the one hand he earnestly contends that *inconceivability* of the opposite is the decisive test of original truths, holds that these very axioms are our earliest inductions from experience. Moreover, he holds that the capacity of induction itself is not only the result of processes of association, but these descend from one generation to another with an augmented tendency, till they acquire that irresistible force which excludes the conceivability of other relations. All these writers may be said to belong to the school of Locke, but they receive only one or two of his leading doctrines and interpret them in a narrow spirit, and apply them to explain conceptions and beliefs to which Locke never thought of applying them.

Dr. Reid and the Scottish School.

§ 255. 5. Dr. *Thomas Reid*, with Hutcheson, Oswald, and Beattie, was aroused by the skeptical conclusions derived by Hume and Berkeley from the doctrines of Locke, to combat his principle as it had till then been interpreted—that all ideas are obtained from sensation or reflection—and to assert for the mind itself an independent power or source of knowledge. This power was called by him *Common Sense*, and to it was referred our belief in the original and fundamental elements of all knowledge. Reid was especially earnest in asserting the necessity of *first principles* as the foundations of knowledge in general and of every special science in particular. Of these principles there is a great variety—*logical, grammatical, mathematical, moral, æsthetical, metaphysical*, as well as those *facts* given in the experiences of sense and consciousness. All these are discerned by that power which he called *common sense*, and occasionally *judgment*. The nature and the conditions of this faculty he did not exactly define, nor its relations to other powers, nor the laws of its acting, nor the character and place of its products. He was content to assert that there must be a source of this kind of knowledge independently of experience, and that these first truths are to be received upon its authority. *Dugald Stewart* followed Reid in insisting upon "*fundamental laws of human belief*," and "*original elements of human knowledge*." He, however, subjected to analysis some of those truths which were asserted by Reid to be original, and allowed to the law of association an influence which Reid had not recognized. *Brown* deviated materially from Reid and Stewart in attaching greater importance, in his analysis of our conceptions, to the laws of association. He resolved the relation of cause and effect into that of invariable antecedence and succession. He occasionally refers to some original belief or tendency to belief as necessary to explain our actual experience. He also distinctly recognized a faculty or power called *relative suggestion*, which of itself originates or discerns certain original relations; making it, like Reid's *judgment*, to be the originator of and voucher for these original relations or categories. His system is not always congruous or consistent with itself, inasmuch as he attributes greater authority at one time to the associational, and at another to the intuitional element.

In France, *Royer Collard* and *Jouffroy* followed in general the method and the doctrines of Reid, with a more analytic scrutiny and a more systematic arrangement of the original data of knowledge. Each of these writers made some important improvements upon the doctrines of their teachers.

Maine de Biran followed out the doctrine of Locke in respect to Reflection, and attempted to find in Reflection the source of some important first truths. He went further than Locke in this direction and borrowed from Leibnitz some important modifications of Locke's teachings in respect to the nature of power and the essential activity of the mind as a discoverer of original and independent truth. *Cousin* sought to unite Reid, Collard and Kant.

These writers might perhaps be more properly grouped together as belonging to a separate school—the *Scottish*, or the *Scottish and French* School. But a more careful study of the doctrines of Locke reveals the fact that in the latter part of the *Essay*, when he came to analyze and account for the ideas of relation, particularly of such primitive relations as *substance, cause*, and *adaptation*, he departs from the doctrines which he was supposed to have laid down in the preceding chapters. He certainly did not place that construction upon them which many of his disciples imposed after his time. In accounting for these original ideas, he seems to ascribe them directly to the intellect itself, and to an original power to discern, and an original necessity to receive them as true. In short, without asserting, in form, any new source of ideas, and without in the least abandoning his previous teachings—while in reply to the objections which were brought against him for inconsistency, he earnestly defends his own consistency with himself—he does in fact take the same ground with Reid and the Scottish School. Cf. (T. E. Webb. *Intellectualism of Locke.*)

If this is a correct interpretation of Locke's real opinions, then Reid and his disciples are properly connected with the school of Locke, notwithstanding their earnest polemic against some of the doctrines which they supposed him to teach.

Kant and his School.

§ 256. 6. From Hume and Reid, who were antagonist disciples in the school of Locke, we pass to the speculations of *Kant*, and consider his views of *first principles* and the *categories*. Kant, like Reid, was aroused by the skepticism of Hume to investigate the foundations of knowledge. He saw that if the solution given by Hume of the relation of causation were accepted and applied to others which are as original and fundamental, then scientific knowledge would be impossible, and religious faith would be unsupported by any rational foundations. He therefore set himself to the work of examining, by critical analysis, the intellectual powers, to ascertain, if possible, whether knowledge *a priori* is possible, and if so, what must be its original elements and authority. The results of his critical inquiries were as follows: The human intellect may be considered as *Sense, Understanding*, and *Reason*, and to each of these powers or modes of action, there are elements *a priori*. To the *Sense, space* and *time* must be assumed as *a priori* conditions. If these are not thus assumed, neither perception nor consciousness could possibly gain the knowledge appropriate to each. Moreover, unless the knowledge of both space and time is *a priori*, the mathematical sciences would be impossible.

The *Understanding* is the power of generalizing and of logical reasoning. To this, certain forms of conception are also necessary as its *a priori* conditions, such as *substance and attribute*, and *cause and effect*. Without these forms *a priori*, the processes of the *Understanding* would be impossible and their products would be untrustworthy.

The *Reason* is the power by which we give unity to our knowledge of both material and spiritual phenomena, as well in the several portions of each, as when these portions are mutually connected and related with one another. To this unifying process, there must be assumed, as necessary presuppositions, certain ideas *a priori*, viz.: *the soul, the external world*, and *God*.

The *a priori* elements of our knowledge, according to Kant, are the receptivities of *space and time* for the *Sense; the forms or categories* for the *Understanding;* and *the ideas* for the *Reason*. That these elements are assumed and applied in all our higher knowledge, was shown by Kant to follow necessarily from the analysis of that knowledge which is gained by the intellect, and indirectly from the direct analysis of the operations of its several powers. These were the positive results of his psychological analysis.

But Kant raised another inquiry. Are these *a priori* and necessary assumptions themselves worthy of confidence? Are they true, and do they hold good of the nature of things, or do they simply arise from the constitution of the human intellect—a change in which might involve a change in these necessary relations and in the knowledge which is built upon them? To these questions of his own asking, Kant makes the following reply: These assumptions have for man a *regulative force*, but perhaps only *a relative truth and validity*. That is, while man must act in his intellectual processes under the belief that these principles are primary and universal, and thus admit them as giving law to his own intellect, and as grounding and explaining all his knowledge, he is not authorized thereby to assume that they hold good as the laws of those minds which may be supposed to be constituted differently from the human, or that they hold true of the knowledge which such minds acquire. On the one hand, we cannot deny that they do hold true for other beings and their knowledge; and on the other, we cannot deny that they do not. For aught that we know, it may be true, that other beings might be so constituted as not to assume these principles, or to know by means of the relations which they involve. We cannot affirm that there are such beings. We cannot deny that there may be. We cannot conceive how there should be. We cannot imagine intellectual processes that do not run back into these relations and principles, nor can we conceive of any knowledge which is not held together by these relations, but we have no rational ground for denying that both are possible.

This is the last result of the critical examination to which Kant subjected the intellectual faculty. These views have had extensive currency among the philosophers of Germany and England, and the assertion of them has wrought like leaven, to stimulate inquiry and to excite to counter assertions. Many who would not accept them have found it difficult to show their groundlessness or their untruth, in part or in whole. Many philosophers who have followed Kant in his analysis of the foundations of our knowledge, have felt themselves constrained to enter a special protest against these views, or to seek to vindicate a different theory.

The only part of Kant's theory with which we are here concerned is the suggestion which he makes, that the relations and principles which we find to be original and assume to be true for our own thinking and knowledge, are not necessarily true and valid for all thinking and all knowledge.

Concerning this we observe:

(1.) It is a question of Speculative Philosophy or Metaphysics, and not at all a question of Psychology. Psychologically considered, the views of Kant do not differ materially from those of other philosophers so far as the proposition is concerned, that certain truths must be received as universal and necessary, and that these are given to the mind *a priori*. It is one chief object of his *Critique* to show that such principles are not obtained by experience, but must be assumed in order to make experience possible, as without them we could have neither experience nor science.

That which he subjoins to this ascertained result of psychological analysis, is the suggestion that this may be true in human psychology only, and not in the psychology of other knowing beings. Whatever may be the probability or reasonableness of this suggestion, it is in no sense a psychological fact. It is purely a philosophical thesis, to be urged and defended on speculative grounds.

(2.) This metaphysical suggestion or thesis is unsupported by any grounds of analogy or probability. The facts which suggested the thesis are the known changes in the objects of sense-perception, which are connected with known changes in the organism of the percipient or in the medium by which this percipient apprehends. These changes are most conspicuous in vision. An object seen through a colored lens, be it red or green or blue, is seen to be red or green or blue. In like manner, the color of objects is, to a limited extent, affected by changes in the physical condition of the eye. Some men, through disease, see objects colored as they are not in reality. Others are incapable of seeing any differences of color, or at best, only a few varieties.

Upon analogies derived from these facts, Kant justifies himself in asserting that there may or might exist created or finite minds which know other relations than those of *time, space, substance, causality*. To this it is enough to reply that the facts from which these suggestions are derived are phenomena of the corporeal organism—while the acts and objects to which they are applied by way of analogy pertain to the pure intellect. We know moreover of the phenomena of the organism, that the corporeal organism is a factor which, with material conditions, not only *presents* the object for the mind to perceive, but makes it to be what it is to a certain extent, so that the object changes with its changing factors and conditions. But to these thoughts or intellectual relations no such conditions are required. Certainly the objects are not known to change with any conditions. So far as these relations are applied to material objects it makes no difference what the objects are. Many are equally applicable to spiritual beings, and their phenomena, products, and trustworthiness cannot be weakened or set aside by analogies derived from material beings and phenomena.

All positive grounds for applying any analogies of the kind are found to be wanting.

(3.) The suggestion of Kant is inconsistent with, and overthrown by, the reach and necessary use of some of these very relations which are brought into distrust. It is open to the charge of being an intellectual *felo de se*. For example, all *the positive ground* for the suggestion, founded upon an analogy which we have seen to be *invalid* because *irrelevant*, rests upon one of these first truths themselves, one of these very original relations, which Kant subjects to metaphysical doubt as to whether it may not be merely contingent upon the human constitution. It is perfectly clear that the question which he raises, is whether knowledge by these relations as a subjective process, and the relations themselves as objective facts, may not be and probably are, *effects* of which the human constitution is *a cause*. We notice also that the reason by which he supports his suggestion is, that we are justified in so interpreting—which we have shown is misinterpreting—certain signs or indications furnished by analogous phenomena. In this argument it will be obvious to all our readers who accept the analysis which we have given of induction, that the assumptions which he contends are only *regulative*, are used and applied by him as though they were *real*. He certainly applies with entire confidence, the relations of *cause* and *effect* as necessarily and really applicable to the constitution of man as viewed by all beings whatever, and wholly omits to notice that he has suggested that these relations necessarily employed in human thinking, are merely contingent upon the accidents of that thinking, and may not belong to the constitution of the soul as viewed or known by any other being, whether creature or Creator.

This is not all. Not only are they used as though they were real, but they are used as real in order to prove that they are only regulative. He reasons thus: Upon the principles to which I must conform as the laws of my human thinking, do I conclude that it is more than probable that these principles themselves are true of human thinking only. How convincing and consistent such reasoning is, it is easy to see.

Hamilton's Positive and Negative Necessity.

§ 257. 7. From Kant to *Hamilton* the transition is natural, because the connection between their views is most intimate. Hamilton holds that our native cognitions are both *Universal* and *Necessary*. The Necessity of a cognition may, however, be of two species. It may be either *Positive* or *Negative*. It may either result from the power of the thinking principle, or from the *powerlessness* of the same to think otherwise. Of *Positive Cognitions* he says: "To this class belong the notion of existence and its modifications, the principles of identity, contradiction, and excluded middle, and the intuitions of space and time." All these are discerned by the mind by a necessity which positively pertains to the objects discerned and in the reality of which the mind absolutely confides.

To the other class belong the relations of *Substance* and *Phenomena*, and of *Cause* and *Effect*. These are necessary through the imbecility of the mind to conceive of existence in any other way than under these relations. This necessity is only a special case of the application of the more *general law of the conditioned;* which in its turn is described as the necessity which constrains the mind to think of every object as a medium between two extremes, each of which is respectively contradictory of the other and so both cannot be true, while yet the mind must think the object under one of the two.

The exposition and discussion of this Law of the Conditioned may be deferred till we consider its application to the special conceptions and relations of Cause and Effect. (Cf. § 297.)

It is enough to say here, that it seems to be in its principle the same with the doctrine of Kant, that certain cognitions are necessary to the mind because of its peculiar constitution, which would no longer be so in case this constitution were changed or other than it is. They are therefore Regulative only, that is, they control the actions of the human mind and their products, because we cannot avoid employing them, knowing all the while that we are obliged to do this because we are finite. They are true relatively, *i. e.*, true only in relation to our limited capacities.

We urge against this substantially the same objections to which the doctrine of Kant is liable, viz.: that we must use these very conceptions which are said to be merely Regulative and Relative, in the very judgments which we form of the mind and these very relations; and again, its tendency is skeptical, like that of Kant. It ought to be regarded with distrust if for no other reason than that it introduces contradictions between the decisions and dicta of the separate activities of the intellect.

The theory of Faith as contrasted with knowledge.

§ 258. 8. To meet, or rather, to shut off, the difficulties propounded by Kant, and in part assented to by Hamilton, *Faith* has been proposed as the source of certain original conceptions and primary beliefs. Sometimes *Feeling*, or some act more akin to the emotive than to the intellectual powers, has been urged as the originator and voucher of the primary beliefs, and indirectly of the knowledge which is built upon them. This faith or feeling has most usually had for its object or objects, the *Absolute*, the *Infinite*, or the *Unconditioned*, rather than the ultimate conceptions under which finite existences are thought by the mind and the primary relations by means of which these existences are classified and connected. *God, the Soul, Time, Space, Immortality*—have been usually the objects which it is asserted are received by this original assent of Faith or Feeling. Sometimes the moral relations have been conceived as the direct object of the soul's apprehension, together with God and the soul. The tendency to cut the knot which an intellectual analysis has failed to untie, is most conspicuous as perpetually reappearing in the entire history of modern philosophy. The need of an ultimate and decisive authority for our confidence in the actings of the soul, has often prompted to a *coup de main*, by which some usurping power, under the fairest names, has seated itself in the place of rule, and the usurpation has been acquiesced in, by reason of the temporary peace and order which has followed in the intellectual convictions and the received systems of science, morality and theology.

Descartes, having vainly sought for some criterion of truth which should assure him that his senses did not deceive him, and that his judgment in regard to his spiritual operations might be trusted, found repose in the veracity and benevolence of the Great Creator, of whose existence he was assured by the innate idea which attests both his existence and his perfections. This being given, the cognitions and inferences of the intellectual faculty may be trusted, when they are properly tested by the criteria or norms which the Creator himself has provided.

Kant, after despairing to find in the speculative Reason any warrant for trusting those necessary cognitions which are universal to all men, and assumed *a priori* as the conditions of all experience and all science, finds in the *categorical imperative* of the *Practical Reason* a voucher for the law of Duty. Unconditional faith in Duty was the corner-stone of his system, the only sure foundation which he could find among the ruins into which he had disintegrated the structures of the merely speculative Intellect, and upon which he could rebuild the same and make them compact and safe. Faith in Duty requires faith in God to defend and reward Duty. Hence the same Practical Reason which commands us categorically (*i. e.*, unconditionally, and without asking or finding reasons or grounds) to believe in Duty, commands us to believe there is a true and perfect God. But such a God will not deceive his creatures. If we trust in Him we may confide in the speculative testimony of the Reason which he has constructed and created, concerning those conceptions which it originates and requires; and may assign them the place which they take and hold in our knowledge, not as being merely *a priori* assumptions under which we are obliged to think, but as being fundamental truths which we must accept as real. By the Practical Reason we allow these *forms of thought* by which we must regulate our thinking, to become the representatives of those *forms of being* which control the world of reality.

Jacobi felt the difficulties in which Kant involved himself and the minds of his generation, but was not content with the solution which he furnished. He adopted another, similar in principle, indeed, but slightly varied in its applications. To the power of apprehending that which is primary and unconditionally true, he gave the names, at first of *Faith*, afterwards of *Feeling* and the *Revelation of the*

Divine, and last of all, of *Reason Proper*. The objects which this power apprehends are not moral and religious objects and relations exclusively; but the objects of sense and consciousness with the relations which they involve, as truly as God, the Soul, and Immortality. These are all received by the direct faith of the soul, and this faith and the truth of what it receives is the precondition of all *analysis, inference* and *deduction:* In all these processes we simply analyze and explicate what is given to faith impliedly and as a whole. Jacobi simply asserted these principles to be the foundation truths of all knowledge. He did not show how they could be true or why we believe them. Indeed, he despaired of any such analysis. He did not feel adequate to illustrate them in the detail; he simply rested in their truth.

Schleiermacher recognized feeling—the *feeling of dependence*—as the ground and medium of all the knowledge of the Absolute that we can attain. But we can neither conceive of God nor define our concepts of him. All efforts in this direction, as well as their results, are entirely inadequate and misleading. So far he is at one with Jacobi. With him he makes feeling or faith the ground of our apprehensions of the Infinite and Divine. In respect to our knowledge of and faith in the conceptions that are fundamental to finite knowledge—he would be foremost to assert that these are *a priori* conditions and assumptions of the intellect, and that nature herself is constructed in correspondence with these forms of human thought: we have therefore the amplest ground for trusting the processes that are essential to our higher knowledge and the results to which they conduct us. The relations of finite existence, including those of *space* and *time*, of *substance* and *attribute*, of *cause* and *effect*, were considered by Schleiermacher forms of existence, or *real forms* in contradistinction to the subjective forms of Kant and Fichte and the notion forms of Hegel. These are apprehended by the intellect directly, or, in the phraseology of his system, by the *intellectual function*, to the operations of which, in connection with the *organic function*, all the forms of finite knowledge are to be referred.

Some of the more recent German philosophers, as *Chalybæus*, *Reiff*, and pre-eminently *Lotze*, rest their confidence in the fundamental assumptions of the human intellect, upon ethical grounds. The questions propounded by Kant, viz.: "Suppose after all that the constitution of our nature should itself not be trustworthy when it causes and impels us to think according to these original forms and fundamental assumptions? Suppose that the relations or forms of things, which seem to correspond to the relations or forms by which we think should prove to be unreal?" they answer thus: "We must believe that nature is benevolent in her indications and therefore true. We assume that goodness and veracity regulate both the objective relations of the universe which we study and the subjective constitution of the intellect which interprets it. For these reasons we rely upon the categories of both thought and being, and learn to think in accordance with them, trusting the results which we gain.

As Hamilton (as we have seen): in his views of the extent and limits of our knowledge, followed Kant and Schleiermacher, so he borrowed from both the required solution. While he asserts that we cannot *think* the *infinite* and *unconditioned*, because *to think* is *to limit* and *to condition*, he concedes that we *know* the same. When he is asked how? he replies, by *faith:* we must *believe* in the *Infinite*. The extremes of our knowledge, between which we form our concepts—and out of the relations of which we form our concepts—we must believe exist and are related to one another. The fact of their necessary existence we receive by a direct insight, which he calls both faith and knowledge. He borrows from Kant conceptions that are appropriate to the Practical Reason—so far at least as ethical distinctions, moral liberty and a personal God are concerned. From Jacobi he adopts the term *faith*. With the doctrine of Schleiermacher the details of his theory of the Unconditioned are closely allied. Cf. Hamilton (*Met., Lec.* 38; also *Appendix, Letter to Calderwood*).

That which gives plausibility to the doctrine that Faith or Feeling is the ultimate ground of this kind of knowledge is that it is not received by any act of conscious assent to propositions, of which the elementary concepts are first distinctly apprehended apart and then united, but the mind first believes or knows before it reflectively discriminates its knowledges into their elements. Hence the act is called faith in opposition to and in distinction from judgment, the last being supposed to involve analysis as well as combination. Ethical and religious objects are those which most frequently bring it into exercise, and these invariably excite more or less feeling. Hence the special source of these convictions is conceived as something not intellectual, and is simply called *feeling* at one time, and faith at another. The oversight lies in making these terms to imply that the act is not intellectual. It must be preëminently an intellectual act and power, for it conditions all the special acts and cognitions of which the intellect is capable.

J. G. Fichte.

§ 259. 9. The immediate successor of Kant was *J. G. Fichte*, whose system was proposed as a modification and improvement of that which was taught in the *Critique of the Pure Reason*. Fichte derived all knowledge,—the materials as well as the forms, the *a posteriori* and the *a priori*,—from the activity of the Ego. Every thing which the mind knows, being as well as relations, so far as it is known, is the work of the Ego, and is evolved from its own creative activity.

So far as the categories of thought are concerned, Fichte endeavors to show that each one of them is necessarily involved in the several concrete creative acts by which the Ego constructs for itself the known universe. Its first act is to affirm its own being. But in this it must apply and evolve the law or relation of identity, A=A. Its second act is to affirm the non-Ego. But this in like manner involves the law of contradiction, (A) is not (non-A). The third is to recognize the indivisible Ego as opposed to a divisible non-Ego. This involves the reciprocal activity of each on the other, and this implies the relation of Causative efficiency. The other relations are all evolved in a similar way by the productive activity of the Ego, together with the non-Ego which this activity calls forth. *Time* and *space, substance* and *attribute, reality, possibility* and *necessity,* etc., etc., are all accounted for by the creative activity of the Ego, as it proceeds from the simpler to the more complex processes and products of human knowledge.

Schelling's view of the categories.

§ 260. 10. *Schelling* followed Fichte—by the effort to mediate between him and Kant—so far as to provide for a common origination and relationship for the subjective and objective. His *intellectual intuition* recognizes at first the indifference of both, from which it develops as correspondent to one another the forms of thought and the forms of being. The authority for the categories in this double application must be in that intuition which affirms them to be common to the two. In his later philosophy, which was modified to avoid and displace the logical idealism of Hegel, Schelling assumes the reality of concrete and actual being, and teaches the mind's competence to originate and affirm necessary and original relations only in their application to, and by occasion of supposed concrete knowledge. For this reason he asserted for these *a priori* relations and for philosophy itself, what he called only a *negative* value.

Hegel's theory of pure thought.

§ 261. 11. *Hegel* substituted *thought* for Schelling's *intellectual intuition*, i. e., that mental activity which produces and is concerned with the concept or logical notion; but he made a fatal mistake in conceiving that *thought*, viz., *abstract thinking*, could be explained independently of *concrete* knowledge and actual being, and that the former could explain the latter by the relations of pure or abstract thought. He was therefore compelled, by logical consistency, to endeavor to evolve and explain every form of actual being by the development or evolution of the notion from within itself.

The categories or the original and necessary relations of knowledge, according to Hegel, are all the relations which are necessarily evolved in the process by which simple, *i. e.*, abstract being is developed into the several forms of thought and existence, and through them all, till the absolute is attained, *i. e.*, till the process is complete and with it the cycle of the original relations or categories which are required for its evolution.

Herbart's theory.

§ 262. 12. According to *Herbart*, some of the categories are the products of the action and reaction of ideas. They are not the necessary laws or forms of the mind's knowledge, but are the growth and result of its psychological functions as determined by the laws which govern the formation and mutual action of the results of the impressions made upon the soul by matter, and the soul's reaction against them. These results are perceptions or representations. Concepts, or general notions, arise only when a number of similar objects have been perceived. These different elements in their struggle for reappearance crowd one another out of view, and only those are apparent which, being alike, reinforce one another, and so survive the struggle. The conceptions of Space and Time are series of reproduced objects, the parts of which are more or less indistinct, as they stand related to *the here* and *the now*. A thing or being and its attributes, is either an original whole analyzed into its constituent parts, giving the attribute of quality, or a whole with its attendant series of time and space accompaniments giving the attribute of quantity. The successful connection of these attendant parts or accessory series is affirmation—the unsuccessful is negation: both these involve the two corresponding forms of judgment or the apprehension of relations.

The relations of *substance to attributes* and of *cause and effect* are inconsistent with the logical laws of *identity and contradiction*, which are assumed by Herbart to be original and independent laws of thought. To remove these inconsistencies is the object of his metaphysical system. This he essays to do by "*the method of relations*." It would seem that the logical laws are the only categories, properly considered, which Herbart accepts, for the reason that these logical criteria are applied by him as the fixed rules and original measures by which every other relation is tried and tested.

CHAPTER III.

FORMAL RELATIONS OR CATEGORIES.

The category of being.

§ 263. Following the classification of categories or intuitions which we have adopted and explained (§ 250), we begin with those which we have defined as *formal*. These are also called Logical, for the reason that Logic has to do with the concept as such, *i. e.*, the *pure concept* and its necessary relations. The concept as such consists of those elements, and those only, that must be conceived as present in every object when thought of. That is, it must embrace those elements only which are common to every such object, whether it is a real or an imagined being. These elements, while they belong to things as well as to concepts, are yet essential to the concept and the other entities of pure logic, and hence are referred pre-eminently to the power of thought.

We begin with *being*. This will be readily acknowledged to be the most extensively applied of all the concepts, and therefore *fundamental*. Everything which we know, we know to exist. To know is impossible and inconceivable, if it does not involve the certainty that that which is known, *exists* or *is*. Being is the correlate of knowledge.

In what sense fundamental.

Hence, this concept is apparently fundamental to all others. It belongs to every object with which the mind has to do in knowledge, and it belongs to each with equal propriety—to Him whom we call, in the poverty of our language, the Being of beings, and to the most transient and trivial creation of the humblest of His creatures; to the universe in the most comprehensive meaning of the term, and to the mathematical point, which is the product of the thought of a moment.

We sometimes dignify the being which is independent and permanent with the assertion that this only or truly has being, or only and truly is; but this is by a metaphor only, and does not in the least affect the proper import of the term or of the concept for which it stands. The *positive existence* of the object, but neither its dignity nor its duration, is expressed by the word.

§ 264. Being is the most *abstract* of all possible concepts. After every property or relation which we know of an object is set aside from any existing thought or thing, there remains the affirmation; *this is*. This resulting concept cannot be thought away. For this reason it is called logically the first or the most elementary of all concepts. As it is the last which we reach by *analysis*, it is the first with which our *synthesis* begins.

The most abstract of all the categories.

Psychologically, the knowledge of being in the concrete, precedes that of being in the abstract. We know individual beings before we know being as a concept.

Logically, or, more properly, *metaphysically*, the concept of being is the first and most fundamental of all the concepts, because it is the most extensively applied, and is the highest of our generalizations (§ 249). But it cannot be understood as a concept, except by means of individual objects. To begin with the concept in the abstract, excluding that knowledge which interprets and makes it clear, is literally to begin with nothing. To attempt to develop from it actual being, is to give an example by failure, of the truth, *ex nihilo nihil fit!* Hegel begins the development and explanation of our real knowledge with the concept of being in the abstract, and seeks to construct and develop from this the conception and knowledge of real existence, and the relations which it involves. In doing this, he is obliged to interpret his meaning by a tacit assumption of that which he formally ignores and denies—*i. e.*, to draw upon direct and presented knowledge for the interpretation of the conceptions and relations which he professes to develop and account for. The attempt is vain; the method is false; the solution is impossible.

The knowledge of being is expressed by judgments or propositions, the subjects of which are known individually. We tacitly assert or think of every such object; *it*, or *this*, *is* or *exists*. From these we generalize the concept—*being*. Being or existence is not, however, an attribute or a relation, though it is conceived or treated as such when it is thus generalized. It is obvious that being must be assumed in order that an attribute or relation may be known.

§ 265. Being cannot be defined—*i. e.*, resolved into any more elementary constituents. It can be *de-*

Is indefinable and indeterminate.

scribed, however, by the conditions or circumstances under which it is present to the mind: When we ask, What is being? we cannot answer in the way of definition. But inasmuch as whenever we know we apprehend being, by referring to the act of knowing we understand, though we cannot define, the import of the concept; *i. e.*, we explain the concept, being, by the act which involves and supposes it.

It was said (§ 196) that all concepts are founded on attributes or relations generalized, and that the only difference between nouns and adjectives arises from their use and not their meaning; the same content being present in every case—a content of attributes only. How, then, it might be urged, is it possible that there should be any concept of being at all, if being is not only not an attribute, but is the direct contrast of an attribute and must be supposed to make an attribute conceivable or possible? This inquiry has in part been answered. In order to be turned into a concept, being is treated as an attribute; it is predicated of the individuals to which it belongs and thus is made to suggest itself as essential to any relation. It is worthy of notice also that some fixed permanent attribute as of *standing*, etc., is usually selected to image or represent *beingness*.

Simple being is a concept wholly *indeterminate*. It stands for itself and for nothing besides. It is supposed in every other. It must be assumed to determine every other. We must begin with being, before we can add a single characteristic to make it definite.

This is what Hegel had in mind in his assertion: *Being or entity is equal to nothing*, *i. e.*, it is equivalent to a notion without content. As an abstract conception, it has no relations to any other concept, and consequently no attributes; it is wholly undefined. "Being, the undetermined, immediate object of knowledge, is in fact *nothing*, no more nor less. Nothing is [has] the same determination, or rather, absence of determination with, and, for that reason, is equivalent to, simple entity. Hegel, (*Logic*, vol. i., p. 22; *Encyc.*, p. 406.)

But though being, as a concept, and in its relation to other concepts, is indeterminate, it is not without signification. The concept is taken from and affirmed of and interpreted by, individual beings which we actually know by direct knowledge.

§ 266. From *being* we pass to *relation;* both existence and relationship being involved in the act of knowing. By relations, individual objects, as well as concepts, are distinguished and connected. But relationship involves diversity in the *concept* produced, and negation as the judgment by which diversity is affirmed.

Relationship. Diversity and similarity.

Two entities—*i. e.*, objects apprehended—are essential to the apprehension of a connecting relation. But if the two are known they must be distinguished—*i. e.*, known as different from each other, in order that they may be again connected.

It follows that the relation which is the most extensive of all others, is the relation of *diversity or difference.*

In every act and object of knowledge two relations are supposed, those of diversity and of similarity. If there is more than one concrete Being, one is diverse from the other. If both are alike Beings, *i. e.*, are comprehended under the concept Being, they must be alike at least in that they are both knowable. In brief, diversity and similarity—*i. e.*, logical or formal *sameness*—are everywhere present. This truth is asserted in the proposition, that every act of knowledge is at once an act of analysis and of synthesis. In every single act of knowledge we separate—*i. e.*, distinguish—in order that we may combine. We can only unite so far as we separate, and we unite by similarity.

The relation of difference or diversity is expressed by the proposition, *this being is not that.* A is not B, or B is not A; the color is not the taste, the taste is not the color; the pictured moon is not the mind, the mind is not the moon which it pictures. I am not the object seen or tasted, etc., etc.

It will be remembered that these propositions are all individual propositions, and that none of them are or can be general. The individual goes before the general in these propositions of relations, as in all others.

From the recognition and affirmation of relations in general are evolved what are called *relative concepts or notions.* From the negative proposition which expresses the relation of diversity are produced what are termed *negative concepts.*

Relative notions. Negative notions.

No sooner is A distinguished from B, than we can apply to it the negative notion of *not-B.* In the same way reciprocally, the notion *not-A* can be affirmed of B. These two notions are purely *relative.* The whole content or import which they express, is limited to the single relation in which they stand to the other object, which other object, A or B, as the case may be, is supposed to be positively known.

In like manner, *other relative notions* may be formed, as if we take a substance and it puts us to sleep, we conceive the unknown something which produces this *sleep-making;* that is, we need know it no further than by its relation to this effect. The only notion which we have of it may be purely relative to the known effect.

The negative relation, as indeed any relative notion, is at first apprehended as individual, and then generalized. No sooner is A pronounced to be *not-B,* than we proceed to apply this to C, D, E, F, etc., as well as to A—indeed, to all objects except B itself. We need know nothing more of them than that they *are,* to be justified in classing them all as *not-Bs,* or in affirming of them the negative concept thus generalized. This is the ground of the division of all real and conceivable things by *dichotomy,* as it is called.

It will be observed, however, that negation expresses a relation between two actual beings, or two beings treated or conceived as real. It supposes two positives known or conceived, each of which is thought as related negatively to the other.

The concept *nothing—nonentity*—is a purely relative concept. All being or entities, whether real or imaginary, are grouped under the most general of all concepts. To this is attached the relation of negation. What is expressed, is the proposition that the concept is exhaustive, and that it is impossible to conceive or believe in any thing beside. By a fiction of speech and of thought this proposition is contracted into the concept *nothing—nonentity*—as though there were a really existing object negatively related to being. To form it we group all known or knowable objects under the general concept of being and attaching to this the negative particle, make *not*-being=*no* thing=*nothing.*

When Hegel asserts that the concept *being* or *entity* equals *nothing* in its import, he has in mind that it is a concept which cannot be analyzed into any constituent concept or thought element: it is therefore unrelated to any other; it is undetermined: it has no notional or formal content. So far from being true that this concept has no import, no concept has an import so extensive. Its import is reached in the various forms of direct knowledge, which furnish the material and meaning to every concept, and a reference to which is supposed every time the concept *being* is used.

Hegel reasons that, because the concept *being* is the *summum genus* among concepts, it is the originator of all other concepts: not only so, but by the law of self-evolution, it is the originator of things or actual beings. The failure of the attempt, and the absurdity of the theory on which it rests is manifest when the effort is made to cross over from the notion world to the real world; when the effort is essayed to evolve time and space, matter and spirit from concepts only. The effort seems to be successful only because the real world with its relations is ever ready at hand behind the concept world which symbolizes it, to furnish the signification which is required. Real being, and real relations are very easily confounded with the generalized concepts of the same. The two are easily interchanged, and it is by a kind of intellectual juggling or slight-of-hand that any success appears to be attained, or any conviction is produced.

Substance and attribute formally conceived.

§ 267. Diversity or negation is applied *to a being as distinguished from its relations, to one relation as distinguished from another relation,* and also *to one being as distinguished from another by means of its relations.* We distinguish or separate objects from one another whether material or spiritual: *first, in real knowledge,* by *intuition or direct inspection; next,* in *thought knowledge,* by employing relations for this purpose, and especially those similar relations by which beings are grouped under concepts.

This introduces us to the category of *substance and attribute,* so far as it is *merely formal.* Whenever a being is thought of, *i. e.,* is distinguished from another being by the number and the extent of its relations, then we have the relation of *substance and attribute in its pure or abstract form.* A *substance* formally conceived is *a being distinguished by certain relations.* An *attribute* is *one of the relations which thus distinguishes a being.*

Every concept whenever it is complex, as having a definite *content,* implies the relation of a *whole* constituted of and separable into *parts.* This implies the relation of *more* or *less.* The *extent* of a concept, as applicable to more or fewer objects, and therefore as higher or lower, implies the same relation. The relations of wholes and parts and of greater or less are properly *formal* relations, as involved in the very nature of the concept. They are relations of formal or logical quantity, which is distinguished from mathematical quantity by characteristics subsequently explained.

The relation of diversity with its several applications suggests the relation of *identity.* In affirming that A is not B, or is diverse from B, we imply that A is identical with itself. That the mind comes to the distinct recognition of this relation at an early period of its development, and makes frequent application of it afterwards, is too obvious to need confirmation. That the relation is original, and is intuitively discerned, is equally clear.

If a concept is known as identical, it is of course implied that the individual beings to which it belongs have similar relations in common. These individuals cannot be distinguished, except by means of the relations of time and space, which are conceivable as possible of either, but not of both together. One concept is distinguished from another by the relations which make the *content* and determine the *extent* of the one and the other.

Many hold, that the first object to which identity is applied is the soul itself, as distinguished from the diverse states of which it is conscious. As the Ego distinguishes itself from its changing states, it knows that the states are varying, but the Ego is the same. In doing so, it must compare itself at one time with itself at another, or itself in one state with itself in another.

Identity again may be affirmed of a material object, as of a house, a ship, a tree, or a horse. In such cases the objects are perceived at different times at least, and are often changed in form, appearance and properties. The test or standard of identity may be real and natural, or it may be conventional and factitious. But the relation itself is not thereby altered.

Identity may also be applied to a purely mental product. Often it is interchanged with similarity, when it is applied to a *concept*, *e. g.*, I have a similar image of the same object which I previously imagined or perceived. It is not necessary that the concept should be formed by all men from the same individuals, but it is meant that the similarity between the individual objects is so perfect that one individual may be substituted for another in forming it, and that it may be applied to one as freely and as properly as to another. When it is thus applied it concerns the relations of *content* and *extent*, and signifies that the same definitions and divisions are applicable in every case.

The logical axioms of Identity, etc.

§ 268. *To guard against using concepts in different senses in any of the processes of thought, the law of identity, the law of contradiction and the law of excluded middle are set forth as the three fundamental laws of thought, i. e., of formal thought.* These respect the identity and diversity of concepts only. They are the axioms of logical thinking, but not necessarily the rules for every form and mode of knowledge. They are such practical rules as have been found necessary from the dangers to which men are exposed from the various forms of expression in which concepts and their relations are phrased.

The *law of identity* is designed to avoid the twofold danger of supposing, on the one hand, because the diction is altered, that the concepts, propositions, and reasonings are changed, or on the other, that, because the phraseology is similar, the meaning is the same.

Complex concepts only can be tried and tested by this law; and these can be tested both in their content and extent. The law applied to the content asserts that a concept is, for purposes of logic, the same with the sum of its constituting elements: A= (a, b, c, d, and e); *i. e.*, all these being taken together, the one is convertible with the other. When applied to the relation of extent, it asserts that the concept as genus is identical with the total of its contained species or subordinate parts. To make the logical law of identity the mere meaningless truism,—A is A, *i. e.*, that a concept in the same form of diction is identical with itself,—is inept and absurd.

The logical axiom or *law of contradiction: A is not not-A*, is only a generalized application of the intuition of difference to any concept whatever, taken in both extent and content. A thing or a concept is not another, it is not any one of the things or concepts from which it differs, nor all of them united. This truth, expressed as a *rule*, requires that the concept "should never be confounded with or substituted for either."

The *law of excluded middle is, every B is either A or not-A*. This is another application of the intuitions of difference and identity when generalized. When A has been distinguished from not-A, it is at once discerned that these two concepts divide the extent of all conceivable existences into two classes. This truth is then stated as a principle; which is ready to be used as a law whenever it is required to guard or correct our thinking.

Much evil has resulted from the error of taking these three logical laws as the original and the only laws of our knowledge. It was entirely natural for philosophers who were practiced in the schools of formal logic to suppose that everything which man believes to be true could be demonstrated by the methods and after the principles of the syllogism. The tenacity with which this persuasion has been adhered to is most remarkable in the history of all systems and schools of thought. For a long period after the revival of philosophy it seemed that man would never cease to attempt to give a logical demonstration for the very axioms and principles on which all demonstration must rest. Logical proof was required for all knowledge, for the belief in a material world, for our confidence in memory, for the distinction between the facts of experience and the illusions of the imagination; in short, for everything known or believed by man,—and to logical proof these three laws of thought were assumed as the axioms. Hence, the attempt was persistently made to found upon these laws *the whole structure* of *human* knowledge, and to deduce or demonstrate from them, the validity of this knowledge in all its forms and applications.

CHAPTER IV.

MATHEMATICAL RELATIONS: TIME AND SPACE.

We proceed to consider the mathematical categories; or those relations which involve the belief in time and space. These relations are of the most extensive application. They all must in a sense be recognized in every act of consciousness and perception. By means of these, material and spiritual objects are parted and united, are individualized and generalized. They suggest the space and time which are infinite and absolute—the correlates of limited time and limited space. In order to relieve the treatment of the subject as much as possible, we will consider them first under their more familiar aspects and relations, and afterwards in those which are more recondite and difficult. We begin with

I. *Extension as given in Sense-Perception; or the relations of matter which introduce and require the knowledge of Space.*

Development of the several relations of extension.

§ 269. All matter is known as extended. The beings or objects of which we become cognizant in the use of the muscular and sensorial apparatus are extended. The percepts and things which are presented to the sensorium as eye and ear and hand, are perceived as extended.

It is not meant that this extension in one or all of its dimensions is known at first as separable from the matter to which it pertains and of which it is affirmed; but as belonging to matter and affirmable of it. All extended objects are known as extended, at least in two dimensions. We cannot conceive the eye and the hand to rest upon or to move along any so-called object without the apprehension of an extended surface. A ball or cube when followed by the eye or grasped by the hand is known to return upon itself, and both are sooner or later known as extended in three dimensions or directions, *i. e.*, as *high*, *broad*, and *deep*. This extension is first known as *outer*, *i. e.*, as enclosing matter. But when the child peeps into a box, or surveys from within, the walls, floor and ceiling of the apartment with which it is familiar, it distinguishes the surfaces which are inner or enclosed by matter, from those which are outer and enclose matter.

After the process of perception is complete by a synthesis of percepts and their relations, the mind proceeds to analyze these elements, and to think of them separately from any single substance. But after disposing of all the qualities apprehended by sense-perception, it still finds a *residuum* in the relations belonging to the inner and outer surfaces of matter as already described. The hand experiments upon these surfaces, and finds them *rough* or *smooth*, etc. The eye discerns them as variously colored, as *light* or *dark*, etc. But no one of the senses finds what we call their *extension*. There is no sense-perception to which this is appropriate, and over against which this may be set as a quality. Moreover, this very property involves the recognition of a void, to which it is also conceived to have constant relation.

What is this void which we call space? What is that property in matter which requires the recognition of space? We may find further aid in answering these questions, if we consider first the attributes and relations which involve the kindred questions in respect to time.

II. *Of Time as apprehended in consciousness; or, the relations of events which introduce and involve the knowledge of Time.*

§ 270. *Every psychical act or state,* whether apprehended more or less distinctly as a part of the whole series; and the entire series viewed as an unbroken whole, are known as *continuing or enduring.*

Duration, how related to the acts of the soul.

How soon, or whether it is by the gradual discipline or the instant application of the powers that psychical phenomena are separated into distinct events, we need not inquire. Whenever they are distinguished, the whole and the parts are known as continuous or enduring. An act that is literally instantaneous, a psychical state beginning and occupying no time at all, is absolutely inconceivable. What we call instants are not timeless, but the least knowable or appreciable portions of time. As every object of sense-perception—whether many as one, or one of many

—must be known as extended, so it is with the phenomena of consciousness. Continuance,or duration,belongs to each and to all.

But there are two distinct classes of psychical objects given to consciousness; *first*, the energy of the *ego* by which it manifests its continued, unbroken, and identical life; and *second*, the special activities which change every instant. As the subject of changing activities—the soul knows itself to be living and acting continuously. It also knows itself as acting and suffering in states that change as continuously. Some of these states may seem also to coincide with others, as one continuous or successively repeated act of knowledge may run side by side with two or more diverse states of feeling.

Upon this continually existing and proceeding life of the soul, all its special activities and states are *projected*, as it were; as one portion of extended matter is perceived over against the background of other matter more extended than itself. These activities thus connected are known to exist in a series involving the relations between one another of *now*, *before*, and *after*. These relations are applied first of all to the individual activities of the soul. But just as we speak of portions of matter as *here*, *there*; *before*, *behind*; *within*, and *without*; so we apply these time relations to the states of the soul. As we find one portion of matter included by or including other portions, so we can cut off a single portion of the continuous life of the soul by voluntary or involuntary effort, and contemplate those states which are included within, or are excluded from it.

Time may be conceived as *void* of psychical phenomena; as space is void of material beings and acts. Not that time can be absolutely void, but portions of the soul's existence can be considered as such, in the sense explained. But it is not at all essential to the knowledge of events in the relations of time, that time should be distinctly conceived as void. We can know events as past, present, and future, by considering each of them as successive phenomena of the continued life of the soul.

We have to do thus far only with time-relations in the concrete, and as given in consciousness. By consciousness as here used it is obvious we do not intend merely the power or the act by which the soul knows its own states as present and immediate. In this sense we cannot be conscious of duration. We must include some use of the representative power in respect to past and future events, as well as the belief that what is represented, was or will be actual. Consciousness must be enlarged to this extent of meaning, before it can connect objects in the relations of time.

III. *Of the mutual relations of Extended and Enduring objects.*

§ 271. Material objects, as we have seen, are apprehended by sense-perception as extended. Spiritual acts and states are known in consciousness as enduring. But sense-perception and consciousness occur in fact, as two elements in the same psychical energy or state. As a consequence, the relations of extension and duration are intimate and interchangeable, and the conceptions and language originally derived from and appropriate to the one, are appropriated to the other.

The mind discerns extended and enduring objects together.

First: The relations of time are transferred from the activities and phenomena of spirit, to the activities and phenomena of matter.

Duration or continuance is, as we have seen, originally discerned of the activities and phenomena of the spirit. To these the relations of time are directly and properly applied. When these relations are affirmed of more than one object, whether of matter or spirit, the intervention of the memory of the observer is required. We cannot say of the trotting of a horse, of the flight of a bullet, or of any other motion, that it continued so many seconds or minutes, without supposing the observer who is all the while looking on, to translate the objects really taking place into *objects* as *perceived* by *himself, i. e.*, into *results of acts of his own*, each *enduring so much time.* Material acts or phenomena must be connected by the soul's subjective activity that they may be recalled. Moreover, whatever may take place in the series of objective or material acts; that which is unobserved is totally omitted in the estimate of time: to the mind as enduring it is, as though it had not been at all. The relation of time can neither be applied, nor thought of as applied to any material acts or events, except through the medium of the duration of some person who has first applied to them his own spiritual experiences either in fact or imagination. Every such application, when fully translated or explicated, is made as follows.

While I was thinking or observing for so long a time the, horse trotted or the bullet sped for the same space of time.

Second: But though duration, as a spiritual experience, is the ultimate standard or measure; the actual measures of the duration even of spiritual phenomena,—are taken from the objective or material world. The reason is obvious. Any standard furnished from individual and spiritual experience must be so indeterminate to one's self as to be useless, and, moreover, must be wholly inaccessible to every one besides. Though, in our ultimate analysis, we say to ourselves, "While I was thinking and feeling so and so, the pendulum vibrated, the horse ran, the bullet sped so or so long," yet it is practically impossible for us to fix and render familiar any individual or often repeated series of thoughts and feelings, so as to use it as a standard even for ourselves. Even if we could do this for ourselves, we could not bring it within the reach and use of others. But two individuals, and a great number of individuals, can observe the same vibrating pendulum, the same advancing and retreating shadow on the dial, or the same rising and setting sun, and can use these as standards to measure all phenomena whether internal or external.

Third: the language of duration is taken from material and extended objects, for a similar reason. In fact and from necessity, all the relations of time are expressed in terms originally appropriate to material objects, and the relations of extension which they involve. *Long, short, before, after*, etc., were first applied to material objects, and from them transferred to the relations of time. As will be seen hereafter, this is but a single example of the necessity by which the language and terms of every kind that are applied to spirit and its relations must be derived from space-objects and space-relations.

Material objects are not only known to be extended, but as measuring one another, *i. e.*, as susceptible of quantity. Quantity supposes the inquiry, How much? How many? or, How great? It has for its answer, So much, So many, So large—referring at once to some object which as a unit or standard measures a whole. The extended material universe, as at first vaguely and confusedly conceived, is unbroken, having only superficial extension. By the process of sense-perception it is soon broken into separate objects, each of which may be compared with the whole, in respect to breadth and the other relations.

As extended objects divide and measure one another, so one or more separate acts or states of the soul which follow one another in a series, may be contemplated as dividing, and yet making up this whole, the whole of time being constituted by the continued activity of the soul during these its different acts. Measure in the general sense, as applied to spirit objects and material objects, implies the relation of *whole and parts.* This relation, as we have seen, is involved in the analysis and synthesis of sense-perception, consciousness, representation, and thought, etc., and is essential to the very process and product of knowledge in every form, and hence belongs among the *formal* relations. Measure, in the more exact sense, we need not say, supposes *number.*

IV. *Of extended and enduring objects as Imaged or represented:* or, *space and time objects as enlarged and measured by the Imagination.*

Limitations of sense-perceptions.

§ 272. Only a small portion of the material universe is apprehended *through the senses* by any single act of the mind. The hand can cognize an object of only equal extent with itself. The eye has a far wider, but still a very limited range. All beyond either, is apprehended and measured by the representative power. Even within the limits to which the eye reaches, and upon those very objects which the eye seems to command, the representative power is largely employed in estimating extent in the dimensions of distance and size.

That which is before the eye is the utmost which the eye can in any sense be said to perceive, and much even of this extent is estimated by the eye of the mind. The objects within the reach of the hand and the direct inspection of the eye, we measure by selecting some one as a unit, in the manner explained. Those beyond these bounds, we measure in a similar way, with this difference only, that the material measured, and the standard by which it is measured, are furnished by the imagination only, working upon the suggestions or occasions which perceived objects furnish. We seem to perceive the real height of the lofty tree that shoots up from the horizon against the sky, while it is but a mote to the eye; we think we perceive the width of the stream that threads the distant meadow with a silvery line, but these estimates are possible only by the aid of the picture-making power, that brings its objects by the side of the tree under which we stand, or upon the margin of the stream where we sit. We have already learned, in considering the acquired perceptions, that it is only by the aid of the imagination that we supply the defects of the senses, and interpret their indications.

Beyond these we use the imagination.

§ 273. We are dependent upon *the imagination* alone for our estimates of distance and size beyond the limits of actual perception. These estimates vary with the actual knowledge which we have gained of such objects by inspection and can recall by the memory, and with the practice which results from the frequent application of definite standards by the representative power. The adult surpasses the child immeasurably in this power. The man of various observation and of disciplined powers excels the man of limited knowledge and of untrained habits; the modern, instructed and taught as he is, presents a very striking contrast to the wisest of the ancients.

A child between three and four years old, of no inferior intelligence, and of good opportunities for instruction and thought, was once asked how far distant the sun sets, and answered promptly, In the next field. This child had walked and driven for miles in every direction from its home, and would have remembered, if prompted by leading questions, that all the roadways along which it had gone were bordered by adjacent houses, fields, and gardens, like those within sight, but it had never learned to combine these objects by imagination or to measure such a whole by the unit of a familiar standard so as to estimate their relative dimensions.

The conceptions and estimates of the uncultivated man are very like those of the immature child, especially if such a man is confined by his habits of life to a single narrow valley or a limited range of travel. Every thing beyond these limits is confused and unmeasured. The horizon of his actual perceptions, or the slightly enlarged horizon of his expeditions for hunting and war, includes all that he knows or soberly imagines. He may at times fill the blank vacuity beyond, with objects that are monstrous, horrid, and grotesque—objects that are terrific to his unintelligent fears, or are bewildering to his insane expectations ; but he fixes on few or none which hold definite or rational relations to others as measures or bounds. The spatial world formed by both child and savage, is well represented by the rude maps of the early geographers, in which the countries actually traversed are drawn with a certain degree of definiteness, though the near is out of all proportion to the remote ; but the regions beyond are a blank bounded by an uncertain line, along which uncouth monsters are placed, or the unknown and measureless water or desert shuts in the picture.

The child and savage neither think nor care how large are the sun and the stars, or how many are the steps, the miles, or leagues, which would be required to reach them. In this way, and in this only, can we explain the very inadequate conceptions on these subjects which the early astronomers accepted.

Measures of time-objects are imaginary.

§ 274. Our estimates of time-objects, like those of space-objects, are largely the work of the representative faculty. The passing and present acts and states of our own spirits, and the coincident operations and phenomena of the material world, are the only time-objects of which we have direct cognizance. Past objects are gone. Future objects do not yet exist. Present objects alone directly confront the mind. The past must be recalled by memory, the future must be anticipated in the imagination, *i. e.*, both must be re-presented to the mind, so as with the present to complete the series of time objects.

Different capacities in different men.

To measure past events, we must be able to recall them in their order, so as to have before us the material which we are to estimate. But men differ greatly in their capacity to revive past objects in their fulness and order. If the capacity to recall with success be possessed, time and effort must be added that any past series may be restored, so as to be estimated and measured. Some self-discipline and practice are required that a measure may be prepared from our inner experience which shall be ready for use, and also that the same standard shall be uniformly applied.

Differences in the estimates of time.

Differences in both these particulars in different persons, and in the same persons at different times, account for the singular differences which are so notorious in our estimates of time. No fact is more generally accepted, than that two series of events may occupy the same length of time as measured by the clock, and may seem to vary very greatly from one another as measured by the mind. If we are waiting impatiently for the arrival of a friend or a railway train; or if we are listening to a tiresome conversation or a tedious lecture, the time seems very long. On the other hand, if the conversation is interesting, or the pastime is absorbing, the time flies swiftly along. The child cannot believe that the hour has come which calls him from his play, to school or to bed. A trip by a steamer seems much longer than a trip by railway, when the time is the same. Each are sensibly shortened if the tedium is beguiled by spirited conversation. A week spent in the daily routine of regular employment, goes quickly by; while a week of constant traveling, filled up by a rapid succession of exciting objects, often seems surprisingly long. The years of childhood glide slowly away. Every day and every month stretches to an interminable length, because our present enjoyment brings no disappointment, and because it stands between us and some future happiness which the mind is impatient to grasp. The years of our busy middle life slip hastily by, though we would fain delay their flight, because we are too busy to measure the passing years.

The constructions and measurements of space and time which we have thus far considered, do not involve definite relations of number and magnitude. They are made for practical use and convenience, and require only general impressions of their time or space relations, or a ready reference to some familiar object or series as a standard of measurement. The mind judges the time spent in one occupation to be about as long as the time spent in another. 'It took me about as long, or twice or half as long, as to do this or that familiar act. The distance from A to B is equal to the distance from C to D; or it may be greater or less.' But when we say London is 3 or 4,000 miles from New York, or, the moon is 238,650 miles distant from the earth; or, Washington and Napoleon were born and died so many years after the birth of our Lord, we apply measurements of a different character, by means of definite standards of both space and time.

It is interesting to notice in this connection, the history of the progress made by the human race in the standards of both time and space. The savage measures time by the budding of the oak, or the return and departure of birds or other game. By and by he marks the coming and going of the moon. Then rude devices like the clepsydra or the sand-dial are introduced. Last of all, the scientific observer employs the chronometer and the astronomical clock.

So, in standards of length, the mind has passed from the use of parts of the body, to measurements by the aid of the pendulum, or a portion of a circle of the earth, in order to find an accurate and trustworthy standard.

Whence standards for both space and time are derived.

Standards of both space and time are derived from material objects, real or imagined. No images can be formed of space or time as such, or of what are sometimes called pure or empty space and time, but only of those objects or events which hold a relation to either or to both. When these are pictured or imaged, they carry with them those relations which the originals necessarily involve, and from which they cannot be severed in reality or in thought (§ 206). Thus, for a standard of space, the words *yard*, or *rod*, or *mile*, may call up some visible or tangible object most indefinitely pictured, or with the words, a *minute*, an *hour*, a *day*, or *year*, some series of events that have required a remembered period, or a part of such a period. Both these are pictured, not for their own sake, but for the sake of the time or space which they suggest. But these standards are concepts as well as images, and they cannot be completely understood, even as images, till they are considered also as concepts. This leads us to consider

V. *Space and time objects as Generalized;* or, *the Concepts of the relations of objects to time and space.*

How the relations of space and time objects are generalized.

§ 275. Different individual objects and events hold *similar* space and time relations, whether they are presented to sense and consciousness, or are represented to the imagination. Space-objects may be alike in relative position, distance, form, size, etc., etc. Time objects may be alike in coexistence, in antecedence or subsequence, in their relative place in the order of occurrence, and in the intervals by which they are separated from one another or from any other event. The mutual relations which exist between time and space objects may also be common to any number of both classes.

These relations are as readily generalized as are the attributes of material or spiritual things. It is as easy to generalize the forms and sizes of objects as their color or their taste; the *beforeness* and *afterness* of a spiritual act, as any one of its qualities of knowledge or feeling.

It is true there is this difference: these relations are in their nature incapable of being directly picturable to the imagination, like the properties of matter and spirit. In order to represent them at all, we must first picture the objects which hold them and so recall or suggest the relations themselves. But as concepts these generalized products are as easily formed and comprehended as any others.

The words by which these relations are named and known, are as truly generic as the terms usually called common. All of them, it is true, have a more or less direct relation to an individual place and time, and seem therefore to be less general than the other appellatives; but they are all capable of being equally attributed to many individual objects, and hence are as truly generic as they. We cannot say *here, there, now, before* and *after*, without implying that an individual observer occupying an individual place at an individual portion of time apprehends the object in this very relation, but it is possible that many objects at different times may be *here* or *there*, and *v. v. now* or *then, before* or *after*, i. e., at the same time, in different places. Hence the *hereness* and *thereness*, the *nowness* and *thenness*, the *beforeness* and *afterness* may be common to many individuals, and like sensible or spiritual qualities, may be affirmed or predicated of all. These objects may be grouped under, or classified by means of these general relations. The terms which denote these, take their place side by side with other common terms. Very many adjectives of time, as *prior, later, present, past,* and *future,* and of space, as *long, short, high, deep,* and *broad,* and of form, as *circular, triangular, square, spherical,* and *conical,* and of motion, as *swift, slow,* etc., will occur as belonging to these classes of words. All these classes of terms, like all other notion words, require some image to explain and illustrate them to the mind. But they are peculiar in this, that every object in nature and in spirit has some relation to time and space, and hence it is indifferent what one is cited to exemplify these universal relations.

VI. *Of Mathematical Quantity; the process by which its concepts are evolved, and their relations to time and space.*

§ 276. These concepts naturally divide themselves into two classes, the *concepts of magnitude* and the *concepts of number*, or the concepts which are respectively related to space and time. We begin with those which imply the existence of space, as being the most easily explained and understood; *i. e.*, with *geometrical concepts* or *concepts of pure magnitude.*

Two classes of mathematical concepts. The geometrical.

Of these the most familiar are, *the point, the line, the surface, the triangle, the square, the rectangle, the rhomboid, the solid, the cube, the sphere, etc.*

These terms stand for both images and concepts, in other words, for the products of the imagination and of thought. As images they are individual, as concepts they are general.

The creative imagination idealizes not only the sensible and spiritual properties of these objects and phenomena, but it idealizes their space and time relations, § 181. It transforms the perceived edge with its actual breadth and ragged outline into the ideal line which has neither breadth nor undulation. It smooths the undulating surface into an evenly lying geometrical superficies. In the same way it refines the blunted corner of a die or cubical block into the mathematical point which is idealized as having place but no extent in any direction. These relations cannot themselves be thus imaged, but an object itself can be imaged with these relations thus idealized. Every such object is at first individual. But when the relation is generalized, we have a concept in place of an image, holding the same relation to the concrete and individual which belongs to any other concept. These concepts, like all other concepts, need to be imaged and illustrated by concrete objects. Only in this way can their import be understood, and their validity established. All geometrical conceptions are dependent upon the assumption of the space-relations of objects. Without these space-relations they have no meaning. They presuppose the belief in these space-relations, as actually belonging to every material existence. They rest upon the belief in that absolute and infinite space which limited space presupposes and involves. Space, with the space-relations of objects, is the ever-assumed background upon which all geometrical constructions are projected, and over against which all its processes are interpreted.

Postulates of Geometrical Quantity.

The reality and the validity of these conceptions rest entirely upon the mind's own power to construct and comprehend them. The mind knows that it can construct these concepts, and knows what they are when constructed. Geometry postulates that every student may make these concepts for himself. Its language is confident, "*draw a line*," "*conceive or construct a plane*," "*think of a point*." It lays the foundations for its reasonings in these postulates. It defines the meaning of these constructions by analyzing their relations to one another and to the space to which they all have a common relation. It illustrates, or, as we usually say, *demonstrates* any relations unknown before by referring to new constructions as exemplified in some material substance, for example, in a cube or sphere, a cone, a dot, a chalk line, a rough surface on a blackboard or paper bounded by marks—which are no mathematical entities but serve to represent them and hold the attention to the constructions they represent. In the so-called demonstrations of Geometry one figure is supposed to be drawn in connection with another. Additional figures are placed by the side of those with which we begin, or those already drawn are so divided as to enable the mind to bring into comparison figures that had been inaccessible and incommensurable. As it is with the original and simpler definitions, or postulates, so is it with these complex constructions: space is supposed as the necessary attendant of each and of all, making possible the original constructions and the evolution of the new relations, which the mind discerns ultimately so soon as the requisite figures and connecting lines have been prepared and combined. As has already been shown, § 229, the nerve and force of the geometrical demonstration rests more upon these successive intuitions than upon that element which is properly deductive.

The concepts of number.

§ 277. *The concepts of number* are conditioned upon those relations of objects to time which are involved in the mind's continued activity in uniting them as parts into wholes.

To number, some object must be selected which shall serve as the unit, *i. e.*, which can be conveniently repeated as a recurring part of a whole of extended objects, or of a continued series of mental states.

These constituent parts are numbered when the mind connects each with the next by relations to its own activity in time. That with which it begins is called first. The next, when connected with the one taken first in time, is second. When another is thus connected, we have the third, and so on. Thus we count, or number. The act seems so simple as scarcely to admit or require explanation. It is obvious, however, that this act is only possible as we connect and contemplate objects in relation to a consecutive series of mental acts—that is, a series of mental acts following each other in time.

We find, then, that the relation of number requires that objects should first be connected as wholes and parts, and then contemplated in an arrangement which depends entirely upon the time-relations of the mind that views them. In other words, number depends upon those relations of time which we assume and know to be inseparable from the soul's own subjective activity.

When a series of mental states is itself measured and numbered, it must be remembered that in reflective consciousness this series itself is made objective to the mind. It is treated or viewed as though it were a series or whole of material objects. It is contemplated by a series of acts wholly subjective, involving as spiritual acts, the attribute of duration to themselves, and as *successive*, the relation of number in the objects which they unite and measure as wholes and parts.

Whatever objects are numbered must be arranged in a continued series. This is possible only by the recognized relation of such objects to the mind's continued action in contemplating them. They must also be viewed reciprocally as wholes and parts, as the mind gathers the objects, when thus arranged, into a group, which it breaks into parts, reuniting these parts with each other at its will, and making its units larger or smaller as choice or chance directs. To both these relations time is the necessary condition,—to the continued subjective act of the mind in connecting objects into a series, and to the arranging of them as wholes and parts.

In other words: To the act of counting, time must be assumed as both the subjective and objective condition; but the relations by which objects are viewed or connected in the act of counting when abstracted, generalized, imaged and symbolized, are the relations of number.

These relations can be applied to any objects whatever—to material objects, to spiritual objects, to acts or states of the mind itself, to the very acts of the mind in numbering; in short, to any objects whatever, whether of direct or reflex cognition. Any series of objects can be used as the symbols or images of number. Thus a row of marbles, of kernels of grain, or a series of marks is usually selected. Such objects can be readily interchanged, and they are chosen because they suggest little more than their numerical relations. For convenience of recording and recalling the results of the processes of counting, arbitrary symbols have been selected. Thus, for two objects made one by a single addition, we employ the symbol of two marks, as in the Roman system, II,—later, the Arabic character 2; then III Rom., 3 Ar.; then instead of five marks we use V and 5; instead of four and six, V diminished by I going before and increased by I following, or the Arabic characters 4 and 6, etc., etc.

The principal concepts of number are *the unit, the sum, the difference, the multiple, the divisor and the ratio.*

These concepts cannot be defined so readily as they can be imaged and exemplified. To explain and illustrate their import we must go back to the several acts which represent them. Their meaning is originally taught and successively enforced by directions to select certain objects and proceed with them thus and thus, *i. e.*, they rest upon postulates as truly as do the concepts of geometry. They assume that the mind can perform certain thought-processes which result in certain thought-products. The *psychological condition* of these processes is the arrangement of objects in a series, whether material or spiritual. Their logical condition is the reality of time-relations, and of time itself as making these relations possible. That number depends upon and implies time, is obvious still further from the language which we continually use in our definitions and analyses. We say, add this so many *times;* ten taken twice, *i. e.*, *two times* ten, is twenty; ten divided one *time* by two, or diminished *once* by three, is respectively five and seven.

The application of number to magnitude.

§ 278. The *application of number to magnitude*, or of the concepts of discrete to those of continuous quantity, depends on the mutual relations of time and space objects which have already been explained, § 273. We take any portion of space as a whole, we divide it into parts, we number these parts, we discern ratios between them. We express the powers of curves by their equivalent formulæ of lines as symbolized by numbers, etc., etc., creating all those conceptions and performing those processes which modern analysis has discovered and applied.

VII. *Of the application of mathematical conceptions to Material phenomena.*

Why, and how mathematical concepts are applicable to material objects.

§ 279. Pure Geometry deals only with ideal constructions in ideal Space, and pure Arithmetic and Algebra with ideal concepts conditioned on ideal Time. The possibility of applying these ideal creations to material things and phenomena is explained by the fact that the concepts of number and magnitude are all generalized from the relations of concrete objects and events to both space and time. In the order of time and acquisition we know applied number and applied magnitude before we know pure number and pure magnitude. The latter are always explained by the former. On the other hand, when we apply the concepts of pure mathematics to material substances, we find that those properties which were left out of view in forming them must be brought into view to modify our ideal inferences. In estimating the velocity of bodies we think of them only as capable of constant force and of accelerated motion. When we compare the results of our mathematical processes we find that they do not hold good of real phenomena, because they assumed what rarely if ever actually occurs, *i. e.*, a force that is entirely constant and equable. Or perhaps, they omitted to recognize the increase of resistance consequent upon the increase of velocity. Thus in *Mechanics*, bodies are viewed as attracted by gravitation, as held together by cohesion, as impelled by a natural or artificial agency, as capable of both force and motion, as acquiring and losing velocity. For the purposes of this science gravitation is idealized as a constant force manifested in motion, the rapidity of which is inversely as the square of the distance. The nature of gravity itself as a material agent, is not considered, nor that of *inertia;* nor is the resistance of intervening media, but only the simple fact of motion, or a tendency to motion, with certain constant relations to space and time. In like manner cohesion is defined by the relations involved in the phenomena of motion. So the laws and properties of bodies in motion or in pressure are expressed by space and time relations. Whether bodies do in fact move or tend to move with regularity in these relations so that their motions can be measured or computed, are facts that can be ascertained by observation and induction only.

For example: Newton's great laws in respect to the causes and continuance of force and motion are all generalized observations of facts of sense enforced on grounds of high probability. In other words, they are grounded upon induction. These laws or facts being assumed, we reason and compute with respect to the direction and rate of bodies in motion, with respect to the pressure and weight of bodies tending to move, and with respect to the results of bodies conspiring together in motion, just as we can reason or compute with respect to a sizeless or weightless point that is supposed to move in a breadthless line. That is, we apply to these material objects the concepts, relations and rules of the pure mathematics. But when we compare the results of our computations and demonstrations with bodies actually existing and phenomena actually occurring, we find that the two do not coincide. When we inquire why the prophecy given by our demonstration or computation is not fulfilled by the *facts* of the velocity, weight or pressure of the material bodies with which we come in contact, we account for the discrepancy by those elements or properties which we were obliged wholly or partially to disregard, such as inertia, resistance, friction, and the like. In many cases these are so unimportant that we subject them to no estimate, but take the result as exact enough for our purposes. In other cases, as in gunnery, astronomy, and the working of machinery, we seek to express the value and effect of these very elements in special mathematical formulæ, and subject them to mathematical computations similar to those which we had applied to the prime forces.

VIII. *Of the relation of space and time concepts to Motion.*

Time and space relations, can be still further generalized.

§ 280. It is obvious that the space and time relations of objects when thus generalized become *universals* of a very wide extension. The inquiry naturally suggests itself whether these relations can be generalized still further, and so be included under relations of a still wider extension, as well as subordinated under one another.

We find the medium of such generalization in the capacity of material objects for motion. Every material thing can be moved. The eye and the hand learn to separate the objects of perception from the great universe with which they are at first united, by the circumstance that they are moved and movable. The limiting surfaces, edges and corners of such objects are determined and traced out by the moving of the hand or the eye along or up to their several limits. Every act of motion brings with it the possible suggestion of some one of the relations of space.

We find, moreover, that there is not a single relation of space which cannot at once be brought before the mind, and, as it were, suggested by motion. Each one of these can, in a certain sense, be expressed and defined in terms and concepts of motion.

Even the relations of position can be expressed by means of motion. The meaning of here and there, above and below, behind and before, are all definable by acts of motion—to and from, this way and that way,—joined with counter or resisting motions, which stop their progress. When the question is asked of a child, What do you mean by any one of these terms? he invariably replies by explanations by motion. He says, in effect, Move an object in this or that direction, and then arrest it, and it will be *here* or *there*, *before* or *behind*, *above* or *below*.

The relations of time can also be generalized by means of the motions of material objects. A moving body suggests duration as truly as it does extension, when all its import is received; the act of starting suggests *then* as truly as it does *there;* the act of stopping suggests *now* as well as *here*. It may have come to do so by a secondary and transferred meaning; but it does so in fact and by a universal and inevitable connection.

Even when time is thought or affirmed of mental acts and events, it is still represented by motion in space. Hence by a natural consequence, when time is affirmed of processes (or states) that are purely spiritual, its relations are represented in language and thought by motions that are corporeal. It follows that motion furnishes all the materials for a common generalization of both space and time objects, and *for the comprehension and arrangement of time and space relations in the same logical system.*

This explains why mathematical *entities* or *quanta* are so naturally defined by means of motion; a fact confirmed and illustrated by many such definitions. These definitions always rest upon, and can be expressed by postulates, and these postulates always suppose an act or acts of motion. In geometry we say, draw a line; terminate or bisect a line, giving a point; move a line and it gives a surface. In arithmetic and algebra we say *count*, that is, unite as wholes, or *add*, *subtract*, *multiply*, and *divide;* all of which terms suggest or supposes some images taken from spatial motion.

Extended and enduring objects are limited.

§ 281. The extended and enduring objects which we have thus far considered, are *limited* objects, and the relations to space and time which they involve are also *limited.* Whether they are presented by sense-perception or consciousness, whether they are represented to the imagination or generalized in thought, they are necessarily limited. The so-called dimensions of extension—length, breadth, and thickness,—and the various relations of duration, can only be affirmed of finite beings and activities. If affirmed of the Infinite, it is of its relations to the finite. Even mathematical relations can be conceived of only as limited or definite quantities. These, as we have seen, presuppose some objects imagined to exist in space, or series of such objects connected by acts continuous in time, of which certain attributes and relations are affirmed, *i. e.*, *they invariably presuppose limited objects.*

The infinite and indefinite have therefore no place in mathematics. What is called the mathematical infinite is either a quantity as yet not measured or numbered, *i.e.*, a quantity in respect to which these processes have been begun but are not yet completed; or a quantity so nearly commensurable with another that it may be substituted for it. The so-called infinite quantities of the mathematics are quantities not yet actually or proximately defined, *i. e.*, mensurable but not yet measured or defined. They should be carefully distinguished from what, in distinction from them, may be called the actual infinite or unconditioned. The conception of the mathematical infinite or indefinite may be rendered possible by the real infinitude of time and space, but in import the two are wholly diverse, if indeed we can be said to have any concept at all of the latter.

IX. *Of Space* and *Time as infinite* and *unconditioned.*

§ 282. The several attributes of limited extension and duration, involve relationship to and questions concerning *space and time.*

Extension and duration distinguished from, but related to space and time.

These attributes and properties, when considered collectively are called collectively, *extension and duration.* The appropriate names of the entities to which these properties involve relations, are *space* and *time.* Thus distinguished, extension and duration, *i. e.*, extension and duration in the concrete, or the extension and duration of individual objects, are known *by experience;* while space and time, as soon as they are apprehended at all, are known *a priori,* i. e., to be the necessary and fundamental conditions of all actual existences and events as extended and enduring.

It is not asserted that in applying these attributes to objects of experience the mind necessarily adverts to the relations to time and space which they imply, but only that when the mind gives attention to them, it cannot fail to discover that these relations are implied, and with them the existence of time and space. To make this discovery the mind may need to make the experience of many objects of sense and consciousness. It may need the discipline of many acts of attention to separate and analyze what is at first known confusedly and without discrimination.

In order fully to appreciate the time and space relations of objects and events to one another as well as to time and space themselves, the imagination may need to be called into exercise. One material object may need to be annexed to another and still others to these, before space can be fully understood in all the relations which it involves to the extended objects thus believed or supposed to exist, or to other extended objects besides. In like manner, many events must be experienced, in order that the common relations of all these and of all conceivable enduring objects to time, may be distinctly apprehended, and clearly distinguished from the time which is common to them all. The psychological conditions of knowledge are clearly distinguishable from the essence and the evidence of the objects that are known. The one describes the *subjective conditions* that render it *possible* for an individual to employ and apply his mind in such a manner as to discern a fact or truth. The other describes *objectively*, *what* in its nature is knowable by all individuals under these subjective conditions, and the evidence, if there be any, by which it is known.

They limit objects and events.

§ 283. Extension and duration are also the *limits* or the grounds of the limits of objects and events. These pertain not to space and time, but to objects and events as related to *Space* and *Time*, and therefore and by this means, to one another.

When, for example, I perceive a box either inclosing or inclosed by what we call a void, and affirm that what is without is not that which is within, or conversely; both that which is within and without are conceived as matter with surfaces mutually coinciding, but yet dividing or limiting the one from the other. If I conceive of the outmost limit of the universe of matter and ask what is beyond, immediately as I ask the question I attach the limiting surface to other matter which is conceived to be beyond, and the outlines of which I begin to trace by the constructive motion of which the imagination is capable. Of this outline, one portion, viz., the limiting surface already described, is fixed. The others are not yet drawn; the mind has no occasion even to conceive them drawn, and it rests in the knowledge or belief that it might complete them in any way in which it chooses. But as soon as they should be completed they must necessarily be conceived as inclosed by or with matter, for the simple reason that an extended surface of that which has no actual being cannot be conceived or thought of.

In a similar way the instant which terminates or limits an event, is the beginning of another as yet inchoate or incomplete. So the beginning of an event already past, is the end of the event that was transacted before it.

What we call Space and Time are those entities which can be occupied, as we say, by beings and events, *i. e.*, which render their actual existence possible, and which in rendering them possible, also make it possible that they should be limited from one another, *i. e.*, distinguished from one another by their common relations to space and time.

It follows that: *Relations of place do not belong to space, but they belong to bodies perceived or imagined to exist in space. Relations of time do not belong to duration, but to events occurring in, i. e., presupposing time.*

§ 284. Space and time are *unlimited*, simply because the conception of *limitation is inapplicable* to them, because by its very nature it is only applicable to and affirmable of extended matter and occurring events,—when we attempt to apply it to Space and Time we can only do it by means of objects and events. This attribute is therefore simply negative. It denies that the relation of limitation which pertains to bodies and acts can pertain to Space and Time.

In what sense Space and Time are unlimited.

It is important to notice this distinction in order that we may preserve ourselves from many of the alleged incompatibilities which are conceived to be involved in the attempt to know or conceive of Space and Time.

Thus Hamilton (*Met.* 38) urges that we are under the necessity of conceiving space and time either as an absolute maximum or an absolute minimum, and that it is impossible to do either, because the mind, as soon as it has fixed the limits to the ultimately great or the ultimately small, will immediately overstep or go beyond the limits which it had just established, and will find itself continually baffled in its impotent efforts to grasp or conceive either.

In the same strain, Kant had urged that the mind, in its attempts to conceive of space and time, must continually set up two incompatible propositions—which he calls *Antinomies*—both of which cannot be true, and yet one of which would seem to be necessary. Both overlook that the *maximum* and *minimum* which we attempt to conceive are not *space* and *time*, but bodies and events as limited in space and time. The maximum and minimum in the case are not space and time, nor are they concepts of either, but they are concepts of bodies and events as related to and limited by space and time. They are limited concepts, and in their very nature logically inapplicable to objects which cannot be limited. To attempt to think of time and space under any such concepts, however great or small, is to make an effort which will involve certain and constant contradiction and inconsistency. To attempt to picture time and space to the imagination is impossible, for we can only picture objects and events with definite properties and characteristics. Even when we lay aside all properties except what we call their time and space relations, what we picture or imagine are still limited objects in space and time—objects with some defined limits of extension and duration, but not space and time themselves. It is true that every time we picture or image such objects we must think of their relations to their correlates, time and space; but time and space, in themselves, can neither be imaged nor pictured.

§ 285. Again, Space and Time cannot be *generalized* or *apprehended by or under concepts*. Concepts suppose definite attributes of objects limited by and individualized in Time and Space. But Time and Space are withdrawn from these conditions of generalization, for they are necessarily supposed as the conditions and correlates of

Space and Time cannot be generalized under higher concepts.

all individual existences and their attributes. Even the relations of extension and duration, by which individual objects are possible, cannot be intelligible except by means of these very entities which are the necessary correlates to these universal properties of all individual existences. These related properties are generalizable, but the entities themselves to which they are related cannot be generalized.

Space and Time cannot in the ordinary sense of the term *be defined.* If we cannot form concepts of these entities by means of generalized attributes or relations, it is manifest that we cannot define these concepts, because to define is simply to state the attributes into which a concept thus formed can be resolved, § 214. They are not simple concepts, for simple concepts pertain to single indecomposible attributes or relations, § 197, and no one will for an instant believe or contend that the import of either space or time is exhausted by any single property or relation.

What is demonstrated to be necessary from the nature of the case, is confirmed by fact and experiment whenever we make the trial. Whenever we endeavor to define these entities we find ourselves employing concepts which presuppose that they are already known. Every concept that we use is an attribute or relation of some object or event which exists in space or time, and which implies some relation of the same to one or both. We fall, therefore, continually into the circle of using in our definitions terms that presuppose that to be known which we attempt to define or describe.

They are known as the conditions of their limited correlates.

§ 286. Space and time are known by *intuition* as the *necessary conditions* of the existence and the conception of all objects and events. Every object and event, as has been already explained, has properties or attributes which imply the existence of these entities. In *knowing* that these objects exist, we know that time and space exist as their *actual* conditions. In *conceiving* of these objects or events as real or possible, we must conceive of them as related to space and time, and, of course, must recognize time and space as their *logical* conditions.

While, then, it is true that we can neither generalize nor define time and space, because the very attributes which we must employ imply both, it is true, on the other hand, that we cannot generalize or define any object whatever without recognizing both, and, therefore, time and space must enter as the material into all our concepts. Again:

Though time and space cannot be defined or conceived by the relations of objects and events which imply time and space, yet, on the other hand, as the *correlates* of all such objects, they can be explained to the mind by means of the limited relations which imply their real existence. So far is it from being true that, because space and time are known by intuition, they are known out of relation to limited objects and events; that rather it is only possible to know them by means of such relations. On the other hand, they are only known as implied in the relations which are called collectively the extension and duration of such concrete realities; on the other, they cannot be generalized nor defined by means of any such relations, because all imply their existence.

It has already, § 247 (5), been asserted that the distinct recognition of these correlates, is, as it were, the *fifth* or last stage of the mind's attainment in cognition; which is reached only by the few who are trained to habits of speculative analysis and discrimination. If this is so, then it is obvious that the number of thinkers is very small who have any occasion to ask the question, whether space and time can be defined, or whether they are known out of relation to or by means of their relations to the concrete. But the persons who have occasion to ask these questions can certainly comprehend that the very relations which cannot possibly be used to define time and space, because they imply them, may, for this very reason, be the only medium of bringing them before the mind for the uses of thought.

What are space and time? Conclusion.

§ 287. *What then, are space and time? Are they substances, qualities,* or *relations?* Or, are they *the forms* or *subjective conditions of knowledge by sense* or *consciousness?* Or is it impossible to ascertain what they are? These questions will force themselves upon the attention of a few; and they require an answer.

Are they *substances?* That they are material things with sensible qualities will scarcely be imagined or contended by any one. No one would honestly believe or seriously urge that they can be heard, or smelled, or seen, or tasted, or touched. Space and time are not perceived in such ways or by such means, and hence cannot be classed with material substances. Nor are they spiritual beings. They have none of the properties of spirits. They cannot think, or feel, or will. Nor can they be apprehended by consciousness in the special sense of the term. Neither time nor space is a spiritual substance.

They are not *qualities* or *properties* of *spirit* or *matter*. Dr. Samuel Clark maintained that space and time are attributes or modes, and that inasmuch as both are infinite, there must be an Infinite Being to which they belong. James Mill, in his *Analysis of the Human Mind*, chap. xiv. asserts that they are simply abstract terms which stand for collective conceptions of those attributes of extension and duration, which belong to individual beings and acts. But it needs no further discussion to prove that they are and can be neither. Nor are they simply *relations*, as Leibnitz maintained. This philosopher defined space as an order of co-existence,' and time as 'an order of succession.' (*Third letter to Dr. S. Clark*, § 4, ed. Erd.,p. 752.) Using extension as its equivalent, he defines space as the order of possible co-existences; and time as the order of inconstant possibilities. (*Reply to Bayle*, ed. Erd.,p. 189.) *Calderwood* defines time as "a certain correlation of existences," and distinguishes his own view from that of Hamilton, who calls it "the image or concept of a certain correlation of existences." (*The Phil. of the Infinite*, 2d ed., c. v.)

It is evident from what has been said already, that space and time are neither *relations* nor *correlations*, but *correlates* to beings and events. Extension and duration are the relations or correlations in question; but these involve space and time as *realities*.

Again: Space and time are not *forms of intuition* [*i. e.*, presentation] in the sense suggested by Kant, that is, not subjective forms only. This philosopher taught that if we distinguish the matter apprehended by perception and consciousness from the forms of this matter, then space is the form of sense-perception or external intuition, and time is the form of consciousness. There is a sense in which this doctrine is true. Extension is the form of material objects in the sense that all such objects are perceived as extended, and none can be apprehended except under the form or condition of being an extended object. When all the matter which is given in the various sensible qualities is thought away, the relations of extension remain. The same is true of the matter furnished in consciousness as distinguished from its relations of duration.

But the doctrine as further expounded by Kant is open to two exceptions. *First:*

He fails to distinguish between extension and duration as relations, and the correlates space and time which they involve. He does not notice that these very relations, after or under which all objects and their concepts are and must be formed, do in their very nature involve the intuitive knowledge of space and time as realities, and that to suppose that they are only forms is to exclude and eliminate that which is given and affirmed by their very nature. *Second:* The suggestion or the assumption that they depend on the subjective constitution of the human intellect is unwarranted by positive evidence and is contradicted by the testimony of the intellect itself. The supposition that intellects of another order might possibly exist, which could know objects without the relation of space and time, is without proof and against proof (§ 259.) In other words, that which makes it possible and necessary for extension and duration to be the forms of perception and consciousness is the fact that the objects of these two modes of knowledge are in reality related to the entities space and time.

§ 288. St. Augustine is reported to have said—" What is time? If not asked, I know, but attempting to explain, I know not." This, in one view, is correct. We know by intuition that time and space exist, but to explain or define what they are, is not so easy. It may relieve our embarrassment in part to explain why we cannot answer the question in one sense, and why we can in another. If, in answering the question *what,* it is expected or required that we should class them with objects limited by space or time, or objects having material or spiritual properties, or objects holding relations to space and time, in other words, that we should class them with beings, qualities, or relations, in the ordinary acceptation of these terms, then it is obvious that we cannot answer this question at all. We cannot say *what* they are. But we know *that* they exist, *i. e.,* there exist realities which answer to the terms. Their existence is implied in the existence of every limited object and property, because every such object and property is related to them. We cannot believe or know that the one exists without knowing that the other exists also. But can we in *any sense of the word, what,* explain the nature of that which we know exists? We can, so far as to say that they are entities to which all these limited objects are related, and which are, therefore, correlates to them. If they are correlates to all limited objects, they are known and described by their relations to them. By their very nature they are entities to which these objects bear these relations, and by their relations to these objects they are known and thought of. They cannot be said to be defined in the sense in

Conclusion.

which limited objects are defined, but they can be suggested or described to the mind as the necessary correlates of limited existences, by means of their relations to them.

These relations to both space and time are represented in thought and language by means of *motion*, as has already been explained, and hence it follows that space and time are set forth in thought and language by the same medium.

We conclude that in whatever sense space and time are unconditioned, infinite, and absolute, they are not so in any such sense as to exclude the possibility of being related to the finite. By means of these relations they can be both conceived and known. (§ 342.)

CHAPTER V.

CAUSATION AND THE RELATION OF CAUSALITY.

Causation as a law, and as a principle.

§ 289. From the formal and mathematical intuitions we come to those which are *real*, i. e., which are required to explain the attributes which are respectively distinctive of material and spiritual *beings*. Into these real relations all the actually existing properties and powers of matter and spirit are resolved. Under the laws which regulate their operation, the effects and purposes that describe the universe are accomplished. We shall consider first, the *relation* of *causality* or *causation*.

The relation of causality is sometimes called *the Law*, at other times *the Principle* of causality, causation, or cause and effect. Causation as *a law* is viewed as a relation actually prevailing in or ruling over the finite universe of physical and spiritual being. Causation as *a principle* is placed first or highest with reference to the other concepts or truths which depend upon or are derived from it—either relatively or absolutely, according as the truth is received as original or derived. The first of these appellations is *objective* and *real*, and indicates its universal prevalence among objects actually existing. The other is *subjective* and *logical*, and designates the place which the relation or the proposition in which it is expressed holds in the systematic arrangement of our knowledge, (cf. § 246).

Causation as a *law* may be stated thus: Every finite event is a caused event, or, more briefly, is an effect. Causation, as a *principle*, may be thus expressed: Every finite event may be accounted for by referring it to a cause as the ground or reason of its existence.

It is scarcely necessary to observe, that the proposition, *every effect must have a cause*, is purely and simply identical. It is mere tautology, expanding in the predicate what had been implied in the subject. The term effect, in its import, implies a cause by a logical necessity. To say an effect must be caused, is as reasonable as to say, a caused event is caused, or, $xy = x \times y$.

§ 290. Causation, both as law and principle, is affirmed of events. But what is *an event?* An *event* is something which is known to be, which was not; or which begins to be or to occur. Events are, therefore, finite, *i. e.*, limited by relations of space or time. Their existence or occurrence implies change. Something is here and now which was not. Of these changes it is affirmed that they were caused.

Event defined.

In the *material world*, events are changes of place or relative position, motions in space, changes of form, changes of properties in respect to existence or intensity. They are often called phenomena, *i. e.*, manifestations to the senses or the consciousness of some power or agency.

Events or phenomena are more numerous and conspicuous in *the vegetable* and *animal* sphere. There is growth, change of form and of structure, the manifestation of new colors, odors, and above all, there is constant motion. In the *mental* or *spiritual* sphere, new thoughts, new feelings, new purposes, pass before the observant eye of consciousness faster than they can be accounted for. But besides phenomena of these classes in the world of life, which appear in acts, states, or qualities, more or less lasting, there are still others in the existence and production of new and separate beings, material and spiritual.

Besides these, there are conditions or states more or less permanent which require to be accounted for, such as the equilibria of forces, or pressure, as illustrated in the action of gravitation or electricity, the tendencies of fluids at rest or in motion. All these, so far as the law of causation is concerned, come under the class of events or phenomena.

Many of these so-called events and phenomena are a *combination of several,* and made up of many units. But whether simple or complex, each one of them is caused. If the question be raised, What is a single event, or the simplest phenomenon? we have only to reply, that any change, the least extensive in space, or the briefest possible in time, which can be discerned by human observation, is a single event.

Cause distinguished from conditions.

§ 291. Again: we distinguish between the causes of an event and the conditions of its actually producing the effect. The stroke of a hammer is the cause of the fracture of a stone, of the flattening of a leaden bullet, of the heating of a bit of iron. The conditions of the effect would, in such a case, be said to be the properties of the stone, the bullet, or the iron.

In any such case the effect is frequently said to be the resultant of the joint action of the striking hammer and the resisting stone, lead, or iron. This doctrine is thus generalized by Mill: "The real cause is the whole of these antecedents (or conditions), and we have, philosophically speaking, no right to give the name of cause to one of them exclusively of the others." (*Log.*, B. iii., c. v., § 3). To the same effect, says Hamilton: "Every effect is only produced by the concurrence of at least two causes (and by cause, be it observed, I mean every thing without which the effect could not be realized)." (*Met., Lec.* 3.) In common life a distinction is made between the efficient and patient cause, the last being used for the object, *i. e.*, that on which the causal agency shows a result, or upon which it is exerted. It is obvious that that whose activity is most obvious or demonstrative, is called the efficient. The patient or recipient often exhibits no force or energy, as, the cohesion of the stone, lead, or iron in the cases supposed.

Sometimes the objects in their matter and chief elements are said to be the same, but the force or causal agency is applied under diverse conditions of quantity, time, or distance: as a chemical agency is doubled; the gravitating force operates with a varying energy; a wave of light acts with twice a given rapidity. These conditions are called in scientific language, *the laws* of the acting of the forces or powers—*the causal agents*—of nature.

The relation cannot be resolved into a time-relation.

§ 292. Of causation both as a law and a principle we assert that the relation is *original and independent.* Those who regard it as secondary and derived usually resolve it into some relation of time. The history of speculation abounds in such attempts. This is not surprising. The relations of time pertain to all objects whatever. If objects are connected by the relation of causality, the same objects must also be united to observation either as co-existent or as successive. The most conspicuous advocates of this disposition of the causal relation are *David Hume, Dr. Thomas Brown,* and *John Stuart Mill.*

Hume defines a cause as *a constantly precedent, and an effect as a constantly subsequent event.* The necessity by which conjoined objects are connected as cause and effect, arises from their being united in the mind's own experience, and the consequent fact that the thought or observation of the one determines the mind to a lively idea of the other. They are discovered to be thus related by the constant conjunction of the two.

Dr. T. Brown agrees with Hume that the relation of cause and effect is nothing more than the constant and invariable connection of two objects in time,—the one as antecedent and the other as consequent. Brown differs from Hume in holding that two objects need only be conjoined in a single instance in order to be known as cause and effect respectively, while the theory of Hume requires that they must be frequently conjoined in order to be causally connected. Indeed the interest and meaning of Hume's causal connection depends upon the tendency of the mind to think of those objects together which have been observed to be conjoined in fact. Brown contends that the only use of repeated observations is to enable the mind to analyze or separate complex objects into their ultimate elements; a single conjunction of any two clearly distinguished objects, in his view, gives their causal connection.

"A cause, therefore, in the fullest definition which it philosophically admits, may be said to be, that which immediately precedes any change, and which, existing at any time in similar circumstances, has been always, and will be always, immediately followed by a similar change. Priority in the sequence observed, and invariableness of antecedence in the past and future sequences supposed, are the elements, and the only elements, combined in the notion of cause. By a conversion of terms, we obtain a definition of the correlative *effect;* and *power,* as I have before observed, is only another word for expressing abstractly and briefly the antecedence itself and the invariableness of the relation."—T. BROWN, *Inquiry, etc.*, Part I., Sec. 1. Cf. *Lectures*, Lec. vii.

The Theory of Hume and Brown has, in its essential features, been reproduced and defended by John Stuart Mill. It is fully and fairly stated in his own language in the following extracts from his System of Logic.

"The law of causation, the recognition of which is the main pillar of inductive philosophy, is but the familiar truth, that invariability of succession is found by observation to obtain between every fact in nature and some other fact which has preceded it." * * "To certain facts, certain facts always do and as we believe always will succeed. The invariable antecedent is termed the cause; the invariable consequent, the effect; and the universality of the law of causation consists in this, that every consequent is connected in this manner with some particular antecedent, or set of antecedents. Let the fact be what it may, if it has begun to exist, it was preceded by some fact or facts, with which it is invariably connected."—B. III., c. v., § 2.

"I have no objection to define a cause, the assemblage of phenomena, which occurring, some other phenomenon invariably commences or has its origin. Whether the effect coincides in point of time with, or immediately follows, the hindmost of its conditions, is immaterial. At all events it does not precede it; and when we are in doubt between two coexistent phenomena, which is cause and which effect, we rightly deem the question solved if we can ascertain which of them preceded the other."—B. III., c. v., § 6.

In other words, causation does not imply production, dependence, efficiency or force, but simply uniform succession or constant conjunction. All events or begun existences are or may be presumed to be invariably preceded by certain events, more or fewer, in a set or assemblage, each one of which is as truly a cause as any other.

Against these views of Mill and others, we contend that the relation of causation cannot be resolved into any relations of Time. Our reasons are these. It is conceded by Mill, that in some cases, no interval of antecedence or succession can be discerned between the cause and the effect. To set aside the force of this undeniable fact, he contends that though this is true, yet all those cases in which we have occasion to resort to the law of causation, are cases of begun existence, in which the cause is obviously before the effect. He insists therefore that "*practically*" his view of the nature of causation cannot be controverted. This we grant, so far as to allow that in most instances in which we have occasion to discover a cause or predict an effect, the event is a begun existence. In other words, practically every caused existence is a begun existence, and every cause *precedes* its effect, and every effect *follows* its cause: or, which is the same thing, the relations *before* and *after* usually attend the relation of causality. This is simply the truism that all events, *i. e.*, all *begun* existences, or phenomena, occur in time; or, stated in another manner, that all finite phenomena are subject to time-relations. But it is one thing to assert, which is all that Mill does in this passage, that we can determine causes and effects by meams of their constantly attending relations of time, and quite another to show that the two relations are identical.

That they are not identical is proved by the fact that, without the assumption of the relation of causation as distinct and real, logical deduction would be impossible. This has been shown in the analysis of deduction already given. Induction also would be unmeaning. It is idle to contend that the force of the reasons and laws by which we explain and predict events is exhausted by resolving them into uniform antecedences and successions in time. This has been already shown under Induction. It will be more conclusively proved when we consider in its place the explanation of Induction given by Mill in his own theory of the nature of the causal relation, § 294. This explanation not only fails to satisfy the mind in respect to induction, but it reacts against his underlying and assumed construction of the causal relation. But aside from these considerations, we contend that the very statement of the proposition is its own sufficient refutation. The human mind clearly distinguishes the relations of time from the relations of *causality* and *production*. The intelligent and universal use of the whole vocabulary of terms appropriate to each of these classes of relations is but the constant attestation that this distinction is made universally and necessarily by the mind; in other words, that causation cannot be resolved into any relation of time.

The principle of causality intuitively evident.

§ 293. The relation of causality being established, we assert that the mind *intuitively believes that every event is caused*, *i. e.*, is produced by the action of some agent or agents, which, with respect to the effect, are called its cause or its causes.

The reasons for this view are the following:—

(*a*) We explain the occurrence of events in common life, on the assumption of this truth. To explain phenomena is to refer to the beings or agencies which have occasioned them. When these producing agents are discovered, and the modes and laws of their action are referred to or unfolded, the process of explanation is complete.

(*b*) When an event has occurred which is not yet accounted for, and the mind is aroused to the effort to solve or explain its occurrence, it believes just as firmly that it can be accounted for in the way described, as if the explanation had been in fact attained. It is as confident before as after the cause has been determined, that its occurrence depends upon some cause or causes. Upon this confidence rest all the inquiries and experiments which it sets on foot.

(*c*) The mind not only explains the past, but it relies upon the future, on the ground of its faith in causation. It provides for or secures future results by availing itself of the causes which it knows will produce them. It employs these agents in all its plans and experiments with entire certainty of the results which they will effect. It predicts these results with confidence so soon as it is certain of all the causes which are or may be put into action.

(*d*) In these explanations and experiments the mind is impelled by a special emotion, always present and powerful. Curiosity is more than an interest and desire to know an event as a fact: it impels to the knowledge of its causes and laws, of its origin and growth. The existence of a strong and apparently original emotional capacity of this sort confirms the view that the relation itself is original as a law of existence, and that the belief in it is a fundamental principle of the mind's knowledge.

What the mind unconsciously assumes to be true in practical life, it distinctly and consciously applies in all the *methods and processes of thought and of science.* We have seen that deductive reasoning has no meaning unless the relation of causality is assumed, and that induction in its researches after the forces and laws of matter and of spirit, makes the same assumption. Science, in all its processes, investigates the properties, the powers, the forces, the attributes, and the laws, of all existing objects. But properties, powers, forces, and attributes are all of them terms which directly assert or indirectly imply that there is a causal energy or activity in these objects. The laws of matter and of spirit have no import, and can admit no application, except as causal agencies are affirmed which these laws measure or formulate. Except as the causal relation is believed and assumed, scientific knowledge can have no import, and scientific inquiries would be meaningless and impossible.

Moreover: the belief in the relation of causality is wrought into and expressed by the structure of language. There are words which express causal activity, words which express the reception of such activity, and words which express the changes which are wrought in objects by means of causal activity. The grammar of every language furnishes proof of this, both in its etymology and its syntax.

These considerations prove decisively that our belief in causation is an intuitive principle which meets all the criteria of *universality*, *necessity*, and *originality*.

Counter theories. The belief not acquired by induction or association.

§ 294. This opinion is disputed by many. Various counter theories have been devised to account for the *universal* or, as the case may be, for the *very general* application of causation. The first of these counter theories which we notice is, that the belief in the universality of causation is, like other general beliefs, acquired by induction. This is the doctrine of J. S. Mill.

"With respect to the general law of causation it does appear that there must have been a time when the universal prevalence of that law throughout nature could not have been affirmed in the same confident and unqualified manner as at present. There was a time when many of the phenomena of nature must have appeared altogether capricious and irregular, not governed by any laws, nor steadily consequent upon any causes." * * "The truth is, as M. Comte has well pointed out, that (although the generalizing propensity must have prompted mankind from almost the beginning of their experience to ascribe all events to some cause more or less mysterious) the conviction that phenomena have invariable laws, and follow with regularity certain antecedent phenomena, was only acquired gradually; and extended itself as knowledge advanced, from one order of phenomena to another, beginning with those whose laws are most accessible to observation."—B. III., c. xxi., § 3.

"I apprehend that the considerations which give, at the present day, to the proof of the law of the uniformity of succession, as true of all phenomena without exception, this character of completeness and conclusiveness, are the following: First, that we now know it directly to be true of far the greater number of phenomena; that there are none of which we know it not to be true, the utmost that can be said being that of some we cannot positively, from direct evidence, affirm its truth," etc., etc. "Besides this first class of considerations there is a second, which still further corroborates the conclusion, and from the recognition of which the complete establishment of the universal law may reasonably be dated." "When every phenomenon that we know sufficiently well to be able to answer the question, had a cause on which it was invariably consequent, it was *more rational to suppose* that our inability to assign the causes of other phenomena arose from our ignorance, than that there were phenomena which were uncaused, and which happened accidentally to be exactly those which we had hitherto had no sufficient opportunity of studying. It must, at the same time, be remarked, that the reasons for this reliance do not hold in circumstances unknown to us, and beyond the possible range of our experience. In distant parts of the stellar regions, where the phenomena may be entirely unlike those with which we are acquainted, it would be folly to affirm confidently that this general law prevails, any more than those special ones which we have found to hold universally on our own planet. The uniformity in the succession of events, otherwise called the law of causation, must be received not as a law of the universe, but of that portion of it only which is within the range of our means of sure observation, *with a reasonable degree of extension to adjacent cases.* To extend it further is to make a supposition without evidence, and to which, in the absence of any good ground from experience for estimating its degree of probability, it would be ridiculous to affect to assign it."—B. III., c. xxi., §§ 4, 5.

Closely allied to this is the doctrine of Hume: that the belief is the result of association. Indeed, Mill blends the two in one, inasmuch as he makes induction to be the result of repeated or inseparable associations. This doctrine expressed by Hume is as follows:

"The first time a man saw the communication of motion by impulse, as by the shock of two billiard-balls, he could not pronounce that the one event was *connected*, but only that it was *conjoined* with the other. After he has observed several instances of this nature, he then pronounces them to be *connected.* What alteration has happened to give rise to this new idea of *connection?* Nothing but that he now *feels* these events to be *connected* in his imagination, and can readily foretell the existence of one from the appearance of the other.—*Inquiry, etc., Lec.* vii., p. 2.

"Necessity is something that exists in the mind, not in objects; nor is it possible for us ever to form the most distant idea of it considered as a quality in bodies. Either we have no idea of necessity, or necessity is nothing but that determination of the thought to pass from causes to effects, and from effects to causes, according to their experienced union. 'A cause is an object precedent and contiguous to another, and so united with it that the idea of the one determines the mind to form the idea of the other, and the impression of the one to form a more lively idea of the other.'"—*Human Nature*, B. I., *Lec.* xiv.

(1.) The advocates of each of these theories *overlook the real question at issue.* The belief to be explained or accounted for, is, that *every event has a cause.* The belief which the advocates of this theory seek to account for, is the belief that to each particular event or class of events, some definite cause *has been or may be actually assigned.* That this last can only be the product of experience is obvious. That this is the belief in support of which they adduce illustrations and arguments is evident from the passages which we have quoted from Hume and Mill. That this is not the belief which is in question, needs no illustration or argument.

(2.) No simple experience of actual events can establish the application of its results any further than the range of *actual events of which we have had this experience.* But in both Generalization and Induction, we go far beyond our actual experience. When, from the observation of a few objects or a few events, we generalize a concept or a law which we apply to objects or events more or less like them, we use the belief that what we have observed will prove true of what we have not observed. Whether what we have observed are called simple uniformities of antecedence and succession, or uniformities of causation, makes no difference with the nature of the act by which we pass from the known to the unknown.

Mill himself most pertinently observes : "We believe that fire will burn to-morrow because it burned to-day and yesterday; but we believe on precisely the same grounds that it burned before we were born, and that it burns this very day in Cochin-China. It is not from the past to the future [only or as such] as past or future, that we infer, but from the known to the unknown ; from facts observed to facts unobserved; from what we have perceived, or been directly conscious of, to what has not come within our experience."

He also admits, in the passages already quoted, that we do not limit ourselves to experience. In asking why, when we cannot assign a definite cause for an event, we yet believe it to be caused, he says it is "*more rational to suppose* that our inability to assign the causes of other phenomena arose from our ignorance than that there were phenomena which were uncaused." While then he insists that we have no warrant from experience in applying the results of experience "to circumstances unknown to us and beyond the possible range of our experience," and contends that "the law of causation must be received not as a law of the universe, but of that portion of it only which is within the range of our means of observation," he is careful to subjoin "*with a reasonable degree of extension to adjacent cases.*" It would be difficult to give a meaning to the phrases "*it is more rational to suppose,*" and "*with a reasonable extension to adjacent cases*" without finding in them a real, though reluctant homage to the intuition, "*Every event must be caused.*"

(3.) *Induction assumes this belief* as already present to or ready to be applied by the mind. Mill concedes that Induction itself has axioms. He says, "whatever be the best way of expressing it, the proposition that the course of nature is uniform, is the fundamental principle, or general axiom of Induction." The proposition that "the course of nature is uniform" must mean that the unknown uniformities of succession or causation correspond to those which are known. If this is a general axiom or fundamental principle of Induction, it would seem that it cannot be gained or derived by means of Induction. And yet Mill contends that *the axiom which is necessarily assumed to give meaning and reality to the process of Induction is acquired by means of the process to which it is a necessary pre-condition.*

(4.) The resolution of this belief into *tenacious or inseparable associations*, or as Hume more bluntly expresses it, into "*custom or habit,*" is more *palpably untenable* than the other theory or form of this theory.

The resolution of the objective reality of this connection into a mere subjective association of the two terms fails to satisfy the mind, because it does not account for what is believed. How the mind comes to think of the one when the other is observed or thought of, is a very different question from this, 'how or by what relation does the mind believe that the objects thus thought of together, are connected in fact?'

It is a mere truism to say that objects observed or thought of together will be conjoined by association. The fact that the mind is constantly determined to one thought by the presence of another, is very different from the fact that the two things thought of, are necessarily determined the one by the other. If the two are viewed simply as psychological experiences, even the subjective law by which the objects concerned are presented to the mind in constant conjunction, is clearly different from the subjective belief that the *objects so presented are united causally.*

The philosopher who directly, like Hume, or indirectly, like

Mill, resolves the principle of causality into the law of association, complicates rather than simplifies the problem. For he imposes upon himself the obligation to show that the objective world of fact corresponds to the subjective world of ideas. This he must show by deduction, induction or intuition: but deduction and induction both rest upon intuition; consequently even the theory which attempts to dispense with intuition must in the final analysis rest upon it, in one form or another, as the ultimate arbiter.

Not resolvable into outward or inner experience, or both. Locke and De Biran.

§ 295. The two next theories resolve the principle of causality *into the observations of experience*, ascribing it to our sense-perceptions of the phenomena of matter, or to our conscious experience of the phenomena of the soul, or, again, to both of these conjointly.

Locke seems to advocate, in different passages of his Essay, every one of these theories. The following passages may be fairly taken to represent each of the three:—

"In the notice that our senses take of the constant vicissitude of things, we cannot but observe that several particulars, both qualities and substances, begin to exist; and that they receive this their existence from the due application and operation of some other being. From this observation we get our ideas of cause and effect. That which produces any simple or complex idea, we denote by the general name, cause, and that which is produced, effect. Thus finding in that substance which we call wax, fluidity, which is a simple idea that was not in it before, is constantly produced by the application of a certain degree of heat, we call the simple idea of heat in relation to fluidity in wax, the cause of it, and fluidity, the effect."—*Essay*, B. II., c. xxvi., § 1.

"A body at rest affords us no idea of any active power to move; before it is set in motion itself, that motion is rather a passion than an action in it. For when the ball obeys the stroke of a billiard-stick, it is not any action of the ball, but bare passion."

"The idea of the beginning of motion, we have only from reflection on what passes in ourselves, where we find by experience, that barely by willing it, barely a thought of the mind, we can move the parts of our bodies which were before at rest. So that it seems to me, we have from the observation of the operation of bodies by our senses, but a very imperfect, obscure idea of active power, since they afford not any idea in themselves of the power to begin any action, either motion or thought. But if from the impulse bodies are observed to make one upon another, any one thinks he has a clear idea of power, it serves as well to my purpose, Sensation being one of those ways whereby the mind comes by its ideas; only I thought it worth while to consider here by the way, whether the mind doth not receive its idea of active power clearer from reflection on its own operations, than it does from any external sensation." B. II., c. xxi., § 4.

Locke's view has been understood to be, that by simple observation and experience of material or spiritual events, we know that they are connected as causes and effects, and that on the ground of the experience thus given in sense and consciousness, we believe, conclude, or infer, that all events are so connected. To the theory as thus interpreted the reply is decisive: *First,* that simple experience of the known can of itself furnish no warrant for a belief concerning the unknown, unless we apply or assume some *a priori* principle or original intuition; *Second,* sense-perception and consciousness are usually so defined as to include the discernment of the relations of space and time. But the relations of space and time are *a priori,* and are discerned by intuition. It cannot then be urged that sense and consciousness as forms or acts of simple experience, are the source or sources of our belief of causation. Experience is *a posteriori,* and excludes any *a priori* element.

Royer Collard and Maine de Biran, two distinguished philosophers of the modern French school, have each introduced important modifications of the theory of Locke.

Royer Collard, *Fragmens de Leçons* (*Œuvres de T. Reid,* T. iv., p. 296), contends that our experience of psychical phenomena alone gives us direct knowledge of the causal relation, inasmuch as mental states are, by their very nature, known to be caused by the *ego.* We know by consciousness that we are causes, and these are the only causes which we do know. But we know that every event is caused, *as a self-evident and intuitive truth.*

Maine de Biran, (*Œuvres,* T. iv.,) expands this general statement into a refined theory which he explains with great subtlety, and defends with equal boldness as follows:—

The soul, in all its higher states and elements of states, is not receptive but active. As active, it is the originator or producer of effects. These effects are of two sorts: those which are purely psychical, and those which are external as they affect the body and originate motion. In those states which are purely psychical, and in the other states so far as they are such, consciousness distinguishes between *the ego, the ego in action,* and *the result of the acting of the ego.*

(*a.*) The *ego,* discerned or *apperceived,* is not the soul as a substance, but only the individual *ego.* (*b.*) The *ego* thus apperceived is known neither as out of action, nor as prepared for action, but *as acting—these acts in all cases being individual.* (*c*.) This activity is also causal or productive action. In its very nature and essence it is known as passing into effects.

In other words, De Biran holds that the relation of causation is gained by the soul through conscious observation of the ego in action. In answer to the more important question, How does it know that every event has a cause? De Biran would reply: On occasion of the individual apperception described, we extend the causative relations to objects other than ourselves, *by a principle of natural induction or analogy.*

His theory, stated in a single proposition, is that we believe all events external to our own experience to be caused, because we explain all such events by natural induction, after the likeness or analogy of that spiritual causation of which we are directly cognizant in ourselves.

The theory of De Biran may be admitted, that we gain our first knowledge of the causal relation from the experience of personal and individual causality, without involving his second position, viz: that, by natural induction, we make a universal application of our individual experience to every possible event. The so-called natural induction of De Biran must rest upon or involve an intuition, equivalent to the *a priori* principle, every event must have a cause. Otherwise it is impossible to see what warrant we have to transfer what is true of our individual experience to the whole spiritual and material universe. The fact that, *psychologically*, we have the earliest and most complete exemplification of the causal relation in our spiritual experience, does not in the least explain, *philosophically*, why it is that we believe this relation to be of universal application.

From the fact assumed or believed that the soul derives its first notion of cause from its conscious activity, the inference has been derived that causation is predicable of spirit only; that a material cause is contradictory in conception and impossible in fact. This inference has been held in two forms.

(1.) It has been inferred, *first*, that the conception of a material cause is self-contradictory; because, forsooth, our knowledge of the causal relation is derived from our own psychical activity. Spirit alone, it is contended, is essentially active and causal, and in spirit, will is that only which is active. Matter is incapable of force; it presents the appearances of antecedent and successive phenomena, but behind these appearances there is no force except what spirit imparts.

Against this view the following objections are decisive: (*a.*) The soul finds in its own positive psychical experience, evidence that "force and power are" not "applicable only to will;" for it finds spiritual energies that are neither intelligent nor voluntary. When it seeks and strives to fix its attention, to recall forgotten objects, and to control its rebellious desires, it contends against actual forces which are not uniformly regulated by intelligence or controlled by the will. There are 'secondary causes' within the soul at least, if there are not in matter.

(*b.*) It does not follow, because we derive the notion of causation or force from the conscious activities of an intelligent will, that the relation itself involves either intelligence or will. Let it be conceded that at first the soul, by a not unnatural illusion, refers every event which it does not produce by its own activity to some spiritual agent other than itself. It soon learns to correct such judgments. It learns that a spirit does not directly blow upon the trees or agitate the sea, for it finds the agitation of the air interposed; it then discovers that this agitation is occasioned by heat; then that heat is dependent upon the sun, or some other agent.

(*c.*) According to this theory, the universe of matter and of spirit, except so far as it is capable of intelligence, is unreal and impossible. Matter without qualities or powers, is inconceivable; but qualities and powers involve force, *i. e.*, causal energy. The exercise of power is also inconceivable, except by beings capable of voluntary energy.

For these reasons we reject the theory. We distinguish intelligent and voluntary activity from simple causal energy. We distinguish *causal* from *creative* force, *i. e.*, origination under conditions furnished by another being from origination without such conditions. We distinguish primary from secondary causes.

(2.) The *second* inference derived from the position that the activity of spirit furnishes the notion of causation, is, that there is but one agent in the universe, and He is the Creator; that causation is conceivable of neither created matter nor created spirit, and the apparent activities of both are held to be varied manifestations of His single force, in phenomena successive to one another. If this doctrine were true, it could not be legitimately derived in the way prescribed by this theory, which makes the notion of causality to be furnished from a created or finite agent, and yet infers it to be inapplicable to any other than a being which is infinite and uncreated.

Malebranche (*Rech. de la Ver.*, p. 2, c. 3,) advocates the theory in question, but not on these grounds, but as an inference from his general theological and philosophical position, that God is the only agent, and that in him we perceive as well as produce every object in the universe.

The theory which resolves causality into a relation of concepts.

§ 296. A class of theories of historical importance comprehends all those which resolve this relation between things into some *a priori* relation between *concepts*—in other words, into some logical axiom, principle, or relation. The fallacy common to all these consists in inverting the order of nature and of reason. A correct estimate of logical relations and principles would show that they are all dependent upon some assumed reality of things. Among such realities, the relation of causality is prominent and fundamental. It cannot be derived from the laws of identity and contradiction, which as we have shown concern concepts only and are designed to hold the mind to consistency in their use.

It has not been uncommon with the philosophers of the later German Schools to seek to resolve the principle of causality into the principle of the sufficient reason *viewed as a logical axiom.* This follows from not clearly determining and carefully keeping in mind the relation of the *ratio essendi* to the *ratio cognoscendi* in the principle of the sufficient reason itself. Because the *logical reason* is more general or extensive in its application than the *real cause,* they have resolved cause into reason, instead of explaining reason by means of the relation of cause. We have already shown, under Deduction, that the syllogistic process, and indeed all logical reasoning supposes the *ratio essendi,* i. e., real causal action, or that which may be conceived as such, and that without this all deduction is meaningless and inconclusive, (§§ 221, 2.)

This inversion of the real order of the dependence of these conceptions may be traced to Wolff and Kant. Kant sanctioned it by the suggestion that is fundamental to his system, that the forms of thought are not necessarily representative of the forms of being. Kant makes the relation of causality to be a metaphysical relation of that explicability of one concept by another which is required by the logical faculty, instead of a real relation of things.

It has been carried to its furthest extreme by Hegel in the fundamental position of his philosophy which he boldly attempted to make universal, viz., that all the so-called relations of being may be developed from and are resolved into relations of thought, so that the actual world is but the necessary evolution of the relations that belong to the logical concept. The relation of the reason to its consequent, and by consequence, of cause to effect, is only a special application of that law of identity; misinterpreted by his logic, which is properly applied only in the sphere of abstract thought.

Hamilton's theory of causation.

§ 297. Another theory called *a priori* is the theory advanced by Sir William Hamilton, (*Met.,Lec.* 39, 40.) This theory derives our conceptions of, and our belief in, this relation, not from a *power,* but an *impotence* of mind; in a word, it resolves it into the more general "*principle of the conditioned.*" The *law of the conditioned* is, that the "conceivable has always two opposite extremes, and that the extremes are equally inconceivable. That the conditioned is to be viewed not as a power, but as a powerlessness of mind is evinced by this —that the two extremes are contradictories, though neither alternative can be conceived or thought as possible, one or other must be admitted to be necessary."

This general powerlessness gives the special relation of causality, when applied to the two positive forms under which every object is and must be conceived, viz., *existence* and *time.* By the necessity of the *first,* the mind cannot but think of every object as existing. It cannot, if it tries, think of anything as not existing. By the *second* the thing existing is not now what it was a moment *before.* We cannot think of any object as non-existing in the present. No more can we think of the same as non-existent in the past. We cannot think of its absolute commencement in the past, nor can we think of its absolute termination in the future. We can neither think of its absolute non-commencement nor of its infinite non-termination. "This gives us the category of the conditioned as applied to the category of existence under the category of time."

By this application of the principle of the conditioned, the principle of causality is gained. For the law of causality is simply this, that when an object appears to commence in time, we cannot but suppose that the complement of existence which it contains has previously existed; "in other words, that all we at present come to know as an effect, must previously have existed in its causes."

According to this theory, the cause or causes of an object are the sum of the constituent elements of its being, existing at a previous time in a different form; the effects are the same as existing in another form at a subsequent time. This applies to every form of causation, even to the creation of the universe. For creation is not a springing of nothing into something; "it is conceived, and is by us conceivable merely as an evolution of a new form of existence by the fiat of the Deity."

The objections to this explanation of the relation of causation, as taught by Hamilton, are the following:

(1.) It is not true that it is an original and necessary belief, that the complement of existence is not changed with the changes of phenomena. For example, when a pile of fuel is consumed by fire, and only an inconsiderable residuum of ashes remains, men do not necessarily and instinctively assert that the total of the original constituents of the fuel is undiminished. So far is this from being true, that, on the other hand, they are slow to accept the evidence furnished by the more careful experiments of science, that the products, when analyzed and gathered after combustion, equal the elements of the substance before it was burned.

(2.) The asserted impossibility to think an object as non-existent is a *logical*, not a real impossibility. We cannot think any thing not to be *in thought*, because, while we think of it, it must exist for us as thought. Even when we think of it as not existing, whether in the present or in the past, we must first think of it as existing in thought, and to this object as existing in thought we deny existence in fact. If we think of a centaur or a hippogriff, we must think of it as being. If, because we cannot *think* of an object actually existing to be non-existent, we may infer that the complement of its existence does not change, we may also infer that, because we think of a centaur and a hippogriff as existing, they both in fact exist.

(3.) The theory is utterly inadequate to explain *psychical causality*. The operations of the soul are, as we have seen, eminently causal. From our conscious experience of this class of actions the first notion of causality is derived. Whether the effects in question are produced by the action of the soul within itself, and are purely psychical, or whether they are wrought in the nervous organism by the soul; whether they are wrought upon matter by the soul, or upon the soul by matter; in each of these cases the theory fails to satisfy. There is no complement of existence appearing in different forms at different times. Whether the effect is psychical, physiological, or material, is not conceived as the same constituents under a new form. It is what the terms denote it to be—a product, an effect, a result of activity, a consequent of the powers and activities which are required for and appropriate to the result.

(4.) It is still more incongruous with any right notion of *crea-*

tive causality. The creation of matter or of mind implies the production or origination into existence of that which did not previously exist in any of its constituents. It is called by Hamilton, "the evolution of a new form of existence by the fiat of the Deity." But evolution ought, in consistency with his theory, to signify the changing of the materials already existing under one form, into some new form of the existence already in being. This would require either that we believe in the co-eternity of matter with God, and that we restrict the agency of the Deity to the exercise of a merely plastic or formative energy, or it would involve the pantheistic view, that in the spiritual nature or constitution of God there was also present a material substance, from which by a new evolution of divine activity, the created universe emerged, as a new form of the matter which had from eternity existed in God. From spirit as such, from a pure spiritual essence, it cannot be conceived that matter should be evolved, in any consistency with the theory of Hamilton as defined by himself.

Conclusion. Our position reaffirmed.

The various attempts to resolve the relation of causality into some other relation either *a posteriori* or *a priori* having failed to be satisfactory, we return with greater confidence to the original position which we have already explained and defended, that it is *original* and *intuitive*. The various applications of the relation and principle of causality in the processes of the intellect, as well as its significance as an assumption fundamental to our higher knowledge, illustrate and enforce its importance.

CHAPTER VI.

DESIGN OR FINAL CAUSE.

From the principle or relation of causation we pass by a natural transition to the principle of design or adaptation, or, as it is usually termed, *of final cause*. This in an eminent sense, is a synthetic relation, and is contrasted with the relation of causality as analytic. The movement of the latter is from the individual to the general, from the less to the more comprehensive. The movement of adaptation and final cause is from the general to the particular and the individual. It unites constituent elements into constituted wholes.

Terms explained. Formal, material, efficient, and final causes.

§ 298. The term *final cause* is thus explained: Aristotle and the schoolmen divided all possible or conceivable causes into four; the *material, formal, efficient*, and *final*. The *material causes* are those material elements or principles of which any existence is composed, whether the matter is bodily or spiritual. The *formal cause* is the property or aggregation of properties which constitute its *essence* or *logical content* (in Aristotelian phraseology, *its form*). In these two senses, the word cause is equivalent to element or constitutive principle, each differing according as that which is constituted is *matter* or *form*.

The *efficient cause* corresponds with the cause of modern philosophy, except that it was formerly appropriated to the most conspicuous or prominent of the agents or conditions that produce a result; whereas, in modern usage, the term is extended to all those agents which, in combination, originate an effect.

The *final cause* was the design or end which was conceived as impelling and directing the action of a number or succession of agencies, till it was actually brought to pass. The significance of this appellation can be understood by an example. If I form a purpose, the event or result when made actual, will be the *end* of a series of events or actions. Hence the *end*, by a secondary signification, is made to signify a *purposed result* or a design, and the adjective *final* receives the same import. This purpose is called *a cause*, because it is conceived when formed as causing those events or acts which are necessary to its realization. Hence the appellation, *final cause*,—*i. e.*, a cause, which, beginning as a thought, works itself at last into a fact as an *end* or *final result*.

Aristotle called the formal cause τὴν οὐσίαν καὶ τὸ τί ἦν εἶναι, the material cause τὴν ὕλην καὶ τὸ ὑποκείμενον, the efficient cause ὅθεν ἡ ἀρχὴ τῆς κινήσεως, and the final cause τὸ οὗ ἕνεκα καὶ τἀγαθόν. *Met.* 1. I. 83 a 27, a 29, a 30, a 31.

Design and adaptation, how related.

§ 299. The relation of design supposes that agencies exist or may exist, which might cause a result to be. *The result* is called the *end* or *final cause*. The *capacity* of *these efficient causes when combined to produce the effect* is called *their adaptation or fitness for it*. This adaptation may be considered *subjectively* or *objectively*. If it is viewed as *arranged* or *known* by the designer, it is considered subjectively, *i. e.*, it is a *design*. But whether it is known or not, the capacity for or the possibility of it exists and remains to be discovered. It pertains to actually existing forces and laws of nature, and is a relation which may be affirmed of such causes. A series or combination of causes, viewed as *fitted* for an end are called *the means*—literally the intermediate agencies—between the end as thought and the end as produced.

The relation assumed as necessary and *a priori*.

§ 300. The position which we assert and defend is that this relation is believed *a priori to pervade all existence, and must be assumed as the ground of the scientific explanation of the facts and phenomena of the universe.* We do not inquire whether this relation is exemplified in our experience as a psychological fact, but whether it lies at the ground of all our knowledge as *a necessary relation* of *things,* and *a first principle or axiom* of *thought*—whether, in other words, the *principle of adaptation* ranks with the principle of efficient causation *as a necessary* and *a priori truth.*

We assert that the relations and laws ascertained by asking the questions *why* and *how,* are not the only relations conceivable, but that others hold the same place in our knowledge, viz., those which the question *what for* requires as its answer. Among his four causes, Aristotle gave the highest preeminence to the οὗ ἕνεκα or the *what for.* Was Aristotle right in assuming that the *end* is as important to be known as the *definition,* the *constituents* and the *origination* of a being or a phenomenon?

Reasons. The mind impelled to connect objects by this relation.

§ 301. Our reasons for the truth of this position are the following:

(1.) The mind is impelled to seek, and is satisfied when it finds that any objects or events are related as means and ends. Whatever these objects may be which are connected under this relation—whether they are individuals that fill only single points in space and endure but for a moment of time, or classes of beings that pervade the universe by their agency, and endure with energy unwasted from generation to generation—the mind inquires, *for what* do these exist and act? and if it can find an answer, it accepts it with rational satisfaction.

It asks the question and accepts the answer in a way precisely analogous to that in which it inquires and learns, By what agency and under what law does any thing exist and act? It asks as pressingly and as persistently, concerning what it may find in this relation, as concerning what it can know under the relation of causation. When it receives a probable answer, it welcomes it with a more complete and a higher satisfaction than a similar explanation by efficient causes and their laws. This ground of analogy would lead us to believe that the two relations are both original and intuitively assumed.

§ 302. (2.) The relations under which this axiom requires that objects should be connected, is *higher* than any of those which arise under the category of efficient or blind causative force.

The relation is higher than that of efficient causation.

The relation of means to ends supposes that of cause and effect. We must first suppose causes or agents to exist, before we can suppose them to be applied or employed as means. But when forces and their laws are ascertained, and by them unity and order and dependence are established among the otherwise disconnected beings and events of the universe, the mind takes a step higher in its aspirations, seeking to rearrange under more elevated relations the objects united under the lower. The one class being presumed, and in part at least successfully established, the mind believes that a higher is possible, and proceeds to discover it. Subjectively viewed, this relation gives a higher satisfaction. Objectively regarded, it stands higher in rational value than that of efficient causation, which is only a stepping-stone and preparation for it.

§ 303. (3.) The principle has been of *essential service in scientific discovery*. Should it be conceded that the appropriate sphere of science proper is to develop and establish the so-called powers and laws of nature, and that the discovery of adaptations lies without its sphere, it would still be true that the belief that the universe is full of such adaptations, is of essential service in suggesting powers and laws previously undeveloped and undetermined. The history of scientific discovery abounds in confirmations of this truth.

The principle has been of essential service in scientific discovery.

When Harvey observed at the outlet of the veins and the rise of the arteries, certain curiously constructed valves,—those in the one, opening inward towards the heart, and in the other, opening outward away from the same, he was persuaded that the arrangement indicated an end which supposed activities and laws which he proceeded to ascertain and determine.

Further illustrations of the value of this principle in scientific discovery will be given when we treat of its application to the several sciences. Cf. § § 316 seq.

The Foundation of the Inductive Philosophy.

§ 304. (4.) *The entire superstructure of the Inductive Philosophy rests upon the principle in question.*

It has already been shown that the Inductive method rests on several assumptions. They are such as these: nature is uniform in her operations and laws; the indications or signs of less obvious powers and laws may be confided in; the analogies of nature are important means of suggesting facts and laws, and of inciting to experiment and discovery; the simplest relationships, the fewest agencies, and the most economical uses of forces are always presumed. These and other like axioms of the student of nature are but varied applications of the principle in question, viz., *that in the universe objectively considered, there is an intelligent and wise adaptation of powers and laws to rational ends, and that the same is true of the relation of the universe to the knowing mind.*

It is not sufficient for the philosopher to say that without these assumptions, the science of nature itself would be impossible, inasmuch as the conception of science requires that powers should be fixed, and laws should be uniform, and indications and analogies should be trustworthy—that were science not to assume the truths of these maxims she would commit suicide. To this it is pertinent to reply, What if science itself should be impossible? What is the imperative necessity for science? Every reply to these questions implies that the adaptations of nature to the methods and impulses of the knowing mind are such as indicate at least that class of designs in the structure of the universe which the possibility of science requires.

Required to explain the phenomena of organic existences.

§ 305. (5.) It is also needed to *explain those phenomena of organic existence,* which the relations of efficient causes are entirely incompetent to resolve or even to define. An organic being, or an organism, can only be defined as a being of which each organ acts for the integrity and well-being of every other organ, and all act together for the life of the whole. More abstractly, and in the terms of the relation in question, an organism is a being in which each of the parts and the whole are respectively means and ends for one another. We find it, in fact, to be true, that in any living being, whether plant or animal, the elements or organs act together so as to promote the action of each other, and of the whole. If the appropriate function of each organ is performed, the function of every other is also fulfilled, and when all together are exerted they are the conditions of the growth, the development and the remaining functions of the plant or animal. In the animal, the action of the lungs is necessary to that of the heart, and the action of the heart to that of the lungs, the action of both to the action of the stomach, and the action of the stomach to that of both these, and the mutual action of these and the remaining organs, to the health and life of the whole body.

The elements or agents of which these organs are composed, have their well ascertained mechanical and chemical properties, and when these are combined in inorganic substances, their results follow the laws which control them. But when they are combined in living beings or their organs, these powers and laws do not explain in the least degree these compounds or their functions. The materials or agents which form the heart, the lungs or the brain, do not at all explain the peculiar substance, form, or functions of these organs; much less do they account for the singular capacity which they possess of producing a whole, on which they depend for their own existence as a living heart, lungs and brain,—which in its turn as a living whole is dependent on each of these.

All that we can do, is—within the sphere of the mechanical and chemical relations of the constituent elements—to observe the resultant products into which they are transmuted; but the laws by which these results are produced, are mostly hidden from view. The Inductive philosophy, with its efficient causations, is here wholly at a loss: It cannot explain how the agents which form the vegetable or the animal cell should impart to that least microcosm the wonderful power of developing a new cell from within itself, or of adding cell after cell to its substance. Much less can it explain why or how it is that one cell is the rudiment of a plant and another that of an animal—that one expands into this plant, and another into that; one into this animal and another into that. All this is totally unknown. The principle of life and the conditions of life are only names for causes that cannot be explained by such methods. The effects cannot even be described, much less explained by the relations of efficient causation.

Under these circumstances we resort to the relation of adaptation and the assumption of design in order to define and explain the phenomena. After no other relation than this can we explain the fact that dead matter is transmuted into living particles, and that aggregates of these particles are developed into living organs, which act together so long as the being lives of which they are parts. By no other law than that of design can we explain how each class of living beings works for itself, having a form, habits, tastes, and instincts peculiar to itself, and how each individual of each class is an end to itself, having an individual form, size, and other peculiarities more or less marked, according to its rank and place in the scale of being.

Relation of final to efficient causes in the higher orders of being.

§ 306. Two facts are here suggested touching the relation of final to efficient causes. The *first* is that the higher we rise in the scale of being, the less we know of the relations of *efficient causes;* while those of *final cause* are *more and more various and conspicuous.* In unorganized matter we have occasion chiefly to apply efficient causes and their unvarying laws. As we ascend into the regions of life, we are more and more baffled in our attempts to detect the elementary forces and to determine their unvarying laws, but are more and more gratified at seeing the relations of adaptation become more and more conspicuous. The *second* is, *That one of these relations does not displace the other,* and the discoveries in respect to the one neither compel nor allow us to dispense with the search after the other. On the contrary, the more complete is our analysis of efficient forces and our determination of their laws, the greater is the opportunity to notice how the structure whose constituents are resolved by analysis, is controlled by manifest fitness and adaptation. Each newly discovered element and determined law opens an opportunity for some adaptation as yet unobserved.

Objections: (1.) Men mistake in their judgments about final causes.

§ 307. To the doctrine that the belief in design is intuitive, the following are urged as objections:

(1.) Men *mistake in discovering or assigning ends,* and the mistakes into which they fall are irrational and foolish; whatever man in his selfishness and limitation may think important to himself, he thinks must have been designed in the economy of nature, and thus is exposed to the danger of setting up his narrow and interested judgments as the real adaptations and intents of the Creator.

It is sufficient to reply that, if men mistake in assigning the ends of phenomena, they do the same in interpreting their efficient causes. We do not raise the question whether men can discover particular ends with infallible certainty, but whether they intuitively believe there are ends to which all beings and agents are adapted, and for which they are designed.

(2.) Our interpretations can neither be tested nor confirmed.

§ 308. (2.) It may be objected that we have no means of *testing and confirming our inductions* in respect to ends, while in respect of causes and laws we are provided with tests, rules and methods which are universally acknowledged to be amply sufficient. "In ordinary cases the methods of agreement, of difference, and of concomitant variations are acknowledged to be ample: In special exigencies artificial experiments may be instituted to supplement the deficiencies of simple observation: But in ascertaining ends we have no such methods, tests, or experiments."

We reply: It will be found on closer inspection, that the methods appropriate to the two are more nearly alike than would be at first imagined. It has been already shown, that the end or purpose, in its relations to the means of its realization, may be conceived of as an efficient force carried back from the end to the beginning of the series of causes and effects, which drives them to their issue by a constant energy. If this be so, the question, What is the particular end of a combination or series? may be answered by the methods appropriate to efficient causes. It may in some cases be less easy to conjecture the probable end than it is to conjecture the probable cause, inasmuch as many such ends might in a given case, be supposed to be equally compatible with the effects. But, on the other hand, in other departments of nature, as the *organic* and *historical,* the ends and adaptations flash upon the mind without the need of inquiry or tests of any kind, while in these very departments the efficient forces usually elude the most subtle analysis, and refuse to yield to the most exact and rigorous methods.

(3.) This relation derived from conscious experience.

§ 309. (3.) It may be still further objected that the adaptation of means to ends is an actual relation, of which *we are aware from our own conscious activity,* and it is simply by a fiction that we transfer it to other, *i. e.*, to material objects.

To this objection we reply, that the activity of our own souls and the relations which are instanced or exemplified in our conscious mental and moral functions, hold precisely the same relation to efficient as to final causes. The most complete knowledge, we may say the only complete knowledge, which we have of power or efficiency, is gained through or by means of the active energy of our own spirits. By this, we in a certain sense image, cf. § 206, this abstract relation whenever we have occasion to affirm it of impersonal or material agents. In doing so, we use examples, associations, and language taken from our personal activity. It is not true, however, that we affirm this relation of all the objects in the universe, because we have happened to experience its agency in our own spirits. It is by an intuition that we affirm it to be necessary to a rational construction of the universe. But this very objection itself suggests an argument in defence of the propriety of making a similar application of final cause. The power of adapting means to ends is one with which we ourselves are very familiar in our own conscious experience. We propose ends. We devise and arrange, *i. e.*, adapt means to bring them to pass. We interpret the actions of others by supposing that they are directed by such intentions and adaptations. We interpret the results of their actions when they are fixed and made permanent in structures controlled by the same relation. It is a fair *argumentum ad hominem* to use in reply, when it is objected that we interpret the universe by a relation derived from our uniform and personal experience, that in this experience we have an agency directed in part, at least, by design. The agency is spiritual, which first proposes ends and then adapts forces for their achievement. It is certainly possible or supposable that the results of a similar agency should pervade the universe, and make themselves manifest in discoverable adaptations. To assume or employ it in the explanation of phenomena is not necessarily unphilosophical.

The relation unquestioned in some applications.

§ 310. (4.) It may be objected still further, that if we recognize final cause as a principle, we introduce into the universe, for the explication of its phenomena, *two principles*, of which the one may at

(4.) Two principles introduced into philosophy which may possibly conflict.

times conflict with the other. In so doing we weaken confidence in the processes and axioms of pure science, and in the stability of the laws and the order of nature. Science, it is contended, must assume not only the *stability* but the *supremacy* of its own laws, and it can neither recognize nor respect any other.

It may be urged in reply that the principle of final cause, is so far from weakening our rational confidence in the stability of the laws of nature or disturbing our faith in the axioms of science, that it confirms both. What science blindly assumes, this rationally accounts for and makes necessary. It gives a reason for the order of nature and the principles of knowledge; and the only reason which can be suggested, viz., the adaptation of such order to the uses and ends of the human intellect, and of human science. As we have shown already, it furnishes the only solid foundation for the assumptions of induction.

But it will still be objected; if efficient causes and physical laws are to acknowledge themselves thus indebted to final causes, they must also confess their subjection to the same, and be ready to stand aside and be suspended whenever the principle of final cause shall require. In other words, the order of nature may be broken whenever the requirements of final cause shall so direct, whenever the claims of the so-called reason of things, or of alleged moral and religious interests may demand an inroad upon this regularity, either in special acts of creation or exertions of miraculous agency. This we assent to, but, we find no reason on this account to reject the principle or its asserted supremacy, but an additional reason for accepting both. If the principle of final cause will not only render the service of sustaining our confidence in the stability of the laws of nature in all ordinary circumstances, but will also account for such extraordinary deviations from this order as may be required in the history of man, then for this double service it deserves to be esteemed of greater value and authority. [Cf. Locke, *Essay*, B. iv. c. xvi. § 13.]

(5.) The search after final causes has hindered discovery.

§ 311. (5.) It is objected still further, that the search after final causes has seriously *hindered the advancement of science*, by turning aside the attention and interest of observers from their appropriate duty, which is the investigation and determination of efficient causes and their laws.

Lord Bacon, it is said, was so alive to its evil influence as to utter his memorable and oft-repeated caution in the words: "Causarum finalium inquisitio sterilis est et tanquam virgo Deo consecrata nihil parit."—*De Aug. Scient.*, III. 4. *Descartes* was still more strenuous in the same opinion, as appears from these assertions: "Totum illud causarum genus quod a fine peti solet in rebus physicis nullum usum habere existimo; non enim absque temeritate me puto posse investigare fines Dei."—*Med.* iv. 20. "Ita denique nullas unquam rationes circa res naturales a fine quam Deus aut natura in iis faciendis sibi proposuit discernimus, quia non tantum debemus nobis arrogare ut ejus consiliorum nos esse participes putemus."—*Princ. Phil.*, p. I. 28.

To this objection we reply: That what Bacon intended was that the attention of the inquirer should not be diverted from the investigation of efficient causes by the suggestion of ends or adaptations, for the appropriate sphere of the interpreter of nature is to develop agents and laws that are unknown, or newly to confirm and exemplify those already established. In this he was right. But, that Bacon himself believed that nature is penetrated and illumined by the higher relations of design is evident from this among similar intimations: "I had rather believe all the fables in the Legend and the Talmud and the Alcoran, than that this universal frame is without a mind." * * "For while the mind of man looketh upon second causes scattered, it may sometimes rest in them and go no further; but when it beholdeth the chain of them, confederate and linked together, it must needs fly to Providence and the Deity."—*Essays*, xvi.

When Bacon says that the inquiry after final causes is without fruit, he must mean 'practical fruit,' or the production of direct advantage to the interests of man. It is, in fact, so far from being barren, that as we have already seen, § 303, the consideration of ends has been fruitful in the suggestion of undiscovered agencies as their means, and has thus proved itself a most important agent in such discoveries. It has been more efficient in leading to the *prudens quæstio*, the *sagacious guess*, or the *ingenious hypothesis*, which has so often opened the way for decisive experiments. If our doctrine is correct, that the methods and rules of induction themselves rest upon the belief in design, then final cause is so far from being barren that she deserves to be honored as the *Alma Mater* of the *Inductive Philosophy* itself.

§ 312. (6.) It is objected again, that what are called the adaptations of nature, are only the *necessary conditions of existence* and its phenomena.

(6.) The adaptations of nature are only the conditions of existence.

When, for example, the eye is said to be adapted to the light, and both to the production of vision, this says the objector, is only another phrase for saying that the eye as we find it, acting with the light as we find it, produces its pictures upon the retina, and these acting with the intellect and sentient organism, produce the sense-perceptions which we call vision. What are called the ends of nature, to which her forces are said to be adapted, are simply the effects of which these forces are the necessary and actual conditions. The fish, we say, is adapted in its structure and its instincts to the water, and the water exists with relation to the fish, but the truth is that there could be no fish without water, for without water, the existence and conception of the fish are impossible. We know what appears, *i. e.*, what is made manifest, and we know it under the single relation of the forces which cause it to be. This is the only relation under which we can regard it. As to whether other effects might or might not have been produced from these causes in different conjunctions and intensities, we have no means of deciding. Whether other effects may not be produced in future we cannot say. All that we know is what *has been*, and *now is*, and by what means. These *have been*, and *are*, and occur under the operation of these very causes and laws. We inquire concerning the actual conditions of things, not concerning possible designs.

In reply to this class of objections, we need only say that they apply, not to the position that the belief in final cause is a first principle, but to the doctrine that this belief is derived from observation and required by experience. If the principle is intuitive and *a priori* (in the sense explained, § 246), we bring it with us to the explanation of the facts. We do not derive it from experience by an *a posteriori* method, but we apply it to experience by one that is purely *a priori*. It is true, if facts and phenomena were inconsistent with the principle, we should be embarrassed by the discrepancy of the two. But no incompatibility is urged, only that final causes are not proved by experience. It is conceded that the explanation by efficient causes is not inconsistent with that by final causes, inasmuch as it is through effects actually produced that we infer these effects were intended and provided for.

But we take issue with the position that we find nothing more than the conditions of existence. We find not simply the conditions of mere existence in the causes of effects produced, but the conditions of well-being, or adaptations to a highly artificial, elevated, and refined existence and enjoyment; and these in forms so manifold as to be entirely consistent with the *a priori* principle which we bring to the explanation of the facts. The illustrations of this assertion can only be gathered from the study of individual examples.

(7.) Adaptation is limited to organic existence.

§ 313. (7.) It may be objected again: that adaptation can only be traced in fact in a limited class of phenomena, viz., those of organized existence, whereas were it necessarily presumed it might be discerned in all kinds of being, the inorganic as truly as the organic.

It is sufficient to reply that examples can be found in every kind of object-matter as will be shown in another place. They are more striking within the region and sphere of life, indeed, but are not less real beyond that sphere. Besides, this axiom is the foundation on which rests the structure of the inductive method, which is as often applied to inorganic as to organic being. This makes it necessary to apply it to every kind and style of existence.

(8.) We are not warranted in affirming it of all kinds of existence.

§ 314. (8.) It might also be urged that we cannot trace or interpret adaptations on a scale sufficiently extensive to warrant our affirming that they exist throughout the whole universe of being. "We may, indeed, guess at them within a limited range of observation. But it is presumptuous to assume that we can trace the adaptations and discover the ends of the entire universe."

If this were admitted to be true, it would not hold against the principle that ends exist, and that adaptations to them regulate all the things that are. It is for the principle that we contend, not for infallibility in the application of it to individual cases.

The same law holds good of final causes as of efficient

causes. That both exist, and both control the universe is known to the human mind by the necessity of its nature. The discovery of instances and examples of each is accomplished by experience and induction. Both can be traced by observation in but few classes of objects, and within that portion of the universe only which comes under our eye or ear, or the report of our fellow-men.

But the one can be traced as far as the other. What is connected with its fellow as adapted to an end under this relation, is an efficient agent or force. If we can trace gravitation as far as the utmost verge of material being, we can also affirm that it was designed to hold the masses in their relative positions and their paths of motion. The principle of final cause moreover is absolutely required to warrant the extension of the relations of efficient causes observed within a limited sphere, throughout those regions of which observation and testimony can give only an uncertain and incomplete report.

(9.) Adaptation cannot be affirmed of an unlimited Being.

§ 315. (9.) Last of all it may be said, that the recognition of this as a first principle would require us to ascribe intention and adaptation to an unlimited Being, whereas it supposes certain forces or powers already given or existing, and the problem arises how to dispose of these so as to attain or produce the designed result. Such a problem can never, it is contended, be presented to an unlimited Being, who, by the very supposition, is not confined to forces or agencies which already exist, but can produce effects by a fiat of creative will. Moreover, the supposition would introduce into such a mind and order the reverse of the rational. It would make the production of agencies go before the disposition of them to an end. It would make blind force precede wise forecast.

None of these inferences are warranted. Because in the order of design thought must recognize the possible adaptations of forces, it does not follow that the forces must exist in order to be thought of as existing, or in order that certain adaptations should be determined on. Both, indeed, may be objects of design, the existence of the forces *and* their adaptations; or, rather, the existence of the forces because of their adaptations to accomplish some end or ends of thought. Even the human mind, impotent as it is to create, sometimes imagines to itself, *i. e.*, creates in thought some new agent in the world of matter or of spirit, and revels in contriving the variety of uses to which it might make it subservient. How much more readily may that Being whose thoughts can in any instant become powers, laws, and facts!

The principle is illustrated and confirmed, by its applications (1.) to metaphysics.

§ 316. But the most instructive view which we can take of this principle is to contemplate the *variety of its applications*. It has already been observed that First or Intuitional Truths, are never apprehended in actual application as general propositions. They can only be discerned in the concrete, as they actually connect individual things or phenomena. Thus we cannot discern causation or adaptation as *universal* and *a priori;* we only discern an event or being as causative or caused, as a means or an end. When we appeal to the use which is made of these relations in the sciences as proof that they are fundamental and intuitive, we expect to find that these sciences constantly assume these relations to be valid, by explaining phenomena by means of them. The constant repetition of this relation and the important uses to which it is applied add incidental strength to the positive arguments for regarding it as an intuition of the intellect.

1. The first application which we notice is that which is made by metaphysical science itself. We have already insisted on its importance in sustaining the metaphysical axioms of Induction. Upon this we need not dwell.

Its application in the formation and arrangement of those general conceptions which are at once the materials and the conditions of all science, is of equal consequence, though perhaps not equally obvious.

(*a.*) The principle of final cause regulates the formation of concepts. We have already seen that so far as the form of the concept is concerned, it is by abstraction or analysis that we separate the qualities or attributes of existing beings, and by synthesis unite them into new and generalized products. These processes regulate the form but not the import of the concept. We are not at liberty to select any attributes which analysis gives us and to unite them into any complex notion which they might form. Some are adapted by logical compatibility to be conjoined, while others are not so fitted.

But again: not all the attributes which are logically compatible are, in fact, united into concepts by any earnest thinker. The centaur, the mermaid, the hippogriff, are logically possible, but not actual. Why? Because the properties or attributes which constitute them are not adapted to exist together in the same being, and, of course, except for the service of the fancy, are never combined. There is something in these properties, or in what they represent, which fits them to co-exist, or they could not with any reason be combined in a concept which connects the rational and real; which represents things as actual or possible, or contemplates them as designed to be under existing powers or laws.

(*b.*) The same principle must be assumed in the arrangement of a system of concepts as genera and species.

It is evident, that as we might make as many concepts as the varied aggregations of single attributes would allow, so these might be arranged into as many genera and species as the fertile law of permutation and combination would permit. Any one attribute might be taken as generic without regard to its actual extent in nature; with this any other might be combined as a *differentia* without regard to the compatibility of the two as provided by the adaptations of nature's laws. It is contended by some, that in the classifications which we actually make, we are guided by mere convenience, that we can make any attribute generic which we please, provided it be more extensive than its differentia in its actual prevalence, but that there are no such things as real *genera* and *species;* these terms having no meaning in such an application. If we assume that there are no affinities or adaptations in properties and laws, and no ends to which the powers of nature are adapted, and which are designed to be permanent, this view is correct. But the moment we assume that such adaptations exist, and that they can be discovered by observation and induction, then the belief in permanent classes is justified and explained.

(*c.*) This relation is essential to an intelligible conception and definition of an individual.

(*d.*) The principle is of the greatest value as a criterion of truth and a rule of certitude. When skepticism suggests that every principle may be questioned, and every observation of fact may be mistaken; that the objective creation may be a shifting phantasmagoria, and the subjective mind but a lying glass of opinion; then the thought of the inconceivable non-adaptation of such a universe to any rational end even of knowledge, restores our confidence in the testimony of the senses, the experiences of consciousness, and the inductions of thought. We try all these indeed by one another, after the tests which experience and science have discovered, but we *trust* them only when they conspire to ends that are worthy of rational order in a universe adapted to be known by a being who is manifestly designed to know, and to confide in his knowledge when it is properly tested and proved.

Applied in geometrical construction and deduction.

§ 317. 2. In the Mathematics even, the presence of this relation is recognized.

In pure geometry it may be applied more frequently than would be anticipated. The circle is adapted to prove a great variety of theorems, and to solve many problems, as is exemplified in any treatise on geometry. If we are required to construct two triangles on the same base, the angles of which at the apex of each shall be right angles, it can readily be done by describing a half-circle on this line as a diameter, and any number of triangles can at once be drawn so as to fulfill the required conditions. We discern in a portion of space bounded by a half-circle, this capacity or adaptation, that waited long to be discerned.

The relations of pure number open as wide a field of inherent fitnesses to serve the ends of the student. It is upon the faith that additional adaptations remain to be discerned that the mathematician prosecutes his work of inventive discovery.

The adaptations of the mathematics to the service of physics are if possible still more striking. No projectile was ever thrown in an exact parabola; yet the theory of this curve is adapted to explain the direction and motion of every body that is launched into the atmosphere. The theory of the lines in which bodies tend to move, and the rates in which bodies move in fact when impelled, is adapted to regulate the mechanics of bodies as they fall to the earth, and the motions of the orbs which revolve in the heavens. It also explains the phenomena of the pressure of fluids. The relations of number solve the mystery of chemical combinations: they explain the symmetry of agreeable forms and the harmony of musical sounds. They enable us to discern a common law in the arrangement of the leaves upon the stem of every tree, and in the placing of the planets along the lines which stretch out from the sun.

On the first thought, it would seem that in extension and number it would be impossible to find so great a variety of possible adaptations. But on reflection, we find that their capacity of multiform application is the only key to the perfection of the sciences of matter and the reduction of its forces to unvarying laws.

We have urged that the belief in final cause must be intuitive, because we could not otherwise confide in the axioms of induction. But we see in the provision for the possibility of mathematical science, and of its universal application to material phenomena as the indispensable condition of their laws, another example of design where we had least expected its manifestations, viz., in those time and space relations which render the mathematics possible.

§ 318. 3. Geology and Paleontology both assume the truth and applicability of the principle of final cause.

Applied in geology, etc;

Geology was at first content to explain the formation of the crust of the globe by analyzing its parts into their constituent elements, and recording the order in which the rocks had been compacted and broken down, and the strata had been formed and deposited. In these investigations it proceeded as a science of observation, watching and recording the operations of the forces of nature according to laws already ascertained.

But, aided by paleontology, geology has proposed to itself a higher problem, and contemplated facts under more elevated relations. It has traced a plan and order of development resting on the assumption of a series of ends subordinated to one another, and terminating in a habitation equally adapted to man's higher and lower nature. It has ventured to recall the successive phases of organic life by reproducing extinct species of plants and animals amid the lakes, marshes and jungles in which they sported and from which they subsisted, and to arrange these phases in the order of time and of a more and more perfect development. The assumption which has directed these bold essays and enabled the geologist successfully to apply the hints furnished by facts observed, is, that an order of fitness and progress has been followed from the first, and every epoch has prepared the way for the next succeeding; the adaptations of each being complete in animals, plants, and scenery. Following the same clue, this science has found in each previous epoch, not merely the materials of the one which succeeded, but the existence of a less perfectly developed form of life. This series terminates with man, who not only represents the highest type of life, but shows that he is the end for which all others are designed, by the fact that he alone can comprehend the import of the plan and recognize the relations of the parts to the whole, and of the whole to himself.

Geology, by the very aims which it proposes, and the splendid results which it has achieved, gives its tacit yet fervent assent to the original authority of the intuition of final cause.

Applied in geography and history;

§ 319. 4. Philosophical Geography gives a similar testimony. This science, as conceived and perfected by Ritter, takes the earth where geology leaves it, and shows how each continent and country was fitted for the part which it has played in the world's history, by its structure, surface, soil, and climate, by its mountain barriers to repel, and its coasts and harbors to invite; by its river-systems to bind remoter portions, and its insular situation to facilitate defence. It shows that every part of the earth was not only adapted from the first to receive and develop the race which was allotted to it, and to become the scene of events which have made it memorable, but to transmit the results of these achievements to neighboring countries and other races, and even to transfer them to remote parts of the earth and a later and better civilization. By referring intellectual and moral influences to favoring physical conditions, it enables us to find an adaptation to important moral results, even in the physical arrangements of the earth.

The Philosophy of History also must assume that the events of human history, have occurred in obedience to definite laws regulating constant forces. Whatever these forces may be called— or whatever may be the law of their action, the historian cannot seek to interpret or explain them without believing that there are definite aims toward which these forces tend, and after which they are regulated.

§ 320. 5. Comparative Anatomy rests upon the same intuition. It could have no meaning, as it would have no truth without it. It is a science of similar adaptations, not only of organs to functions, but of analogies of form and feature and inner structure to the completeness of a progressive plan, and even to the achievement of an æsthetic effect and the expression of an æsthetic import. Give this science a bone, and it will draw or model the animal, tell you how large he was, how formed, on what he lived, what were his habits and disposition, what the length of his life,—just so far as it reads the adaptations that gather and cluster around this fragment of a skeleton, which except as thus interpreted were only a broken and abraded fossil.

also in comparative anatomy and physiology.

6. In Physiology, special and general, similar relations are more numerous and manifest. The departments of animal and vegetable life abound, or rather overflow with examples of fitness and adjustment. The nicer the analysis of elements and of organs, and the more subtle the detection of offices and functions, so much the more exquisite are the discerned relations of adaptation of each to each. Not only is there seen a fitness of one organ to another,—as of the lungs to the heart,—and to the common end of all, but there is a fitness of every organ to the element in and by which it acts,—as of the lungs to the air and of the eye to the light. The more we learn of the structure of the one and of the properties of the other, the nicer are the adaptations which we discern between the two.

The adaptations of the body of man to the functions and uses of the rational soul, are still more striking; but we here approach, if we do not cross, the line which divides physiology from Anthropology.

§ 321. 7. In Anthropology we trace these higher adaptations. The human hand does not differ more strikingly from the hand of the monkey than the mind of the monkey from the mind of man. The mind of man has endeavored to discover and combine the powers of nature, and to devise the appliances of art. Whatever the mind has prompted the hand to construct, the hand has been able to frame, either through the seemingly exhaustless versatility of its flexible organism, or by the tools and machinery with which it

Applied in anthropology;

has contrived to supplement its powers. So wonderful has been this service, that it has been questioned, whether the human intellect or the human hand has been the most conspicuous in shaping human destiny and in developing human history. The hand has also by the economy of nature been fitted to be the medium of conveying varied intellectual and emotional expression to the intellect and heart, which have been as mysteriously fitted to receive and interpret its indications. The hand invites and repels, commands and forbids, soothes and enrages. It appeases with its gentle waving, and smites with its ferocious energy. It adores with the uplifted arm, it blesses with the outspread palm; it blasphemes with aimless and impotent motions, and curses with its downward stroke.

In the provisions for and the capacities of language.

But there is no adaptation of the mind and body that gives to both united, an interest which at once so fascinates and baffles our prying scrutiny, as that exhibited in the agency of both in the production, use, and development of language. There are two conditions of language, the bodily and the mental. The bodily are also two, the mouth and the ear, to which the hand and the eye are accessory. But for expression the mind must also furnish the material through its required capacities and development. Language is impossible until the mind observes and generalizes and affirms. The mind must first think the material and spiritual universe with which it comes in contact, into the thought-world which its powers and laws fit it to create, before it can give to it expression by language. This adaptation of the vocal and the spiritual to each other, and of the possible elaboration of the one to the possible refinement of the other, go far beyond any observed fitness of the eye to the light, or of the ear to the agent of sound. Not only are these two parts of the complex body and soul fitted to expand side by side with one another, but the expression of thought in language reacts with wondrous energy on the development and refinement of thought itself, so that it is not only true that the developed thought finds itself able to employ language in its service, but it is also true that the thought in order to be developed, must express itself in language. Man not only speaks because he thinks, but he speaks in order that he may think, *i. e.*, think with clearness, precision and progress. The two are not merely so adapted that the one can expand side by side with the other, but it is difficult to say which is the most dependent on the other.

The celebrated Galen says, in his treatise concerning the human body, that by the variety and accordant action of its adjustments, it seems to utter an anthem of praise to its Maker. But the philosopher who reflects on the mystery of human language, in the subtlety of the elements involved, the variety of the conjunctions, the delicacy of the structure, and the capacities for growth and development, might find, as he watches in the lispings of infancy the feeble beginnings of such splendid results, a new meaning in the familiar words "Out of the mouths of babes and sucklings thou hast perfected praise."

§ 322. 8. In Psychology the manifestations of final cause are more frequent and obvious than in either physiology or anthropology.

Application to psychology.

It is now and then difficult for consciousness to analyze its operations under the relations of efficient causation, or to trace each product back to the separate force from which it springs into being. But the adaptations of these operations and products to one another, and to the manifest ends of the soul's culture and well-being are often so obvious and remarkable, that they partially settle questions that would otherwise remain unsolved. For example, in considering the acquired perceptions, we find that animals possess from the beginning, a capacity of judging of distance and size which man is forced to acquire by slow and painful effort. We question whether our observations can be trusted, whether there is not some error or oversight in the analysis of the phenomena. The consideration of the end to be accomplished by this arrangement relieves the difficulty. Man, we observe, needs the discipline required by the slow process of acquiring what the animal knows at the beginning. The consideration of adaptation removes the similar difficulties suggested by the question, "why the range of instinct is so much wider and more unerring in the lower animals than it is in man, the highest of all?" When we consider the diversity of the destiny and ends of the two we accept with less hesitation the evidence which observation furnishes.

Above all, psychology acquaints us with the rational faculty as that pre-eminent power which proposes ends and devises means for their accomplishment. It acknowledges that this is the highest of the intellectual powers, that it is lawfully supreme, that in the service of this power we investigate causes and determine their laws. In the subjection and adaptation of the lower powers to this highest of all it finds confirmation of the propriety of assuming the relation of adaptation in all our interpretations of nature. If "on the earth there is nothing great but man, and in man, there is nothing great but mind," it is emphatically true that in the mind there is nothing great but the reason which proposes and discovers ends, and is itself an end to the lower actings of the intellect.

Applied and assumed in ethics.

§ 323. (9.) Ethics, the science of duty, which is so closely allied to, if it is not a department of psychology, is founded entirely upon the intuition in question. Its subject-matter is derived from the ends of human existence and human activity. The comprehensive and fundamental question which it asks, is, for what kind of activities is the human soul adapted by its constitution, and what must man be and do to fulfil these ends of his being? In these inquiries, it rests on the single assumption that man is fitted for one kind of activity rather than for another, and that the action for which he is fitted is right, while the action for which he is not fitted is wrong. It asks, how shall these adaptations be discovered? By what faculty or capacity, one or more, are they discerned and responded to? What are the tests or criteria by which they are distinguished? What external actions or duties must we perform in order most effectually to fulfil these several ends of our being?

Corresponding to the power of apprehending duty, is the faculty of will or choice qualifying man to fulfil the ends for which he exists. The existence of this power, its importance to human development and responsibility, and the necessity that it should be defended in its integrity, explain the necessity of moral trial, and the possibility of moral evil—under the one relation of the ends which the possession of this power and the exposures which it involves are adapted to fulfil.

The adaptations with which ethics has to do, are chiefly psychical, and suppose a spiritual organism in the soul—a system of internal adaptations in the several powers with which it is endowed, which indicate our duties and our obligations. These all look toward moral perfection. To this the soul is adapted and to it it tends and is impelled. Without this intuition and faith in its truth, ethics can have no meaning and duty no authority. If reason as proposing ends is the highest ruling power in man, then the reason, when it discovers and proposes the highest moral ends, exercises its loftiest functions, and reigns sovereign over the inner and outer world by a self-justified authority.

Application to theology.

§ 324. 10. In Theology, or the science of God, whether natural or revealed, this principle is of supreme importance. The most of the so-called demonstrations of the being of God, find their material or grounds of proof in the indications of design that are furnished in the material universe.

The common argument for the Divine existence.

These arguments are usually stated somewhat thus: Design proves or implies a designer; The universe abounds in design; Therefore the universe implies or proves a designer: or, order and adaptation imply a designer; The universe abounds in order and adaptation; Therefore a designer exists.

The major premise in this argument is obviously assumed or received as *a priori*. The minor is a statement of fact grounded on observation or induction. Those who employ this argument would not accept the view, that the belief that adaptation prevails throughout the universe is a first truth or axiom of thought. They rest their belief upon observation, and they search through the universe to discover instances of the presence of this relation. Having observed a sufficient number, they generalize them by induction, and then apply, as the minor premise of their syllogism, the proposition which they have established by this cumulative evidence.

We have sought to prove that the proposition affirming final causes is a first principle or intuitive truth; that it is not in any sense dependent on observation, but is an original and necessary belief or category; that so far from being derived from induction, it is the necessary ground on which induction itself must rest for its validity and application.

Those who accept the relation of final cause as necessary and *a priori* may be grouped under two leading classes or divisions, according as the adherents of each reject or accept the belief of a personal God. The one class believe in an immanent force, which involves no relation to any thing beyond the universe as a whole. They fully accept the truth that design rules throughout nature. They find examples of the relation of final cause everywhere present. But they insist that these do not necessarily carry the thoughts out of nature: Final cause or design is a force in nature itself, *immanent* in each separate object, and in all existing objects taken as an organism of parts mutually related and connected.

Those who hold this doctrine, concede that adaptation prevails in nature, and must be assumed to explain its powers and operations; also, that it works in every case as though a personal mind had contrived these ends and the relations which they involve, and continues to direct them. But they urge that we are not compelled to ascribe this adaptation to a personal being, but may refer it to an impersonal, unconscious, unthinking force, as blind and unintelligent as the efficient forces that act by mechanical laws.

The second class contend that the necessary correlate to *adaptation* is a designing mind: They conceive of adaptation as the objective relation to which thought is an essential supplement. Adaptation does not prove or indicate design, but it rationally implies it; if, therefore, the adaptation is real, so is the designing mind. In assuming the one truth by an *a priori* necessity, you must accept the other. The belief in adapted things both logically and really carries with itself the belief in adapting thought and an adaptive thinker. The mind need not necessarily think of the two at the same instant, or in the same connection. The attention may be so concentrated upon the adaptation objectively considered, its ingenuity, the variety of the means employed, the intricacy and order of the combinations required, that it does not consciously refer to the correlate, but this fact does not prove that it is not necessarily involved. For example: in a machine of human devising, an ingenious mind can discern very many adaptations, without adverting to the mind which produced them, or distinctly recognizing the fact that it proceeded from any thought; but as soon as it raises the question and reflects on the relation, it cannot but assent to the additional truth.

The application of this principle in the service of Natural Theology raises still another question; viz., What relation has efficient to final causation in the universe? Does each require its separate principle or agent, or do both united direct us to one? Does the adapting agent simply take the efficient forces and laws of the universe as it finds them, and arranging them as best it may, bring out of them the wisest results to which its sagacity may adapt them, or does it also originate the forces which it arranges and combines? The one view gives the eternity of matter, with its hindrances and limitations and possibilities of evil, making the Deity a Demiurgos or Plastic energy. The other makes the originator and the arranger to be the same power and the same mind. The one view is the cruder theism of Ancient Philosophy, the other the purer theism of the Jewish and Christian Scriptures.

It would carry us too far from our appropriate theme to argue here the question between the two. The discussion of it belongs to a treatise on Natural Theology. Psychology suggests that the analogy of the human soul, which combines in itself—under limits—a creating force and an adapting or designing force, furnishes a decisive argument in favor of the conclusion, that the creator and thinker is One Being.

CHAPTER VII.

SUBSTANCE AND ATTRIBUTE: MIND AND MATTER.

Uses and Etymology of the Terms.

§ 325. We return again to the relation of *Substance and Attribute*, and to the important application of it in the determination of the definitions of *Mind and Matter* and of *Real and Phenomenal Being*. The relation is so fundamental and so much discussed in Psychology and Philosophy, as to require a careful consideration.

The substance or substratum with which we have to do, is the *Real substance or substratum*. As such it should be carefully distinguished from the *logical substance or subject*. A logical subject is any thing which is conceived in thought as a substance with attributes, whether it does or does not exist in fact. Thus any *abstractum* can be treated in thought and described in language as though it had real being, and were endowed with real attributes. The concepts *power, goodness, responsibility, representation, republic, wages, wealth*, or any other abstract notion, may be conceived in thought and treated in language as having properties or qualities which are affirmed of each as though it were a real being. *Real substance* ought also to be distinguished from the *grammatical subject*. The grammatical subject is any word which is used in language as though it denoted a logical subject.

The concepts, substance and attribute, cannot be resolved by the etymology of the terms which designate them. The words *subject, substance, substratum*, have a common derivation which literally imports something *standing or lying under*, and implies that there is something placed above or upon it which may be removed. This suggests the impression that the attributes are superinduced upon the substance, as folds or wrappings are thrown over or around a nucleus or core within. This prompts to the effort to lay off the covering, to separate the wrappings from the nucleus which they invest, to scale off the laminæ or folds, and find the substance or substratum within or beneath, bare of all qualities and relations. But the attempt to lay aside qualities in order to find their subject is soon discovered to be vain. It is as though one should cut down the trees in order to find the forest. It is found to be impossible to discover an actually existing subject without attributes. The simplest and barest object in the universe—that which in its nature is the most uninteresting and the most undistinguished—as the *mote* in a sunbeam, the minutest perceptible *grain* of *sand*, the *atom* or *molecule* which the physicist cannot perceive, the *monad* of which the metaphysician confidently speculates—must always be conceived as having place and form, and as involving the relations of extension and force.

The *etymology* and *use* of the terms *attribute*, *quality*, *property*, and *accident* do not give us any greater satisfaction as to the nature of the distinction. The term *attribute* simply directs the attention to the fact that we *attribute to*, or *affirm of*, a being, something which we distinguish from itself; but it does not in the least explain what we distinguish or that from which it is distinguished. *Quality* is a term of classification merely, *i. e.*, it signifies that the being is of a certain class, without explaining why it belongs to the class in question. *Property* indicates, that what we thus attribute or affirm belongs peculiarly or *properly* to the being or substance, and *accident* that it belongs to it *occasionally*. These different words are only different names for the same conception, as differently used. But their *etymology* or *application* throw no light upon the conception itself, or how it originates, or why it is distinguished from its correlate *substance*.

We learn, moreover, that we can no more find an attribute without substance, than we can find a substance without attributes. We cannot separate length from something which is long, nor color from something colored, nor thought from a thinking being, nor joy from a rejoicing being. The two conceptions are never parted in the world of real existence. They are not merely correlated by a logical necessity, but they are always inseparably conjoined in actual existence.

Substance and attribute in the abstract.

§ 326. But though substance and attribute do not exist apart, they can be conceived of and defined as abstracted from one another. Abstractly considered, the concept, substance, is less general than that of simple being. Being has already been explained as every object that is, or that is conceived to be, knowable or known. But every thing that is known is not only known to be, but is also known as related.

Hence, with every act of knowledge, the concept of *being as related,* at once arises and becomes universally applicable to every object that is known. Certain of these relations may be used to distinguish, define, and explain these knowable objects. *Any being with relations so discerned and applied as to distinguish it from other beings,* is conceived as *a substance, i. e.: a substance* is *a being distinguishable and definable by a complex of relations.*

The conception of attribute arises in a similar way. As soon as an object is discerned in a definite relation to another object, this relation can be affirmed of or attributed to this object. When one relation or more is applied to define or distinguish any one of these beings, it becomes an attribute, as used in this generic and technical sense. *Every relation by which a being is known or distinguished* is *an attribute.*

It deserves to be noticed here that there are also as many different substances as there are beings distinguishable in kind by combinations of relations. *An individual substance* is known only by the individual relations which it shares with no other. *The substance is not, however, made up or constituted, by its relations. It is not the same thing as a collection of attributes. It is distinguished and defined only by these relations.* From this it is manifest that the *category of substance and attribute* is not simple and original like the other relations or categories which we have considered, but is *complex and derived.* Any one of these relations, when employed for the ends of recognition or description, for definition or classification, for reasoning or explanation, in short, for knowledge of any sort, whether common or scientific, becomes an attribute. Any existing thing, when it is sufficiently permanent or oft-recurring to require to be known by attributes, is called a substance.

There are two classes of objects-matter to which this category is most frequently applied, *spiritual substances* and *corporeal substances. Abstract ideas,* or *abstracta,* follow the analogy of real beings, and so do grammatical subjects. Mathematical entities do the same so far as this relation is concerned. We shall consider the two classes which are here named, and begin with spiritual or mental substance.

Spiritual or mental substance.

§ 327. Here we encounter, at the outset, the objection or difficulty that a mental or spiritual being cannot be a substance at all. This difficulty is merely verbal, and is of purely casual association, arising simply from the fact that substance more frequently implies material existence. Dismissing this objection as merely verbal and superficial, we proceed to inquire in what sense spirit is a substance and what are its distinguishing attributes, especially in the form which it assumes as the human soul.

Our previous inquiries have taught us that the prominent attributes of the substance which we call the human soul are its capacities *to know, to feel,* and *to will.* But to know, to feel, to will, are operations or modes of activity and suffering. These capacities are energies, simply causative of certain effects, or which involve energies that are causative. These three attributes obviously fall under the category or *relation of causation.*

The power of the soul *to be conscious* of its acts and states is also a capacity for *causal* efficiency, which like the others is known by its exercise and its results.

But we know more of the substance of the soul than that it is the cause or recipient of those effects which we call its states. The truth is established by consciousness that the soul knows these acts and states to be *its own, i. e.,* to be caused or suffered by the individual *ego,* or *self.*

These states and products of the soul's causal activity, are transient and changing, while the *ego* is *permanent* and *enduring.* As the cause or recipient of these changes the soul is *identical* with itself. They are diverse, the soul is one. The psychical attributes therefore require the categories of *identity, diversity* and *time,* as well as that of *causation.*

Besides the attributes of the soul which are revealed in consciousness, it is capable of acts or processes of which it is conscious only of *the results* in its psychical experience. The capacities for these results, whether the results are dependent on psychical or material conditions, are also causative, and are therefore properly classed among causative attributes.

Besides the relations of *causation* there are relations of *design* which pertain to the soul. These are conspicuous both in the relations of one power and act of the soul to another, and also in the relations of the soul to the external world and to the body which connects it with that world. All of these relations are attributes of the soul; some of these, however, are so necessary to an adequate conception of its nature as to deserve to be considered as *essential* and distinguishing.

We find then, that those relations of the individual *ego* by which it is usually defined, are its capacities to do and to suffer, to know and attain its end or destiny, and these attributes are all found in the Categories of Causation and Design. When to these we add its relations of Identity and Time we complete the cycle of its attributes. From this analysis we derive the following definition: *That Substance which we call the Human Soul, is an identical enduring self, capable of spiritual acts and states in the succession of time, which are adapted to certain ends with respect to the universe of being.* The relation of substance and attribute, when thus applied, is that of a being and of certain distinguishing or essential relations, as of time, identity, causation, and design, which appertain to the being.

Material substance defined.

§ 328. A material substance, again, is, like spiritual substance, a being discerned or discernible by intuitive or direct knowledge and also definable by a sufficient variety and number of relations to distinguish it from other beings. These relations are discerned by sense-perception and consciousness, and are generalized by thought. *A Material Substance may be defined: a being occupying definite limits in space, and productive of specific sensations in the sentient soul, on occasion of which it is perceived or known to exist.*

First of all, it is related to space in trinal extension. It might be urged that, in one sense, the spectrum cast by the camera on a screen, or the rainbow flung athwart a cloud are material substances, with only superficial or binal extension; but material substance, in the ordinary sense, has threefold extension, or, as we say, extension in three dimensions.

Corporeal substance has a *second* relation to space, viz., that of space-occupying or space-filling. This is often called the *solidity* or *impenetrability of matter*, but should be carefully distinguished from that power of matter to awaken the sensation of hardness, which is also called solidity. The first is a relation to space which is tested and expressed by the application of motion. The second is the capacity of the body to excite a specific sensation on occasion of touch.

The *third* class of relations which belong to corporeal substance are its powers variously to affect, through the senses, the body as animated and ensouled, and also the soul itself as a sentient agent. Every material substance has power to produce certain so-called impressions on the so-called organs of sensation, *i. e.*, upon the body as organized to receive these impressions. Of these effects, the vibration of the tympanum, and the formation of the image on the retina, are examples. These may occur without sensation, as is manifest in cases of disease, of mental excitement, and the use of anæsthetic agents. But the condition of any of these effects, is *a living body*. Consequent upon these are those effects upon the sensitive or sentient soul which are called sensations, or sensations proper. The condition of the last is a body living and *ensouled*. In sensation, or the sense-element of the complex act called sense-perception, the soul is purely receptive, or *passive*, and the material substance is active. Its various powers to produce these sensations are all comprehended under the *category* or *relation of causation*.

On the condition of the experience of these sensations the being which causes them is known to exist as a *Non-ego*. But the possibility of being perceived is in itself no attribute of matter in the sense of being a causative power. To perceive is an act of the mind. The causative energy and the capacity which fits for this act, both pertain to the mind alone. Matter, so far as it is perceived, acts neither upon the body nor the soul. Matter is, *i. e.*, exists, and is known by the mind to be. It is not correct to say, that it is known *only* as the cause of the sensations which the soul suffers or receives; making it to be known *only* as the *unknown cause* of a *felt* effect. Rather, it is known *to be and also as causing* these sensations. As existing, it is known to have other relations than its power to cause sensations. Space is a reality, and so are the spatial relations of matter as known. To define matter with J. S. Mill (Logic, I., c. 3, § 7,) as "the external cause to which we ascribe our sensations," is to overlook entirely these relations of matter to space, and misinterpret the act of knowledge itself. To say that "matter may be defined as a permanent possibility of sensation" (Exam. of Ham., c. xi.), is to fall into a more serious error.

Besides the relations of material substances to space and to the animated and ensouled body, there is a class of relations which it holds to other bodies. These are its powers to produce effects in or upon them. They comprehend all the properties of matter, whether mechanical, chemical, or organic, which have as yet been discovered, or which science may in future unfold. That all these attributes are comprehended under the *causal relation* is too obvious to need illustration or proof.

The relations of matter thus far considered are those of *space and causation.* We define material substance by means of these as a being having a definite form or outline (*involving relations to space or other bodies existing in space*), occupying exclusively some portion of space (*involving space-relations*), and productive of specific sensations in the sentient soul on occasion of which it is known to be (*involving relations of causation and objective reality*).

We repeat the remark, that this complex or collection of relations do not *constitute* a material substance. They simply *indicate* that it *is* a material substance. They are relations which *define* and *distinguish* it as such. They constitute its *logical essence* only. The same is true of the element *being* which is implied in such definition. *Being*, like every other simple notion, cannot be defined; but it does not follow, as we have already seen, that it cannot be known and understood.

Space occupation and identity of matter.

§ 329. A material substance has been defined as exclusively occupying a portion of space. It is not required that this portion of space should be of any definite size or dimensions. A grain of sand is a material substance; so is a large mass of sand-stone: so is a portion of water or the indefinitely expanded atmosphere. All that is required is, that the mass, be it greater or smaller, should be so fixed and held together in its parts as to occupy continuously their defined limits. The continuity of parts is of more importance than the continuity of definite outline. This continuity or coherence of parts is maintained in different substances by different agencies. The constituent parts may be held together by simple mechanical aggregation under the force of cohesive attraction. They may be held more closely by the polar force of crystalline arrangement. They may be united still more intimately under the laws of chemical affinity. They may be combined and assimilated into the forms and products of organic existence; or the substance may be conceived as an ultimate molecule, or a simple cell. Every being that is one and continuous, of whatever size, in whatever form, or held together by whatever bond of union, is a material substance.

A certain continuity in time, or permanence, is either required as a defining characteristic of substance, or is implied in its definition. This integrity of the whole is presumed as having continued and as likely to continue for some considerable period, or the being indicated would scarcely be called a substance. It certainly would not be worth noticing by defining attributes if it did not so remain.

The *relative permanence* of material substance explains the conception of its identity. Identity in such a substance may pertain to the constituent elements only, or to the form only, or to the uniting force, or it may be applied to the connection of one part with another in a series of changes which involve a total alteration of both constituents and form. Thus if the same particles remain united in the same form by mechanical aggregation, the substance is eminently the same; the only diversity in such a case being that of relation to other substances—a diversity of time or place or both. Should the constituents remain the same and the form be changed, it might be called the same, provided the constituents were viewed as more important than the form. If the external form is changed by growth or development, as in plants or animals, the continuously acting force is regarded as making them a substance. If the parts of a knife or a ship are displaced and replaced by successive removals and substitutions while the form and functions are retained, the substance is still called the same by a loose analogy taken from living agents and their gradual accretion and growth.

The production of new substances.

§ 330. We have seen that a change in form and structure or in both, involves the production of a new substance, because it involves the production of relations which clearly distinguish such a substance. A living being, as an animal, consists in part of certain material particles or elements. If a succession of changes or decompositions and recompositions could go on before our eyes, so that we could trace the same particles back through every form in which they can possibly exist, through plant, mineral, earth, air, water, etc., in every possible form of chemical and crystalline combination, till we had reached the ultimate molecules or elements of all and of each, we should evolve a series of substances, one after another, in a consecutive order of gradation.

But the simplest elements, the ultimate particles, would still be substances with attributes which they must continue to retain and from which they could never in fact be parted. Those who seek an interior substance, divested of attributes—the nucleus of the outer—are misled by a secondary use of the word.

The real Essence or the *Thing in itself.*

The "*underlying substance*" of the schools, the "*thing in itself*" of Kant, are mere names, which signify either being in the abstract or being in the concrete. If it is being in the abstract, then it must be synonymous with matter as knowable, *i. e.*, it is *a concept* only, which can be separated from its relations in thought but never in fact. If it is being in the concrete, then this must be known *with* its relations and never apart from them. In either case the substance or thing *in itself*, cannot be known *by itself.*

A material substance not necessarily independent.

§ 331. A material substance is not necessarily *independent or self-subsistent.* This was insisted on by Spinoza, who defines substance to be "*that which exists and is conceived by itself.*" "*Per substantiam intelligo id quod in se est et per se concipitur; hoc est id cujus conceptus non indiget conceptu alterius rei a quo formari debeat.*" *Ethica*, p. i., def. 3. From this definition the inference was direct and irresistible, that no finite substance is possible, because every so-called finite material substance is produced or sustained by other material beings, and is dependent on them; or, on the other hand, there is but one such substance, and that is the total of all which exists—the universe; this totality being conceived as absolute and independent.

Locke falls into a similar manner of speaking when he describes the constitution "which every thing has within itself without any relation to any thing without." Similar to this is the doctrine of Whewell, that substance is indestructible. "The supposition of the existence of substance is so far from being uncertain, that it carries with it irresistible conviction, and substance is necessarily conceived as something which cannot be produced or destroyed." *Hist. of Scient. Ideas,* vol. ii., p. 32.

Our analysis has shown that a material substance is so far from being independent of other beings and forces, that, properly speaking, no material substance is in any sense independent, or can be conceived to be so. Every material substance is what it is by the productive or sustaining force of all other beings and forces in the universe. It is also conceived and defined to be what it is by its relations to these forces, the expressed and the implied. It cannot exist and cannot be defined except by these relations to other beings and agencies. If material substance is dependent, it is not necessarily indestructible. Should the forces which sustain it be withdrawn, or their action be changed, it would cease to be, or cease to be the same substance that it was.

And yet we constantly assume that material substances are permanent,—not the ultimate particles alone, but even the continuous forms in which they exist and perpetually reappear. If we did not assume this, we should not define the constituents of either,—we should neither form them into concepts, nor apply these concepts for the ends of knowledge. What is the nature and what are the grounds of this assumption? They are none other than that the agencies and laws which sustain and produce these substances will remain, in order to accomplish certain ends for which they exist. In other words, it is only by *relations of orderly design* that we can explain or vindicate that belief in the permanence of the material structure as to its forms of being and their constituents which is received as an axiom in all physical or inductive philosophy. That this permanence or indestructibility is not essential or necessary, that it cannot be viewed as an axiomatic truth, appears from the broader and deeper axioms on which it rests. These axioms involve the relation of design and belief in a designer.

Dogmas that seem to deny permanence.

There are philosophers who deny that there are any permanent forms or elements of material substance. Such believe that nothing is fixed, either in substance or attributes; that every thing in the universe is in a perpetual flux, that the law of development controls all existence, so that one form and species of being is evolved from another—the more complex from the more simple—in endless progression. One relation of permanence in nature must, however, be assumed by all these philosophers, and that is, *the permanence of this law or principle of development itself.* If it may be assumed from the limited facts and observations adduced, that the law of development has prevailed in all the ages, and has evolved one form of being after another, by a steady progress and in regular order, then the permanence of the law of development itself must be referred to a fixed purpose and design of nature. The law of development cannot, therefore, drive the fact of design out of the universe, nor dispense with the assumption of design as one of the axioms of science.

The reciprocal relations of material and spiritual substance.

§ 332. Our analysis of matter and spirit has shown that many of the attributes of both can only be explained and understood by means of one another. The one can be defined and known only by the other. To understand and describe the one we must make use of the other. But the two are in some important respects very unlike. In order to show this with success, we must first consider the difference between *the direct and reflex knowledge* of both.

Mind and matter directly and indirectly known.

The mind knows both matter and mind by direct and reflex knowledge. By direct knowledge in sense-perception, it knows matter as a being. By direct knowledge in consciousness, it knows itself as the agent which knows matter and is also the subject of certain sensations. It knows both these objects in certain relations. It knows matter not only to exist, but to be diverse from itself the knower, and to be *extended, i. e.,* to have space relations; it knows itself to exist, and *enduringly* to feel and act, *i. e.,* to have time relations.

By indirect or reflex knowledge the soul is considered as *sentient* as well as *percipient.* As *sentient* it receives or suffers certain effects, viz.: the sensations of which matter is the cause. As *percipient* it knows by consciousness its own subjective states as thus caused, and by sense-perception the being that causes them; also that this being is not itself, and is extended in space. The being having these capacities to cause these effects in itself as a sentient it defines as *matter.*

In other words, in sense-perception, the intellect knows something more than subjective or spiritual effects, viz., specific sensations, as of touch, sight, etc., for which it assumes an unknown cause. It not only knows itself directly and those acts and objects that are purely spiritual, but it knows material objects also, and by its prerogative as an agent competent to know. If it did not know them directly as beings, it could not know them as extended or as diverse from itself, or even as causal agents.

We say, then, without reserve, that *the mind in sense-perception, knows matter or material being as truly and as directly as in consciousness it knows the ego, or mental being.*

The qualities of matter as primary and secondary.

§ 333. The qualities of matter have been divided into *the primary* and *the secondary.* The *primary* include its relations to space, as *extension and space-occupancy, or impenetrability.* The *secondary* include its causative relations to the sentient soul, as *hardness, color, smell, taste,* etc. [Dugald Stewart divides these relations into *Mathematical affections* and *Primary* and *Secondary* qualities. Hamilton recognizes three classes, the *Primary,* the *Secondary,* and the *Secundo-primary: the primary* including the relations of extension; the *secundo-primary,* resistance, gravity, repulsion, and inertia; and the *secondary,* the capacities to cause sensations.]

The principle of this two-fold division is obviously just and the application of it is easy. The analysis already made has shown that these two classes of attributes are clearly distinguished in fact. The relations of matter to space do not in their logical content, as distinguished by the mind, involve the recognition of any sensation. On the other hand, the powers of matter to produce certain sensations of touch, sight, smell, taste, and sound, can only be known by considering the sensations themselves as caused by these powers. Of the first class we have direct and positive knowledge. Of the second our knowledge is indirect and relative, it being explained by the effects produced.

There is still another element in matter which does not fall within either class, and this is matter as *existing* in distinction from its relations to other matter, to the sentient spirit, or to space or time. This is known by direct mental apprehension, in connection with felt sensations and on condition of the excited or *impelled* sensorium. Matter is known as being and also as causing these sensations: not as though its being were only known by relation to these sensations; but it is directly known as being and also as related to these sensations which it causes. Every thing which is known as related, is known to be; consequently, the matter which is known as related to the sensations must also be known to exist.

Two questions remain to be considered in respect to these two classes of qualities. (*a.*) Are the primary qualities distinguished from the secondary in being alone essential to the conception of matter, as Locke and others assert? (*b.*) Do the primary qualities alone give us a knowledge of matter as it really is and as distinguished from a relative knowledge?

§ 334. (*a*) Are the primary qualities alone essential to matter? The primary qualities are essential to the conception of matter, so far as they are required and sufficient to define and distinguish this kind of being from every other. It is of course implied that such relations are always true of this kind of existence—that they are always present and never absent in a single individual. This being assumed, we have only to ask for a sufficient number of relations to serve the purposes of definition. It is obvious that for this purpose no other relations of matter are necessary than its relations to space. These are always present, and for the purposes of defining the concept, matter, these only are required.

It is contended that they are essential, and therefore primary, in another sense, viz., in being adequate as they exist in different forms and varieties to account for all the secondary qualities, so that color, taste, heat, electricity can all be resolved into the number, position, and motion of homogeneous molecules. It is obvious however that this is not a logical, or psychological, or even a metaphysical problem, but one that is purely physical—a problem which can be solved by extensive observations of every species of matter and a more penetrating insight into its powers and laws than has yet been reached. Its solution must be left with the physicists, to whom it properly belongs.

Real and phenomenal or relative knowledge.

§ 335. The second question, (*b*) involves the more comprehensive inquiry, *Is our knowledge of either matter or spirit real, or only phenomenal?*

The *real*, in the language of recent philosophy, is opposed both to the *phenomenal* and the *relative.* It is used in the first connection by Kant, and in the second by Hamilton.

We have seen that the knowledge of the primary qualities of matter is more direct than that of the secondary qualities, because to the apprehension of the latter the reflexive consideration of the soul as sentient and percipient is required. But the knowledge of *the relations* of matter,—as indeed of the mind,—in one sense of the term, must necessarily be a *relative* knowledge, whether the relations are primary or secondary.

But besides this knowledge of the mutual *relations* of matter and spirit, we have also a knowledge of both *directly* as beings—of matter by *perception* and of spirit by *consciousness*. Is this direct knowledge *real* knowledge? This question is important, and has been so much discussed in modern speculation as to require special consideration.

The phenomenal, as contrasted with the real, may be understood in two senses. It may mean that that which *appears to one sense is not what it appears* to be to another; as when a stick, thrust in the water, appears to be bent, but is not so in fact; or, when the rainbow appears to be, but is not, a solid arch. In cases like these, the inference is drawn that one percept, as that given to the eye, is the sign of another, that which is appropriate to the touch. We infer that what we see with the eye is, or will prove solid, or, as we say, real, to the touch. In this sense, that which is known by the sense of touch, is held to be *real*, while what is apparent to or inferred from vision or any other sense is *phenomenal*.

The *phenomenal, in the second sense,* is anything *manifested to direct observation*—either of sense or consciousness—as distinguished from *the elements into which it is resolved, and the powers or laws by which it is explained.* For example, the rainbow, as apprehended by the eye, is a phenomenon; but the light reflected from rain-drops at a certain angle from the sun, is said to be the reality. The rain-drop, again, as a phenomenon, is a portion of water definite in form, relations to the light, and appearance to the eye. Water, again, when chemically analyzed, is the product of certain agents in certain proportions, etc. etc. The reality of light is an ether capable of certain undulations.

According to this use of these contrasted terms, every thing apprehended by the senses, all that is known as most solid and real in the world of matter, is only phenomenal. That only is *real* which is discovered by *science* of the elements and laws into which these phenomena are resolved, and by which they are explained. Any thing which remains to be thus explained and resolved, is phenomenal, relatively to the agents and laws which explain it.

Under this contrast, that which is directly and constantly known, which interests our feelings, which is most important, and, in one sense, is most permanent, is pronounced unreal; and that only is called real which is reached by special and artificial analysis, and is expressed by recondite relations. Of the analysis which attains to reality so understood, we are never certain that we have reached the end. The real agents behind these shifting changes which we call the phenomenal universe of material being, may not yet have been ascertained; and after all that science has discovered, we are still forced to ask, What is reality, and shall we ever be able to lay hold of it? The phenomena of the mind, again, are what appears to consciousness, as contrasted with the powers and relations into which they may be resolved and by means of which they may perhaps be explained. The states and operations of the mind, the products themselves, nay, even the *ego* itself, of all of which we have direct knowledge in consciousness—all these are phenomena.

According to Kant and Hamilton, reality, or the thing in itself, can never be known. It is transcendental to our knowledge; we only know that it is. We cannot even know the truth of its relations; for the relations or categories by which the understanding judges, do not connect realities, but only phenomena. Even the relations of space and time do not apply to realities, but only to phenomena. And even if both the forms of the understanding and of intuition could be applied to things as well as to phenomena, these forms may themselves be only subjective: that is, the phenomenal products of the human agent have an existence relative only to the constitution of the human being. (Cf. §§ 256, 7.)

The real as thus opposed to the phenomenal is called by Kant the *noumenon* or the thing in itself. This cannot be discerned by the senses, nor can it be apprehended by consciousness. It ever flits from our grasp, and leaves phenomena only in our possession as shadows which never satisfy, but simply point to something which we never can reach. This real we cannot know by the intellect. It is true, the Reason, as distinguished from the Understanding, must assume it to exist, in order to regulate its operations and conclusions, but even the Speculative Reason does not know that it in fact exists. It is only the Practical or Moral Reason which commands us to believe that it exists in the three forms of Matter, the Soul, and God. This knowledge is called relative, because it is dependent on the constitution of the soul, and the ultimate relations by which we connect the objects which we know. If these were changed, all our present knowledge would be changed with them. It is therefore relative to these powers and dependent on the relations according to which they connect objects in knowing.

In the language of Hamilton: "Our whole knowledge of mind and of matter is relative—conditioned—relatively conditioned. Of things absolutely or in themselves—be they external, or be they internal—we know nothing or know them as incognizable; and become aware of their incomprehensible existence only as this is indirectly and accidentally revealed to us through certain qualities related to our faculties of knowledge, and which qualities, again, we cannot think as unconditioned, irrelative, existent in and of themselves. All that we know is therefore phenomenal —phenomenal of the unknown.

"Our knowledge is relative: 1st, because existence is not cognizable absolutely and in itself, but only in special modes; 2d, because these modes can be known only if they stand in a certain relation to our faculties; and 3d, because the modes, thus relative to our faculties, are presented to and known by the mind only under modifications determined by the faculties themselves." *Met., Lec.* 8.

To secure ourselves against this distrust of our capacity to know the real, we have endeavored to distinguish between objects as perceived by sense and consciousness, and as known in higher relations. Things and facts given in experience are, *as phenomena,* just what they appear to be. But when we view them in their *relations to causes and laws,* we call those real whose causes are permanent and always active. These are constant, ever-present, and enduring effects. If the causes are occasional and short-lived, their effects are said to be unreal. The universal light and the wakeful eye co-operate to produce and prepare for the perceiving mind the reality which we call the visible universe. Let this light be dimmed, or the eye be dimmed (one or both), and the colored universe is an actual reality no longer. But inasmuch as its conditions or causes are ever ready to produce this phenomenal being, it is said to be real or a reality. It is only as we assure ourselves that these conditions are permanent, because they are sustained by the agencies and the designs of the living God, that we find the profoundest import as well as the sufficient ground of reality.

But when we hear Kant and Hamilton inquire, May not the intellect which perceives, also create the objects which it beholds, with a similar liability to change as the sensorium *i. e.*, Is not existence with its categories, itself a phenomenon dependent upon changeable forces, and therefore relative to the powers and forms of the intellect? We answer, No. Every analogy fails by which we interpret the realities of the *knowing* by the phenomena of the *sentient* soul. The soul, as intellect, not only acts in knowing according to the constitution which makes it what it is, but it assumes, and must assume, that its intuitive relations are discerned and affirmed by every intellect whether creating or created, and are therefore the real elements of all trustworthy knowing as a subjective process, and of all valid knowledge as an objective fact. To whatever object-matter this process and its results are applied (whether it be to material or spiritual, or to the thinking agent itself), these categories are absolute and real, and cannot be even supposed to be relative or phenomenal. To suppose them such, is to commit intellectual suicide. It is to introduce constant antagonism into every process which we perform, and the elements of self-destruction into every result which these processes evolve, as well as logical incompatibility and confusion into the language by which both processes and results are expressed. It is to philosophize ourselves into the impossibility of philosophy, and by assumptions which we argue that we may neither assume nor confide in. It is not only to offend against reason by introducing inconsistency into that which in its very nature is self-consistent, but it is to overlook or deny those designs which we must assume that the universe exists to fulfill,so far at least as it is capable of *being known*.

CHAPTER VIII.

THE FINITE AND CONDITIONED.—THE INFINITE AND ABSOLUTE.

The questions concerning the finite and its relations, the conditioned and its dependence upon the absolute, are the most vexed and the most unsettled of any in modern speculation. Can the infinite be conceived or known by a finite intellect? Can the unconditioned be brought under those relations which are appropriate only to the conditioned? These questions we must attempt to answer, if we would analyze all the powers and explain all the products of the human intellect. We can do this more successfully if we consider the finite and the conditioned apart from the infinite and the absolute. We begin with

I. *The finite and the conditioned.*

§ 336. The process of knowledge in all the forms as yet considered, is a unifying and therefore a *limiting* process. Each object which it takes in hand it analyzes into many parts, and discriminates into various elements. The parts it then proceeds to recombine into a completed whole: the elements it blends into a perfected product. It leaves it a completed whole or finished result, which passes into the sum of its possessions as a known, a defined, and therefore a limited or finite object.

To know a limiting process.

Thus, in sense-perception, the objects are perceived by being first separated into distinct percepts, each of which is perfected by a separate act of analytic attention, and which are again united into a completed whole in space.

The units thus constituted may be enlarged by the imagination and memory. Spatial objects may be added one to another, so as to increase the space-unit to the farthest limit; or the imagination may suppose them created where they are not. Memory may add to the present mental states all that have gone before within its own experience. Imagination supplies all that now exist or that might exist in other minds. Each of these forms of the representative power, after its own manner, produces units or finite wholes.

Thought, by its similarities observed, unites the like into new combinations or units. It refers diverse effects to a common cause, acting under similar laws. It subordinates means the most diverse to a single end, by their conspiring and designed adaptation, and thus unites them as preëminently one.

The finite universe, how conceived.

§ 337. We can imagine that all material objects perceivable could be united as one by a single mind endowed with capacities ample enough to grasp so many by a single act. We can also imagine every existing mind as co-operating with every other mind, and can suppose each to know all the powers of these minds, and all their acts. We can believe it possible that these agents and objects should be known in all their knowable likenesses and dissimilarities, in all their causal agencies, in all the laws under which their forces act and the ends to which they are adapted. We can conceive this assemblage of separate objects, material and spiritual, with their several phenomena, to be but an assemblage of effects, produced by other agencies and other beings in previous times, and these by others; each aggregate of beings and forces producing others, under permanent agencies and fixed laws. Moreover, we can conceive these beings, with their powers and laws, as co-existing in space; these successive evolutions, whether of separate beings or new phenomena, as developed in time, as designed for separate ends, and all these ends as conspiring together for a series of ends, constituting in this way an intelligible and orderly system. This assemblage of all objects believed to be existing in space and acting in time, with all the agencies and laws and relations now known or which may be afterward discovered, make up the *finite universe* as knowable, or as conceived by man. It is called the *universe*, because it includes as a whole all the separable objects apprehensible by sense and consciousness. It is the *finite* universe, because each of these objects is limited to a portion of space and period of time, and subjected to all the conditions of existence and of action which their actual forces, laws, and ends prescribe. It exists and acts under the action of these forces, ends, and laws.

The finite universe is *limited and conditioned*. It is limited because it is made up of objects and events which are bounded by one another, and have a limited or definite extension. The existing spirits which we know, exist and act within certain defined spheres of extension. When all these extended beings, and these spheres of spiritual being and action, are gathered into the universe known, its extension is still limited or defined. So far, also, as we trace this universe of beings and phenomena backwards or forwards through the series of its changing developments, its duration is limited by a beginning and end. There is *a first* and *a last* of the series, if it is limited; whether the terms designate a single object or a single act, or collectively designate many objects and acts,

It is also a *conditioned universe*. Every part and element in it *depends* on something other than itself, for what it is and for what it does. It begins to be by the operation of one or more agents acting according to laws, and these agents are the necessary conditions of its existence. It also continues to exist under the operation of *conditions*. These conditions are the causes, laws, and ends of its being, and these prescribe its being, as well as the sphere and the results of its activity. Each part of the universe being thus dependent on productive forces other than itself, the universe itself, as a whole, is said to be conditioned as well as limited. But is this all that we know? Is this all that exists? Besides the limited, is there the unlimited? Besides the conditioned and dependent, is there the unconditioned, the self-existent, and self-active? These questions introduce

II. *The infinite and absolute, and their relations to the finite and dependent.*

§ 338. To understand the import of the questions concerning these much-vexed topics, and to attempt to answer them, it is necessary, first of all, to clear away all uncertainty in respect to the terms which are employed, and to bring the mind to a definite apprehension of the various senses in which they may be interchanged and confounded. We consider, *first* of all, the etymology of the more important of these terms.

The import of the terms infinite and absolute.

We begin with the *infinite*.

Infinite signifies, literally, that which is not bounded or terminated. It is primarily applied to spatial quantity. Every thing which has extent is terminated or bounded by some other object or objects which are also extended. The line or surface which divides one surface or solid from another, is called its *limit,* and the surface or solid, as necessarily thus terminated or terminable, is called finite or limited. In like manner, the mathematical point is conceived as terminating or limiting the mathematical line, and the line itself is limited or finite. By an obvious transference of signification from the objects of space to those of time, the first and last of any succession of events or series of numbers is called its limit, and every series of numbers, numbered objects, or events and portions of time, is finite or limited.

The terms originally appropriate to extension, duration, and number, are still further applied to the exercise of power by material and spiritual agents. The exercise of power by man, whether spiritual or material, is possible only in certain places, at certain times, and with respect to a certain number of objects, or a measured quantity or mass of matter, and thus power itself becomes measurable by the relations of quantity and number as applied to its effects and the means by which they are caused. Man can only accomplish certain effects in *limited* places, times, and number, and hence he is said to be limited in his powers. He can only know and do certain things under all these favoring circumstances, and is therefore a finite being. The word *finite* is, therefore, *originally a term of quantity,* and *secondarily of causal or productive agency. The infinite,* in the general sense, is the *not*-finite. Logically conceivable, there are as many sorts of the not-finite or infinite as there are senses of the finite.

The unconditioned is the non-conditioned.

§ 339. *The unconditioned* comes next in order. Logically, it is the negative of the conditioned, and follows its meaning. The *conditioned* is that which is in any sense *dependent* upon any thing else, either as a *material* of its composition, a *cause* or *means* of its production, or an *object* of its psychical activity. Thus, silver is a condition of a silver spoon; heat is the condition of the melting of iron; and a material world the condition of the act of sense-perception. Every condition has this in common with every other, viz., that that to which it is the condition cannot be what it is without it, whether it is a thing, an act or an effect. It is therefore said to be limited by these conditions. It can neither be, nor be thought of without them. They are necessary to it. They must be given or present with it, and are therefore called its conditions.

The *primary* signification of the conditioned is that of *necessary dependence.* Its *secondary* application is to objects of quantity, thus reversing the process through which the *finite* passes. The *finite* proceeds from a signification of *quantity* to one of *quality.* The conditioned proceeds from *quality* to *quantity.*

Applied to quality and quantity.

The line and surface are the conditions as well as the limits respectively of the surface and the solid, but solely because they are essentially necessary to the conception of each. In the same manner, space and time are the conditions of extension and duration, because they are essential to the possibility of each. These can neither be logically thought of, nor really exist, except as they involve space and time as their conditions. All the limits of objects of quantity are also their conditions, but all the conditions of such objects are not necessarily their limits.

The *unconditioned* is that which is *not conditioned*—i. e., not necessarily dependent on another object for thought, being, or act, as a constituent, cause, or object. Whenever the positive, the conditioned, can be applied, the negative unconditioned, can be logically conceived as its opposite.

The absolute, several senses of.

§ 340. *The absolute* is still another term that is often interchanged with the infinite and the unconditioned. Originally and etymologically, it signifies *freed from,* or severed. This signification is purely negative with reference to that from which the subject is freed. It was also applied to mean *the finished or completed,* even as the Latin word *absolutus,* as is thought, was originally used of the web when ready to be taken from the loom. Both these senses have passed into the modern use of the term, and determined the varieties of its application. *First* of all, absolute and absolutely is applied to any thought or thing as viewed apart from any of its relations—regarded simply by itself. This meaning is near akin to that of an object viewed as *complete within or by itself. Next,* it is applied to that which is complete of itself so far as the relations of dependence are concerned; to that which is necessarily dependent on nothing besides itself. In this sense it is very near in meaning to the *primary* sense of the *unconditioned* already explained. *Still further* it is used in the sense of severed or separated from all relations whatever, or *not related*—i. e., not admitting of any relations. This sense Hamilton and Mansel have transferred also to the unconditioned and the infinite. *Still again:* it is applied to relations of *quantity,* and here the signification of *complete* or *finished* is applied to the greatest possible or conceivable whole, to *the total* of all existence, whether limited or unlimited in extent and duration.

The Hegelian sense.

In the Hegelian terminology, the absolute takes a special signification from the fundamental assumptions of the Hegelian system. When the notion, *der Begriff,* has completed every possible form of development, and, as it were, done its utmost possible by the force of the movement essential to itself, the absolute is reached. This absolute completes every possible form of development, and explains every kind of object conceivable and knowable by the mind, from the undetermined notion with which it begins, up to the highest form of development, when it becomes self-conscious in the human spirit by distinguishing itself from the material universe. The conscious spirit thus evolved, and reflecting in itself all these lower forms of existence, is essential to and completes the development of *the absolute.* This absolute is perpetually reproduced by the lower forces of the universe, and itself perpetually represents all these in its own reflective thinking.

It is obvious from what has been said, that these three terms are all used in applications which are often interchanged, but which should be carefully and sharply distinguished. The infinite, the unconditioned, the absolute, may denote some property or relation of a being, *in the abstract,* or may stand for a *being or entity* which is believed or supposed to be infinite, unconditioned, or absolute. That is, the infinite, etc., may stand for the infinitude, the unconditionedness, the absoluteness of some being—*i. e.,* as an *abstractum* or property of a being; or for that which is infinite, unconditioned, or absolute. One of these acceptations is obviously very different from the other. The one may readily be confounded with the other.

What is not true of the absolute, etc.

§ 341. These concepts and the entities which they represent are not of necessity merely *negative conceptions,* nor are they the products of what is called negative thinking.

We have seen from our analysis of the terms *infinite, unconditioned,* and *absolute,* that they are all originally negative in form, and that this form, strictly interpreted, would denote the absence or the denial of the positive attributes, with which these negatives are combined. From this unquestioned fact the inference has been derived that, because the terms were negative, the concepts are also negative.

But this inference, by whomsoever it is countenanced or made, is manifestly invalid. It does not follow, because a concept is designated by a negative term, that it is not positively conceived; or, because an object is called by such a name, that it is not really known. If the only fact that is prominent before the mind be that an object is not something else—whether it be a being or a quality—it may be designated by a negative term. This term does not deny its real existence, or that it is both knowable and known, for it may assume and imply both. It simply sets forth its contrast with something else. If we see a bat, and say of it, It is not a bird, or, It is not a beast, or if the Sandwich Islanders, for lack of name, had called the *ox* a *not-dog*, the use of a negative appellation would not necessarily authorize us to infer the absence of definite conceptions or of positive knowledge. So, when we gather together the entire sphere of finite being, and stretching our thought beyond, apprehend something which is unlike it and contrasted with it by being *not finite, not conditioned*, and *not dependent,* we do not confess that we cannot conceive it or that we do not know it as something positive and real because we emphasize this single relation of contrast by the use of such negative terms as the infinite, the unconditioned, and the absolute.

Again, these concepts are not "negative" in that they are produced by what is called "*negative thinking.*" This negative thinking is distinguished from the mere thinking of a negative—*i. e.*, thinking a positive in a negative relation—as above explained. Those who teach this, assert that our conceptions of the unconditioned, etc., are necessarily negative, because they are the result of an attempt to think them which is unsuccessful, and which, whenever it is repeated, reminds us of the impotence or imbecility of our faculties.

"Every thing conceivable in thought lies between two extremes, which, as contradictory of each other, cannot both be true, but of which, as mutual contradictions, one must." "Space cannot be conceived by us either as an infinite or a finite *maximum*, or an infinite or finite *minimum*, and yet if it is conceived at all it must be conceived as one of these, and forasmuch as we cannot conceive it under either, we have only a negative idea of space, *i. e.*, an idea which results from an impotent attempt to conceive it. The same is true of time, and even of causation itself."—Hamilton, *Met., Lec.* 38. Mansel illustrates the process of negative thinking still more definitely. "A negative concept, on the other hand, which is no concept at all, is the attempt to realize in thought those combinations of attributes of which no corresponding intuition is possible." "The only negative ideas with which the logician or metaphysician as such is concerned, are those which arise from an attempt to transcend the conditions of all human thought." * * "Such negative notions, however, must not be confounded with the absence of all mental activity. They imply at once an attempt to think and a failure in that attempt."—Mansel, *Proleg. Logica*, chap. i.

Again: The unconditioned, etc. is not necessarily, as a concept or as a being, *exclusive* of *all relations*. It is not *unrelated*, or *the unrelated.*

The doctrine of Spinoza was contrary to this. The maxim on which he rested for the statement and defence of it was *Omnis determinatio est negatio.* Every relation implies a distinction into parts related; the one part cannot be the other: hence, the absolute, as related, cannot be complete or perfect of itself. It cannot be *unconditioned*, for, in order to be related, it must require, or, so far as related must be conditioned upon, that which is related to and is not itself. It cannot be *unlimited*, for, in order to be what it is, or what it is asserted to be in the given relation, it must depend on something out of itself. The unconditioned cannot, therefore, be related. Hamilton gives the following reasons for the same opinion: "A relation is always a *particular* point of view; consequently, the things thought as relative and correlative are always thought restrictively, in so far as the thought of the one discriminates and excludes the other and likewise all things not conceived in the same special or relative point of view." And again: "We conceive God as in the relation of Creator; and in so far as we merely conceive him as Creator, we do not conceive him as *unconditioned*, as *infinite*," etc. (*Letter to Calderwood;* cf. *Mansel, Limits of Rel. Thought, Lec.* 2.)

The proper answer to these representations is the following: It is not at all essential to the conception of the absolute which the human mind requires, or to its reality, that it should exclude all relations, but only a *certain class* of relations, viz., those of *dependent being* or *origination*. *The truly absolute and infinite is not the unrelated as such, but that which is not dependent on any other being for its existence or its activity.*

Again: The unconditioned, etc., is *not the sum of all actual or conceivable being.*

This view of the absolute is closely connected with the preceding. The denial of all relations to the absolute involves the denial of all parts or entities, whether real or thought-parts, which can be related, and this requires the conception of the absolute, as the total of all existences and conceivable things, the *Τὸ ἓν καὶ Παν, the all which is also one.* This position was actually taken by *Spinoza*, who was driven by logical consistency to acknowledge but one being or substance in the universe.

Hamilton (*Letter to Calderwood*) reasons as though this were the only possible conception of the true absolute. Mansel, (*Limits of Rel. Thought, Lec.* 2,) expressly asserts: "That which is conceived as absolute and infinite must be conceived as containing within itself the sum not only of all actual, but of all possible modes of being. For, if any actual mode can be denied of it, it is related to that mode, and limited by it." "The metaphysical representation of the Deity, as absolute and infinite, must necessarily, as the profoundest metaphysicians have acknowledged, amount to nothing else than the sum of all reality."

Of this view of the absolute, we need only say, that it is not the only possible conception, nor is it the most rational conception which can be taken of it. In a gross quantitative sense, we may say that the finite, plus the so-called infinite, equals the absolute, and that the result is in conception and in fact the unconditioned and the infinite, because nothing can be affirmed of it in the way of distinction or relation. But the question at once returns, Is this the absolute and the unconditioned which the mind necessarily receives in thought and believes in fact? This absolute cannot be totality, for it is expressly supplied by the mind in addition to the finite, and in order to account for and explain this. It cannot include that or require that which it itself accounts for and explains.

Again: The absolute, again, is not a concept or entity which is divested of all *interior relations*—a something *entirely one and simple.*

Those who contend that the absolute does not admit the idea of parts, because parts imply division and relationship, are driven by a logical necessity to the conclusion that it must be one and indivisible in respect of parts and relations. Hence it has been inferred that the absolute cannot be a personal being. 'A person distinguishes himself from that which is not himself, his own being from his acts, and both from their objects, whether these be real or spiritual. His acts must be successive to one another also, and thus be separable and distinguishable in time. All these divisible parts and distinguishable relations are,' it is urged, 'entirely incompatible with the concept and reality of the absolute.'

These views are held by those who deny the possibility of personality in God, as well as by those who, like *Kant, Mansel,* and *Hamilton,* believe that God is personal, but deny that, so far as He is believed to be personal, He can be known as the Absolute.

It is enough to say of this view of the absolute, as has been said already, that the absolute does not necessarily exclude the possibility of parts or relations. The absence of necessary dependence upon the finite and the complete dependence of the infinite upon itself, does not imply such a simplicity or oneness of being, as excludes complexness or personality.

The absolute, etc., are knowable.

§ 342. It has been earnestly held that the absolute or the infinite *is unknowable by a finite mind.* Some have held that the mind cannot properly know either, *that* it is, or *what* it is; others that we can know *that* it is, but not *what* it is.

Views of Kant, Hamilton, and Mansel.

Kant, Hamilton, and *Mansel* all hold that we cannot know, though we may believe that the infinite exists, simply because the conception of the infinite is not within the grasp of the finite. Kant teaches that the reason why we cannot know the infinite, is, that our faculties of knowing both the finite and the infinite have merely a subjective necessity and validity, and therefore we cannot trust their results as objectively true. Moreover, if we apply them to the infinite, we are involved in perpetual *antinomies* or *contradictions.* Our only apprehension of the absolute is, therefore, by the practical reason, and comes in the way of a moral necessity through the categorical imperative, which requires us to receive certain verities as true. *Jacobi, Schleiermacher,* and others say, that we reach these by faith or feeling, and not by knowledge. Hamilton asserts that we find ourselves impotent to know them, in consequence of the contradictions which the attempt involves. But he expressly asserts "that the sphere of our belief is much more extensive than the sphere of our knowledge; and therefore, when I deny that the infinite can by us be *known,* I am far from denying that by us it is, must, and ought to be believed. This I have indeed anxiously evinced, both by reasoning and authority." (*Letter to Calderwood.*)

Herbert Spencer dissents in part.

Herbert Spencer reasons against Hamilton and Mansel, to the conclusion that we can know *that* the Infinite exists, but we cannot know *what* it is. He contends that we can know that it is, because, "To say that we cannot know the Absolute is, by implication, to affirm that there is an Absolute. In the very denial of our power to know what the Absolute is, there lies hidden the assumption that it is, etc. Besides that definite consciousness of which logic formulates the laws, there is also an indefinite consciousness which cannot be formulated."—*First Principles*, P. I., c. iv., § 26.

We contend that the absolute is knowable — that man can both know *that* it is and *what* it is. But, first of all, we would define the sense in which it *cannot be known*, either as *that or what.*

(*a.*) It cannot be known by the *imagination*, either as representative or creative. The imagination can only picture that which is limited by space and time, and which is possessed of limited powers of matter or spirit. The absolute and infinite has none of the attributes of matter or spirit, as limited by space and time. It cannot, therefore, be either imaged or pictured.

It would be more exact to say that the analogies between any finite objects and the infinite are so general and attenuated, that the imagination can render no available or efficient service by introducing any images of the finite.

To attempt to image the relation of dependence which exists between the infinite and the finite by such special and limited examples of it as exist between different limited beings, is either superfluous or misleading. The relation may be known as so general, like that of simple entity, as not to need an example; or the use of examples may introduce many extraneous and unimportant circumstances, which may be conceived as essential to the relation in question. Thus, when it is reasoned that self-existence, personality, the creation of another than itself, the possession of a complex nature—one or all, are incompatible with the true infinite and unconditioned, the reasoning is founded on the attempted exemplification of the infinite by the finite, and on the unessential accessories which finite images suggest. Logically expressed, it is a case of *fallacia accidentis.*

The *antinomies of Kant* and *the essential contradictions of Hamilton* each of which seem necessary to the mind, and each of which exclude the other, are all made by the mind itself in the attempt to illustrate the infinite by the finite. The antinomies of Kant are incompatibilities between *an image* and *a relation* which the image exemplifies, or between two images adduced to illustrate different relations, or between two concepts, both of which are not necessary to the mind. The solution of them is to be found in a re-statement of the conceptions between which these incompatibilities are said to exist. Thus, for example, in the alleged antinomy involved in the propositions *the world is in time and space* and is *neither finite nor infinite;* the contradiction lies between a fact or image borrowed from experience, and an alleged *a priori* necessity of thought. But the incompatibility of the one with the other arises from a misconception of what is involved in our conception of the infinite, a confounding of the extended in space with space itself. When Hamilton says we must conceive of space as a bounded or a not bounded sphere, he introduces the image of an object existing in space and limited in space, in order to illustrate space itself, and confounds the one with the other.

We observe still further, (*b.*) that the absolute, etc., though knowable, is not a notion that is the product of reasoning, *inductive* or *deductive*, or that can be *defined* in a *system of logical classification.*

It cannot be inferred by induction, because, as has been shown, it is assumed in the very process of induction, as its necessary condition. (Cf. §§ 237, 240.)

It cannot be deduced by syllogistic reasoning, because, as has been shown, all deduction rests either on a previous process of induction, or on the intuitions of time and space. (Cf. § 226, 7, 8.) But induction requires the absolute as its condition.

Nor can the concept be defined for the ends of logical classification. The infinite is not properly co-ordinate with the finite, for the reason that it must be assumed as the ground of all such classification. Every notion or concept of every finite existence implies the unconditioned, and holds some relation to it, but its relations are not necessarily used in defining the notion for logical or scientific ends. The relations of substance and attribute, as used in such definition and classification, are applicable only to objects, which for their existence and their relations are dependent on the fixed conditions of finite being. They imply the presence of time and space relations, and the limitations of the powers of created beings by the laws which are determined by these relations. The cause and effect, the adaptations and ends, which logic usually recognizes in its operations, are fixed in a similar manner by established forces and laws.

The so-called categories—*i. e.*, the generic relations which are supreme and final in scientific definition and classification—cannot be applied to the infinite, because the infinite is required and assumed for the explanation of these very categories. These categories rest upon the infinite, and presuppose it.

We next affirm *positively* that the absolute is and can be known as the *correlate* which must be necessarily assumed to explain and account for the finite universe.

If the absolute is necessary to explain the finite, then it holds some relations to it. If it is its correlate, it must be connected with it by some relations. What these relations are, it is not needful to inquire. All that we need here to urge, is, it is so far from being true, that because it is absolute it is not related, that, on the contrary, it cannot be the absolute without being known as related. We cannot know *that it is*, without knowing, to a certain degree, *what it is*. If it is necessary to the mind to assume the absolute in order to explain the infinite, then the finite is certainly explained by these relations which it holds to the absolute. These relations must be real, else our knowledge is a fiction. They must be capable of expression in language. The relations between the finite and the infinite need not, of course, be the same as those which exist between the finite and the finite, but they must be real and cognizable relations.

The absolute apprehended by the intellect.

§ 343. The apprehension of the absolute is an *act of knowledge, even when called an act of faith* or *feeling*. (Cf. § 258.)

Hamilton opposes the one to the other, as faith to knowledge, because he affirms that to know is always "to condition;" and therefore if we know the unconditioned, we must condition the unconditioned, and limit the infinite. His doctrine is, that "we believe the infinite, but do not know it to be. The sphere of our faith, is wider than the sphere of our knowledge." But to know as related, is not the same as to condition in the special signification in which the unconditioned and the infinite are opposed to the conditioned and the finite. The knowledge of the unconditioned may be *a priori*, intuitive, and necessary, but it is knowledge nevertheless. It may be higher than any reasoned or logically defined knowledge, but it is still knowledge.

To call it faith, in any but a purely technical and private sense of the word, is to put it out of all relation to knowledge. To contrast it with knowledge in respect to its essential characteristics, is to weaken the very foundations on which both knowledge and science are made to rest. Especially is this the case, if this so-called faith is referred to an impotence of the intellect, and is made to depend on the conscious imbecility and known limitations of the powers. This is so far from being true, that, to know the absolute, is to know in the highest and the most positive sense possible to the mind. For if we cannot assume the infinite, we can neither define nor reason the finite. Without the intuition of the unconditioned, it is impossible to have any grounded science of the conditioned.

Not known exhaustively or adequately.

§ 344. But though we have a real and proper knowledge of the absolute, we do not therefore attain to an adequate and exhaustive, or what is often called an absolute knowledge of it. But this forms no objection to the reality of this knowledge. Indeed, an exhaustive knowledge, even of the finite, is only ideally conceivable, but is in fact impossible. An absolute knowledge of all the relations of an individual object—*e. g.*, a mass of rock, a tree, an animal, or a man, would imply a complete mastery of all the relations which each holds to every other object in the universe, in respect to its properties and ends —in other words, an exhaustive knowledge of the universe itself.

The finite universe infinite to our knowledge.

For man, the unexhausted finite must ever be as the infinite. But the fact that he knows the finite in part, is not inconsistent with the proposition that he knows it in truth. Nor ought the fact that man knows the infinite but in part, to be used to prove that, so far as he knows it, he does not know it as it is. To man there is, in both finite and infinite, a background always unexplored, and which, perhaps, never can be explored by man. If this is so, then the finite is as the infinite to him. The limited forest, into the mazes of which the child has not yet penetrated, the shallow abyss the depths of which he has not ventured to sound, are to him the symbols of infinitude.

In *both finite and infinite*, there is *a common* mystery, which *cannot be overcome*, and that is the *mystery of self-existence.* It does not relieve this mystery, to accept the fact of self-evolved and self-evolving forces and laws; nor does it increase it, to accept the fact of a self-existent creating intelligence whom we assume to explain the order and thought of the finite universe.

The absolute a thinking agent.

We may then positively affirm that the absolute is a *thinking agent.* The universe is a *thought* as well as a *thing*. As fraught with design, it reveals thought as well as force. The thought includes the origination of the forces and their laws, as well as the combination and use of them. These thoughts must relate to the whole universe. If so, it follows that the universe is controlled by a single thought, and is the thought of an individual thinker. If gravitation everywhere prevails, and gravitation is a thought as well as a thing, then the universe, so far as it depends on and is affected by gravitation, is a single thought. But a thought implies a thinking agent, and if the universe is a single thought, it was thought by one thinking agent. That this thinking person should be self-existent, involves no greater mystery than a self-existent thing or system of things.

Must be assumed to explain thought and science.

§ 345. We assume that this Absolute exists, in order that thought and science may be possible. We do not demonstrate his being by deduction, because we must believe it in order to reason deductively. We do not infer this by induction, because induction supposes it; but we show that every man who believes in either, or in both, must assume it, or give up his confidence both in these processes and in their results. We do not demonstrate that God exists, but that *every man must assume that He is.* We analyze the several processes of knowledge into their underlying assumptions, and we find that the one assumption which underlies them all is a self-existent intelligence, who not only can be known by man, but who *must be* known by man in order that man may know any thing besides. In analyzing our psychological processes, we develop and demonstrate an ultimate truth, and that is the truth which the unsophisticated intellect of child and man requires and accepts, that there is a self-existent personal intelligence, on whom the universe depends for the being and the relations of which it consists. We are, therefore, not alone justified—we are compelled—to conclude our analysis of the human intellect with the assertion, that its processes involve the assumption that there is an uncreated Thinker, whose thoughts can be interpreted by the created intellect which is made in His image.

INDEX.

Absolute, (see Infinite;) original meaning of, 545; the Hegelian sense, 546; used in the concrete and abstract, 546.

Abstract thinking, 325; concepts, 332.

Abstraction, 328.

Acquired sense-perceptions, chapter on, 132–150; examples of, 132; defined, *do.*; importance of, 132, 3; many gained very early, 133; of smell and hearing, 133, 4; of sight, 134; of distance, of magnitude, 134, 6; of size, 136; mistaken judgments of both, *do.*; of percepts appropriate to touch, 137, 8; of place of sensations, 139; of control of bodily motions, *do.*; provisions for, 139–141; how controlled, 141–143; involve memory, 145; and induction, 146; infants capable of such inductions, 146, 7; objections, 148; from the case of animals, 148, 9; other acquisitions of the infant, 149.

Activity of the soul, essential to its nature, 18; essential to knowledge, 42; in sense-perception, chapter on, 180; is attested by consciousness, 181; varies in energy, *do.*; success depends on attention, *do.*; differs in different men, 182; shown in innervation of organs, *do.*; directed to different objects, 183; selects and combines, 184; separates single objects in infancy, 185; continued through life, *do.*; illustrated in different men, 185; easily performed, 186.

Adaptation, 500; how related to design, *do.*

Æsthetics, its relations to psychology, 8.

Agassiz, on species, 353; on classification, 414.

Analogy of nature, 393.

Analysis, involved in knowledge, 46.

Analytical reasoning in mathematics, 378.

Anthropology, defined, 2.

Antinomies of Kant, and Hamilton, 475.

Apperception, 62, 3.

Aristotle, division of powers of the soul, 31, 2; theory of sense-perception, 192; enumeration of laws of association, 231; on universals, 340; regarded the middle term as causal, 374, 5; fourfold division of causes, 499.

Arnauld, theory of sense-perception, 195.

Association of ideas, 210; chapter on, 225–254; other terms for, 225; importance and mystery of, 225, 6; method of discussion, 227; division of, *do.*; not explained by bodily organization, 227, 8; defect of all physiological explanations, 228, 9; actual influence of the body, 229; exercised by means of psychical states, 230, 1; vital sensations may act as links of association, *do.*; ideas do not attract one another, 231, 2; crude statements of Hobbes and others, *do.*; relations do not attract ideas, 232; relations stated as three, seven, two, and one, 232–4; law of redintegration, 234; how far satisfactory, 235; objection, 236; the real solution, 237; explains phenomena, 237, 8; associations with sensible objects, 239; of home, *do.*; relations of acquisition and reproduction the same, 239, 40; secondary laws of association defined and named, 241; discussed, 241, 2; apparent exceptions to, 243; Hobbes' often-quoted illustration, *do.*; two theories in explanation, 244; capable of interruption and control, 245, 6; not the only power of the soul, 246; indirectly controlled, 247; relation to habits, question concerning, 248, 9; higher and lower laws of, 250; prevalence of higher, 250; of lower, 251; casual associations, 251, 2; in changes of fashions, 252; the moral influence of, *do.*; influence on language, 253; on philosophy, *do.*

Associational psychology, 38–40; prominent writers, 38; explanation of necessary truths, 39; fundamental error, 39–40; usually materialistic, 40.

Associational school, their views of intuitions, 436.

Astronomy, discoveries in, 397, 9.

Attention defined, 47; beginnings of, 152, 3; Stewart's theory, 178; can be given to two objects at once, 178, 9; is the utmost attention possible to more than one? 179.

Attribute, relations most frequently used as, 168; sensations so used, 170, 1; etymology and meaning of, 525; in the abstract, 525; material; indicate but do not constitute matter, 530.

Auxiliary lines in geometry, 384, 5.

Axioms, mathematical, 382, 3; Analytical and synthetical, 382; are they properly premises? 383.

Bain, A., an associationalist, 38.

Being, correlate of knowledge, 44; varieties of, 45; some more lasting and important, *do.*; contrasted with phenomenon, *do.*; one kind mistaken for another, *do.*; not known apart from relations, *do.*; category of, 446; fundamental in what sense, *do.*; different sorts of, *do.*; the most abstract, 447; how explained, *do.*; concrete known first, *do.*; knowledge of, expressed in propositions, 448; not a relation, *do.*; cannot be defined, *do.*; treated as an attribute, *do.*; indeterminate, *do.*; both spiritual and material, directly known, 449.

Berkeley's view of sensation, 103; theory of sense-perception, 197; doctrine of the concept, 342, 3.

Biran, de, M., views of intuitions, 437; theory of causation, 493–495; how far correct, 494.
Black's, Dr., discovery of carbonic acid gas, 395.
Blind, the, when restored to sight, 137, 8; how they judge of form, size, etc., 138; the reports of, critically noticed, 163–165.
Bodily organism, 97, 8.
Bonnet, theory of vibration, 227.
Brain, the organ of the soul, 38.
Brown, Dr. T., denies consciousness of ego, 69; admits it, 70; theory of tactual and other sensations, 123; theory of sense-perception, 199; of the nature of the concept, 343; of intuitions, 436; theory of causation, 484, 5.
Buxton, Sir T. F., advice on memory, 274.

Categories. (See Intuition.)
Causation, and causality, chapter on, 481–499; as a law principle and distinguished, 481, 2; the principle of, intuitively evident, 484, 6; reasons for, 487; resolved into a time-relation, 484; by Hume, *do.;* by Brown, *do.;* by J. S. Mill, 485; not a relation of time, 495, 6; not explained by induction, 488, 91; nor by association, 489; not gained by experience, inner or outer, 492–495; Locke's views, 492; views of R. Collard and M. de Biran 493; theory of De Biran, 493-5; two positions of, 493; how far correct, 494–5; denied to matter, 494; denied to created spirits, 495; Malebranche, *do.;* theories *à priori*, 495, 6; explained by law of contradiction, 496; Wolf, Kant Hegel, *do.;* Hamilton's explanation by the law of the conditioned, 496–9; objections to, 498, 9; conclusion, true doctrine of, 499.
Cause distinguished from condition, 483; four classes of, 499.
Cerebralists. (See Cerebral Psychology.)
Cerebral Psychology, 36; supposes consciousness, 37.
Clarke, S., definition of space and time, 479.
Classification, how it arises, 335; by children and savages, 336; in science, *do.;* relations to knowledge, 337; significance of, 338; assumes final cause, 514.
Coleridge, S. T., on the arts of memory, 276, 7.
Complex notions, 333.
Concept, formation of, chapter on, 327–339; of material objects, 327; when it begins, *do.;* similarity discerned, 328; involves analysis, *do.;* attributes distinguished, *do.;* called abstraction, *do.;* to prescind, *do.;* comparison, *do.;* generalization, 329; predication, *do.;* assumes substance and attribute, *do.;* appellations concept, 330; and notion, *do.;* not a percept, *do.;* nor an image, *do.;* relative, *do.;* a mental product, *do.;* universal, 331; predicable, *do.;* respects attributes only, 332; concrete and abstract, *do.;* simple and complex, 333; content and extent, 334; mutual relations of the two, 334–337; how far they add to knowledge, 338; theories of nature of, chapter on, 339–340; Socrates and Plato on, 339; Aristotle, *do.;* Porphyry, the Realists, Nominalists, and Conceptualists, 340; Thomas Hobbes, 340, 1; John Locke, G. W. Leibnitz, George B. Berkeley and D. Hume, 341-3; Reid, Dr. T. Brown, Sir W. Hamilton, 343; J. S. Mill, 344; I. Kant, 344; Hegel, 345; nature of, chapter on, 345–357; distinguished from the act, 345; implies substance and attribute, *do.;* is relative, 346; founded on similarity, *do.;* classifies, *do.;* gives import to names, 347; the import explained by individuals, *do.;* nominalists how far right, 347, 8; the conceptionalist, 349, 50; the realist, 350, 2; mistakes of the realist, 352, 3; why language aids thinking, 353, 4; symbolic and intuitive knowledge, 354–357; formed by judgment, 358; how related to it, 359; in mathematics, 380; of space and time objects, 464, 5; mathematical, *do.;* of geometry, *do.;* of number, *do.;* of space and time, 476, 7.
Conceptualists, the, 339; strife adjusted, 349.
Concrete thinking, 325; concepts, 332.
Condillac and school, on the origin of knowledge, 436.
Conditioned. (See Infinite.)
Consciousness, and natural consciousness, chapter on, 61–67; defined, 61; applied to any act of knowledge, 61; a collective term for all the intellectual states, 62; metaphorical uses of, *do.;* proper meaning, *do.;* called inner sense, *do.;* called apperception, *do.;* German equivalent for, 63; called reflection, *do.;* exercised in two forms, *do.;* the two defined, 64; natural consciousness as an act, *do.;* an act of knowledge, *do.;* results in a product, 65; is *sui generis*, *do.;* consciousness, the object, 66, 7; object complex, *do.;* elements threefold, 67; relations to one another, 67, 8; elements not regarded with equal attention, 68; the activity an object, *do.;* also the ego, 70, 1; different views, 68, 9; proof that we are conscious of the ego, 70; unconscious admissions, 70, 1; are we conscious of objects? 71, 2; summary of doctrine of consciousness, 72; object of c. a condition of being, 72; Descartes' doctrine, *do.;* c. involves all the categories, 73; development and growth of c., 74, 5; exercised more or less completely in different persons, 75; capacity for, not developed, 76; latent modifications of, *do.;* capable of degrees, 76, 7; Leibnitz's doctrine of, 77; philosophical or reflective, chapter on, 78–93; characterized by attention, 78; the morbid consciousness in children, hypochondriacs, etc., 78, 9; egoistic consciousness, 79; ethical type, *do.;* in the reflective, attention is persistent, 80; comprehensive, 81; comparative and classifying, 81, 2; interpretive, 82; searches for conditions and laws, *do.;* relations to natural consciousness, 82, 3; imparts new knowledge, 83; in what sense, *do.;* relations of language to each, 84, 5; does not create phenomena, 85; dangers from exact terminology, *do.;* psychology, tried by the language of common life, 85, 6; by the actions, 87; conditions of the successful interpretation of both, 88, 9; why men are so positive in their philosophical opinions, 88; explains slow progress of psychology, 89; explains difficulties in studying psychology, 90–92.
Conservative faculty. (See Memory.)
Content, of notion, 334.
Contradiction, law of, 453.
Copernicus, discovery, 397.
Copula, force of, 361, 3.
Cousin, on origin of knowledge, 425; views of intuition, 437.
Critical or speculative stage of knowledge, 52.

Dalton's discovery of chemical equivalents, 395, 6.
Dana, on species, 353.
Darwin, on species, 352.
Davy's discovery, 396, 7.
Deaf mutes,reason why they cannot speak,142.
Deduction, chapter on, 366–377 ; how related to induction, 367, 8 ; its two forms, 369, 71 ; is not explained by the *dictum de omni et nullo*, 371, 2; but rests on the relation of reason to consequent, 373; this rests on causation, 374, 7; varieties of, chapter on, 378–391 ; various classes of, 378; probable 378, 9; mathematical, 380, 3; not purely deduction, 383, 4 ; examples of, 384, 5 ; immediate or logical, 386, 7 ; distinguished from the process of preparation, 388, 9 ; does it add to knowledge ? 389.
Definition, 361, 3.
Democritus, theory of sense-perception, 191.
Descartes, *cogito, ergo sum*, 72 ; theory of sense-perception, 193, 5 ; on innate ideas, 442 ; on final cause, 509.
Design, or final cause, chapter on, 498–523 ; (see Final Cause ;) how related to adaptation, 499.
Development, of the intellect explained, 51, 2 ; order and stages of, *do.* ; of consciousness, stages of, 74, 6 ; of sense-perception, chapter on, 150–165 ; of touch, 154–156 ; of vision, 156–159.
Dianoetic faculty, 59.
Dictum de omni et nullo, 372.
Discovery and Invention, the conditions of, 409–413 ; attention, 410 ; familiarity, 411 ; constructive imagination, 411, 12 ; wise judgment, 413 ; ready deduction, *do.* ; reference to Divine mind, 414, 15 ; experiment, 415.
Diversity or otherness, relation of, 449 ; proposition expressing it, *do.* ; relation to negation, 450.
Division of the concept, 364.
Dreams, and dreaming, 283–286 ; dreams, the soul active constantly, 283, 4 ; the soul acts with feeble energy, 284 ; with varying energy, *do.* ; representative power active, *do.* ; irregular, *do.* ; the judgment feeble, *do.* ; the reasoning power, 284, 5 ; consciousness feeble, 286 ; estimates of time in, *do.* ; moral responsibility in, *do.* ; the emotion in, *do.* ; the activity of the will in, *do.*
Dugald Stewart. (See Stewart.)
Duration, how related to the soul's acts, 456 ; applied to two objects, 457 ; relations of *do.* ; void, *do.* ; relations to extension, 458 ; transferred to material acts, *do.* ; measures of, whence derived, *do.* ; language of, 459 ; how related to time, *do.* ; affirmed of events, but not of time, *do.*

Ego, the, known in consciousness, 70, 1 ; denied by many, *do.* ; distinguished from the self, 83, 4.
Elaborative faculty, 59.
Empedocles, theory of sense-perception, 190, 1.
Enthymeme, the, 369.
Error, possible of relations only, 45 ; of the senses belong to the acquired sense-perceptions, 144 ; two classes of, 144, 5.
Ethics, its relation to psychology, 7, 8 ; assumes final cause, 520.
Essence, 362.
Event, defined, 482 ; different classes of, 482, 3.
Excluded middle, law of, 453.
Extended objects limited, 43.
Extension known in perception, 106 ; in vision superficial only, 128 ; extra organic, how acquired, 154, 5 ; known in sense-perception, 455 ; blended with matter, *do.* ; the several relations of, 456 ; relations to duration, 458, 9 ; related to space, 473 ; limits objects, 474 ; affirmed of objects not of space, 475.
Extent, of notion defined, 334 ; of mathematical concepts, 382.
Externality, known in perception, 105, 6 ; in touch, 122 ; two meanings of, *do.* ; of the body to the soul, 123, 4 ; of one body to another, *do.* ; extra organic, how acquired, 154–156.
Eye, the structure of, 126 ; single objects seen with two eyes, 129, 30 ; dignity of 131, 2.
Faculties of the soul, 24–34 ; the soul, not parts or organs, 24 ; often so misconceived, 25 ; do not act apart, *do.* ; grounds of belief in, 25–27 ; states like and unlike, 26 ; one dependent on another, *do.* ; distinguishable by a prominent element, *do.* ; more obvious than powers of matter, 27 ; why called human, 28 ; not independent, *do.* ; relations of, important in education, 28, 9 ; history of doctrine of, 31, 2 ; synonyms for, 33, 4 ; of the intellect, how conceived, 53, 4 ; leading faculties named, 54 ; severally defined, 54–60.
Fainting. (See Phantasy.)
Fichte, T. G., on the categories, 444.
Final cause, chapter on, 499–523 ; terms explained, division of causes, 500 ; the relation discerned *a priori*, 501 ; reasons for the position, 501–505 ; the mind seeks this relation, 501 ; acknowledges it to be higher, *do.* ; is of service in discovery, *do.* ; the only basis of deduction, 503 ; explains organic phenomena, 502 ; conspicuous in the highest order of beings, 504 ; does not displace efficient causes, 505 ; objections to the position, 505–513 ; men mistake, 505 ; they cannot test their inductions, 506 ; the relation subjective only, *do.* ; involves two principles, 508 ; hinders discovery, 509 ; Bacon and Descartes on, *do.* ; adaptations are necessary conditions only, 510 ; limited, 511 ; cannot be ascribed to an unlimited Being, 512 ; application of the principle, *do.* ; in metaphysics, *do.* ; in induction, *do.* ; in the formation of the concepts *do.* ; in classification, 513 ; in the notion of an individual, *do* ; as a rule of truth, 514 ; in mathematics, *do.* ; in geology and paleontology, 515 ; in phil. geography, 516 ; in comp. anatomy, 517 ; in physiology, *do.* ; in anthropology, *do.* ; in psychology, 519 ; in ethics, 520 ; in theology, 521 ; two classes of theories of God, 521 ; reasons for accepting a personal God, 522.
Finite and the Infinite, (see Infinite) ; and conditioned, the chapter on, 541 ; result of processes of knowledge, *do.* ; the finite universe how conceived, 542 ; is limited and conditioned, 543.
First principles. (See Intuition.)
First truths. (See Intuition.)
Forgetfulness. (See Memory.)
Forgotten. (See Memory.)
Formal cause, 500.
Formal categories, 446 ; chapter on, 446–455.
Forms, of thought and being, 324 ; of knowledge, Kant and Hamilton error, concerning, 538.

Franklin's discovery of electricity, 394, 5.
Functions of the soul defined, 33.

Galileo, discovery by, 398.
Gassendi, illustration of memory, 263.
Generalization, 319, 22.
Geography, Phil., assumes final cause, 329.
Geology, assumes final cause, 516.
Geometrical reasoning, (see Mathematical quantities); constructions of, 302, 3; figures, construction of, 304; quantities measurable, 305; example of, 335; concepts, how formed, 466; rests on what assumptions, 467; postulates of, *do.*
God, belief in, assumed in inductive and scientific knowledge, 408.

Habit, relation to association, 248; theory of, *do.; often supposes a difficulty, do.;* bodily, *do.;* mental, 249; emotional, *do.*
Hallucinations, 216; case of Nicolai, 293; not purely physical, *do.;* how explained, *do.*
Hamilton, Sir Wm., division of faculties, 32; consciousness of Ego, 69; theory of extra-organic perception, 156, 7; doctrine of latent modifications, 244; on the nature of the concept, 343; Hamilton's dictum of the syllogism, 372; on origin of knowledge, 424; positive and negative necessity, 441; theory of causation by law of the conditioned, 495–498, *sqq.;* of primary, secondary and secundo-primary qualities, 535; on the real and phenomenal, 538, 9; negative thinking, 546–550; on the Infinite, 549.
Harvey's discovery prompted by final cause, 502.
Hauser, Casper, how the world looked to, 162.
Hearing, sense-perceptions of, 113–116; organ, 113; varieties, how far distinguishable, 114, 15; condition of language, 115; expresses feeling, *do.;* dignity, 116; acquired perceptions of, 133, 4.
Hegel, on the nature of the concept, 445; on the categories, 445; being equals nothing, 448; error, *do.*
Herbart, doctrine of faculties, 32; theory of sense-perception, 204.
Herbert Spencer, (see Spencer,) an associationalist, 38; doctrine of necessary truths, 39–40.
Hobbes, crude views of association, 231, 2; often-quoted illustration, 243.
Hume denies consciousness of ego, 68; enumeration of laws of association, 232; doctrine of the concept, 342; on tuitions, 436; theory of causation, *do.*

Ideals, nature of, 301; varieties of, 304, 6; related to individual experience, 306, 7.
Identity, law of, etc., do not explain deduction, 371, 2; category of, 452; affirmable of spirit and matter, *do.;* logical law of *do.;* concerns concepts, *do.;* guards against what, 546; founded on real identity, misapplication of by Hegel and others, *do.;* of material substance, *do.*
Image, technical name for objects of representation, 209; relation to concept, 349; of space and time objects, 457.
Imagination, a modification of representation, 211; mathematical, 212; poetic, 213; philosophical, *do.;* the, chapter on, 295–318; materials and conditions for, 295, 6; space and time, 295; thought-relations, 296; material qualities, *do.;* spiritual, *do.;* how far can it modify these materials? 296–301; its combining office, 301; idealization of space and time objects, the mathematical imagination, 301, 3; psychical idealization, 303, 4; capable of growth and culture, 307, 9; constantly exercised, *do.;* the poetic, 309, 12; the philosophic, 312, 14; the ethical, 314, 15; the religious, 316–318.
Imaging, of concepts, 349; of space and time objects, 460; of the infinite, etc., 316–318.
Individual, notion of, rests on final cause, 499.
Induction, relation to psychology, 35; how related to deduction, 367, 8; chapter on, 391–416; loosely defined, 392; the so-called purely logical, *do.;* proper induction, 393; very frequent, *do.;* how differs from simple judgment, *do.;* importance of a correct theory, of, *do.;* in science, 394–399; why in science more difficult, 400–402; the problem of, difficult, 403; involves certain assumptions, 404–408; three rules of induction, 408; conditions of successful hypothesis, 410–413; relation of experiments, 415.
Inductive science. (See Induction.)
Infants capable of induction, 446, 7; condition of the soul in, 150, 2; learns to touch, 162.
Infinite, unconditioned and absolute, chapter on, 541-556; relations to the finite, 541, 2; literal import of infinite, 543; transferred from quantity to quality, 544; variety of senses of, *do.;* the terms used in the concrete and abstract, 546; not negative conceptions, 547; not produced by negative thinking, *do.;* Hamilton and Mansel, *do.;* not unrelated, 548; Spinoza, *do.;* Hobbes' doctrine of, *do.;* not the sum total of being, 549; totality not infinite, *do.;* not a matter of quantity, *do.;* not one and simple, *do.;* is knowable, *that* and *what* it is, 550; Herbert Spencer's doctrine of, *do.;* cannot be imagined, 551; Kant's antinomies explained, *do.;* not known by reasoning or induction, 552; not defined for classification, *do.;* holds relations to the finite, *do.;* known by knowledge, and not by faith or feeling, *do.;* not known exhaustively, 554; self-existence common to the finite and infinite, 555; is a thinking person, 555; relations to space and time, 556.
Innate Ideas, doctrine of, 435.
Inner sense. (See Consciousness.)
Insanity, 294, 5.
Intellect, growth and development of, 51, 2; rules for culture of, 52, 3; faculties of, 54; learns to control the body, 141, 3; its state before sense-perception, 150.
Intuition and Intuitive knowledge, 58, Part IV., 419–556; defined and enumerated, chapter on, 419–433; not gained by ordinary processes, 422; referred by some to a special faculty, *do.;* various appellations for, *do.;* not first in time, 422; Locke's polemic against, *do.;* first in logical importance, *do.;* in what sense principles, 423; different senses of the word, *do.;* how related to origin of knowledge, 424; stages of the mind's progress in, 425–427; explanation of the limited assent to them, 427; tested by the language and actions of men, 428, 9; three criteria, 429; logically independent, *do.;* divided into three classes, 433; theories of, chapter on, 433–445; of direct men-

tal vision, 434; light of nature, *do.*; innate ideas, *do.*; school of Locke, 435; Condillac, 436; Hume, *do.*; of the associational school, *do.*; Dr. Reid and the Scottish school, 437; the French school, *do.*; Kant and his school, 438; criticism of, 439, 40; Hamilton, 441; of faith, *do.*; Schleiermacher, 443; ethical school, 444; J. G. Fichte, *do.*; Schelling and Hegel, 446; Herbart, *do.*; Trendelenburg, *do.*

Judgment, chapter on, 358-366; forms the concept, 359; how related to the concept, *do.*; psychological and logical, *do.*; the logical judgment, 360; force of the copula, 361; judgment of content, *do.*; essence, 362; judgment of extent, 363; importance to science, 364; propositions of extent and content how related, 365.

Kant, theory of sense-perception, 203; on the nature of the concept, 345; on origin of knowledge, 424; views of categories and intuitions, 438; criticism of, 439; of practical reason, 442; doctrine of space and time, 478; on causations, 495; error concerning forms of knowledge, 538; *the thing in itself*, 532; on the real and phenomenal, 540; antinomies, 550.

Kepler, discovery by, 398; exclamation, 414.

Knowledge defined and discussed, 42-50; how far definable, 42; is action, *do.*; exercised under conditions, *do.*; these various, 43; two classes of objects, 43; preparation of objects, 44; involves certainty, *do.*; being its correlate, *do.*; involves apprehension, of relations, 45; objection, *do.*; involves analysis and synthesis, 46; when the process is complete, 46, 7; these products objects of subsequent knowledge, 47; representative and represented knowledge, *do.*; acts of kn. diverse in energy, *do.*; attention, 47, 8; some objects known more easily than others, *do.*; psychological and philosophical kn., 48; critical stage of kn., 50; direct and reflex, of matter and spirit, 534; direct involves apprehension of being as well as relations, *do.*; reflex, difficnlt to analyze, *do.*

Language, relation to psychological truth, 84; of common life, a test of truth, 85; influenced by association, 253; relation to thought, 226, 7; the study of, 327.

Law, its relations to psychology, 8.

Law and power, 481.

Leibnitz, doctrine of latent consciousness, 76; latent modifications in association, 244; on the sufficient reason, 375; criticism on Locke's doctrine of origin of knowledge, 424; on intuitions, 435; sufficient reason as applied by Wolf, 491.

Light of nature, 434.

Limit and limitation of objects and events, 459.

Limited, the distinguished from the conditioned, 548.

Locke, doctrine of reflection and consciousness, 63; theory of sense-perception, 195, 6; doctrine of knowledge, 196; of association, 231; on the syllogism, 434; on innate ideas, 383; on intuitions, etc., 422; theory of causation, 491; relation to Mill and Hume, *do.*; to de Biran, *do.*; to his own doctrine of knowledge, *do.*; on substance, 532; on primary and secondary qualities, 536.

Logic, its relation to Psychology, 9; to metaphysics, 10.

Logical relation of processes and products, 49; contrasted with psychological, 50; do not always coincide, *do.*

Maas, theory of association, 235.

Malebranche, theory of sense-perception, 195; of causation, 494.

Mansel, H. L., on negative thinking, and on the Infinite, etc., 547-550.

Materialism accounted for, 12-13; arguments in favor of, 14—17; counter-arguments, 17-21.

Materialists, their views of psychology, 36.

Mathematical affections of matter, Stewart's doctrine of, 535.

Mathematical, reasoning, 380-386; its entities or concepts, 380-383; into categories, 419-433.

Mathematical relations, chapter on, 454; quantity, 465; concepts, two classes of, *do.*; application to matter, *do.*; to mechanics and chemistry, 469; to light, sound, and heat, *do.*; suggested and defined by motion, 471.

Mathematics, rests on final cause, 514.

Matter, relations of the soul to, 11-24; phenomena first attended to, 12; prepossessions which it engenders, 13; furnishes language for physical phenomena, 22-24.

Matter and form, in sense-perception, 192, 3.

Matter, its capacity to be perceived not an attribute, 535, 6; known as being, *do.*; its most important relations to the soul as sentient, *do.*

Measurement involves number, 462; involves both number and magnitude, *do.*

Memory a modification of representation, 210, 11; chapter on, 254-278; essential elements in an act of, 254-256; memory technically defined, 256, 7; representation and recognition, 257; spontaneous and intentional, 258; spontaneous defined, *do.*; original differences in, *do.*; relations peculiar to it, 259; its value, *do.*; requires the rational also, 260; the intentional memory defined, *do.*; relations to the knowing mind, 260, 1; recovery of forgotten objects, 262, 3; memory as the power to retain, 263; how accounted for, *do.*; figurative explanations, Gassendi's, *do.*; ready and tenacious, *do.*; forgetfulness, 264; forgotten knowledge recovered, 265, 6; dependence on the bodily condition, *do.*; influenced by the season or the time of the day, 266; sudden loss of memory, 267; how explained, *do.*; varieties of, 268-271; development of, 271, 2; in infancy, childhood, and youth, *do.*; cultivation of the memory, 273-277; fundamental principles, 274; Buxton's advice, *do.*; artificial memory, *do.*; value, objections, 275; when useful, 276; Coleridge's arts of memory, 277; moral conditions of, 277.

Metaphysics, its relations to psych., 9; to logic, *do.*; assumes final cause, 512.

Microcosm, the soul a, 73.

Middle terms, 369-374; invention of, 389.

Mill, James, an associationalist, 38; denies consciousness of ego, 70; admits it, 71; doctrine of association, 232; on intuition, 436.

Mill, John Stuart, an associationalist, 38; doctrine of necessary truths, 39; conscious-

ness of ego, 70; doctrine of association, 232; on the nature of the concept, 344; doctrine of the syllogism, 370; on intuitions and first truths, 486; theory of causation, 487; relation to those of Hume and Brown, *do.*; definition of the soul, 529; definition of body, error in, *do.*
Mind and matter, chapter on, 525–540.
Mnemonics. (See Memory.)
Morell, J. D., perception into classification, 176.
Motion bodily, provision for, by nature, 140; for combined activity, 141; how controlled by the intellect, 141–3; aids sense-perception, 172.
Motion, relation of space and time concepts to, 470; universality of, *do.*; indicates position and rest, 471; suggests time relations, *do.*; mathematical quantities, *do.*; the condition of generalization, *do.*
Müller, J., theory of nerve endings in touch, 121; theory of extra-organic perception, 155; theory of sense-perception, *do.*
Muscular sense-perceptions defined and divided, 109; lowest in rank, *do.*; in touch, 117, 18; first developed, 153.

Names, significance of, 338. (See Words.) Of concepts, advantages of, are sensuous, 353, 4.
Negative notions, 448.
Nerves, reflex action of, 99; afferent and efferent, *do.*; subject to various affections, *do.*; special function in sensation, *do.*
Nervous system described, 98.
Newton, discovery by, 407.
Noetic faculty, 59.
Nominalists, the, 341–47; strife adjusted, 349, 50.
Nothing, Hegel's use of, 450.
Notion. (See Concept.)
Number, how developed, 462; defined, 468; relations, how symbolized, *do.*; concepts of, *do.*; application to magnitude, *do.*

Objects—object- and subject-, 43; material distinguished from percepts, 165; involve two relations, 166; percepts united in space and time, 166–8; involve substance and attribute, 168–171.
Organic sense-perceptions, 110.
Original sense-perceptions defined, 132.
Owen, on species, 353.

Perception. (See Sense-perception.)
Perception, proper, Hamilton's doctrine of, 104; defined, 105–109; an act of knowledge, 105; a *non-ego*, *do.*; an extended *non-ego*, 106; accompanies every sense, *do.*; with varying clearness, 107; in inverse ratio to sensation-proper, *do.*; in different sensations and senses, 108; of touch, 120–125; in vision, 128–131; acquired, 132.
Percepts, how gained, 165; how combined, *do.*; distinguished from things, *do.*; combined into things by two stages, 166.
Phantasy, a modification of representation, 211; chapter on, 278–295; defined, 278; examples of, *do.*; why infrequent, 278; fainting, sleep, etc., 278, 9; several suppositions possible, 279; why probably explicable by laws, 279; depend on laws of representation, *do.*; bodily condition influential, 280; creative power possible in, 281; sleep considered physiologically, 282; prominent phenomena, *do.*; considered psychologically, 283–286; somnambulism, 286–294; insanity, 294.
Phenomenal and real. (See Real.)
Phenomenon defined, 34.
Philosophical consciousness. (See Consciousness.)
Physiology defined, 2; assumes final cause, 517.
Plato, theory of sense-perception, 191; on universals, 339, 40.
Political Science, its relation to psychology, 8.
Porphyry's Questions on universals, 340.
Postulates, 381.
Power and law distinguished, 376.
Powers of the soul. (See Faculties.)
Predicable, 331.
Prescind, to, 328.
Presentation. (See Presentative Knowledge.)
Presentative Faculty defined and divided, 55, 6.
Presentative Knowledge, Part I., 61–206.
Primary laws of association, 227–240.
Primary Qualities, 535, 6.
Principle, various senses of the term, 423, 4.
Probable or problematical reasoning, 378, 9; founded on causes and laws, 279.
Proposition. (See Judgment.)
Psychological contrasted with logical relations, 49.
Psychology defined and vindicated, 1–11; improperly named, 1; properly a science, *do.*; relations to psychology and anthropology, 2, 3; its phenomena peculiar, 3; known by consciousness, *do.*; interest of, *do.*; value of, promotes self-knowledge, 4; teaches self-control, *do.*; promotes moral culture, *do.*; aids in understanding others, 6; indispensable to educators, *do.*; aids in the study and enjoyment of literature, 6, 7; the mother of all the human sciences, 7; relation to ethics, *do.*; to political and social science, 8; to law, *do.*; to æsthetics, *do.*; to theology, *do.*; special relation to logic and metaphysics, 9; why called phil. and met., 10; disciplines to method, *do.*; a branch of physics, 11; why distrusted, 11, 12; its phenomena overlooked, 12; resolved into material agencies, 13; is it a science? 34–41; the materials, whence derived, 34, 5; an inductive science, 34; also the science of induction, 35; objections against psychology as a science, 35, 6; answers, *do.*; views of materialists, 36; of cerebralists, *do.*; views refuted, 37; phrenologists, *do.*; Associationalists, 38–40; *a priori* theory, 40; wherein defective, 41; method of observing and interpreting its phenomena, 80–82; in what sense imparts new knowledge, 83; aided by language, 84; misled by exact terminology, 85; tried by the language of common life, 85, 6; by the actions, how it can interpret both, 87; why men are so positive in their theories of, 88; slow progress and divisions explained, 89, 90; special difficulties of studying, 90, 92

Qualities of matter, primary and secondary, 535, 6; two and threefold classification, *do.*; Aristotle's, Descartes', and Locke's, *do.*; Reid's, Stewart's and Hamilton's, *do.*; the secundo-primary not established, *do.*; Hamilton's locomotive energy, *do.*; are the

primary qualities essential to the notion of matter? 536; do they give real knowledge? 539.
Quantity, relations of, 456; mathematical, 465.

Real and phenomenal, 536–539; contrasted in two senses, *do.;* Kant's doctrine of, *do.;* Hamilton's, 540; their views criticised, *do.;* question not peculiar to philosophers, *do.;* special sense of real, *do.;* relations of the intellect trustworthy, *do.*
Real categories, 540.
Realism, truth, and significance of, 350–352; assert permanent relations, 351; mistakes, 352.
Realists. (See Realism.)
Reason and consequent, relation of, 373, 4.
Reason to. (See Reasoning.)
Reasoning, deductive, chapter on, 366–377; reasoning implies judgment, 366, 7; inductive and deductive, 367; often conjoined, 368; deductive, (see Deduction;) probable, 378–380; mathematical, 380–386; formal, 386, 7.
Redintegration, law of, 234, 5; how far it accounts for the laws of association, 235, 6.
Reflection, as used by Locke. 63.
Reflective consciousness. (See Consciousness.)
Regulative faculty, 59.
Reid, consciousness of ego, 69; defective view of sensation, 103; theory of perceiving externality by touch, 107; theory of sense-perception, 198; on the nature of the concept, 342, 3; on axioms, 383; on intuition and first truths, 437.
Relations involved in knowledge, 45, 6; objects unrelated, 46; relations do not attract ideas, 232, 3; of place in assoc., 233; of time and of both, *do.;* of similarity and contrast, *do.;* of cause and effect, *do.;* of means and end, 234; general relations or categories, (see C.;) formal relations, chapter on, 446–454; mathematical, chapter on, 454–480.
Relative notions, 449.
Repetition, in sense-perception, excites interest, 173–5.
Representation, defined, 56; its objects, *do.;* conditions, 57.
Representation and R. Kn., Part II., 206–295; defined, 206; not limited to sensible objects, *do.;* a creative power, 207; appellations for, 207, 8; objects of, 208; individual, *do.;* involve relations, 209; no technical names for objects, of, *do.;* conditions and laws of, 210; divisions of, 210–213; interest and importance of, 213, 14; object of, chapter on, 215–224; why needs discussion, *do.;* three heads of inquiry, 215; psychical, *do.*; transient, *do.;* not spectrum or hallucination, 216; intellectual, *do.;* relation of object to its original, 217; comparable to no other, *do.;* does not resemble its objects, *do.;* contradictions involved, *do.;* no resemblance in memory or recognition, *do.;* mental pictures less exciting, 219; consist of fewer elements, *do.;* recalled slowly in parts, 220; objects of imagination, 221; usefulness of representative objects to thought, *do.;* less distracting than realities, 222; more easily compared, *do.;* and generalized, *do.;* serviceable in action, 224; conditions and laws of Rep., chapter on, (see Association of Ideas,) 225–254.
Representative faculty. (See Representation.)
Retention, 262.
Retina, image on, 126.
Royer-Collard, on causation, 437.

Schema, nature and service of, 222, 3.
Schleiermacher, theory of sense-perception, 205; on intuitions and the categories, 443.
Science, classifications of, 336, 7; nomenclature of, 337; related to common knowledge, 365; defined, *do.;* when complete, 367.
Scientific knowledge. (See Science.)
Secondary laws of association, 240–243.
Secondary Qualities, 535, 6.
Secundo-primary qualities, 535.
Sensation proper, defined, 102; experienced in the soul, *do.;* connected with an organism, 103; Reid's view of, *do.;* Berkeley's, *do.;* Hamilton's, 104; involve relations of place, *do.;* differ in kind and degree, 105; definiteness of place, *do.;* inversely to perception proper, *do.;* muscular, 109; organic, 110; special, 111; of taste, 112; of hearing, 113; of gentle touch, 117; acute and painful of, 118; of temperature, *do.;* of weight, 119; muscular in touch, *do.;* of vision, 126.
Sense-perception, 93–205; conditions and process, chapter on, 93–109; defined, 95; called earliest into action, *do.;* seems easy to understand, *do.;* why difficult, 94; what it is not, 94, 5; example of, in an orange, *do.;* what it is, 96; eight topics of inquiry, 97; conditions of sense-perception, 97–98; bodily organism, 98; nervous system, *do.;* sensorium, 99; appropriate objects a condition, 100; action of object on sensorium, *do.;* process of sense perception, 101–109; psychical, not physiological, *do.;* classes of sense-perceptions, chapter on, 109–131; three named, 109; muscular, *do.;* organic, 110; special, *do.;* smell, 111; taste, 112; hearing, 113–116; *q. v.;* touch, 116–126; *q. v.;* sight, 126–132; *q. v.;* acquired sense-perceptions, chapter on, 132–150, *q. v.;* development and growth of, chapter on, 150–165; interest of the problem, 150; perplexing to the imagination, 150; data for solving it, 151, 2; products of, chapter on, 165–179; conditions of perception of things, 171; energy by contrast, etc., 172; motion, *do.;* repetition, 173; need of, explained, 173–175; familiarity, 175; repetition not recognition, 176; continuance of time, 177; activity of the soul in, chapter on, 180–187; summary and review of theory of, 187–189, theories of, chapter on, 189–205.
Sensorium described, 99; known as extended, 122.
Sensory. (See Sensorium.)
Sight, sense of, 126; organ of, 126–131; conditions of, 126; image on the retina, function of, 127; as sensation, 127, 8; as perception, 128–130; place of the object as originally seen, 130; dignity of vision, 131; acquired perceptions of, 134, 35; *e.;* why and how its percepts are projected in space, 157–159; percepts of, combined with those of touch, 159–162.
Simple notions, 333.
Sleep. (See Phantasy.)
Smell, sense-perceptions of, 110; organs, 111; acquired perceptions of, 133, 4.
Socrates, on universals, 339.

Somnambulism, three species of, 286, 7; natural, 287; activities required in, *do.;* magnetic, *do.;* representation in excess, *do.;* also some sense-perceptions, 288; acute but limited, *do.;* the sense-organs used, *do.;* extraordinary intellectual activities, 289; state usually forgotten, 291; when remembered, *do.;* alternate states, *do.;* artificial somnambulism, 292; hypnotism *do.;* relation to somnambulism, *do.;* control of one mind by another, 293.

Soul, the, signification of the term, 1, 2; original designation, 2; secondary meanings, *do.;* relations of, to matter, 11-24; phenomena of, resolved into matter, 12; phenomena at first overlooked, 13; arguments for the material structure of, 14-17; for its spiritual essence, 17-21; its phenomena real, 21; cannot be judged by material analogies, *do.;* described in language of physical origin, 22, 3; consequent dangers, 23, 4; faculties of, (see Faculties;) unity of, higher than any other, 29, 30; does not exclude complexness, 31; powers of the soul threefold, *do.*

Sound, sense-perceptions of, 113-116.

Space, a condition of imagination, 295; void, how first known, 454; inclosed and inclosing space, *do.;* these relations analyzed, *do.;* objects as imaged, 457; relation to motion, *do.;* as infinite, *do.;* in what sense unlimited, *do.;* cannot be generalized, 463; nor defined, *do.;* known by intuition, *do.;* correlate of the extended, *do.;* not a substance, *do.;* nor a quality, *do.;* nor a relation, or correlation, 476; nor a form, *do.;* in what sense knowable, *do.;* conclusion respecting, 478.

Space and Time, chapter on, 454; objects generalized, *do.;* their relations individual and general, 458.

Species, in sense-perception, scholastic doctrine of, 193; nature and permanence of, 350-353.

Spectra, 216; 293, 4.

Speculative or critical stage of knowledge, 52-419.

Spencer, Herbert, an associationalist, 38-40; doctrine of consciousness, 66; resolves perception into recognition, 176; on the knowledge of the Infinite, 550.

Spinoza's definition of substance, 532; on the Infinite, 548.

Spirit, original meaning of, 1, 2.

Standards of space and time, 464.

States of the soul defined, 34.

Stereoscope, invalid inference from, 129.

Stewart, Dugald, consciousness of ego, 69; theory of attention, 178; theory of sense-perception, 198, 9; explanation of latent modifications of consciousness, 244; on the syllogism, 372; on geom. axioms, 383; on primary and secondary qualities, 535.

Studies, natural order of, 52, 3.

Subject-objects, 43.

Substance and Attribute, relation of, in sense-perception, 168-171; supposes reflex knowledge, *do.;* supposed in the concept, 332; category of, 524; chapter on, *do.;* import of the terms, *do.;* etymology of, *do.;* different theories of, *do.;* Locke on, 532; Hume, Reid, Kant, Whewell, 532.

Substance represented by touch-percepts, 170, 1; distinguished from logical and grammatical subject, 524; etymology of, 524; in the abstract, 525; three classes of, 526; spiritual substance, 525; distinguished by attributes of causation and design, *do.;* spiritual and human defined, 526, 7; J. S. Mill's definition, 529; material defined, 528; related to space in a two-fold way, 530; power to affect the senses, *do.;* matter not causative of perception, *do.;* Mill, Brown, and Kant on, 532; permanently occupies space, 530; not self-subsistent, 531; Spinoza's error and definition, *do.;* Whewell's, 532; belief in permanence founded in design, 533.

Syllogism and Deduction, chapter on, 366-377; parts of, 369; possible changes in, 371; does not rest on the *dictum de omni et nullo*, 372; not identical with induction, 373; explained by relation of reason to consequent, 373, 4; this by causation or its equivalent, 374; sanctioned by Aristotle and Leibnitz, 374, 5; immediate syllogisms, 375, 6.

Symbolic Knowledge, 354-357; can the infinite and spiritual be symbolized? 357.

Synthesis, involved in knowledge, 46.

System, chapter on, 416-418; any arrangement of content or extent, 416; of both united, *do.;* of propositions of either, or both, 417; of less obvious concepts, *do.;* in science, 418; of *abstracta*, *do.*

Systemization. (See System.)

Taste, sense-perceptions of, 112-113; variety, names of, *do.*

Tennyson, on self-consciousness, 75.

Theology, relations to psychology, 8; relations to final cause, 521.

Theories of nature of concepts and universals, (see Concept);—of sense-perception, chapter on, 189-205; universal, 189; reflex influence mischievous, 190; liable to be erroneous, *do.;* pertain chiefly to vision, *do.;* of the earlier Greek philosophers, *do.;* Empedocles, *do.;* Democritus, 191; the Socratic school, *do.;* Plato, *do.;* Aristotle, 192; the schoolmen 193; Descartes, 194; Malebranche, 195; Arnauld, *do.;* Locke, 195, 6; Berkeley, 197; Hume, *do.;* Reid, 198; Stewart, *do.;* Brown, 199; Hamilton, 199-202; Condillac, 202; Kant, 203; Herbert, 204; Schleiermacher, *do.*

Thing in itself explained, 532; Kant's doctrine of. (See Kant.)

Thinking. (See Thought.)

Thinking, and Thought-knowledge, Part III., 319; terms variously applied, 319; relation to higher knowledge, *do.;* dignity of, 320; illustrated by an example 320, 1; thought defined, 321; use of term justified, 322; appellations for the power, 322, 3; forms of, 324; relation to lower powers, *do.;* when does it begin? 325; abstract and concrete, *do.;* difficulty of abstract, *do.;* to language, 326.

Thought, faculty of, defined, 57; its objects, *do.;* its conditions, 58; how far prepared by thought itself, *do.;* certain intuitions assumed in, *do.;* analysis of, involves two general inquiries, 59, 60.

Time and Space, relations of, chapter on, 454; estimates of, 461; objects generalized, 463. (See T. & S.)

Time, a condition of imagination, 295; objects as imaged, 459; measure of, 461; estimates of, *do.;* relation to motion, 470; time-relations generalized and suggested by motion, 471; as infinite, 473; in what sense unlimited, 475; cannot be generalized, *do.;* not defined, *do.;* is known by intuition, *do.;* correlate of the enduring, 476; not a substance, 477; nor a quality, *do.;* nor a relation or correlation, *do.;* nor a form, 478; in what sense knowable, *do.;* conclusion respecting, 479.

Touch, sense, of, 116–126; organ, 116, 17; conditions of, 117; variety of sensations, *do.;* gentle touch, 118; involving violence, *do.;* of temperature, 118, 19; of pressure, 119; muscular, *do.;* perception proper of, 120, 1; of extension, *do.;* conditions and act, 121, 2; of extension direct, not indirect, 122; perception of externality in two senses, 122; of the body to the soul, 123, 4; of one body to another, 124; the leading sense, *do.;* called general sensibility, 125; furnishes terms for the intellect, *do.;* percepts of, combined with those of sight, 157, 8.

Unconditioned, (see Infinite,) primary and secondary sense of, 544.

Universal, 226; theories of, nature of. (See Concept.)

Universe, the finite, how conceived, 542.

Vibrations of nerves supposed to account for representation, 97.

Vision. (See Sight.)

Weber, E. H., experiments on touch, 116.

Whewell, erroneous definition of substances, 532.

Wolf, on causation, 495.

Worcester, Marquis of, discovery of steam, 412.

Words, importance of, 353, 4; no substitute for intuition, 354, 5; operate by suggestion, 356.

The Illustrated Library of Wonders

TWENTY VOLUMES READY, CONTAINING OVER

ONE THOUSAND BEAUTIFUL ILLUSTRATIONS.

The extraordinary popularity achieved by the "ILLUSTRATED LIBRARY OF WONDERS" is based upon the following points: 1st. Its design is entirely unique. No similar collection has before been undertaken in this country, and the novelty of the plan is only rivalled by its acceptability. 2d. The subjects discussed are widely diverse, and they are treated by writers who are not only thoroughly familiar with them, but who have been specially selected for their ability to write in a popular and entertaining style. Each volume is moreover a complete treatise in itself upon the particular subject to which it is devoted, and comprises the latest developments in each department of investigation and discovery. 3d. The different volumes are profusely illustrated by designs from the best artists, most carefully executed and specially adapted to the elucidation of the text. They are handsomely printed upon tinted paper, and every care has been taken in their mechanical production to make them an ornament to the FAMILY or SCHOOL LIBRARY, or acceptable for use as PRIZES or PRESENTS, for all of which purposes they are unexcelled.

The volumes may be purchased separately, or in libraries classified according to their subjects, as below:

Each one Volume 12mo. Price per Volume - - - - - $1 50.

Wonders of Nature.

	No. illustrations.
THE HUMAN BODY, -	[illegible]
THE SUBLIME IN NATURE, -	44
INTELLIGENCE OF ANIMALS, -	54
THUNDER AND LIGHTNING, -	39
BOTTOM OF THE SEA, - -	68
METEORS.	

Six Volumes in a neat box, $9 00

Wonders of Art.

	No. illustrations.
ITALIAN ART, - - - - -	28
ARCHITECTURE, - - - -	60
GLASS-MAKING, - - - -	63
LIGHTHOUSES & LIGHTSHIPS,	60
WONDERS OF POMPEII, -	22
EGYPT 3,300 YEARS AGO, -	40

Six Volumes in a neat box, $9 00

Wonders of Science.

	No. illustrations.
THE SUN. BY GUILLEMIN, -	58
WONDERS OF HEAT, - -	93
OPTICAL WONDERS, - - -	71
WONDERS OF ACOUSTICS, -	110

Four Vols. in a neat box, $6 00

Wonderful Adventures and Exploits.

	No. illustrations
ARMS AND ARMOR	
BODILY STRENGTH & SKILL,	[illegible]
BALLOON ASCENTS, - - -	30
GREAT HUNTS, - - - -	22

Four Vols. in a neat box, $6 00

r the twenty volumes named above in a handsome case for $30 00.

Any or all the volumes of the ILLUSTRATED LIBRARY OF WONDERS sent to any address, post or express charges paid, on receipt of the price.

A descriptive Catalogue of the Wonder Library, with specimen illustrations, sent to any address on application.

POPULAR AND STANDARD LIST OF BOOKS

PUBLISHED BY

Charles Scribner & Co.

IN 1869 AND '70.

ALEXANDER, H. C. Life of J. A. Alexander. Two vols. crown 8vo....... $5.00

ALEXANDER, J. W. Forty Years' Correspondence with a Friend. *New Edition.* One vol. 12mo........ 2.50

ALEXANDER, J. A. Sermons. *New Edition,* One vol. 12mo........ 2.50

ANDERSON. Foreign Missions. One vol. 12mo........ 1.50

BOWEN, Prof. FRANCIS. Am. Political Economy. One vol. crown 8vo.. 2.50

BROWNING, Mrs. E. B. Lady Geraldine's Courtship. Illustrated. One vol. small 4to........ 3.00

BUSHNELL, HORACE. Women's Suffrage. One vol. 12mo........ 1.50

CONYBEARE & HOWSON. St. Paul. *Complete Edition.* One vol. 8vo. 3.00

COOLEY, Prof. Le ROY C. A Text Book on Natural Philosophy. One vol. 12mo........ 1.50

——— A Text-Book of Chemistry. One vol. 12mo. 1.25

DAY, Prof. HENRY N. The Young Composer. One vol. 12mo........ 1.00

——— The American Speller. One vol. 12mo........ .25

FISHER, Rev. GEO. P. Supernatural Origin of Christianity *New Edition.* One vol. 8vo........ 3.00

FROUDE, J. A. History of England. Vols. XI. and XII., cr. 8vo., per vol.. 3.00

——— Vols I. to X. inclusive. 12mo., per vol 1.25

HAMILTON. Reminiscences of J. A. Hamilton. One vol. 8vo........ 5.00

HARPER, MARY J. Practical Composition. One vol. 12mo........ .90

HEADLEY, J. T. The Adirondacks. *New Edition.* One vol. 12mo........ 1.75

HURST, JOHN F. D.D., History of the Church in the Eighteenth and Nineteenth Centuries. Two vols. 8vo........ 6.00

ILLUSTRATED LIBRARY OF WONDERS. Thirteen vols. Illustrated, per vol........ 1.50

JERNINGHAM'S (Mrs.) Journal. One vol. 16mo........ .75

LANGE'S (Prof. J. P., D.D.) COMMENTARY. Romans. One vol. 8vo. 5.00

——— Proverbs. One vol. 8vo. 5.00

LIFTING THE VEIL. One vol. 12mo........ 1.25

LORD, Dr. JOHN. Ancient States and Empires. One vol. crown 8vo.... 3.00

MAINE, H. S. Ancient Law. *New Edition.* One vol. crown 8vo........ 2.50

Mc ILVAINE, Prof. J. H. Elocution. One vol. 12mo........ 1.75

MOMMSEN, Dr. THEODOR. The History of Rome. Three vols. cr. 8vo., per vol........ 2.00

MULLER, MAX. Chips from a German Workshop. Two vols. cr. 8vo... 5.00

PORTER, Prof. N. The Human Intellect. One vol. 8vo........ 5.00

POUCHET, F. A. The Universe. One vol. royal 8vo........ 12.00

SEAMAN, E. C. The Am. System of Government. One vol. 12mo........ 1.50

SHEDD, WILLIAM G. T., D.D. A History of Christian Doctrine. *New Edition.* Two vols. 8vo........ 5.00

——— A Treatise on Homiletics and Pastoral Theology, *New Edition.* One vol. 8vo........ 2.50

SONGS OF LIFE. Illustrated. One vol. small 4to........ 6.00

STANLEY, ARTHUR PENRHYN, D.D. Lectures on the History of the Jewish Church. *New Edition.* Two vols. crown 8vo........ 5.00

——— Lectures on the History of the Eastern Church. *New Edition.* One vol. cr. 8vo........ 2.50

——— Sinai and Palestine. *New Edition.* One vol. cr. 8vo........ 2.50

TRENCH, R. C. Studies in the Gospels. *New Edition.* One vol 12mo .. 2.50

WOOD, Rev. J. G. Bible Animals. One vol. 8vo........ 5.00

WOOLSEY, T. D., D.D. Essay on Divorce. One vol. 12mo........ 1.75

www.ingramcontent.com/pod-product-compliance
Lightning Source LLC
LaVergne TN
LVHW021226110826
845150LV00002B/259

* 9 7 8 1 4 2 5 5 6 3 7 9 0 *